Lecture Notes in Computer Science 4047

Commenced Publication in 1973
Founding and Former Series Editors:
Gerhard Goos, Juris Hartmanis, and Jan van Leeuwen

Matthew Robshaw (Ed.)

Fast
Software Encryption

13th International Workshop, FSE 2006
Graz, Austria, March 15-17, 2006
Revised Selected Papers

 Springer

Volume Editor

Matthew Robshaw
Telecom Research and Development
38-40 rue du General Leclerc, 92794 Issy les Moulineaux, Cedex 9, France
E-mail: matt.robshaw@francetelecom.com

Library of Congress Control Number: 2006928940

CR Subject Classification (1998): E.3, F.2.1, E.4, G.2, G.4

LNCS Sublibrary: SL 4 – Security and Cryptology

ISSN	0302-9743
ISBN-10	3-540-36597-4 Springer Berlin Heidelberg New York
ISBN-13	978-3-540-36597-6 Springer Berlin Heidelberg New York

Springer is a part of Springer Science+Business Media

springer.com

© Springer-Verlag Berlin Heidelberg 2006

Typesetting: Camera-ready by author, data conversion by Scientific Publishing Services, Chennai, India
Printed on acid-free paper   SPIN: 11799313   06/3142   5 4 3 2 1 0

Preface

Fast Software Encryption (FSE) 2006 is the 13th in a series of workshops on symmetric cryptography. It has been sponsored for the last five years by the International Association for Cryptologic Research (IACR), and previous FSE workshops have been held around the world:

1993 Cambridge, UK	1994 Leuven, Belgium	1996 Cambridge, UK
1997 Haifa, Israel	1998 Paris, France	1999 Rome, Italy
2000 New York, USA	2001 Yokohama, Japan	2002 Leuven, Belgium
2003 Lund, Sweden	2004 New Delhi, India	2005 Paris, France

The FSE workshop is devoted to research on fast and secure primitives for symmetric cryptography, including the design and analysis of block ciphers, stream ciphers, encryption schemes, analysis and evaluation tools, hash functions, and message authentication codes.

This year more than 100 papers were submitted to FSE for the first time. After an extensive review by the Program Committee, 27 papers were presented at the workshop. Of course, the program would not have been complete without the invited speaker, and the presentation by Eli Biham on the early history of differential cryptanalysis was particularly appreciated by workshop attendees.

We are very grateful to the Program Committee and to all the external reviewers for their hard work. Each paper was refereed by at least three reviewers, with papers from Program Committee members receiving at least five reviews. The local Organizing Committee at Graz worked very hard and we particularly thank Melanie Blauensteiner, Christophe De Cannière, Sharif Ibrahim, Florian Mendel, Norbert Pramstaller, Christian Rechberger, and Michaela Tretter-Dragovic for their generous efforts and strong support. In Paris we are indebted to Henri Gilbert and Helena Handschuh, who shared their valuable experience from FSE 2005, and to Côme Berbain and Olivier Billet for proofreading and preparing the FSE pre-proceedings.

To close, we thank the IACR secretariat, Kevin McCurley, and Shai Halevi for their help with the registration process and we thank the IACR for their support of FSE. We are grateful to K.U. Leuven for their web-based review software and we thank both France Telecom and Siemens, Munich for their financial support of FSE 2006.

Matt Robshaw	France Telecom R&D	FSE 2006 Program Chair
Vincent Rijmen	Graz University of Technology	FSE 2006 General Chair

FSE 2006
March 15-17, 2006, Graz, Austria

Sponsored by the
International Association for Cryptologic Research (IACR)

Program and General Chairs

Matt Robshaw ....France Telecom R&D   Program Chair
Vincent Rijmen ....Graz University of Technology   General Chair

Program Committee

Kazumaro Aoki ....NTT, Japan
Steve Babbage ....Vodafone, UK
Anne Canteaut ....INRIA, France
Carlos Cid ....Royal Holloway, University of London, UK
Joan Daemen ....STMicroelectronics, Belgium
Orr Dunkelman ....Technion–Israel Institute of Technology, Israel
Helena Handschuh ......Spansion, France
Thomas Johansson ......Lund University, Sweden
Antoine Joux ....DGA and University of Versailles, France
Charanjit Jutla ....IBM Watson Research Center, USA
Xuejia Lai ....Shanghai Jiaotong University, China
Stefan Lucks ....University of Mannheim, Germany
Mitsuru Matsui ....Mitsubishi Electric, Japan
Willi Meier ....FH Aargau, Switzerland
Kaisa Nyberg ....Helsinki University of Technology and Nokia, Finland
Elisabeth Oswald ....Graz University of Technology, Austria
Bart Preneel ...........K.U.Leuven, Belgium
Håvard Raddum ....University of Bergen, Norway
Matt Robshaw ....France Telecom R&D, France
Phillip Rogaway ........U.C.Davis, USA and Mah Fah Luang Univ., Thailand
Moti Yung ....RSA Security and Columbia University, USA

Sponsors

France Telecom R&D
Siemens, Munich

External Reviewers

Frederik Armknecht	Daniel Augot	Elad Barkan
Mihir Bellare	Côme Berbain	Dan Bernstein
Eli Biham	Alex Biryukov	John Black
Nick Bone	Christophe de Cannière	Pascale Charpin
Frédéric Didier	Claus Diem	Martin Feldhofer
Henri Gilbert	Louis Granboulan	Johann Groszschaedl
Shai Halevi	Philip Hawkes	Martin Hell
Christoph Herbst	Tetsu Iwata	Nathan Keller
John Kelsey	Alexander Kholosha	Lars Knudsen
Ted Krovetz	Ulrich Kühn	Simon Künzli
Joe Lano	Cédric Lauradoux	Stefan Mangard
Florian Mendel	Marine Minier	Sean Murphy
Anderson Nascimento	Philippe Oechslin	Matthew Parker
Ludovic Perret	Raphael C.-W. Phan	Norbert Pramstaller
Christian Rechberger	Vincent Rijmen	Greg Rose
Markku-Juhani O. Saarinen	Tsuneo Sato	Eric Schost
Tom Shrimpton	Hervé Sibert	Dirk Stegemann
John Steinberger	Michael Steiner	Stefan Tillich
Jean-Pierre Tillich	Hiroki Ueda	Charlotte Vikkelso
Johan Wallén	Dai Watanabe	Ralf-Philipp Weinmann
Doug Whiting	Go Yamamoto	Kan Yasuda

Table of Contents

Proposals

Hash Functions II

Modes and Models

Implementation and Bounds

Stream Ciphers II

Cryptanalysis of Achterbahn

Thomas Johansson[1], Willi Meier[2,*], and Frédéric Muller[3]

[1] Department of Information Technology, Lund University
P.O. Box 118, 221 00 Lund, Sweden
thomas@it.lth.se
[2] FH Aargau, 5210 Windisch, Switzerland
w.meier@fh-aargau.ch
[3] HSBC-France
Frederic.Muller@m4x.org

Abstract. We present several attacks against the Achterbahn stream cipher, which was proposed to the eSTREAM competition. We can break the reduced and the full version with complexity of 2^{55} and 2^{61} steps.

Extensions of our attacks are also described to break modified versions of the Achterbahn stream cipher, which were proposed following the publication of preliminary cryptanalysis results.

These attacks highlight some problems in the design principle of Achterbahn, *i.e.*, combining the outputs of several nonlinear (but small) shift registers using a nonlinear (but rather sparse) output function.

1 Introduction

The European project ECRYPT recently decided to launch a competition to identify new stream ciphers that might be suitable for widespread adoption. This project is called eSTREAM [3] and received 35 submissions, some of which have already been broken.

Among these new algorithms, a challenging new design is Achterbahn [5]. It is a relatively simple, hardware-oriented stream cipher, using a secret key of 80 bits. In this paper, we present several attacks which break the cipher faster than a brute force attack. Our results provide new directions to break stream ciphers built by combination of several small, but nonlinear shift registers, like Achterbahn.

2 Description of Achterbahn

2.1 General Structure

Achterbahn uses 8 small non-linear registers, denoted by $R_1, \ldots, R_8$. Their size ranges from 22 to 31 bits (see Table 1). The total size of the internal state is

---

[*] The second author is supported by Hasler Foundation www.haslerfoundation.ch under project number 2005.

M.J.B. Robshaw (Ed.): FSE 2006, LNCS 4047, pp. 1–14, 2006.

Table 1. Length of non-linear registers in Achterbahn

Register	Length
R_1	22
R_2	23
R_3	25
R_4	26
R_5	27
R_6	28
R_7	29
R_8	31

211 bits. At the t-th clock cycle, each register produces one output bit, denoted respectively by $y_1(t), \ldots, y_8(t)$. Then, the t-th output bit $z(t)$ of the stream cipher Achterbahn is produced by the filtering function F as

$$z(t) = F(y_1(t), y_2(t), y_3(t), y_4(t), y_5(t), y_6(t), y_7(t), y_8(t))$$
$$= y_1(t) \oplus y_2(t) \oplus y_3(t) \oplus y_4(t) \oplus$$
$$y_5(t)y_7(t) \oplus y_6(t)y_7(t) \oplus y_6(t)y_8(t) \oplus y_5(t)y_6(t)y_7(t) \oplus y_6(t)y_7(t)y_8(t).$$

We can observe that F is a sparse polynomial of degree 3. There are only 3 monomials of degree 2 and 2 monomials of degree 3. In the full version of Achterbahn, the input of F is not directly the output of each register, but a key-dependent combination of several consecutive outputs[1]. In the reduced version of Achterbahn, the input of F is directly the output of each register.

Each register is clocked similarily to a Linear Feedback Shift Register (LFSR), except that the feedback bit is not a linear function, but a polynomial of degree 4. Details of this clocking are not relevant in our attack. We refer to the original description of Achterbahn for more details [5].

2.2   Initialization

The internal state of Achterbahn is initialized from a secret key K of size 80 bits and from an initialization vector IV of length 80 bits.

First, the state of each register is loaded with a certain number of key bits (this number depends on the register length). Then, the rest of the key, followed by the IV, is introduced sequentially in each register. More precisely, this introduction consists simply in XORing the auxiliary input to the feedback bit during the register update. At some point, one bit in the register is forced to 1 to prevent the all-zero state. Before the encryption starts, several extra clockings are applied for diffusion purpose.

---

[1] The number of consecutive outputs involved in this linear combination varies from 6 for R_1 to 10 for R_8.

2.3 Evolutions of Achterbahn

In September 2005, some preliminary cryptanalysis results were announced on the eSTREAM website [6]. These results allow to break the reduced version of Achterbahn with 2^{56} computation steps and the full version with complexity of 2^{73} computation steps.

After the publication of these results, the designers of Achterbahn proposed to modify the output filter F of Achterbahn in order to strengthen the cipher [4]. This is a natural idea, given the nature of the published attacks. The first suggestion, that we will refer to as **Achterbahn-v2** in this paper, uses a new combining function F' instead of F, where

$$F'(y_1(t), \ldots, y_8(t)) = F(y_1(t), \ldots, y_8(t)) \oplus y_5(t)y_6(t) \oplus y_5(t)y_8(t) \oplus y_7(t)y_8(t).$$

Another alternative suggested in [4] is to replace F by F'' defined as

$$F''(y_1(t), \ldots, y_8(t)) = y_1(t) \oplus y_2(t) \oplus y_3(t) \oplus \sum_{4 \leq i < j \leq 8} y_i(t)y_j(t) \oplus$$

$$\sum_{4 \leq i < j < k \leq 8} y_i(t)y_j(t)y_k(t) \oplus \sum_{4 \leq i < j < k < l \leq 8} y_i(t)y_j(t)y_k(t)y_l(t).$$

We refer to **Achterbahn-v3** for the cipher instantiated with F''.

3 Weaknesses of Achterbahn's Design

3.1 General Observations About the Design

Combination of several small Linear Feedback Shift Registers (LFSR) is a well-known method for building stream ciphers. The output of the registers are generally combined with a function F, in order to produce one keystream bit (see Figure 1). A popular example is the algorithm E0 [1], which is used in the Bluetooth technology[2]. Unfortunately such constructions have some problems, that

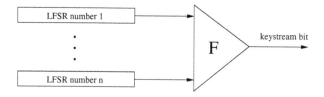

Fig. 1. Stream Cipher built by Combination of LFSR's

originate from the linearity of the LFSR's. For instance, correlation attacks [8,9] exploit linear approximations of the function F to attack the whole stream

---

[2] E0 has the particularity that the function F uses a small auxiliary memory.

cipher. Another method is algebraic attacks [2] that take advantage of low degree polynomial equations satisfied by F.

Criteria that should be satisfied by the boolean function F, in order to counter such attacks have been widely studied. However there appears to be limitations that cannot easily be removed. To improve the designs, it is often suggested to replace linear registers by nonlinear registers. This idea is the bottomline of Achterbahn's design.

3.2   Linear Complexity of Achterbahn

If the linear registers of Figure 1 are replaced by nonlinear registers, one may expect to counter many problems arising from the linearity of LFSR's. A usual tool to analyze such constructions is the **linear complexity**. For a binary sequence, it is defined as the length of the shortest LFSR that could generate the sequence.

For a LFSR of length n bits, the linear complexity of its output sequence is $L = n$, provided its feedback polynomial is properly chosen. For a nonlinear register, it is not always easy to compute the linear complexity of its output sequence, but clearly it cannot exceed its period. In the case of Achterbahn, the keystream bit b is computed by

$$b = F(y_1, \ldots, y_8).$$

Then, it is well-known that the linear complexity of the keystream sequence is at most

$$L = F(L_1, \ldots, L_8),$$

where L_i denotes the linear complexity of each single register and F is now seen as a polynomial on the integers, with its coefficients $\in \{0, 1\}$. This observation shows that **it would be insecure to combine the small nonlinear registers using a linear function**. Indeed, in this case, the linear complexity L of Achterbahn would be bounded by 8×2^{31} since 31 is the length of the largest register.

For Achterbahn, F is not linear, but its algebraic degree is 3. The original paper [5] does not contain an exact proof of the linear complexity of the 8 nonlinear registers, but it is reasonable to assume that $L_i \simeq 2^{n_i}$ where n_i denotes the length of register R_i. With this assumption, the linear complexity of Achterbahn's outputs is :

$$L \leq 2^{28} \times 2^{29} \times 2^{31} = 2^{88}.$$

If we apply the Berlekamp-Massey algorithm [7], we can expect to distinguish this sequence if we analyze 2^{89} known output bits. Since the running time of Berlekamp-Massey is about L^2, this attack is way above the complexity of a brute-force attack.

3.3 Ideas for Improvement

These observations about the linear complexity were taken into account by the designers of Achterbahn (see page 20 of [5]). However, we should also consider that several refinements are possible :

- The output function is **sparse**. Indeed $z(t)$ is computed by a simple filter, which is almost linear. For instance, when $y_6(t) = 0$, only one nonlinear term remains. If $y_5(t)$ is also equal to 0, the output function becomes purely linear.
- Each single register has a **small period**. This is unavoidable due to the small size of each register (31 bits for the largest one, R_8).
- Each register is **autonomous**. Therefore when we guess its initial state, we know its content at all stages of the encryption.

Our idea is to guess the initial state of two registers (R_5 and R_6). Then we **select** particular positions in the output sequence, for which

$$y_5 = y_6 = 0.$$

All nonlinear terms in F cancel out, so the linear complexity of this subsequence is much smaller than for the whole Achterbahn. Finally, we test if several **parity checks**, resulting from the low linear complexity are satisfied or not. Hence, we can determine when the initial guess on R_5 and R_6 is correct.

Several tricks are needed in order for the attack to work properly. In particular, it is important to find low-weight parity checks. The details of this attack are given in the next section. Attacks in the same vein can also be mounted in the case of Achterbahn-v2 and Achterbahn-v3.

4 Cryptanalysis of Reduced Achterbahn

4.1 Preliminary

Our starting point is to observe that when $y_5(t) = 0$ and $y_6(t) = 0$, the output function becomes purely linear, so

$$z(t) = l(t) = y_1(t) \oplus y_2(t) \oplus y_3(t) \oplus y_4(t).$$

Although its period is rather large, $l(t)$ has a very **low linear complexity** L as pointed out in Section 3.2. Indeed, L is bounded by

$$L \leq 2^{n_1} + 2^{n_2} + 2^{n_3} + 2^{n_4} \simeq 2^{26}.$$

By definition, l can be generated by a LFSR of length L, so it will satisfy some **parity checks** involving L consecutive bits at most. Actually, it can be demonstrated that **sparse parity checks** are satisfied, which will prove to be crucial in the rest of our attack.

4.2   Construction of Sparse Parity Checks

We denote by T_i the period of register R_i. From [5], we can see that

$$T_1 = 2^{22} - 1,$$
$$T_2 = 2^{23} - 1,$$
$$T_3 = 2^{25} - 1,$$
$$T_4 = 2^{26} - 1.$$

Let

$$ll(t) = l(t) \oplus l(t + T_1).$$

Because the period of the first register is T_1, this expression does not contain any term in y_1. Similarly, define

$$lll(t) = ll(t) \oplus ll(t + T_2),$$
$$llll(t) = lll(t) \oplus lll(t + T_3).$$

Here $llll(t)$ contains no term in y_2 or y_3, so it is a combination of bits coming from the register R_4 only. Thus it satisfies

$$llll(t) = llll(t + T_4).$$

In other terms, we have the following relation on the bits $l(i)$,

$$
\begin{aligned}
0 = {} & l(t) + l(t + T_1) + l(t + T_2) + l(t + T_3) + l(t + T_4) \\
& + l(t + T_1 + T_2) + l(t + T_1 + T_3) + l(t + T_1 + T_4) \\
& + l(t + T_2 + T_3) + l(t + T_2 + T_4) + l(t + T_3 + T_4) \\
& + l(t + T_1 + T_2 + T_3) + l(t + T_1 + T_2 + T_4) + l(t + T_1 + T_3 + T_4) \\
& + l(t + T_2 + T_3 + T_4) + l(t + T_1 + T_2 + T_3 + T_4).
\end{aligned}
$$

This is the basic **parity check** on $l(t)$ that we will use in our attack. We can observe that it is the XOR of 16 different bits from the sequence $l(i)$. They all belong to a time interval of length

$$T_{max} = T_1 + T_2 + T_3 + T_4 = 113246204 \simeq 2^{26.75}.$$

Such parity checks are satisfied by the keystream sequence, under certain constraints on the outputs of the registers R_5 and R_6 (several bits $y_5(i)$ and $y_6(i)$ must be equal to 0).

We split the attack in two phases. First, we **precompute** particular states of R_5 and R_6 for which $z(t) = l(t)$. Then we look at a given keystream sequence and test when the parity check is satisfied. This information is used to **identify** one of the precomputed states of R_5 and R_6.

4.3   Precomputation

The goal of the precomputation step is to identify particular state values of R_5 and R_6 for which the parity checks will be satisfied. For that, we need $y_5(t)$ and $y_6(t)$ to be both equal to 0 for the 16 positions that appear in the previous parity check. Consider the case of register R_5 first. We are looking for states of R_5 at time t such that the corresponding outputs satisfy :

$$y_5(t) = 0,$$
$$y_5(t + T_1) = 0,$$
$$y_5(t + T_2) = 0,$$
$$y_5(t + T_3) = 0,$$
$$y_5(t + T_4) = 0,$$
$$y_5(t + T_1 + T_2) = 0,$$
$$y_5(t + T_1 + T_3) = 0,$$
$$y_5(t + T_1 + T_4) = 0,$$
$$y_5(t + T_2 + T_3) = 0,$$
$$y_5(t + T_2 + T_4) = 0,$$
$$y_5(t + T_3 + T_4) = 0,$$
$$y_5(t + T_1 + T_2 + T_3) = 0,$$
$$y_5(t + T_1 + T_2 + T_4) = 0,$$
$$y_5(t + T_1 + T_3 + T_4) = 0,$$
$$y_5(t + T_2 + T_3 + T_4) = 0,$$
$$y_5(t + T_1 + T_2 + T_3 + T_4) = 0.$$

If we enumerate the 2^{27} possible states of R_5 and clock the register T_{max} times, we can find all states that satisfy the above equations[3]. The expected number of solutions is

$$2^{27} \times 2^{-16} = 2^{11},$$

since there are 16 binary constraints to satisfy simultaneously. The complexity of this stage is about $2^{27} \times T_{max} = 2^{53.75}$. It is possible to do it more efficiently if we store the whole sequence of outputs from R_5, but this step will prove not to be the bottleneck of our attack. Similarly, we can find 2^{12} states of R_6 that satisfy the same 16 constraints. The corresponding time complexity is $2^{54.75}$. To summarize, we can enumerate

$$2^{12} \times 2^{11} = 2^{23}$$

favorable states for the registers R_5 and R_6. We store these 2^{23} states in an auxiliary table.

---

[3] We could envisage degenerated registers, for which these equations can never occur simultaneously. However, this is not the case in Achterbahn, and such degenerations would probably lead to other types of attacks.

In addition, for each favorable state, we clock R_5 and R_6 until we reach another favorable state. In the auxiliary table, we store the distance from each favorable state to the next one. This information will be useful in the next section. In average, we need 2^{32} clockings per favorable state, resulting in a time complexity of $2^{23} \times 2^{32} = 2^{55}$ steps.

4.4   Identification

We suppose that we are given a certain sequence of 2^{40} keystream bits. To simplify what follows, we start by computing the parity checks on the keystream bits,

$$
\begin{aligned}
pc(t) = {} & z(t) + z(t + T_1) + z(t + T_2) + z(t + T_3) + z(t + T_4) \\
& + z(t + T_1 + T_2) + z(t + T_1 + T_3) + z(t + T_1 + T_4) \\
& + z(t + T_2 + T_3) + z(t + T_2 + T_4) + z(t + T_3 + T_4) \\
& + z(t + T_1 + T_2 + T_3) + z(t + T_1 + T_2 + T_4) + z(t + T_1 + T_3 + T_4) \\
& + z(t + T_2 + T_3 + T_4) + z(t + T_1 + T_2 + T_3 + T_4),
\end{aligned}
$$

for $t = 0 \ldots 2^{40} - T_{max}$.

It is very likely that R_5 and R_6 are in a favorable state, for at least one of the first 2^{32} positions in the sequence. We call t_0 such a position. Then we must have $pc(t_0) = 0$. This is only one bit of information, which is not sufficient to identify a favorable state.

Therefore, we enumerate all positions t_0 from 0 to 2^{32} and all the 2^{23} favorable states. Suppose we have $pc(t_0) = 0$ (otherwise we discard immediately the candidate). Then we use the auxiliary table to search for the next favorable state. Suppose the table says it will occur at the position $t_1 > t_0$. Then we jump to the position t_1 in the keystream sequence and check if $pc(t_1) = 0$. If it is not the case, we discard this candidate. Otherwise, we iterate the process.

Since we have 2^{40} keystream bits and the distance between two favorable states is about 2^{32}, we might be able to iterate up to $2^8 = 256$ times the process with success. This is sufficient to identify a favorable state, while a false alarm is very unlikely.

With our "early abort" strategy, we need to test only an average of 2 parity checks for each of the $2^{32} \times 2^{23} = 2^{55}$ candidates. So the time complexity of this phase is about 2^{56} steps.

4.5   Retrieving the Key

We have identified the value of the state of R_5 and R_6 at a certain position t_0 in the output sequence. We would like to retrieve the key from this information, so a natural idea is to backtrack the updating of these registers. This is easy to do until we reach the initial state, since the update is invertible.

Next, we want to backtrack the initialization process of Achterbahn. During the extra clockings for diffusion and during the IV introduction, there is no difficulty to backtrack, since we can always predict the feedback bit. Unfortunately, we can no longer backtrack during the phase where the key was introduced.

Then, our idea is to perform a **meet-in-the-middle attack**. We split the key in two halves of 40 bits each. On the one hand, we guess the first 40 bits from the key and predict the state of R_5 and R_6 after the introduction of these 40 bits. On the other hand, we guess the last 40 bits from the key and backtrack the introduction of these bits from the known state of R_5 and R_6. We search for a match between the two lists of 2^{40} elements[4].

We should observe $2^{40} \times 2^{40} \times 2^{-55} \simeq 2^{25}$ matches since the lengthes of R_5 and R_6 sum up to 55 bits. Each of them provides a key candidate, which is easy to test by producing several keystream bits. To summarize, from one known state of R_5 and R_6, we can retrieve the secret key with time and memory complexity of 2^{40}.

4.6   Analysis

Both the precomputation and the identification phase of our attack have a time complexity of about 2^{56} steps. In addition, we need to store about 2^{40} (parity checks of) keystream bits and an auxiliary table of size 2^{23} after the precomputation phase.

The key recovery phase can be achieved using different trade-offs between time and memory. It is possible to do it with time and memory of 2^{40}. But a more reasonable trade-off could be with time 2^{50} and memory 2^{30}.

4.7   Cryptanalysis of Full Achterbahn

If we want to attack the full Achterbahn, we must take into account the key-dependent linear combination used to compute the outputs of each register. This additional feature preserves the period of each registers, as well as the properties of the function F, so the observations on parity checks are unchanged. However, when looking for the favorable states of R_5 and R_6, we must guess in addition the $8 + 9 = 17$ key-dependent taps.

Depending on our guess on these key-dependent taps, we obtain a different set of favorable states. Therefore we must repeat 2^{17} times the second phase of our attack, and the whole complexity for attacking the full Achterbahn is about 2^{73} computation steps.

5   Another Cryptanalysis of Achterbahn

In this section, we propose another attack technique against Achterbahn, based on approximating its output function by a linear expression.

---

[4] One bit is forced to 1 in each register to avoid the "all zero" state. The update is therefore not invertible, but we can easily guess the value of the erased bit, which has a negligible impact on the time complexity.

5.1   Linear Approximations of the Output Function

Reconsider Achterbahn's output function given in Section 2,

$$z(t) = F(y_1(t), y_2(t), y_3(t), y_4(t), y_5(t), y_6(t), y_7(t), y_8(t))$$
$$= y_1(t) \oplus y_2(t) \oplus y_3(t) \oplus y_4(t) \oplus$$
$$y_5(t)y_7(t) \oplus y_6(t)y_7(t) \oplus y_6(t)y_8(t) \oplus y_5(t)y_6(t)y_7(t) \oplus y_6(t)y_7(t)y_8(t).$$

We use the notation $l(t) = y_1(t) \oplus y_2(t) \oplus y_3(t) \oplus y_4(t)$ to refer to the linear part of F. it is easy to observe that F verifies the following linear approximations,

$$z(t) = l(t) \oplus y_5(t) \text{ with probability } 10/16,$$
$$z(t) = l(t) \oplus y_6(t) \text{ with probability } 12/16,$$
$$z(t) = l(t) \oplus y_7(t) \text{ with probability } 12/16,$$
$$z(t) = l(t) \oplus y_8(t) \text{ with probability } 10/16.$$

In particular, we focus on the second approximation,

$$z(t) = l(t) \oplus y_6(t), \tag{1}$$

with probability $\frac{12}{16} = 0.75 = 0.5\,(1 + 0.5)$. Therefore the **bias** of this linear approximation is $\varepsilon = 0.5$.

5.2   Using the Sparse Parity Checks

Similarly to Section 4.2, we can construct parity checks satisfied by the sequence of bits $l(t) \oplus y_6(t)$. Such a parity check will involve 32 keystream bits (instead of 16 like in Section 4.2) distant from at most

$$T_{max} = T_1 + T_2 + T_3 + T_4 + T_6 = 381681659 \simeq 2^{28.51}$$

positions. This parity check is not directly satisfied by the output sequence of Achterbahn since $l(t) \oplus y_6(t)$ is only an approximation of the output function. However we can sum up 32 times the linear approximation (1) over different values of t, which has the effect of multiplying the biases. Therefore, the parity check is satisfied by the sequence $z(t)$ with probability

$$0.5\,(1 + \varepsilon^{32}) = 0.5\left(1 + \frac{1}{2^{32}}\right).$$

Therefore if we consider a sequence of 2^{64} output bits and evaluate all the parity checks, we will detect this bias. This allows to distinguish Achterbahn's outputs from truly random sequences. In addition, this attack is not affected if we add key-dependent taps to each register, so its complexity is the same for the reduced and for the full Achterbahn.

5.3   Guessing One Register

A natural extension of the previous distinguishing attack consists in guessing the initial content of register R_1 (there are 2^{23} candidates). Then, we can eliminate the term $y_1(t)$ in the previous linear approximation. Consequently, the weight of the parity check drops from 32 to 16, bringing the bias from 2^{-32} to 2^{-16}.

For the correct guess, we detect a bias by looking at 2^{32} keystream bits, while there is no bias for incorrect guesses. Once the correct guess has been identified, it is straightforward to repeat the process to target other registers. To summarize, this attack costs about 2^{55} computation steps and requires 2^{32} keystream bits. For the full Achterbahn, the number of guesses for R_1 is 2^{29} instead of 2^{23} increasing the complexity of the key recovery from 2^{55} to 2^{61}.

6   The Case of Achterbahn-v2

6.1   Time-Memory Trade-Off

We reconsider the attack described in Section 4 in order to break the reduced version of Achterbahn-v2. Because of the linear complexity arguments, one can still construct the sparse parity checks satisfied by the linear part $l(t)$ (F and F' have both the same linear part). The criteria chosen by the designers is that it is no longer possible to cancel out the nonlinear part

$$nl(t) = z(t) \oplus l(t) \tag{2}$$

by guessing 2 registers only, like for the "basic" Achterbahn. However, $nl(t)$ depends only on the initial state of the registers R_5, R_6, R_7 and R_8, which represents $27 + 28 + 29 + 31 = 115$ unknown bits. So we can apply the usual **time-memory-data trade-off** for stream ciphers :

- **Precomputation step:** Pick at random $2^{57.5}$ initial states for the registers R_5, R_6, R_7 and R_8 Then evaluate the parity check of Section 4.2 on the sequence of $nl(t)$ bits. We do this for 115 parity checks and we store the resulting vector of 115 bits in a table. Finally, the $2^{57.5}$ entries of this table are sorted according to the stored value.
- **Identification step:** Analyze a sequence of $2^{57.5}$ keystream bits and for each encountered position, evaluate the first 115 parity checks.

The parity check evaluated on the $z(t)$ bits is equal to the parity check evaluated on the $nl(t)$ bits, since it cancels out on the $l(t)$ bits (see relation (2)). Therefore, when we find a match, we learn the state of R_5, R_6, R_7 and R_8 during the encryption process. It turns out that a match is indeed expected here, due to the birthday paradox.

Besides, it is clear that the state of the 4 remaining registers, R_1, R_2, R_3 and R_4 as well as the secret key could be further retrieved, with an analysis similar to the one described in Section 4. We estimate the cost of this attack to $2^{57.5}$ in time and data complexity.

There is a technical detail to mention. Evaluating each parity check (in the pre-computation step) requires to handle some $nl(t)$ bits which are located $T_{max} \simeq 2^{26.75}$ positions apart. To deal with this, we suggest to first compute the contribution of each separate register to the parity checks, for each possible state. This requires about $T_{max} \times 2^{31} \simeq 2^{57.75}$ steps for the longest register, *i.e.*, R_8. Then for each of the $2^{57.5}$ candidates, evaluating the 115 parity checks just requires several table look-ups and several XOR's. Arguably, the basic step in our attack costs about as much as testing one key in an exhaustive search.

This attack applies to the "reduced" Achterbahn-v2. Considering the case of the "full" Achterbahn-v2, we observe that each register has an "extra" entropy of $8, 9$ or 10 bits, due to the secret feed-forward. This sums up to 36 new unknowns. An exhaustive search over these unknowns is impossible, since it would bring the complexity above exhaustive search. Hence, we propose an alternative strategy in order to break the full Achterbahn-v2.

6.2   Breaking the Full Achterbahn-v2

We propose a different extension of the attack of Section 4. The modification is that we target positions where $y_5(t) = y_6(t) = 1$ instead of 0. For these selected positions,

$$F'(y_1(t), \ldots, y_8(t)) = y_1(t) \oplus y_2(t) \oplus y_3(t) \oplus y_4(t) \oplus y_7(t) = \lambda(t)$$

So F' is reduced to a linear term $\lambda(t)$ with 5 terms (instead of 4 terms like $l(t)$) We view the bit $y_1(t) \oplus y_2(t)$ as the output of a single register with period $T_1 \cdot T_2$, so again we can write sparse parity checks of weight 16, involving terms located

$$T_{max} = T_1 \cdot T_2 + T_3 + T_4 + T_7 \simeq 2^{45}$$

positions apart. T_{max} is now much larger than previously, but as already pointed out in Section 4.3 the precomputation of "favorable" states for the register R_5 (or for R_6) can be done without clocking it T_{max} times (T_5 clockings are sufficient, taking into account its periodicity). Similarily, the increase of T_{max} does not change the time complexity of the identification step described in Section 4.4.

To summarize, the full Achterbahn-v2 can be attacked with the same complexity as the attack against the full Achterbahn described in Section 4.7, except that the data complexity is increased to $T_{max} = 2^{45} + 2^{40} \simeq 2^{45}$ known keystream bits.

7   The Case of Achterbahn-v3

Achterbahn-v3 has a new output function F'' which is not as sparse as its predecessors. However, F'' is approximable by a sparse linear function. Then the attack proposed in Section 5 can be applied. First, we observe that

$$z(t) = F''(y_1(t) \ldots y_8(t)) = y_1(t) \oplus y_2(t) \oplus y_3(t) \oplus y_4(t),$$

with probability $\frac{9}{16} = 0.5 \cdot (1 + \frac{1}{8})$. Then we apply the attack of Section 5 with the register-guessing trick. We guess the initial state of register R_1 (complexity 2^{22}), then we evaluate the parity check (of weight 8 since only 3 terms remain). The resulting bias is

$$\varepsilon = \left(\frac{1}{8}\right)^8 = 2^{-24},$$

so 2^{48} keystream bits are needed to detect the correct initial state of R_1. The time complexity of this attack is $2^{22} \times 2^{48} = 2^{70}$ for the reduced Achterbahn-v3. For the full Achterbahn-v3, there is an auxiliary factor of 2^6 to take into account for the secret feed-forward.

8   Conclusion

We proposed several attacks against Achterbahn and its modified versions. Table 2 summarizes all these cryptanalysis results. In spite of the nonlinear update, the fact that all registers are small and autonomous allows us to envisage several new attacks. Our idea is first to observe that a linear output function would give a low linear complexity and therefore an easily brekable cipher. Then we suggest to approximate the output function by a linear expression, and we build parity checks that the linearized version of Achterbahn should satisfy.

Table 2. Summary of cryptanalysis results against Achterbahn

Type of Attack	Technique	Target	Complexity	Data
Key recovery (Sec. 4)	Linear Complexity	reduced Achterbahn	2^{56}	2^{40}
		full Achterbahn	2^{73}	2^{40}
Distinguisher (Sec. 5)	Linear Approx.	reduced Achterbahn	2^{64}	2^{64}
		full Achterbahn	2^{64}	2^{64}
Key recovery (Sec. 5)	Linear Approx.	reduced Achterbahn	2^{55}	2^{32}
		full Achterbahn	2^{61}	2^{32}
Key recovery (Sec. 6)	Time-Memory Linear Complexity	reduced Achterbahn-v2	$2^{57.5}$	$2^{57.5}$
		full Achterbahn-v2	2^{73}	2^{45}
Key recovery (Sec. 7)	Linear Approx.	reduced Achterbahn-v3	2^{70}	2^{48}
		full Achterbahn-v3	2^{76}	2^{48}

Following the publication of some preliminary results, the designers of Achterbahn suggested to modify the output filter. However, we pointed out that some attacks of the same nature are still possible. It is interesting to notice that our attacks are independent of the feedback of the nonlinear registers, so it illustrates some problems of the design itself, rather than an unfortunate instantiation.

References

1. Bluetooth.   Bluetooth Specification, November 2003.   available at http://www.bluetooth.org.
2. N. Courtois and W. Meier. Algebraic Attacks on Stream Ciphers with Linear Feedback. In E. Biham, editor, *Advances in Cryptology – Eurocrypt'03*, volume 2656 of *Lectures Notes in Computer Science*, pages 345–359. Springer, 2003.
3. eSTREAM - The ECRYPT Stream Cipher Project
   `http://www.ecrypt.eu.org/stream/`.
4. B. Gammel, R. Göttfert, and O. Kniffler. Improved Boolean Combining Functions for Achterbahn. eSTREAM, ECRYPT Stream Cipher Project, Report 2005/072, 2005. `http://www.ecrypt.eu.org/stream`.
5. B. Gammel, R. Göttfert, and O. Kniffler. The Achterbahn Stream Cipher. eSTREAM, ECRYPT Stream Cipher Project, Report 2005/002, 2005. `http://www.ecrypt.eu.org/stream`.
6. T. Johansson, W. Meier, and F. Muller. Cryptanalysis of Achterbahn. eSTREAM, ECRYPT Stream Cipher Project, Report 2005/064, 2005. `http://www.ecrypt.eu. org/stream`.
7. J. Massey. Shift-Register Synthesis and BCH Decoding. *IEEE Transactions on Information Theory*, 15:122–127, 1969.
8. W. Meier and O. Staffelbach. Fast Correlations Attacks on Certain Stream Ciphers. In *Journal of Cryptology*, pages 159–176. Springer, 1989.
9. T. Siegenthaler. Correlation-immunity of Nonlinear Combining Functions for Cryptographic Applications. In *IEEE Transactions on Information Theory*, volume 30, pages 776–780, 1984.

Cryptanalysis of Grain[*]

Côme Berbain[1], Henri Gilbert[1], and Alexander Maximov[2]

[1] France Telecom Research and Development
38-40 rue du Général Leclerc, 92794 Issy-les-Moulineaux, France
[2] Dept. of Information Technology, Lund University, Sweden
P.O. Box 118, 221 00 Lund, Sweden
{come.berbain, henri.gilbert}@francetelecom.com
movax@it.lth.se

Abstract. Grain [11] is a lightweight stream cipher proposed by M. Hell,
T. Johansson, and W. Meier to the eSTREAM call for stream cipher
proposals of the European project ECRYPT [5]. Its 160-bit internal
state is divided into a LFSR and an NFSR of length 80 bits each. A
filtering boolean function is used to derive each keystream bit from the
internal state. By combining linear approximations of the feedback function of the NFSR and of the filtering function, it is possible to derive
linear approximation equations involving the keystream and the LFSR
initial state. We present a key recovery attack against Grain which requires 2^{43} computations and 2^{38} keystream bits to determine the 80-bit
key.

Keywords: Stream cipher, Correlation attack, Walsh transform.

1 Introduction

Stream ciphers are symmetric encryption algorithms based on the concept of
pseudorandom keystream generator. In the typical case of a binary additive
stream cipher, the key and an additional parameter named initialization vector
(IV) are used to generate a binary sequence called keystream which is bitwise
combined with the plaintext to provide the ciphertext. Although it seems rather
difficult to construct a very fast and secure stream cipher, some efforts to achieve
this have recently been deployed. The NESSIE project [24] launched in 1999 by
the European Union did not succeed in selecting a secure enough stream cipher.
Recently, the European Network of Excellence in Cryptology ECRYPT launched
a call for stream cipher proposals named eSTREAM [5]. The candidate stream
ciphers were submitted in May 2005. Those candidates are divided into software
oriented and hardware oriented ciphers.

---

[*] The work described in this paper has been supported in part by Grant VR 621-
2001-2149, in part by the French Ministry of Research RNRT X-CRYPT project
and in part by the European Commission through the IST Program under Contract
IST-2002-507932 ECRYPT.

M.J.B. Robshaw (Ed.): FSE 2006, LNCS 4047, pp. 15–29, 2006.

Hardware oriented stream ciphers are specially designed so that their implementation requires a very small number of gates. Such ciphers are useful in mobile systems, e.g. mobile phones or RFID, where minimizing the number of gates and power consumption is more important than very high speed.

One of the new hardware candidates submitted to eSTREAM is a stream cipher named Grain [11] which was developed by M. Hell, T. Johansson, and W. Meier[1] as an alternative to stream ciphers like GSM A5/1 or Bluetooth E_0. It uses a 80-bit key and a 64-bit initialization vector to fill in an internal state of size 160 bits divided into a *nonlinear feedback shift register* (NFSR) and a *linear feedback shift register* (LFSR) of length 80 bits each. At each clock pulse, one keystream bit is produced by selecting some bits of the LFSR and of the NFSR and applying a boolean function. It is well known that LFSR sequences satisfy several statistical properties one would expect from a random sequence, but do not offer any security. Their combination with NFSR sequences is expected to improve the security. However, NFSR based constructions have not yet been as well studied as LFSR based constructions. The claimed security level of Grain is 2^{80}, and it was conjectured by the authors of Grain that there exists no attack significantly faster than exhaustive search.

In this paper, we describe two key recovery attacks against Grain. The proposed attacks exploit linear approximations of the output function. The first one requires 2^{55} operations, 2^{49} bits of memory, and 2^{51} keystream bits, and the second one requires 2^{43} operations, 2^{42} bits of memory, and 2^{38} keystream bits.

This paper is organized as follows. We first describe the Grain stream cipher (Section 2) and we derive some linear approximations involving the LFSR and the keystream (Section 3). We then present two techniques for recovering the initial state of the LFSR (Section 4). Finally, we present a technique allowing to recover the initial state of the NFSR once we know the LFSR initial state (Section 5).

2   Description of Grain

Grain [11] is based upon three main building blocks: an 80-bit linear feedback shift register, an 80-bit nonlinear feedback shift register, and a nonlinear filtering function. Grain is initialized with the 80-bit key K and the 64-bit initialization value IV. The cipher output is an L-bit keystream sequence $(z_t)_{t=0,\ldots,L-1}$.

The current LFSR content is denoted by $Y^t = (y_t, y_{t+1}, \ldots, y_{t+79})$. The LFSR is governed by the linear recurrence:

$$y_{t+80} = y_{t+62} \oplus y_{t+51} \oplus y_{t+38} \oplus y_{t+23} \oplus y_{t+13} \oplus y_t.$$

---

[1] The design of Grain was also submitted and recently accepted for publication in *the International Journal of Wireless and Mobile Computing, Special Issue on Security of Computer Network and Mobile Systems*.

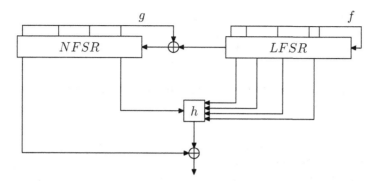

The current NFSR content is denoted by $X^t = (x_t, x_{t+1}, \ldots, x_{t+79})$. The NFSR feedback is disturbed by the output of the LFSR, so that the NFSR content is governed by the recurrence:

$$x_{t+80} = y_t \oplus g(x_t, x_{t+1}, \ldots, x_{t+79}),$$

where the expression of nonlinear feedback function g is given by

$$
\begin{aligned}
g(x_t, x_{t+1}, \ldots, x_{t+79}) = {} & x_{t+63} \oplus x_{t+60} \oplus x_{t+52} \oplus x_{t+45} \oplus x_{t+37} \oplus x_{t+33} \oplus x_{t+28} \\
& \oplus x_{t+21} \oplus x_{t+15} \oplus x_{t+9} \oplus x_t \oplus x_{t+63}x_{t+60} \oplus x_{t+37}x_{t+33} \\
& \oplus x_{t+15}x_{t+9} \oplus x_{t+60}x_{t+52}x_{t+45} \oplus x_{t+33}x_{t+28}x_{t+21} \\
& \oplus x_{t+63}x_{t+45}x_{t+28}x_{t+9} \oplus x_{t+60}x_{t+52}x_{t+37}x_{t+33} \\
& \oplus x_{t+63}x_{t+60}x_{t+21}x_{t+15} \oplus x_{t+63}x_{t+60}x_{t+52}x_{t+45}x_{t+37} \\
& \oplus x_{t+33}x_{t+28}x_{t+21}x_{t+15}x_{t+9} \\
& \oplus x_{t+52}x_{t+45}x_{t+37}x_{t+33}x_{t+28}x_{t+21}.
\end{aligned}
$$

The cipher output bit z_t is derived from the current LFSR and NFSR states as the exclusive or of the masking bit x_t and a nonlinear filtering function h as follows:

$$
\begin{aligned}
z_t &= x_t \oplus h(y_{t+3}, y_{t+25}, y_{t+46}, y_{t+64}, x_{t+63}) \\
&= h'(y_{t+3}, y_{t+25}, y_{t+46}, y_{t+64}, x_t, x_{t+63}) \\
&= x_t \oplus x_{t+63}p_t \oplus q_t,
\end{aligned}
$$

where p_t and q_t are the functions of $y_{t+3}, y_{t+25}, y_{t+46}, y_{t+64}$ given by:

$$
\begin{aligned}
p_t &= 1 \oplus y_{t+64} \oplus y_{t+46}(y_{t+3} \oplus y_{t+25} \oplus y_{t+64}), \\
q_t &= y_{t+25} \oplus y_{t+3}y_{t+46}(y_{t+25} \oplus y_{t+64}) \oplus y_{t+64}(y_{t+3} \oplus y_{t+46}).
\end{aligned}
$$

The boolean function h is correlation immune of the first order. As noticed in [11], "this does not preclude that there are correlations of the output of $h(x)$ to sums of inputs", but the designers of Grain appear to have expected the NFSR masking bit x_t to make it impractical to exploit such correlations.

The key and IV setup consists of loading the key bits in the NFSR, loading the 64-bit IV followed by 16 ones in the LFSR, and clocking the cipher 160 times in a special mode where the output bit is fed back into the LFSR and the NFSR. Once the key and IV have been loaded, the keystream generation mode described above is activated and the keystream sequence (z_t) is produced.

3 Deriving Linear Approximations of the LFSR Bits

3.1 Linear Approximations Used to Derive the LFSR Bits

The purpose of the attack is, based on a keystream sequence $(z_t)_{t=0...L-1}$ corresponding to an unknown key K and a known IV value, to recover the key K. The initial step of the attack is to derive a sufficient number N of linear approximation equations involving the 80 bits of the initial LFSR state $Y^0 = (y_0, \ldots, y_{79})$ (or equivalently a sufficient number N of linear approximation equations involving bits of the sequence (y_t)) to recover the value of Y^0. Hereafter, as will be shown in Section 5, the initial NFSR state X^0 and the key K can then be easily recovered.

The starting point for the attack consists in noticing that though the NFSR feedback function g is balanced, the function g' given by $g'(X^t) = g(X^t) \oplus x_t$ is unbalanced. We have:

$$Pr\{g'(X^t) = 1\} = \frac{522}{1024} = \frac{1}{2} + \epsilon_{g'},$$

where $\epsilon_{g'} = \frac{5}{512}$. It is useful to notice that the restriction of g' to input values X^t such that $x_{t+63} = 0$ is totally balanced and that the imbalance of the function g' is exclusively due to the imbalance of the restriction of g' to input values X^t such that $x_{t+63} = 1$.

If one considers one single output bit z_t, the involvement of the masking bit x_t in the expression of z_t makes it impossible to write any useful approximate relation involving only the Y^t bits. But if one considers the sum $z_t \oplus z_{t+80}$ of two keystream bits output at a time interval equal to the NFSR length 80, the $x_t \oplus x_{t+80}$ contribution of the corresponding masking bits is equal to $g'(X^t) \oplus y_t$, and is therefore equal to y_t with probability $\frac{1}{2} + \epsilon_{g'}$. As for the other terms of $z_t \oplus z_{t+80}$, they can be approximated by linear functions of the bits of the sequence (y_t). In more details:

$$z_t \oplus z_{t+80} = g'(X^t) \oplus y_t \oplus h(y_{t+3}, y_{t+25}, y_{t+46}, y_{t+64}, x_{t+63})$$
$$\oplus h(y_{t+83}, y_{t+105}, y_{t+126}, y_{t+144}, x_{t+143}).$$

To find linear approximations of the term $h(y_{t+3}, y_{t+25}, y_{t+46}, y_{t+64}, x_{t+63})$, we can restrict our search, since the restriction of $g'(X^t)$ to input values such that $x_{t+63} = 0$ is balanced, to input values such that $x_{t+63} = 1$, which amounts to finding linear approximations of $p_t \oplus q_t$.

We found a set of two best linear approximations for this function, namely:

$$L_1 = \{y_{t+3} \oplus y_{t+25} \oplus y_{t+64} \oplus 1; y_{t+25} \oplus y_{t+46} \oplus y_{t+64} \oplus 1\}.$$

Each of the approximations of L_1 is valid with a probability $\frac{1}{2} + \epsilon_1$, where $\epsilon_1 = \frac{1}{4}$.

Now the term $h(y_{t+83}, y_{t+105}, y_{t+126}, y_{t+144}, x_{t+143})$ is equal to either q_{t+80} or $p_{t+80} \oplus q_{t+80}$, with a probability $\frac{1}{2}$ for both expressions. We found a set of 8 best simultaneous linear approximations for these two expressions, namely:

$$L_2 = \{\ y_{t+83} \oplus y_{t+144} \oplus 1; \qquad y_{t+83} \oplus y_{t+105} \oplus y_{t+126} \oplus y_{t+144} \oplus 1;$$
$$y_{t+83} \oplus y_{t+126} \oplus y_{t+144}; \ y_{t+83} \oplus y_{t+105} \oplus y_{t+126};$$
$$y_{t+83} \oplus y_{t+105}; \qquad y_{t+83} \oplus y_{t+105} \oplus y_{t+144} \oplus 1;$$
$$y_{t+105} \oplus y_{t+144}; \qquad y_{t+105} \oplus y_{t+126} \oplus y_{t+144} \oplus 1;\ \}.$$

Each of the 8 approximations of L_2 has an average probability $\epsilon_2 = \frac{1}{8}$ of being valid.

Thus, we have found 16 linear approximations of $z_t \oplus z_{t+80}$, namely all the linear expressions of the form

$$y_t \oplus l_1(y_{t+3}, y_{t+25}, y_{t+46}, y_{t+64}) \oplus l_2(y_{t+83}, y_{t+105}, y_{t+126}, y_{t+144}),$$

where $l_1 \in L_1$ and $l_2 \in L_2$. Each of these approximations is valid with a probability $\frac{1}{2} + \epsilon$, where ϵ is derived from $\epsilon_{g'}$, ϵ_1, and ϵ_2 using the Piling-up Lemma:

$$\epsilon = \frac{1}{2} \cdot 2^2 \cdot \epsilon_{g'} \cdot \epsilon_1 \cdot \epsilon_2 = \frac{5}{4096} \simeq 2^{-9.67}.$$

The extra multiplicative factor of $\frac{1}{2}$ takes into account the fact that the considered approximations are only valid when $x_{t+63} = 1$. The LFSR derivation attacks of Section 4 exploit these 16 linear approximations.

3.2   Generalisation of the Attack Method

In this Section, we try to generalise the previous approximation method. The purpose is not to find better approximations than those identified in Section 3.1, but to derive some design criteria on the boolean functions g and h'. However in the previous approximation, we used the fact that the bias of g depends on the value of x_{t+63}, so that the approximations of g and h' are not correct independently. We do not take this phenomenon into account in this Section. Therefore, we only provide a simplified picture of potential generalised attacks.

The function $g(X^t, Y^t)$ operates on $w(g) = w_L(g) + w_N(g)$ variables taken from the LFSR and the NFSR, where $w_L(g)$ is the number of variables taken from the LFSR and $w_N(g)$ the number of variables taken from the NFSR. Let the function $A_g(X^t, Y^t)$ be a linear approximation of the function g, i.e.

$$A_g(X^t, Y^t) = \bigoplus_{i=0}^{w_N(g)-1} d_i x_{t+\phi_g(i)} \oplus \bigoplus_{j=0}^{w_L(g)-1} c_j y_{t+\psi_g(j)}, \quad c_j, d_i \in \mathbb{F}_2, \quad (1)$$

such that the distance between $g(\cdot)$ and $A_g(\cdot)$ defined by:

$$d_g = \sharp\{x \in \mathbb{F}_2^{w(g)} : A_g(x) \neq g(x)\} > 0,$$

is strictly larger than zero. Then, we have

$$\Pr\{A_g(x) \neq g(x)\} = \frac{1}{2^{w(g)}} d_g,$$

i.e.

$$\Pr\{A_g(x) + g(x) = 0\} = 1/2 + \epsilon_g,$$

where the bias is:

$$\epsilon_g = 1/2 - 2^{-w(g)} d_g.$$

Similarly, the function $h'(X^t, Y^t)$ can also be approximated by some linear expressions of the form:

$$A_{h'}(X^t, Y^t) = \bigoplus_{i=0}^{w_N(h')-1} k_i x_{t+\phi_{h'}(i)} \oplus \bigoplus_{j=0}^{w_L(h')-1} l_j y_{t+\psi_{h'}(j)}, \quad k_j, l_i \in \mathbb{F}_2. \quad (2)$$

Recall, $z_t \overset{p}{=} A_{h'}(\cdot)_t$ with some probability p. Knowing the expressions (1) and (2), one can sum up together $w_N(A_g(\cdot))$ expressions of $A_{h'}(\cdot)$ at different times t, in such a way that all terms X^t will be eliminated (just because the terms X^t will be cancelled due to the parity check function $A_g(\cdot)$, leaving the terms Y^t and noise variables only). Note also that any linear combination of $A_{h'}(\cdot)$ is a linear combination of the keystream bits z_t.

The sum of $w_N(A_g(\cdot))$ approximations $A_{h'}(\cdot)$ will introduce $w_N(A_g(\cdot))$ independent noise variables due to the approximation at different time instances. Moreover, the cancellation of the terms X^t in the sum will be done by the parity check property of the approximation $A_g(\cdot)$. If the function $A_{h'}(\cdot)$ contains $w_N(A_{h'})$ terms from X^t, then the parity cancellation expression $A_g(\cdot)$ will be applied $w_N(A_{h'})$ times. Each application of the cancellation expression $A_g(\cdot)$ will introduce another noise variable due to the approximation $N_g : g(\cdot) \rightarrow A_g(\cdot)$. Therefore, the application of the expression $A_g(\cdot)$ $w_N(A_{h'})$ times will introduce $w_N(A_{h'})$ additional noise variables N_g. Accumulating all above and following the Piling-up Lemma, the final correlation of such a sum (of the linear expression on Y^t) is given by the following Theorem.

Theorem 1. *There always exists a linear relation in terms of bits from the state of the LFSR and the keystream, which have the bias:*

$$\epsilon = 2^{(w_N(A_{h'})+w_N(A_g)-1)} \cdot \epsilon_g^{w_N(A_{h'})} \cdot \epsilon_{h'}^{w_N(A_g)},$$

where $A_g(\cdot)$ and $A_{h'}(\cdot)$ are linear approximations of the functions $g(\cdot)$ and $h'(\cdot)$, respectively, and:

$$\Pr\{A_g(\cdot) = g(\cdot)\} = 1/2 + \epsilon_g, \qquad \Pr\{A_{h'}(\cdot) = h'(\cdot)\} = 1/2 + \epsilon_{h'}.$$

This theorem gives us a criteria for a proper choice of the functions $g(\cdot)$ and $h'(\cdot)$. The biases ϵ_g and $\epsilon_{h'}$ are related to the *nonlinearity* of these boolean functions, and the values $w_N(A_g)$ and $w_N(A_{h'})$ are related to the *correlation immunity* property; however, there is a well-known trade-off between these two properties [27]. Unfortunately, in the case of Grain the functions $g(\cdot)$ and $h'(\cdot)$ were improperly chosen.

4   Deriving the LFSR Initial State

In the former Section, we have shown how to derive an arbitrary number N of linear approximation equations in the $n = 80$ initial LFSR bits, of bias $\epsilon \simeq 2^{-9.67}$ each, from a sufficient number of keystream bits. Let us denote these equations by:

$$\bigoplus_{i=0}^{n-1} \alpha_i^j \cdot y_i = b^j, j = 1, \ldots, N.$$

In this Section we show how to use these relations to derive the initial LFSR state Y^0. This can be seen as a decoding problem, up to the fact that the code length is not fixed in advance and one has to find an optimal trade-off between the complexities of deriving a codeword (i.e. collecting an appropriate number of linear approximation equations) and decoding this codeword.

An estimate of the number N of linear approximation equations needed for the right value of the unknown to maximize the indicator

$$I = \sharp\left\{ j \in \{1, \ldots, N\} \ \middle| \ \bigoplus_{i=0}^{n-1} \alpha_i^j \cdot y_i = b^j \right\},$$

or at least to be very likely to provide say one of the two or three highest values of I, can be determined as follows.

Under the heuristic assumption that for the correct (respectively incorrect) value of Y^0, I is the sum of N independent binary variables x_i distributed according to the Bernoulli law of parameters $p = Pr\{x_i = 1\} = \frac{1}{2} - \epsilon$ and $q = Pr\{x_i = 0\} = \frac{1}{2} + \epsilon$ (resp. the Bernoulli law of parameters $Pr\{x_i = 1\} = \frac{1}{2}$ and $Pr\{x_i = 0\} = \frac{1}{2}$, mean value $\mu = \frac{1}{2}$, and standard deviation $\sigma = \frac{1}{2}$), N can be derived by introducing a threshold of say $T = N(\frac{1}{2} + \frac{3\epsilon}{4})$ for I and requiring: (i) that the probability that I is larger than T for an incorrect value of Y^0 is less than a suitably chosen false alarm probability p_{fa}; (ii) that the probability that I is lower than T for the correct value is less than a non detection probability p_{nd} of say 1%. For practical values of p_{fa}, the first condition is by far the most demanding. Setting the false alarm rate to $p_{fa} = 2^{-n}$ ensures that the number of false alarms is less than 1 in average.

Due to the Central Limit Theorem, $\frac{\sum x_i - N\mu}{\sqrt{N}\sigma}$ is distributed according to the normal law, so that:

$$Pr\left\{ \frac{1}{N} \sum x_i - \mu > \frac{3\epsilon}{4} \right\} = Pr\left\{ \frac{\sum x_i - N\mu}{\sqrt{N}\sigma} > \frac{3\sqrt{N}\epsilon}{4\sigma} \right\} \tag{3}$$

can be approximated by $\frac{1}{\sqrt{2\pi}} \int_{\lambda}^{+\infty} e^{-\frac{t^2}{2}} dt$, where $\lambda = \frac{3\sqrt{N}\epsilon}{2}$. Consequently, if N is selected in such a way that $\frac{3\sqrt{N}\epsilon}{2} = \lambda$, i.e.

$$N = \left(\frac{2\lambda}{3\epsilon} \right)^2,$$

where λ is given by:

$$\frac{1}{\sqrt{2\pi}} \int_{\lambda}^{+\infty} e^{-\frac{t^2}{2}} dt = p_{fa} = 2^{-n},$$

then inequality 3 is satisfied.

A naive LFSR derivation method would consist of collecting N approximate equations, computing the indicator I independently for each of the 2^n possible values of Y^0 and retaining those Y^0 candidates leading to a value of I larger than the $N(\frac{1}{2} + \frac{3\epsilon}{4})$ threshold. This method would require a low number of keystream bits (say $\frac{N+80}{16}$) but the resulting complexity $N \cdot 2^{80}$ would be larger than the one of exhaustive key search.

In the rest of this Section, we show that much lower complexities can be obtained by using the fast Walsh transform algorithm and a few extra filtering techniques in order to speed up computations of correlation indicators. Former examples of applications of similar Fast Fourier Transform techniques in order to significantly decrease the total complexity of correlation attacks can be found in [4] [9] [16].

4.1   Use of the Fast Walsh Transform to Speed Up Correlation Computations

Basic Method. Let us consider the following problem. Given a sufficient number M of linear approximation equations of bias ϵ involving m binary variables y_0 to y_{m-1}, how to efficiently determine these m variables? Let us denote these M equations by $\sum_{i=0}^{m-1} \alpha_i^j \cdot y_j = b^j, j = 1, \ldots, M$. For a sufficiently large value of M, one can expect the right value of $(y_0, \ldots, y_{m-1})$ to be the one maximizing the indicator:

$$I(y_0, \ldots, y_{m-1}) = \sharp\left\{ j \in \{1, \ldots, M\} \;\middle|\; \sum_{i=0}^{m-1} \alpha_i^j \cdot y_j = b^j \right\}$$

$$= \frac{M}{2} + \frac{1}{2} \cdot S(y_0, \ldots, y_{m-1}),$$

where:

$$S(y_0, \ldots, y_{m-1}) = \sharp\left\{ j \in \{1, \ldots, M\} \;\middle|\; \sum_{i=0}^{m-1} \alpha_i^j \cdot y_i = b^j \right\}$$

$$- \sharp\left\{ j \in \{1, \ldots, M\} \;\middle|\; \sum_{i=0}^{m-1} \alpha_i^j \cdot y_i \neq b^j \right\}.$$

Equivalently one can expect $(y_0, \ldots, y_{m-1})$ to be the value which maximizes the indicator $S(y_0, \ldots, y_{m-1})$. Instead of computing all of 2^m values of $S(y_0, \ldots, y_{m-1})$ independently, one can derive these values in a combined way using fast Walsh transform computations in order to save time.

Let us recall the definition of the Walsh transform. Given a real function of m binary variables $f(x_1, \ldots, x_{m-1})$, the Walsh transform of f is the real function of m binary variables $F = W(f)$ defined by:

$$F(u_0, \ldots, u_{m-1}) = \sum_{x_0, \ldots, x_{m-1} \in \{0,1\}^m} f(x_0, \ldots, x_{m-1})(-1)^{u_0 x_0 + \ldots + u_{m-1} x_{m-1}}.$$

Let us define the function $s(\alpha_0, \ldots, \alpha_{m-1})$ by:

$$\sharp \{ j \in \{1, \ldots, M\} \mid (\alpha_0^j, \ldots, \alpha_{m-1}^j) = (\alpha_0, \ldots, \alpha_{m-1}) \wedge b^j = 1 \}$$
$$- \sharp \{ j \in \{1, \ldots, M\} \mid (\alpha_0^j, \ldots, \alpha_{m-1}^j) = (\alpha_0, \ldots, \alpha_{m-1}) \wedge b^j = 0 \}.$$

The function s can be computed in M steps. Moreover, it is easy to check that the Walsh transform of s is S, i.e.

$$\forall (y_0, \ldots, y_{m-1}) \in \{0,1\}^m, W(s)(y_0, \ldots, y_{m-1}) = S((y_0, \ldots, y_{m-1})).$$

Therefore, the computational cost of the estimation of all the 2^m values of S using fast Walsh transform computations is $M + m \cdot 2^m$; the required memory is 2^m.

Improved Hybrid Method. More generally, if $m_1 < m$, one can use the following hybrid method between exhaustive search and Walsh transform in order to save space.

For each of the 2^{m-m_1} values of $(y_{m_1}, \ldots, y_{m-1})$, define the associated restriction S' of S as the m_1 bit boolean function given by:

$$S'(y_0, \ldots, y_{m_1-1}) = \sharp \left\{ j \in \{1, \ldots, M\} \;\middle|\; \sum_{i=0}^{m_1-1} \alpha_i^j \cdot y_i = \sum_{i=m_1}^{m} \alpha_i^j \cdot y_i \oplus b^j \right\}$$
$$- \sharp \left\{ j \in \{1, \ldots, M\} \;\middle|\; \sum_{i=0}^{m_1-1} \alpha_i^j \cdot y_i \neq \sum_{i=m_1}^{m} \alpha_i^j \cdot y_i \oplus b^j \right\}.$$

It is easy to see that if we define $s'(\alpha_0, \ldots, \alpha_{m_1-1})$ as

$$\sharp \left\{ j \in \{1, \ldots, M\} \;\middle|\; (\alpha_0^j, \ldots, \alpha_{m_1-1}^j) = (\alpha_0, \ldots, \alpha_{m_1-1}) \wedge \sum_{i=m_1}^{m} \alpha_i^j \cdot y_i \oplus b^j = 1 \right\}$$
$$- \sharp \left\{ j \in \{1, \ldots, M\} \;\middle|\; (\alpha_0^j, \ldots, \alpha_{m_1-1}^j) = (\alpha_0, \ldots, \alpha_{m_1-1}) \wedge \sum_{i=m_1}^{m} \alpha_i^j \cdot y_i \oplus b^j = 0 \right\},$$

then S' is the Walsh transform of s'.

Therefore, the computational cost of the estimation of all the 2^m values of S using this method is $2^{m-m_1}(M + m_1 \cdot 2^{m_1})$. If we compare this with the former basic Walsh transform method, we see that the required memory decreases from 2^m to 2^{m_1}, whereas the time complexity increase remains negligible as long as $m_1 << log_2(M)$.

4.2 First LFSR Derivation Technique

In order to reduce the LFSR derivation complexity when compared with the naive method of complexity $N \cdot 2^n$, we can exploit more keystream to produce more linear approximation equations in the unknowns y_0 to y_{n-1}, and retain only those equations involving the $m < n$ variables y_0 to y_{m-1}, i.e. which coefficients in the $n - m$ variables y_m to y_{n-1} are equal to 0.

Thus a fraction of about 2^{m-n} of the relations are retained and we have to collect about $N2^{n-m}$ approximate relations to retain N relations. This requires a number of keystream bits of:

$$\frac{N2^{n-m} + 80}{16}.$$

As seen in the former Section, once the relations have been filtered, the computational cost of the derivation of the values of these m variables using fast Walsh transform computations is about $m2^m$ for the basic method, and more generally $2^{m-m_1}(N + m_1 2^{m_1})$ if fast Walsh transform computations are applied to a restricted set of $m_1 < m$ variables.

Thus, the overall time complexity of this method is:

$$N2^{n-m} + m2^m,$$

and more generally:

$$N2^{n-m} + 2^{m-m_1}(N + m_1 2^{m_1}).$$

Once the m variables y_0 to y_{m-1} have been recovered, one can either reiterate the same technique for other choices of the m unknown variables, which increases the complexity by a factor of less than 2 if $m \geq \frac{n}{2}$, or test each of the 2^{n-m} candidates in the next step of the attack (NFSR and key derivation).

An estimate of the number N of equations needed is given by

$$N = \left(\frac{2\lambda}{3\epsilon}\right)^2,$$

where λ is determined by the condition $\frac{1}{\sqrt{2\pi}} \int_{\lambda}^{+\infty} e^{-\frac{t^2}{2}} dt = 2^{-m}$. This condition ensures that the expected number of false alarm is less than 1.

The minimal complexity is obtained for $m = 49$. For this parameter value, we have $\lambda = 7.87$ and $N = 2^{24}$. The attack complexity is about 2^{55}, the number of keystream bits needed is around 2^{51}, and the memory needed is about 2^{49}.

4.3 Second LFSR Derivation Technique

An alternative method is to derive new linear approximation equations (of lower bias) involving $m < n$ unknown variables y_0 to y_{m-1} by combining the R available approximate equations of bias ϵ pairwise, and retaining only those pairs of relations for which the $n - m$ last coefficients collide. One obtains in this way about $N' = R^2 \cdot 2^{m-n-1}$ new affine equations in y_0 to y_{m-1}, of bias $\epsilon' = 2\epsilon^2$. The allocation of the m variables maximizing the number of satisfied equations can be found by fast Walsh computations as explained in the former Section.

The number N' of relations needed is about $\left(\frac{2\lambda}{3\epsilon'}\right)^2$, where λ is determined by the condition $\frac{1}{\sqrt{2\pi}} \int_{\lambda}^{+\infty} e^{-\frac{t^2}{2}} dt = 2^{-m}$. The required number R of relations of bias ϵ is therefore $R = (N'2^{n-m+1})^{\frac{1}{2}}$, and the number of keystream bits needed is about $\frac{R+80}{16}$. The complexity of the derivation of the N' relations is $\max(R, N') = \max((N'2^{n-m+1})^{\frac{1}{2}}, N')$.

Once the N' relations have been derived, the computational cost of the derivation of the values of these m variables using fast Walsh transform computations is about $m \cdot 2^m$ for the basic method, and more generally if fast Walsh transform computations are applied to a restricted set of $m_1 < m$ variables it costs $2^{m-m_1}(N' + m_1 \cdot 2^{m_1})$.

Thus the total complexity of the derivation of the m LFSR bits is:

$$\max((N'2^{n-m+1})^{\frac{1}{2}}, N') + m2^m,$$

and more generally:

$$\max((N'2^{n-m+1})^{\frac{1}{2}}, N') + 2^{m-m_1}(N' + m_12^{m_1}).$$

The minimal complexity is obtained for $m = 36$. For this parameter value, we have $\lambda = 6.65$ and $N' = 2^{41}$. The attack complexity is about 2^{43}, the number of keystream bits needed is about 2^{38} and the memory required is about 2^{42}.

5   Recovering the NFSR Initial State and the Key

Once the initial state of the LFSR has been recovered, we want to recover the initial state $(x_0, \ldots, x_{79})$ of the NFSR. Fortunately, the knowledge of the LFSR removes the nonlinearity of the output function and we can express each keystream bit z_i by one of the following four equations depending on the initial state of the LFSR:

$$z_i = x_i, \qquad z_i = x_i \oplus 1,$$
$$z_i = x_i \oplus x_{63+i}, \quad z_i = x_i \oplus x_{63+i} \oplus 1.$$

Since functions p and q underlying h are balanced, each equation has the same occurrence probability. We are going to use the non linearity of the output function to recover the initial state of the NFSR by writing the equations corresponding to the first keystream bits.

The 16 first equations are linear equations involving only bits of the initial state of the NFSR because $63 + i$ is lower than 80.

To recover all the bits of the initial state, we introduce a technique which consists of building chains of keystream bits. The equations for keystream bits z_{17} to z_{79} involve either one bit of the NFSR ($z_i = x_i$ or $z_i = x_i \oplus 1$) or two bits ($z_i = x_i \oplus x_{63+i}$ or $z_i = x_i \oplus x_{63+i} \oplus 1$). An equation involving only one bit allows us to instantly recover the value of the corresponding bit of the initial state. This can be considered as a chain of length 0. On the other hand, an equation involving two bits does not allow this because we do not know the value of x_{63+i} (for $i > 16$).

However, by considering not only the equations for z_i but also all the equation for $z_{k \cdot 63+i}$ for $k \geq 1$, we can cancel the bits we do not know and retrieve the value of x_i. With probability $\frac{1}{2}$, the equation for z_{63+i} involves one single unknown bit. Then it provides the value of x_{63+i} and consequently the value of x_i. Here the chain is of length 1, since we have to consider one extra equation to retrieve x_i. The equation for z_{63+i} can also involve two bits with probability $\frac{1}{2}$. Then we have to consider the equation of $z_{2 \cdot 63+i}$, which can also either involve only one bit (we have a chain of length 2) or two bits and we have to consider more equations to solve. Each equation has a probability $\frac{1}{2}$ to involve 1 or 2 bits. Consequently the probability that a chain is of length n is $\frac{1}{2^{n+1}}$ and the probability that a chain is of length strictly larger than n is $\frac{1}{2^{n+1}}$.

We want to recover the values of $x_{17}, \ldots, x_{79}$. We have to build 64 different chains. Let us consider $L = 63 \cdot n$ bits of keystream. The probability that one of the chains is of length larger than n is less than $= 64 \cdot 2^{-n-1}$ and therefore less than 2^{-n+5}. If we want this probability to be bounded by 2^{-10}, then $n > 15$ and $L > 945$ suffices. Consequently a few thousands of keystream bits are required to retrieve the initial state of the NFSR and the complexity of the operation is bounded by $64 \cdot n$.

Since the internal state transition function associated to the special key and IV setup mode is one to one, the key can be efficiently derived from the NFSR and LFSR states at the beginning of the keystream generation by running this function backward.

6   Simulations and Results

To confirm that our cryptanalysis is correct, we ran several experiments. First we checked the bias ϵ of Section 3.1 by running the cipher with a known initial state of both the LFSR and the NFSR, computing the linear approximations, and counting the number of fulfilled relations for a very large number of relations. For instance we found that one linear approximation is satisfied 19579367 times out of 39060639, which gives an experimental bias of $2^{-9.63}$, to be compared with the theoretical bias $\epsilon = 2^{-9.67}$.

To check the two proposed LFSR reconstruction methods of Section 4, we considered a reduced version of Grain in order to reduce the memory and time required by the attack on a single computer: we shortened the LFSR by a factor of 2. We used an LFSR of size 40 with a primitive feedback polynomial and we reduced by two the distances for the tap entries of function h: we selected taps number 3, 14, 24, and 33, instead of 3, 25, 46, and 64 for Grain.

The complexity of the first technique for the actual Grain is 2^{55} which is out of reach of a single PC. For our reduced version, the complexity given by the formula of Section 4.2 is only 2^{35}. We exploited the 16 linear approximations to derive relations colliding on the first 11 bits. Consequently the table of the

Walsh transform is only of size 2^{29}. We used $15612260 \simeq 2^{23}$ relations, which corresponds to a false alarm probability of 2^{-29}. Our implementation needed around one hour to recover the correct value of the LFSR internal state on a computer with a Intel Xeon processor running at 2.5 GHz with 3 GB of memory. The Walsh transform computation took only a few minutes.

For the actual Grain, the second technique requires only 2^{43} operations which is achievable by a single PC. However it also requires 2^{42} of memory which corresponds to 350 GB of memory. We do not have such an amount of memory but for the reduced version the required memory is only 2^{29}. Since the complexity given by the formula of Section 4.3 is dominated by the required number of relations to detect the bias, our simulation has a complexity close to 2^{43}. In practice, we obtained a result after 4 days of computation on the same computer as above and $2.5 \cdot 10^{12} \simeq 2^{41}$ relations where considered and allowed to recover the correct LFSR initial state.

Finally, we implemented the method of Section 5 to recover the NFSR. Given the correct initial state of the LFSR, and the first thousand keystream bits, our program recovers the initialization of the NFSR in a few seconds for a large number of different initializations of both the known LFSR and unknown NFSR. We also confirmed the failure probability assessed in Section 5 for this method (which corresponds to the occurrence probability of at least one chain of length larger than 15).

7 Conclusion

We have presented a key-recovery attack against Grain which requires 2^{43} computations, 2^{42} bits of memory, and 2^{38} keystream bits. This attack suggests that the following slight modifications of some of the Grain features might improve its strength:

- Introduce several additional masking variables from the NFSR in the keystream bit computation.
- Replace the nonlinear feedback function g in such a way that the associated function g' be balanced (e.g. replace g by a 2-resilient function). However this is not necessarily sufficient to thwart all similar attacks.
- Modify the filtering function h in order to make it more difficult to approximate.
- Modify the function g and h to increase the number of inputs.

Following recent cryptanalysis of Grain including the key recovery attack reported here and distinguishing attacks based on the same kind of linear approximations as those presented in Section 3 [19] [26], the authors of Grain proposed a tweaked version of their algorithm [12], where the functions g and h' have been modified. This novel version of Grain appears to be much stronger and is immune against the statistical attacks presented in this paper.

We would like to thank Matt Robshaw and Olivier Billet for helpful comments.

References

1. M. Briceno, I. Goldberg, and D. Wagner. A pedagogical implementation of A5/1. Available at http://jya.com/a51-pi.htm, Accessed August 18, 2003, 1999.
2. A. Canteaut and M. Trabbia. Improved fast correlation attacks using parity-check equations of weight 4 and 5. In B. Preneel, editor, *Advances in Cryptology— EUROCRYPT 2000*, volume 1807 of *Lecture Notes in Computer Science*, pages 573–588. Springer-Verlag, 2000.
3. V. Chepyzhov and B. Smeets. On a fast correlation attack on certain stream ciphers. In D. W. Davies, editor, *Advances in Cryptology—EUROCRYPT'91*, volume 547 of *Lecture Notes in Computer Science*, pages 176–185. Springer-Verlag, 1991.
4. M. W. Dodd. *Applications of the Discrete Fourier Transform in Information Theory and Cryptology*. PhD thesis, University of London, 2003.
5. ECRYPT. eSTREAM: ECRYPT Stream Cipher Project, IST-2002-507932. Available at http://www.ecrypt.eu.org/stream/, Accessed September 29, 2005, 2005.
6. P. Ekdahl and T. Johansson. Another attack on A5/1. In *Proceedings of International Symposium on Information Theory*, page 160. IEEE, 2001.
7. P. Ekdahl and T. Johansson. Another attack on A5/1. *IEEE Transactions on Information Theory*, 49(1):284–289, January 2003.
8. H. Englund and T. Johansson. A new simple technique to attack filter generators and related ciphers. In *Selected Areas in Cryptography*, pages 39–53, 2004.
9. H. Gilbert and P. Audoux. Improved fast correlation attacks on stream ciphers using FFT techniques. personnal communication, 2000.
10. J.D. Golić. Cryptanalysis of alleged A5 stream cipher. In W. Fumy, editor, *Advances in Cryptology—EUROCRYPT'97*, volume 1233 of *Lecture Notes in Computer Science*, pages 239–255. Springer-Verlag, 1997.
11. M. Hell, T. Johansson, and W. Meier. Grain - A Stream Cipher for Constrained Environments. ECRYPT Stream Cipher Project Report 2005/001, 2005. http://www.ecrypt.eu.org/stream.
12. M. Hell, T. Johansson, and W. Meier. Grain - A Stream Cipher for Constrained Environments, 2005. http://www.it.lth.se/grain.
13. T. Johansson and F. Jönsson. Fast correlation attacks based on turbo code techniques. In *Advances in Cryptology—CRYPTO'99*, volume 1666 of *Lecture Notes in Computer Science*, pages 181–197. Springer-Verlag, 1999.
14. T. Johansson and F. Jönsson. Improved fast correlation attacks on stream ciphers via convolutional codes. In *Advances in Cryptology—EUROCRYPT'99*, volume 1592 of *Lecture Notes in Computer Science*, pages 347–362. Springer-Verlag, 1999.
15. F. Jönsson. *Some Results on Fast Correlation Attacks*. PhD thesis, Lund University, Department of Information Technology, P.O. Box 118, SE–221 00, Lund, Sweden, 2002.
16. A. Joux, P. Chose, and M. Mitton. Fast Correlation Attacks: An Algorithmic Point of View. In Lars R. Knudsen, editor, *Advances in Cryptology – EUROCRYPT 2002*, volume 2332 of *Lecture Notes in Computer Science*, pages 209–221. Springer-Verlag, 2002.
17. B. S. Jr. Kaliski and M. J. B. Robshaw. Linear Cryptanalysis Using Multiple Approximations. In Yvo G. Desmedt, editor, *Advances in Cryptology – CRYPTO '94*, volume 839 of *Lecture Notes in Computer Science*, pages 26–39. Springer-Verlag, 1994.
18. M. Matsui. Linear cryptanalysis method for DES cipher. In Tor Helleseth, editor, *Advances in Cryptology – EUROCRYPT '93*, volume 765 of *Lecture Notes in Computer Science*, pages 386–397. Springer-Verlag, 1993.

19. A. Maximov. Cryptanalysis of the "Grain" family of stream ciphers. In *ACM Transactions on Information and System Security (TISSEC)*, 2006.

20. W. Meier and O. Staffelbach. Fast correlation attacks on stream ciphers. In C.G. Günter, editor, *Advances in Cryptology—EUROCRYPT'88*, volume 330 of *Lecture Notes in Computer Science*, pages 301–316. Springer-Verlag, 1988.

21. W. Meier and O. Staffelbach. Fast correlation attacks on certain stream ciphers. *Journal of Cryptology*, 1(3):159–176, 1989.

22. W. Meier and O. Staffelbach. The self-shrinking generator. In A. De Santis, editor, *Advances in Cryptology—EUROCRYPT'94*, volume 905 of *Lecture Notes in Computer Science*, pages 205–214. Springer-Verlag, 1994.

23. M. Mihaljevic and J.D. Golić. A fast iterative algorithm for a shift register initial state reconstruction given the noisy output sequence. In J. Seberry and J. Pieprzyk, editors, *Advances in Cryptology—AUSCRYPT'90*, volume 453 of *Lecture Notes in Computer Science*, pages 165–175. Springer-Verlag, 1990.

24. NESSIE. New European Schemes for Signatures, Integrity, and Encryption. Available at http://www.cryptonessie.org, Accessed August 18, 2003, 1999.

25. W.T. Penzhorn and G.J. Kühn. Computation of low-weight parity checks for correlation attacks on stream ciphers. In C. Boyd, editor, *Cryptography and Coding - 5th IMA Conference*, volume 1025 of *Lecture Notes in Computer Science*, pages 74–83. Springer-Verlag, 1995.

26. M. Hassanzadeh S. Khazaei and M. Kiaei. Distinguishing Attack on Grain. ECRYPT Stream Cipher Project Report 2005/001, 2005. http://www.ecrypt.eu.org/stream.

27. T. Siegenthaler. Correlation-immunity of non-linear combining functions for cryptographic applications. *IEEE Transactions on Information Theory*, 30:776–780, 1984.

28. T. Siegenthaler. Decrypting a class of stream ciphers using ciphertext only. *IEEE Transactions on Computers*, 34:81–85, 1985.

Cryptanalysis of the Stream Cipher DECIM*

Hongjun Wu and Bart Preneel

Katholieke Universiteit Leuven, ESAT/SCD-COSIC
Kasteelpark Arenberg 10, B-3001 Leuven-Heverlee, Belgium
{wu.hongjun, bart.preneel}@esat.kuleuven.be

Abstract. DECIM is a hardware oriented stream cipher with an 80-bit key and a 64-bit IV. In this paper, we point out two serious flaws in DECIM. One flaw is in the initialization of DECIM. It allows to recover about half of the key bits bit-by-bit when one key is used with about 2^{20} random IVs; only the first two bytes of each keystream are needed in the attack. The amount of computation required in the attack is negligible. Another flaw is in the keystream generation algorithm of DECIM. The keystream is heavily biased: any two adjacent keystream bits are equal with probability about $\frac{1}{2} + 2^{-9}$. A message could be recovered from the ciphertext if that message is encrypted by DECIM for about 2^{18} times. DECIM with an 80-bit key and an 80-bit IV is also vulnerable to these attacks.

1 Introduction

DECIM [1] is a stream cipher that has been submitted to the ECRYPT stream cipher project [4]. The main feature of DECIM is the use of the ABSG decimation mechanism [1], an idea similar to the shrinking generator [3,6]. Another excellent feature is that a 32-bit buffer is used in DECIM to ensure that at each step DECIM generates one output bit.

In this paper, we point out two flaws in DECIM, one in the initialization algorithm, and another one in the keystream generation algorithm. The flaw in the initialization allows for any easy key recovery from the keystreams when one key is used with about 2^{20} random IVs. The flaw in the keystream generation algorithm results in a heavy bias in the keystream, hence the cipher is vulnerable to a broadcast attack.

In Sect. 2 we describe the DECIM cipher. Section 3 presents an key recovery attack on DECIM. The key recovery attack on DECIM is improved in Sect. 4. The broadcast attack on DECIM is described in Sect. 5. Section 6 shows that DECIM with an 80-bit IV is also vulnerable to the attacks. Section 7 concludes this paper.

---

* This work was supported in part by the Concerted Research Action (GOA) Ambiorics 2005/11 of the Flemish Government and in part by the European Commission through the IST Programme under Contract IST-2002-507932 ECRYPT.

M.J.B. Robshaw (Ed.): FSE 2006, LNCS 4047, pp. 30–40, 2006.

2   Stream Cipher DECIM

DECIM uses the ABSG decimation mechanism in the keystream generation in order to achieve high security and design simplicity. The keystream generation process and the key/IV setup are illustrated in Sect. 2.1 and 2.2, respectively.

2.1   Keystream Generation

The keystream generation diagram of DECIM is given in Fig. 1. DECIM has a regularly clocked LFSR which is defined by the feedback polynomial

$$P(X) = X^{192} + X^{189} + X^{188} + X^{169} + X^{156} + X^{155} + X^{132} +$$
$$X^{131} + X^{94} + X^{77} + X^{46} + X^{17} + X^{16} + X^{5} + 1$$

over $GF(2)$. The related recursion is given as

$$s_{192+n} = s_{187+n} \oplus s_{176+n} \oplus s_{175+n} \oplus s_{146+n} \oplus s_{115+n} \oplus s_{98+n} \oplus s_{61+n}$$
$$\oplus s_{60+n} \oplus s_{37+n} \oplus s_{36+n} \oplus s_{23+n} \oplus s_{4+n} \oplus s_{3+n} \oplus s_{n} .$$

At each stage, two bits are generated from the LFSR as follows:

$$y_{t,1} = f(s_{t+1}, s_{t+32}, s_{t+40}, s_{t+101}, s_{t+164}, s_{t+178}, s_{t+187}),$$

$$y_{t,2} = f(s_{t+6}, s_{t+8}, s_{t+60}, s_{t+116}, s_{t+145}, s_{t+181}, s_{t+191}),$$

where the Boolean function f is defined as

$$f(x_{i_1}, ..., x_{i_7}) = \sum_{1 \le j < k \le 7} x_{i_j} x_{i_k} .$$

The binary sequence y consists of all the $y_{t,1}$ and $y_{t,2}$ as

$$y = y_{0,1} y_{0,2} y_{1,1} y_{1,2} \cdots y_{t,1} y_{t,2} \cdots$$

The keystream sequence z is generated from the binary sequence y through the ABSG decimation algorithm. The sequence y is split into subsequences of the form $(\bar{b}, b^i, \bar{b})$, with $i \ge 0$ and $b \in \{0, 1\}$; $\bar{b}$ denotes the complement of b in $\{0, 1\}$. For every subsequence $(\bar{b}, b^i, \bar{b})$, the output bit is b for $i = 0$, and $\bar{b}$ otherwise. The ABSG algorithm is given below

> Input: $(y_0, y_1, ...)$
> Set: $i \leftarrow 0$; $j \leftarrow 0$;
> Repeat the following steps:
> $e \leftarrow y_i$, $z_j \leftarrow y_{i+1}$, $i \leftarrow i + 1$;
>   while $(y_i = \bar{e})$ $i \leftarrow i + 1$;
> $i \leftarrow i + 1$; output z_j; $j \leftarrow j + 1$;

Remarks. The above description of the ABSG and the pseudo-code of ABSG are quoted from [1]. However the outputs of the pseudo-code are the complements

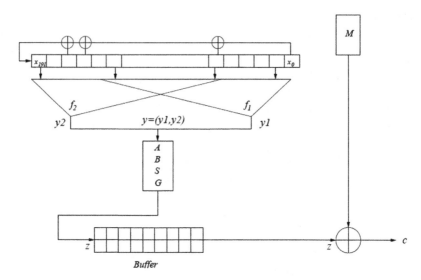

Fig. 1. Keystream Generation Diagram of DECIM [1]

of that of the ABSG algorithm. Anyway, this difference has no effect on the security of DECIM. In the rest of the paper, we assume that the DECIM uses the pseudo-code of ABSG given above.

DECIM is designed to output one bit every two stages. A 32-bit buffer is used to ensure that the probability that there is no output bit is extremely small (2^{-89}).

2.2   Initialization

The secret key K is an 80-bit key. The 64-bit IV is expanded to an 80-bit vector by adding zeros from position 64 up to position 79. The initial value of the LFSR state is loaded as follows

$$s_i = \begin{cases} K_i \vee IV_i & \text{for } 0 \le i \le 55 \\ K_{i-56} \wedge \overline{IV_{i-56}} & \text{for } 56 \le i \le 111 \\ K_{i-112} \oplus IV_{i-112} & \text{for } 112 \le i \le 191 \end{cases}$$

The LFSR is clocked 192 times. After the t-th clocking, $y_{t,1}$ and $y_{t,2}$ are XORed to the $x_{t,192}$ as

$$s_{t+192} = s_{t+192} \oplus y_{t,1} \oplus y_{t,2}.$$

Then one of two permutations π_1 and π_2 is applied to permute 7 elements s_{t+5}, s_{t+31}, s_{t+59}, s_{t+100}, s_{t+144}, s_{t+177}, s_{t+186}. Two bits $y_{t,1}$ and $y_{t,2}$ are input to the ABSG, if the output of the ABSG is 1, then π_1 is applied; if the output of the ABSG is 0 or if there is no output, then π_2 is applied. The two permutations are defined as

$$\pi_1 = (1\ 6\ 3)(4\ 5\ 2\ 7), \pi_2 = (1\ 4\ 7\ 3\ 5\ 2\ 6).$$

3   Key Recovery Attack on DECIM

In this section, we develop attacks to recover the secret key of DECIM. This non-optimized attack applies when the same secret key is used with a number of random IVs, and the first 3 bytes of each keystream are known. The optimized attack is given in the next section.

3.1   The Effects of the Permutations π_1 and π_2

The two permutations in the initialization stage of DECIM provide high non-linearity to the initialization process. However, the permutations also cause some bits in the LFSR to be updated in an improper way. This has a very negative impact on the security of DECIM.

The permutation π_1 is poorly designed. In order to investigate the effects of this permutation, we analyze a weak version by assuming that only this permutation is used in the initialization process, i.e., we replace π_2 with π_1. The values of 140 elements in the LFSR ($s_5, s_6, \ldots, s_{58}$, and $s_{100}, s_{101}, \ldots, s_{185}$) would never be updated by the initialization process. For example, s_{21} would always become s_{192+6}. The details are given below. We trace the bit s_{21}, after 16 steps it becomes s_{16+5} due to the shift of the LFSR. Then it becomes s_{16+177} due to the permutation π_1. After 33 steps, it becomes s_{49+144} due to the shift of the LFSR. Then it becomes s_{49+31} due to the permutation π_1. After 26 steps, it becomes s_{75+5} due to the shift of the LFSR. Then it becomes s_{75+177} due to the permutation π_1. This process repeats and at the end of the initialization process, it becomes s_{192+6}.

The first bit of the keystream is given as $y_{192,2}$; it is computed as $y_{192,2} = f(s_{192+6}, s_{192+8}, s_{192+60}, s_{192+116}, s_{192+145}, s_{192+181}, s_{192+191})$. By tracing the bits of the LFSR during the initialization process, we know that $s_{192+6} \Leftarrow s_{21}$, $s_{192+8} \Leftarrow s_{23}$, $s_{192+116} \Leftarrow s_{132}$, $s_{192+145} \Leftarrow s_{160}$, $s_{192+181} \Leftarrow s_{33}$. If every key and IV pair is randomly generated, then according to the loading of the key and IV, we know that s_{21}, s_{23}, and s_{33} take value 1 with probability 0.75. Thus according to the definition of the function f, the value of $y_{192,2}$ is 0 with probability 0.582. So the first bit of the keystream is heavily biased. It shows that the effect of the permutation π_1 is terrible.

In DECIM, there are two permutations, π_1 and π_2. They are chosen according to the output of ABSG: π_1 is chosen with probability $\frac{1}{3}$, π_2 with probability $\frac{2}{3}$. Due to these two permutations, the number of bits that are not updated by the initialization process is reduced to 54.5 (obtained by running 2^{16} random key and IV pairs). It shows that the permutations π_1 and π_2 which are chosen by the output of ABSG have a negative impact on the security of DECIM.

3.2   Recovering K_{21}

In the initialization process, we monitor the bit s_{21}. s_{21} becomes s_{192+6} with probability $\frac{1}{27}$. If s_{192+6} takes the value 0, and all the other bits in the LFSR at the 192-th step are distributed uniformly, then the value of the first bit of the keystream is 0 with probability $q_0 = \frac{56}{128}$. If s_{192+6} is takes the value 1, and all the other bits of the LFSR at the 192-th step are distributed uniformly, then the value of the first bit of the keystream is 0 with probability $q_1 = \frac{72}{128}$. Denote the probability that the value of the first keystream bit is 0 when $s_{21} = 0$ as p_0, and the probability that the value of the first keystream bit is 0 when $s_{21} = 1$ as p_1. Then $\Delta p = p_1 - p_0 = \frac{1}{27} \times (q_1 - q_0) = 2^{-7.75}$. In an experiment we chose 2^{20} random IVs for $s_{21} = 0$, and another 2^{20} random IVs for $s_{21} = 1$, and we found that $\Delta p = 2^{-7.99}$. The experimental result confirms that the theoretical result $\Delta p = 2^{-7.75}$ is correct.

The above property can be applied to recover K_{21} as follows. Suppose that the same key is used with N random IVs to generate keystreams. For the keystreams with $IV_{21} = 0$, we compute the probability that the value of the first bit is 0, and denote this probability as p'_0. For the keystreams with $IV_{21} = 1$, we compute the probability that the value of the first bit is 0, and denote this probability as p'_1. If $p'_1 > p'_0$, we decide that $K_{21} = 0$; otherwise, $K_{21} = 1$. For $N = (\frac{\Delta p}{2})^{-2} \times 2 = 2^{18.5}$, the attack can determine the value of K_{21} with success rate 0.977.

3.3   Recovering $K_{22}K_{23} \ldots K_{30}$

By tracing the bits in the initialization process, we notice that each s_{22+i} is mapped to $s_{192+7+i}$ with probability $\frac{1}{27}$ for $0 \leq i \leq 8$ (each of them is only mapped by π_1 at s_{t+5}). We know that $s_{22+i} = K_{22+i} \vee IV_{22+i}$, and $s_{192+7+i}$, $s_{192+9+i}$ are used in the generation of $y_{193+i,2}$ for $0 \leq i \leq 10$. In this section, we show that the key bits $K_{22}K_{23}K_{24} \ldots K_{30}$ can be recovered from the keystream.

An attack similar to that given in Sect. 3.2 can be applied to recover the value of K_{23} from the first keystream bits generated from $2^{18.5}$ IVs.

In order to determine the values of K_{22} and K_{24}, we observe the second bit of the keystream. Due to the disturbance of the ABSG, $y_{193,2}$ becomes the second keystream bit with probability 0.5. Thus $\Delta p' = 0.5 \times \Delta p = 2^{-8.75}$. To recover K_{22} and K_{24}, we need $2^{20.5}$ IVs in order to obtain a success probability of 0.977.

In order to determine the value of K_{25}, we observe the second and third bits of the keystream. $y_{194,2}$ becomes the second bit of the keystream with probability $\frac{1}{8}$, and becomes the third bit of the keystream with probability $\frac{1}{4}$. Thus $\Delta p'' = \frac{1}{2} \times (\frac{1}{4} + \frac{1}{8}) \times \Delta p = 2^{-10.165}$. To recover K_{25}, we need $2^{22.3}$ IVs in order to obtain a success probability of 0.977.

We omit the details of recovering $K_{26} \cdots K_{29}$. To recover K_{30}, we observe the fifth, sixth and seventh bits of the keystream. $y_{199,2}$ would become one of these three bits with probability $\frac{77}{256}$. Thus $\Delta p''' = \frac{1}{3} \times \frac{77}{256} \times \Delta p = 2^{-11.068}$. To recover K_{29}, we need $2^{23.5514}$ IVs in order to obtain the success rate 0.977.

3.4   Recovering $K_9K_{10} \ldots K_{19}$

By tracing the bits in the initialization process, we notice that each s_{9+i} is mapped to $s_{192+166+i}$ with probability $\frac{1}{27}$ for $0 \leq i \leq 10$ (each of them is only mapped by π_1 at s_{t+5}). We know that $s_{9+i} = K_{9+i} \vee IV_{9+i}$, and $s_{192+166+i}$ is used in the generation of $y_{194+i,1}$ for $0 \leq i \leq 10$. The attacks given in this section are similar to those given above. We only illustrate how to recover K_9 and K_{19}.

In order to determine the value of K_9, we observe the second bit of the keystream. $y_{194,1}$ becomes the second bit of the keystream with probability $\frac{1}{4}$. Thus $\Delta p^{(4)} = \frac{1}{4} \times \Delta p = 2^{-9.75}$. To recover K_9, we need $2^{22.5}$ IVs in order to obtain a success probability of 0.977.

In order to determine the value of K_{19}, we observe the 8-th, 9-th and 10-th bits of the keystream. $y_{204,1}$ becomes one of these three bits with probability 0.25966. Thus $\Delta p^{(5)} = \frac{1}{3} \times 0.25966 \times \Delta p = 2^{-11.28}$. To recover K_{19}, we need $2^{23.98}$ IVs in order to obtain a success probability of 0.977.

3.5   Recovering $K_{32}K_{33}\ldots K_{46}$

By tracing the bits in the initialization process, we notice that each s_{144+i} is mapped to $s_{192+16+i}$ with probability $\frac{1}{27}$ for $0 \leq i \leq 14$ (each of them is only mapped by π_1 at s_{t+5}). We know that $s_{144+i} = K_{32+i} \oplus IV_{32+i}$, and $s_{192+16+i}$ is used in the generation of $y_{200+i,1}$ for $0 \leq i \leq 14$.

Since for s_{144+i} ($0 \leq i \leq 14$), the key bits are XORed with the IV bits, the attack is slightly modified. For example, if the probability of 0 in the keystream for $IV_{32} = 0$ is higher than the probability of 0 in the keystream for $IV_{32} = 1$, then we predict that $K_{32} = 0$; otherwise, $K_{32} = 1$. We only illustrate how to recover K_{32} and K_{46}.

In order to determine the value of K_{32}, we observe the sixth, seventh and eighth bits of the keystream. $y_{200,2}$ becomes one of these three bits with probability 0.28027. Thus $\Delta p^{(6)} = \frac{1}{3} \times 0.28027 \times \Delta p = 2^{-11.17}$. To recover K_{32}, we need $2^{23.755}$ IVs in order to obtain a success probability of 0.977.

In order to determine the value of K_{46}, we assume that starting from the fourth bit of the sequence y, each bit becomes the output with probability $\frac{1}{3}$. Then $y_{214,2}$ becomes one of the 12th, 13th, $\ldots$, 18th bits of the keystream with probability 0.16637. Thus $\Delta p^{(7)} = \frac{1}{7} \times 0.16637 \times \Delta p = 2^{-13.145}$. To recover K_{29}, we need $2^{26.482}$ IVs in order to obtain a success probability of 0.977.

The attacks given in this section recover 36 bits of the secret key with about 2^{26} random IVs. For each IV, only the first 3 bytes of the keystream are needed in the attack.

4   Improving the Key Recovery Attack

In the above attacks, we deal with the bits affected only by π_1 at s_{t+5} during the initialization (the bits affected by π_2 are not considered in the attack). In order to improve the attack, we have used a computer program to trace all the possibilities for each bit s_i ($0 \leq i \leq 175$) during the initialization process to find out the distribution of that bit at the end of initialization. Then we have searched the optimal attack for that bit. We have performed the experiment, and found that 44 key bits can be recovered with less than 2^{20} IVs, and only the first 2 bytes of the keystream are required in the attack. The experiment results are given in Table 1 in Appendix A.

5   The Keystream of DECIM Is Heavily Biased

The nonlinear function f in DECIM is extremely simple. However this Boolean function is balanced but not 1-resilient. Unfortunately the ABSG decimation mechanism and the buffer in the output function fail to eliminate the bias existing in the output of f, hence the keystream is heavily biased.

5.1   The Keystream Is Biased

We start with analyzing the function f

$$f(x_{i_1}, ..., x_{i_7}) = \sum_{1 \le j < k \le 7} x_{i_j} x_{i_k} .$$

If any bit of the input of f is equal to 1, then f outputs a '1' with probability $\frac{72}{128}$; otherwise it outputs a '1' with probability $\frac{56}{128}$. Thus for $f(x_{i_1}, ..., x_{i_7})$ and $f(x'_{i_1}, ..., x'_{i_7})$, if one bit of one input is always equal to one bit of another input (i.e., $x_{i_a} = x'_{i_b}$ where $0 \le a, b \le 7$), then the outputs related to these two inputs would be equal with probability $(\frac{56}{128})^2 + (\frac{72}{128})^2 = \frac{65}{128}$.

Note that $y_{t,1}$ and $y_{t,2}$ are computed as follows

$$y_{t,1} = f(s_{t+1}, s_{t+32}, s_{t+40}, s_{t+101}, s_{t+164}, s_{t+178}, s_{t+187}),$$

$$y_{t,2} = f(s_{t+6}, s_{t+8}, s_{t+60}, s_{t+116}, s_{t+145}, s_{t+181}, s_{t+191}).$$

Denote $A = \{1, 32, 40, 101, 164, 178, 187\}$, $B = \{6, 8, 60, 116, 145, 181, 191\}$, and denote each element of A by a_i, and each element of B by b_i ($1 \le i \le 7$). Then $y_{t,1} = y_{t+a_i-a_j,1}$ and $y_{t,2} = y_{t+b_i-b_j,2}$ with probability $\frac{65}{128}$ for $1 \le i, j \le 7$ and $i \ne j$. And $y_{t+b_i-a_j,1} = y_{t,2}$ with probability $\frac{65}{128}$ for $1 \le i, j \le 7$. It shows that the binary sequence y is heavily biased.

The heavily biased sequence y is used as input to the ABSG decimation algorithm. It results in a heavily biased output. In the attack, we are interested in those biases in y that would not be significantly reduced by the ABSG Algorithm. Thus we will analyze the bias of $(y_{t+3,1}, y_{t,2})$, $(y_{t+4,1}, y_{t,2})$ and $(y_{t,2}, y_{t+2,2})$ to find out how they affect the randomness of the output of ABSG.

For example, we analyze the effect of the bias of $(y_{t+3,1}, y_{t,2})$. $y_{t+3,1} = y_{t,2}$ with probability $\frac{65}{128}$. Denote the i-th bit of the sequence y by y^i. Thus $y^i = y^{i+5}$ with probability $\frac{129}{256}$. (y^i, y^{i+5}) would affect the bias of the output of the ABSG in two approaches. One approach is that (y^i, y^{i+5}) becomes (z_j, z_{j+2}) with probability $\frac{1}{4}$ (case 1: $y_i = y_{i-1}$, $y_{i+2} \ne y_{i+1}$ and $y_{i+3} = y_{i+2}$; case 2: $y_i \ne y_{i-1}$, $y_{i+1} = y_{i-1}$ and $y_{i+3} = y_{i+2}$). Thus for this approach, the bias of (y^i, y^{i+5}) causes $z_j = z_{j+2}$ with probability $\frac{513}{1024}$. Another approach is that if $y_i = y_{i-1}$ and $y_{i+2} = y_{i+1}$, then (y_i, y_{i+4}) becomes (z_j, z_{j+2}). Note that $y_{i+4} = y_{i-1}$ with probability $\frac{129}{256}$, so $z_j = z_{j+2}$ with probability $\frac{129}{256}$. This approach happens with probability $\frac{1}{4}$. Thus the bias of (y^i, y^{i+5}) causes $z_j = z_{j+2}$ with probability $\frac{513}{1024}$. Combining these two approaches, we know that $z_j = z_{j+2}$ with probability $\frac{257}{512}$.

We continue analyzing the above example since the output of ABSG decimation algorithm should pass through the buffer before becoming keystream.

By analyzing the ABSG decimation algorithm and the buffer, we notice that if (y^i, y^{i+5}) becomes $z_j = z_{j+2}$ after the ABSG decimation algorithm, then it becomes $z'_k = z'_{k+1}$ with probability 0.6135 after passing through the buffer; if (y^i, y^{i+4}) becomes $z_j = z_{j+2}$ after the ABSG decimation algorithm, then it becomes $z'_k = z'_{k+1}$ with probability 0.5189 after passing through the buffer. Thus after passing through the buffer, the two approaches lead to $z'_k = z'_{k+1}$ with probability $\frac{1}{2} + 0.6135 \times \frac{1}{1024} + 0.5189 \times \frac{1}{1024} = \frac{1}{2} + 2^{-9.82}$.

A similar analysis can be applied to the biases resulting from $(y_{t+4,1}, y_{t,2})$ and $(y_{t,2}, y_{t+2,2})$. The bias of $(y_{t,2}, y_{t+2,2})$ would cause $z'_k = z'_{k+1}$ with probability about $\frac{1}{2} + 2^{-10.84}$, and the bias of $(y_{t+4,1}, y_{t,2})$ would cause $z'_k = z'_{k+1}$ with probability about $\frac{1}{2} + 2^{-11.73}$.

Combining the effects of $(y_{t+3,1}, y_{t,2})$, $(y_{t+4,1}, y_{t,2})$ and $(y_{t,2}, y_{t+2,2})$, the bias of $z'_k = z'_{k+1}$ is about $\frac{1}{2} + 2^{-9.82} + 2^{-10.84} + 2^{-11.73} = \frac{1}{2} + 2^{-9.00}$.

Now we verify the above analysis with an experiment. We have generated about 2^{30} keystream bits from DECIM and found that $z'_k = z'_{k+1}$ is about $\frac{1}{2} + 2^{-8.67}$. The experimental result shows that the theoretical result is close to that obtained from the experiment.

5.2   Broadcast Attack

Due to the bias in the keystream, part of the message could be recovered from the ciphertexts if the same message is encrypted many times using DECIM with random key and IV pairs. A similar attack has been applied to RC4 by Mantin and Shamir [5].

Suppose that one message bit is encrypted N times, and each keystream bit is 0 with probability $\frac{1}{2} + \Delta p$ with $\Delta p > 0$. Denote the number of '0' in the ciphertext bits by n_0. If $n_0 > \frac{N}{2}$, we conclude that the message bit is equal to '0'; otherwise, we conclude that the message bit is equal to '1'. For $N = \Delta p^{-2}$, the message bit is recovered with a success probability of 0.977.

Thus if one message is encrypted about 2^{18} times with different keys and IVs, the message could be recovered from the ciphertexts.

6   Attacks on DECIM with 80-bit IV

The keystream generation algorithm of DECIM with an 80-bit IV is the same as DECIM with a 64-bit IV. Thus DECIM with an 80-bit IV still generates heavily biased keystream and it is vulnerable to the broadcast attack.

The initialization process of DECIM with an 80-bit IV is slightly different from the 64-bit IV version. The key and IV are loaded into the LFSR as

$$s_i = \begin{cases} 0 & \text{for } 0 \leq i \leq 31 \\ K_{i-32} \oplus IV_{i-32} & \text{for } 32 \leq i \leq 111 \\ K_{i-112} & \text{for } 112 \leq i \leq 191 \end{cases}$$

Similar to the attack given in Sect. 4, we have carried out an experiment to compute the IVs required to recover each bit. With 2^{21} IVs, 41 bits of the secret key could be recovered. Only the first 2 bytes of the keystream are required in the attack. The experiment results are given in Table 2 in Appendix A.

7   Conclusion

In this paper, we have developed two attacks against the stream cipher DECIM. The key could be recovered easily from the keystream with about 2^{20} random IVs. And the keystream of DECIM is heavily biased. The results indicate that DECIM is very weak.

Recently, the designers of DECIM have proposed DECIM v2 [2]. DECIM v2 is much simpler than DECIM. The initialization of DECIM v2 uses 768 steps of the keystream generation algorithm with the output bit being XORed to the LFSR. The filter is changed and f is one-resilient. DECIM v2 is not vulnerable to the attacks presented in this paper.

Acknowledgements

The authors would like to thank the anonymous reviewers for their helpful comments.

References

1. C. Berbain, O. Billet, A. Canteaut, N. Courtois, B. Debraize, H. Gilbert, L. Goubin, A. Gouget, L. Granboulan, C. Lauradoux, M. Minier, T. Pornin, H. Sibert. "Decim - A New Stream Cipher for Hardware Applications," *ECRYPT Stream Cipher Project Report 2005/004*. Available at http://www.ecrypt.eu.org/stream/
2. C. Berbain, O. Billet, A. Canteaut, N. Courtois, B. Debraize, H. Gilbert, L. Goubin, A. Gouget, L. Granboulan, C. Lauradoux, M. Minier, T. Pornin, H. Sibert. "DECIM v2," *ECRYPT Stream Cipher Project Report 2006/004*. Available at http://www.ecrypt.eu.org/stream/
3. D. Coppersmith, H. Krawczyk, and Y. Mansour. "The Shrinking Generator," in *Advances in Cryptology - CRYPTO'93*, volume 773 of Lecture Notes in Computer Science, pages 22-39. Springer-Verlag, 1993.
4. ECRYPT Stream Cipher Project, at http://www.ecrypt.eu.org/stream/
5. I. Mantin, A. Shamir. "A Practical Attack on Broadcast RC4," in *Fast Software Encryption (FSE 2001)*, LNCS2335, pp. 152-164, Springer-Verlag, 2001.
6. W. Meier and O. Staffelbach. "The Self-Shrinking Generator," in *Advances in Cryptology - EUROCRYPT'94*, volume 950 of Lecture Notes in Computer Science, pages 205-214, Springer-Verlag, 1994.

A   The Number of IVs Required to Break DECIM

Table 1 gives the number of IVs required to break DECIM with a 64-bit IV. 44 key bits can be recovered with less than 2^{20} IVs. Table 2 gives the number of

Table 1. Number of IVs required to recover the key bits (64-bit IV)

	Affected Bits	Amount of IVs ($\log_2$)		Affected Bits	Amount of IVs ($\log_2$)
K_0	$s_{112} \Rightarrow s_{192+60}$	18.95	K_1	$s_{57} \Rightarrow s_{192+122}$	20.83
K_2	$s_{58} \Rightarrow s_{192+116}$	18.80	K_3	$s_{115} \Rightarrow s_{192+104}$	20.46
K_4	$s_{116} \Rightarrow s_{192+105}$	21.41	K_5	$s_{117} \Rightarrow s_{192+106}$	21.54
K_6	$s_{118} \Rightarrow s_{192+107}$	21.67	K_7	$s_{119} \Rightarrow s_{192+108}$	21.72
K_8	$s_{120} \Rightarrow s_{192+145}$	21.21	K_9	$s_{121} \Rightarrow s_{192+110}$	21.92
K_{10}	$s_{10} \Rightarrow s_{192+116}$	17.69	K_{11}	$s_{11} \Rightarrow s_{192+117}$	19.62
K_{12}	$s_{68} \Rightarrow s_{192+6}$	18.88	K_{13}	$s_{69} \Rightarrow s_{192+7}$	20.82
K_{14}	$s_{70} \Rightarrow s_{192+8}$	18.82	K_{15}	$s_{127} \Rightarrow s_{192+116}$	16.66
K_{16}	$s_{128} \Rightarrow s_{192+117}$	18.70	K_{17}	$s_{17} \Rightarrow s_{192+6}$	16.92
K_{18}	$s_{18} \Rightarrow s_{192+7}$	18.82	K_{19}	$s_{19} \Rightarrow s_{192+8}$	16.80
K_{20}	$s_{20} \Rightarrow s_{192+9}$	18.73	K_{21}	$s_{21} \Rightarrow s_{192+6}$	18.59
K_{22}	$s_{22} \Rightarrow s_{192+7}$	20.67	K_{23}	$s_{23} \Rightarrow s_{192+8}$	18.70
K_{24}	$s_{80} \Rightarrow s_{192+146}$	20.80	K_{25}	$s_{25} \Rightarrow s_{192+116}$	17.97
K_{26}	$s_{138} \Rightarrow s_{192+6}$	17.79	K_{27}	$s_{139} \Rightarrow s_{192+7}$	19.87
K_{28}	$s_{140} \Rightarrow s_{192+8}$	17.86	K_{29}	$s_{141} \Rightarrow s_{192+9}$	19.67
K_{30}	$s_{142} \Rightarrow s_{192+10}$	21.46	K_{31}	$s_{31} \Rightarrow s_{192+182}$	18.36
K_{32}	$s_{32} \Rightarrow s_{192+183}$	20.70	K_{33}	$s_{33} \Rightarrow s_{192+113}$	20.97
K_{34}	$s_{34} \Rightarrow s_{192+114}$	21.03	K_{35}	$s_{91} \Rightarrow s_{192+116}$	19.95
K_{36}	$s_{36} \Rightarrow s_{192+116}$	15.55	K_{37}	$s_{37} \Rightarrow s_{192+117}$	17.56
K_{38}	$s_{94} \Rightarrow s_{192+145}$	18.94	K_{39}	$s_{39} \Rightarrow s_{192+104}$	19.62
K_{40}	$s_{152} \Rightarrow s_{192+60}$	16.43	K_{41}	$s_{153} \Rightarrow s_{192+116}$	17.90
K_{42}	$s_{154} \Rightarrow s_{192+117}$	19.93	K_{43}	$s_{43} \Rightarrow s_{192+108}$	20.61
K_{44}	$s_{156} \Rightarrow s_{192+145}$	16.90	K_{45}	$s_{157} \Rightarrow s_{192+146}$	18.96
K_{46}	$s_{46} \Rightarrow s_{192+35}$	20.45	K_{47}	$s_{47} \Rightarrow s_{192+6}$	16.68
K_{48}	$s_{160} \Rightarrow s_{192+145}$	18.68	K_{49}	$s_{161} \Rightarrow s_{192+181}$	15.59
K_{50}	$s_{162} \Rightarrow s_{192+182}$	17.59	K_{51}	$s_{51} \Rightarrow s_{192+116}$	15.62
K_{52}	$s_{52} \Rightarrow s_{192+117}$	17.64	K_{53}	$s_{53} \Rightarrow s_{192+118}$	19.47
K_{54}	$s_{54} \Rightarrow s_{192+119}$	20.05	K_{55}	$s_{55} \Rightarrow s_{192+120}$	20.61
K_{56}	$s_{168} \Rightarrow s_{192+76}$	22.27	K_{57}	$s_{169} \Rightarrow s_{192+103}$	18.43
K_{58}	$s_{170} \Rightarrow s_{192+104}$	18.17	K_{59}	$s_{171} \Rightarrow s_{192+105}$	18.93
K_{60}	$s_{172} \Rightarrow s_{192+106}$	19.11	K_{61}	$s_{173} \Rightarrow s_{192+107}$	19.24
K_{62}	$s_{174} \Rightarrow s_{192+108}$	19.42	K_{63}	$s_{175} \Rightarrow s_{192+109}$	19.58

IVs required to break DECIM with an 80-bit IV. 41 key bits can be recovered with less than 2^{21} IVs. Only the first 2 bytes of the keystream are required in the attack, and the amount of computation required in the attacks is negligible.

We explain Table 1 with K_0 as an example. K_0 is related to s_{112} since $s_{112} = K_0 \oplus IV_0$. s_{112} is mapped to s_{192+60} with probability 0.0318 (this probability is obtained by tracing s_{112} through the initialization process). Thus K_0 could be recovered by observing the first bits of the keystreams. About $2^{18.95}$ IVs are required to achieve a success probability of 0.977.

Table 2. Number of IVs required to recover the key bits (80-bit IV)

	Affected Bits	Amount of IVs ($\log_2$)		Affected Bits	Amount of IVs ($\log_2$)
K_0	$s_{32} \Rightarrow s_{192+183}$	20.70	K_1	$s_{33} \Rightarrow s_{192+113}$	20.97
K_2	$s_{34} \Rightarrow s_{192+114}$	21.03	K_3	$s_{35} \Rightarrow s_{192+115}$	21.13
K_4	$s_{36} \Rightarrow s_{192+116}$	15.55	K_5	$s_{37} \Rightarrow s_{192+117}$	17.56
K_6	$s_{38} \Rightarrow s_{192+118}$	19.43	K_7	$s_{39} \Rightarrow s_{192+104}$	19.62
K_8	$s_{40} \Rightarrow s_{192+105}$	20.37	K_9	$s_{41} \Rightarrow s_{192+121}$	20.30
K_{10}	$s_{42} \Rightarrow s_{192+107}$	20.48	K_{11}	$s_{43} \Rightarrow s_{192+108}$	20.61
K_{12}	$s_{44} \Rightarrow s_{192+109}$	20.77	K_{13}	$s_{45} \Rightarrow s_{192+34}$	20.70
K_{14}	$s_{46} \Rightarrow s_{192+35}$	20.45	K_{15}	$s_{47} \Rightarrow s_{192+6}$	16.68
K_{16}	$s_{48} \Rightarrow s_{192+7}$	18.72	K_{17}	$s_{49} \Rightarrow s_{192+8}$	16.68
K_{18}	$s_{50} \Rightarrow s_{192+9}$	18.66	K_{19}	$s_{51} \Rightarrow s_{192+116}$	15.62
K_{20}	$s_{52} \Rightarrow s_{192+117}$	17.64	K_{21}	$s_{53} \Rightarrow s_{192+118}$	19.47
K_{22}	$s_{54} \Rightarrow s_{192+119}$	20.05	K_{23}	$s_{55} \Rightarrow s_{192+120}$	20.61
K_{24}	$s_{56} \Rightarrow s_{192+121}$	20.63	K_{25}	$s_{57} \Rightarrow s_{192+122}$	20.83
K_{26}	$s_{58} \Rightarrow s_{192+116}$	18.80	K_{27}	$s_{59} \Rightarrow s_{192+12}$	23.00
K_{28}	$s_{60} \Rightarrow s_{192+13}$	23.41	K_{29}	$s_{61} \Rightarrow s_{192+14}$	23.66
K_{30}	$s_{62} \Rightarrow s_{192+15}$	23.78	K_{31}	$s_{63} \Rightarrow s_{192+16}$	24.09
K_{32}	$s_{64} \Rightarrow s_{192+17}$	24.00	K_{33}	$s_{65} \Rightarrow s_{192+18}$	24.19
K_{34}	$s_{66} \Rightarrow s_{192+19}$	24.22	K_{35}	$s_{67} \Rightarrow s_{192+5}$	23.44
K_{36}	$s_{68} \Rightarrow s_{192+6}$	18.88	K_{37}	$s_{69} \Rightarrow s_{192+7}$	20.82
K_{38}	$s_{70} \Rightarrow s_{192+8}$	18.82	K_{39}	$s_{71} \Rightarrow s_{192+60}$	16.77
K_{40}	$s_{72} \Rightarrow s_{192+61}$	18.75	K_{41}	$s_{73} \Rightarrow s_{192+62}$	20.59
K_{42}	$s_{74} \Rightarrow s_{192+63}$	21.11	K_{43}	$s_{75} \Rightarrow s_{192+64}$	21.71
K_{44}	$s_{76} \Rightarrow s_{192+65}$	21.67	K_{45}	$s_{77} \Rightarrow s_{192+66}$	21.85
K_{46}	$s_{78} \Rightarrow s_{192+67}$	21.81	K_{47}	$s_{79} \Rightarrow s_{192+145}$	18.82
K_{48}	$s_{80} \Rightarrow s_{192+146}$	20.80	K_{49}	$s_{81} \Rightarrow s_{192+70}$	22.05
K_{50}	$s_{82} \Rightarrow s_{192+71}$	22.18	K_{51}	$s_{83} \Rightarrow s_{192+72}$	22.40
K_{52}	$s_{84} \Rightarrow s_{192+73}$	22.43	K_{53}	$s_{85} \Rightarrow s_{192+74}$	22.42
K_{54}	$s_{86} \Rightarrow s_{192+75}$	22.43	K_{55}	$s_{87} \Rightarrow s_{192+76}$	22.55
K_{56}	$s_{88} \Rightarrow s_{192+154}$	24.02	K_{57}	$s_{89} \Rightarrow s_{192+155}$	24.04
K_{58}	$s_{90} \Rightarrow s_{192+156}$	24.15	K_{59}	$s_{91} \Rightarrow s_{192+116}$	19.95
K_{60}	$s_{92} \Rightarrow s_{192+117}$	21.97	K_{61}	$s_{93} \Rightarrow s_{192+118}$	23.77
K_{62}	$s_{94} \Rightarrow s_{192+145}$	18.94	K_{63}	$s_{95} \Rightarrow s_{192+146}$	20.91
K_{64}	$s_{96} \Rightarrow s_{192+147}$	22.79	K_{65}	$s_{97} \Rightarrow s_{192+148}$	23.33
K_{66}	$s_{98} \Rightarrow s_{192+149}$	23.77	K_{67}	$s_{99} \Rightarrow s_{192+150}$	23.64
K_{68}	$s_{100} \Rightarrow s_{192+63}$	22.65	K_{69}	$s_{101} \Rightarrow s_{192+4}$	23.12
K_{70}	$s_{102} \Rightarrow s_{192+65}$	23.66	K_{71}	$s_{103} \Rightarrow s_{192+178}$	23.80
K_{72}	$s_{104} \Rightarrow s_{192+179}$	23.77	K_{73}	$s_{105} \Rightarrow s_{192+145}$	20.94
K_{74}	$s_{106} \Rightarrow s_{192+181}$	18.24	K_{75}	$s_{107} \Rightarrow s_{192+182}$	19.97
K_{76}	$s_{108} \Rightarrow s_{192+183}$	21.81	K_{77}	$s_{109} \Rightarrow s_{192+6}$	20.86
K_{78}	$s_{110} \Rightarrow s_{192+7}$	22.83	K_{79}	$s_{111} \Rightarrow s_{192+8}$	20.94

On Feistel Structures Using a Diffusion Switching Mechanism

Taizo Shirai and Kyoji Shibutani

Sony Corporation, Tokyo, Japan
{Taizo.Shirai, Kyoji.Shibutani}@jp.sony.com

Abstract. We study a recently proposed design approach of Feistel structure which employs diffusion matrices in a switching way. At ASIACRYPT 2004, Shirai and Preneel have proved that large numbers of S-boxes are guaranteed to be active if a diffusion matrix used in a round function is selected among multiple matrices. However the optimality of matrices required by the proofs sometimes pose restriction to find matrices suitable for actual blockciphers. In this paper, we extend their theory by replacing the condition of optimal mappings with general-type mappings, consequently the restriction is eliminated. Moreover, by combining known lower bounds for usual Feistel structure, we establish a method to estimate the guaranteed number of active S-boxes for arbitrary round numbers. We also demonstrate how the generalization enables us to mount wide variety of diffusion mappings by showing concrete examples.

Keywords: blockcipher, Feistel structure, optimal diffusion mappings.

1 Introduction

A Feistel structure is one of the most widely used and best studied structures for the design of blockciphers. It was proposed by H. Feistel in the early 1970s; subsequently the structure was adopted in the well-known blockcipher DES [6,7]. During the 30-year of modern blockcipher research history, extensive studies have been made on Feistel structure [10, 13, 16]. Currently, many well-known blockciphers employ the design of Feistel structures [1, 12, 15, 17].

On the other hand, an optimal diffusion which is a linear function with the maximum branch number is widely regarded in the recent blockcipher research; the concept is used in the design of AES/Rijndael and many other cryptographic primitives [2, 17, 14, 5]. However the effect of an optimal diffusion especially in Feistel structure is still needed to be studied.

In 2004, Shirai and Shibutani proposed a novel design concept of Feistel structure which employs plural optimal diffusion matrices in a switching manner. In their design approach, a diffusion matrix in the round function is switched among multiple matrices in a predefined order [21]. We call the matrix switching technique *Diffusion Switching Mechanism* (*DSM* for short) in this paper. Then, Shirai and Preneel has first shown the theoretical explanation of the effects of the

M.J.B. Robshaw (Ed.): FSE 2006, LNCS 4047, pp. 41–56, 2006.

DSM [19]. They proved that the immunity against both differential and linear cryptanalysis would be strengthened due to the fact that difference and linear mask cancellation in characteristics caused by a small number of active S-boxes will never occur.

The theory of the *DSM* opened a new line of research on the Feistel structure. However the optimality condition for matrices in their result sometimes pose restriction to find various matrices suitable for actual blockciphers. For example, our experimental result showed that there are no 8×8 matrices over $GF(2^8)$ satisfying both the optimality and certain practically favorable conditions.

In this paper, we generalize the *DSM* theory by eliminating the conditions of diffusion mappings. This generalization enables us to estimate the guaranteed number of active S-boxes for any types of diffusion mappings if we get knowledge of branch numbers of the mappings. Let a minimum differential branch number among all diffusion matrices used in the Feistel structure be B_1^D, and let a smallest differential and a linear branch number of diffusion matrices composed of two alternate (i.e. i-th and $i + 2$-th) rounds be B_2^D, B_2^L, respectively, and three alternate $(i, i + 2, i + 4$-th) rounds differential branch number be B_3^D. Then, we prove novel extended result on the numbers of active S-boxes such that $R(B_1^D + B_2^D)$ differential active S-boxes for 6R-round, $R(2B_1^D + B_3^D)$ for 9R-round and RB_2^L linear active S-boxes for 3R-round are theoretically guaranteed.

In addition, we show how to estimate the lower bound of number of active S-boxes for arbitrary number of rounds. Kanda has already shown the results on lower bound of the number of active S-boxes for single matrix based ordinary Feistel structure [9]. By combining our results and Kanda's results, lower bounds for any number of rounds can be calculated in a simple manner. Consequently, we can make use of the proved lower bounds for designing Feistel ciphers which hold desirable expected immunity against differential attack and linear attack [4]. We also confirm effects of the generalization by showing concrete example 8×8 matrices for a 128-bit block Feistel structure.

This paper is organized as follows: in Sect. 2, we introduce some definitions used in this paper. Previous works including *ODM-MR design* approach are shown in Sect. 3. We prove in Sect. 4 the extended theorems regarding *Diffusion Switching Mechanism* (*DSM* for short) as our main contribution. In Sect. 5, we discuss the new design approach by presenting some examples and numerical values. Finally Sect. 6 concludes the paper.

2 Preliminaries

In this paper, we treat a typical type of Feistel structure, which is called a balanced Feistel. It is defined as follows [16].

Definition 1. *(Balanced Feistel structure)*
Let b be a block size, r be a number of rounds, and k be a size of round key. Let $k_i \in \{0,1\}^k$ $(1 \leq i \leq r)$ be round keys provided by a certain key schedul-ing algorithm and $x_i \in \{0,1\}^{b/2}$ be intermediate data, and let $F_i : \{0,1\}^{b/2} \times$

$\{0,1\}^k \to \{0,1\}^{b/2}$ be an F-function at the i-th round. The algorithm of a balanced Feistel structure is defined as : (1) Input $x_0, x_1 \in \{0,1\}^{b/2}$, (2) Calculate $x_{i+1} = F_i(x_i, k_i) \oplus x_{i-1}$ $(1 \leq i \leq r)$, (3) Output $x_r, x_{r+1} \in \{0,1\}^{b/2}$.

Then we define SP-type F-functions which are special constructions of a F-function [18, 9].

Definition 2. *(SP-type F-functions)*
Let a length of a round key $k = b/2$. Let m be the number of S-boxes in a round, and n be the size of the S-boxes, with $mn = b/2$. Let $s_{i,j} : \{0,1\}^n \to \{0,1\}^n$ be the j-th S-box in the i-th round, and let $S_i : \{0,1\}^{b/2} \to \{0,1\}^{b/2}$ be the function generated by concatenating m S-boxes in parallel in the i-th round. Let $P_i : \{0,1\}^{b/2} \to \{0,1\}^{b/2}$ be the linear Boolean function. Then SP-type F-functions are defined as $F_i(x_i, k_i) = P_i(S_i(x_i \oplus k_i))$.

Note that we denote the intermediate variables $z_i = S_i(x_i \oplus k_i)$ in this paper.

Definition 3. *((m,n,r)-SPFS)*
An (m, n, r)-SPFS is defined as an r-round Feistel structure with SP-type round function using m n-bit S-boxes, and for which all $s_{i,j}$, P_i are bijective. An $mn \times mn$ matrix M_i $(1 \leq i \leq r)$ over $GF(2)$ denotes a matrix representation of a linear Boolean function P_i where $P_i(x) = M_i x$.

Remark 1. Because of the bijectivity of S-boxes and linear function P in (m, n, r)-SPFS, all F-functions are bijective.

We also give definitions of bundle weight and branch number [5].

Definition 4. *(bundle weight)*
Let $x \in \{0,1\}^{pn}$ represented as $x = [x_0 x_1 \ldots x_{p-1}]$ where $x_i \in \{0,1\}^n$, then the bundle weight $w_n(x)$ is defined as

$$w_n(x) = \sharp\{x_i | x_i \neq 0\} .$$

Definition 5. *(Branch Number)*
Let $P : \{0,1\}^{pn} \to \{0,1\}^{qn}$. The branch number of P is defined as

$$\mathcal{B}r_n(P) = \min_{a \neq 0}\{w_n(a) + w_n(P(a))\} .$$

Remark 2. The maximum branch number is $\mathcal{B}r_n(P) = q+1$. If a linear function has a maximum branch number, it is called an **optimal diffusion mapping** [2]. It is known that an optimal diffusion mapping can be obtained from maximum distance separable (MDS) codes [5].

3   Previous Work

The precise estimation of the lower bound of the number of active S-boxes of blockciphers has been known as one of the practical means to evaluate strength of ciphers, because the lower bound can be used to estimate weight distributions of differential and linear characteristics [11,3,1,18,9,5]. It is shown that the weight distribution is connected with the bound of the expected differential probability or the linear hull probability by Daemen and Rijmen [4].

Recently, Shirai and Preneel proved the following corollary which can be used to estimate the lower number of active S-boxes of a specially designed Feistel structure [21,19].

Definition 6. *Let p be a positive integer, and A, B be $p \times p$ square matrices. Then $[A|B]$ denotes a $p \times 2p$ matrix obtained by concatenating A and B. Similarly, the three matrices case is defined for $[A|B|C]$.*

Corollary 1. *Let E be an (m,n,r)-SPFS blockcipher where $r \geq 6$. If $[M_i|M_{i+2}|M_{i+4}]$ and $[{}^tM_j^{-1}|{}^tM_{j+2}^{-1}]$ are optimal diffusion mappings for any i, j $(1 \leq i \leq r - 4, 1 \leq j \leq r - 2)$, respectively, any $3R$ consecutive rounds $(R \geq 2)$ in E guarantee at least $R(m + 1)$ differential and linear active S-boxes.*

The design approach is called *ODM-MR* (Optimal Diffusion Mappings across Multiple Rounds) design approach. To apply the corollary to practical Feistel structures, we need to use at least three different matrices [19]. For example, let A_0, A_1, A_2 be the matrices which satisfy the following conditions.

1. Choose $nm \times nm$ matrices A_0, A_1, A_2 over $GF(2)$ satisfying the following optimal diffusion conditions:
   (a) $\mathcal{B}r_n([A_0|A_1|A_2]) = m + 1$,
   (b) $\mathcal{B}r_n([{}^tA_0^{-1}|{}^tA_1^{-1}]) = \mathcal{B}r_n([{}^tA_1^{-1}|{}^tA_2^{-1}]) = \mathcal{B}r_n([{}^tA_2^{-1}|{}^tA_0^{-1}]) = m + 1$.
2. Set these three matrices as $M_{2i+1} = M_{2r-2i} = A_{i \bmod 3}$, for $0 \leq i < r$ in an $2r$-round Feistel structure $(m, n, 2r)$-SPFS (Fig.1).

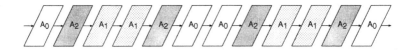

Fig. 1. Example Allocation of Matrices A_0, A_1, A_2

The corollary states that the $(m, n, 2r)$-SPFS with the above settings guarantees $2(m + 1)$, $3(m + 1)$ and $4(m + 1)$ differential and linear active S-boxes in 6, 9 and 12 consecutive rounds, respectively. Fig. 2 illustrates the statement.

In this way, using multiple diffusion matrices in a switching way for round functions makes Feistel structure stronger against differential attack and linear attack. In this paper, we call the new design concept a *Diffusion Switching Mechanism* (DSM) in general. From now on, we will extend the DSM to treat not only optimal diffusion matrices but also any general type matrices.

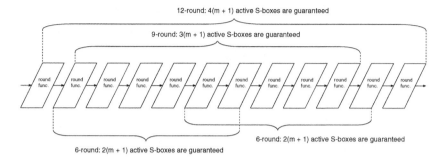

Fig. 2. Guaranteed Active S-boxes by *ODM-MR design*

4 DSM for General Matrices

In this section, we show the extended theory of the DSM theoretically[1]. The following two subsections are devoted to proving three theorems. To ease the proofs, we first introduce an additional definition.

Definition 7. *Consider differential characteristics or linear characteristics. Let D_i and L_i denote the number of differential and linear active S-boxes in the i-th round, respectively. These values are determined by the differences $\Delta x_i, \Delta z_i$ or by the linear masks $\Gamma x_i, \Gamma z_i$. Since all S-boxes are bijective, we have the following relations,*

$$D_i = w_n(\Delta x_i) = w_n(\Delta z_i) \ , \qquad L_i = w_n(\Gamma x_i) = w_n(\Gamma z_i) \ ,$$

where $w_n(\cdot)$ is the bundle weight as defined in Definition 4.

Remark 3. If we have a nonzero input difference for an (m, n, r)-SPFS, we obtain the following conditions:

$$(d0) \quad D_i = 0 \Rightarrow D_{i-2} \neq 0, D_{i-1} \neq 0, D_{i+1} \neq 0, D_{i+2} \neq 0 \ ,$$
$$(d1) \quad D_i = 0 \Rightarrow D_{i-1} = D_{i+1} \ .$$

Similarly, if a nonzero input mask is given, we have

$$(l0) \quad L_i = 0 \Rightarrow L_{i-2} \neq 0, L_{i-1} \neq 0, L_{i+1} \neq 0, L_{i+2} \neq 0 \ ,$$
$$(l1) \quad L_i = 0 \Rightarrow L_{i-1} = L_{i+1} \ .$$

4.1 Proofs for the Lower Bound of Differential Active S-Boxes

In this section we prove Theorem 1 and Theorem 2; the proof is based on three lemmata. Firstly we define a concept of minimum branch numbers for three types of matrices.

---

[1] The composition of the extended version of proofs almost follows that of proofs for *ODM-MR* [19].

Definition 8. *For an* (m, n, r)-*SPFS, minimum branch numbers* B_1^D, B_2^D *and* B_3^D *are defined as follows.*

$$B_1^D = \min_{1 \leq i \leq r} (\mathcal{B}r_n(M_i)) \;,$$

$$B_2^D = \min_{1 \leq i \leq r-2} (\mathcal{B}r_n([M_i | M_{i+2}])) \;,$$

$$B_3^D = \min_{1 \leq i \leq r-4} (\mathcal{B}r_n([M_i | M_{i+2} | M_{i+4}])) \;.$$

Obviously, the following inequality holds.

$$B_1^D \geq B_2^D \geq B_3^D \;. \tag{1}$$

Note that these values can be derived from any given set of diffusion mappings M_i $(1 \leq i \leq r)$. Introducing these values into the proofs means that the constraint of optimal diffusion mappings will disappear. This is an essence of our generalization.

Firstly, Lemma 1 shows relations between D_i of (m, n, r)-SPFS and B_1^D.

Lemma 1. *Let* E *be an* (m, n, r)-*SPFS blockcipher, then* E *has the following condition* (d2).

$$(d2) \quad D_{i+1} \neq 0 \Rightarrow D_i + D_{i+1} + D_{i+2} \geq B_1^D \;.$$

Proof. From the relation between the differences $\Delta z_{i+1}, \Delta x_i$ and Δx_{i+2} in a 3 consecutive rounds, we obtain the following equation.

$$M_{i+1} \Delta z_{i+1} = \Delta x_i \oplus \Delta x_{i+2} \;.$$

Since M_i has a branch number at least B_1^D we have

$$w_n(\Delta z_{i+1}) \neq 0 \Rightarrow w_n(\Delta z_{i+1}) + w_n(\Delta x_i \oplus \Delta x_{i+2}) \geq B_1^D \;. \tag{2}$$

Eq. (2) and the inequality $w_n(\Delta x_i) + w_n(\Delta x_{i+2}) \geq w_n(\Delta x_i \oplus \Delta x_{i+2})$ yield (d2).

□

Remark 4. By combining Remark 3 and (d2), we obtain additional underlying conditions (d3) and (d4).

$$(d3) \quad D_i = 0 \Rightarrow D_{i+1} + D_{i+2} \geq B_1^D \;,$$
$$(d4) \quad D_{i+2} = 0 \Rightarrow D_i + D_{i+1} \geq B_1^D \;.$$

Eq. (d3) and (d4) mean that if a k-th round has no active S-boxes, any 2 consecutive rounds next to the k-th round always contain more than B_1^D active S-boxes.

Next, we show the property of (m, n, r)-SPFS for two matrices case.

Lemma 2. *Let* E *be an* (m, n, r)-*SPFS blockcipher,* E *has the following conditions* (d5), (d6).

$$(d5) \quad D_{i+4} = 0 \Rightarrow D_i + D_{i+1} + D_{i+3} \geq B_2^D \;,$$
$$(d6) \quad D_i = 0 \Rightarrow D_{i+1} + D_{i+3} + D_{i+4} \geq B_2^D \;.$$

Proof. From the relation between 5-round differences,

$$M_{i+1}\Delta z_{i+1} \oplus M_{i+3}\Delta z_{i+3} = \Delta x_i \oplus \Delta x_{i+4} \ .$$

Then,

$$[M_{i+1}|M_{i+3}] \begin{pmatrix} \Delta z_{i+1} \\ \Delta z_{i+3} \end{pmatrix} = \Delta x_i \oplus \Delta x_{i+4} \ .$$

Since $[M_{i+1}|M_{i+3}]$ has a branch number at least B_2^D, and from Remark 3, we see that $w_n(\Delta z_{i+1}) = 0$ and $w_n(\Delta z_{i+3}) = 0$ will never occur simultaneously, we obtain

$$w_n(\Delta z_{i+1}) + w_n(\Delta z_{i+3}) + w_n(\Delta x_i \oplus \Delta x_{i+4}) \geq B_2^D \ .$$

Assuming the cases $\Delta x_i = 0$ or $\Delta x_{i+4} = 0$, we directly obtain (d5) and (d6). □

By using the previously obtained conditions (d0) − (d6), we show the following theorem for differential active S-boxes.

Theorem 1. *Let E be an (m, n, r)-SPFS blockcipher, any 6 consecutive rounds in E guarantee at least $B_1^D + B_2^D$ differential active S-boxes.*

Proof. Consider the total number of active S-boxes in 6 consecutive rounds from the i-th round,

$$\sum_{k=i}^{i+5} D_k = D_i + D_{i+1} + D_{i+2} + D_{i+3} + D_{i+4} + D_{i+5} \ .$$

If $D_{i+1} \neq 0$ and $D_{i+4} \neq 0$, the condition (d2) guarantees that $D_i + D_{i+1} + D_{i+2} \geq B_1^D$ and $D_{i+3} + D_{i+4} + D_{i+5} \geq B_1^D$. Therefore we obtain $\sum_{k=i}^{i+5} D_k \geq 2B_1^D$. If $D_{i+1} = 0$,

$$\sum_{k=i}^{i+5} D_k = D_i + D_{i+2} + D_{i+3} + D_{i+4} + D_{i+5} \ .$$

From (d1),

$$\sum_{k=i}^{i+5} D_k = 2 \cdot D_{i+2} + D_{i+3} + D_{i+4} + D_{i+5}$$

$$= (D_{i+2} + D_{i+3}) + (D_{i+2} + D_{i+4} + D_{i+5}) \ .$$

From (d3) and (d6),

$$\sum_{k=i}^{i+5} D_k \geq B_1^D + B_2^D \ .$$

The case of $D_{i+4} = 0$ is proved similarly from (d1), (d4) and (d5). Combining with (1), we have shown that any 6 consecutive rounds in E guarantee at least $B_1^D + B_2^D$ differential active S-boxes. □

Immediately, we obtain the following corollary.

Corollary 2. *Let E be an (m, n, r)-SPFS blockcipher. Any 6R consecutive rounds in E guarantee at least $R(B_1^D + B_2^D)$ differential active S-boxes.*

The result is compatible with *ODM-MR* by substituting $m + 1$ for B_1^D and B_2^D.

Next, we show the property of (m, n, r)-SPFS for three matrices case.

Lemma 3. *Let E be an (m, n, r)-SPFS blockcipher. E satisfies the following condition (d7).*

$$(d7) \quad D_i = D_{i+6} = 0 \Rightarrow D_{i+1} + D_{i+3} + D_{i+5} \geq B_3^D \ .$$

Proof. First, from the difference relation in 7 consecutive rounds, we obtain

$$M_{i+1}\Delta z_{i+1} \oplus M_{i+3}\Delta z_{i+3} \oplus M_{i+5}\Delta z_{i+5} = \Delta x_i \oplus \Delta x_{i+6} \ .$$

Then,

$$[M_{i+1}|M_{i+3}|M_{i+5}] \begin{pmatrix} \Delta z_{i+1} \\ \Delta z_{i+3} \\ \Delta z_{i+5} \end{pmatrix} = \Delta x_i \oplus \Delta x_{i+6} \ .$$

Since $[M_{i+1}|M_{i+3}|M_{i+5}]$ has a branch number at least B_3^D, and from Remark 3, $w_n(\Delta z_{i+1})$, $w_n(\Delta z_{i+3})$, and $w_n(\Delta z_{i+5})$ cannot be simultaneously 0, we get that

$$w_n(\Delta z_{i+1}) + w_n(\Delta z_{i+3}) + w_n(\Delta z_{i+5}) + w_n(\Delta x_i \oplus \Delta x_{i+6}) \geq B_3^D \ .$$

By assuming $\Delta x_i = 0$ and $\Delta x_{i+6} = 0$, we derive the condition (d7).   □

From the additional condition (d7), we derive the following theorem.

Theorem 2. *Let E be an (m, n, r)-SPFS blockcipher. Any 9 consecutive rounds in E guarantee at least $2B_1^D + B_3^D$ differential active S-boxes.*

Proof. Consider the total number of active S-boxes in 9 consecutive rounds,

$$\sum_{k=i}^{i+8} D_k = D_i + D_{i+1} + D_{i+2} + D_{i+3} + D_{i+4} + D_{i+5} + D_{i+6} + D_{i+7} + D_{i+8} \ .$$

If $D_{i+1} \neq 0$ then $D_i + D_{i+1} + D_{i+2} \geq B_1^D$ from (d2), and Lemma 1 guarantees that the sum of the remaining 6 consecutive rounds $\sum_{k=i+3}^{i+8} D_k \geq B_1^D + B_2^D$. Consequently $\sum_{k=i}^{i+8} D_k \geq 2B_1^D + B_2^D$. Similarly, if $D_{i+7} \neq 0$, at least $2B_1^D + B_2^D$ active S-boxes are guaranteed.

If $D_{i+1} = D_{i+7} = 0$, we obtain

$$\sum_{k=i}^{i+8} D_k = D_i + D_{i+2} + D_{i+3} + D_{i+4} + D_{i+5} + D_{i+6} + D_{i+8} \ .$$

From (d1),

$$\sum_{k=i}^{i+8} D_k = 2 \cdot D_{i+2} + D_{i+3} + D_{i+4} + D_{i+5} + 2 \cdot D_{i+6}$$

$$= (D_{i+2} + D_{i+3}) + (D_{i+2} + D_{i+4} + D_{i+6}) + (D_{i+5} + D_{i+6}) \ .$$

From $(d3)$, $(d7)$ and $(d4)$,

$$\sum_{k=i}^{i+8} D_k \geq B_1^D + B_3^D + B_1^D = 2B_1^D + B_3^D \ .$$

Combining with (1), we have shown that any 9 consecutive rounds in E guarantee at least $2B_1^D + B_3^D$ differential active S-boxes.   □

Immediately, we obtain the following corollary.

Corollary 3. *Let E be an (m, n, r)-SPFS blockcipher. Any 9R consecutive rounds in E guarantee at least $R(2B_1^D + B_3^D)$ differential active S-boxes.*

The result is compatible with *ODM-MR* by substituting $m + 1$ for B_1^D and B_3^D.

4.2   Proofs for the Lower Bound of Linear Active S-Boxes

In this subsection, we will show the proof of the guaranteed number of linear active S-boxes of (m, n, r)-SPFS.

Definition 9. *For an (m, n, r)-SPFS, minimum branch number B_2^L is defined as follows.*

$$B_2^L = \min_{1 \leq i \leq r-2} (\mathcal{B}r_n([^t M_i^{-1} |^t M_{i+2}^{-1}])) \ .$$

Theorem 3. *Let E be an (m, n, r)-SPFS blockcipher. Any 3 consecutive rounds in E has at least B_2^L linear active S-boxes.*

Proof. From the 3-round linear mask relation,

$$\Gamma x_{i+1} = {}^t M_i^{-1} \Gamma z_i \oplus {}^t M_{i+2}^{-1} \Gamma z_{i+2} \ .$$

Then,

$$\Gamma x_{i+1} = [^t M_i^{-1} |^t M_{i+2}^{-1}] \begin{pmatrix} \Gamma z_i \\ \Gamma z_{i+2} \end{pmatrix} \ .$$

Since $[^t M_i^{-1} |^t M_{i+2}^{-1}]$ has a branch number at least B_2^L, and from Remark 3, $w_n(\Gamma z_i)$ and $w_n(\Gamma z_{i+2})$ cannot be simultaneously 0, we obtain

$$w_n(\Gamma z_i) + w_n(\Gamma x_{i+1}) + w_n(\Gamma z_{i+2}) \geq B_2^L \ .$$

By using the notion of L_i, this implies,

$$(l1) \quad L_i + L_{i+1} + L_{i+2} \geq B_2^L \ .$$

□

As a result, we obtain the following corollary.

Corollary 4. *Let E be an (m, n, r)-SPFS blockcipher. Any 3R consecutive rounds in E guarantee at least RB_2^L linear active S-boxes.*

The result is compatible with *ODM-MR* by substituting $m + 1$ for B_2^L in the corollary 1.

5   Discussion

5.1   Comparison of the Results

The statement of the corollaries 2 and 3 are independently applicable. Therefore, it is possible that the both of these corollaries may produce different lowrbounds for the same round numbers.

For example, consider an $(m, n, 18R)$-SPFS, $3R(B_1^D + B_2^D)$ and $2R(2B_1^D + B_3^D)$ differential active S-boxes are lower bounded by the corollary 2 and 3, respectively. Letting two parameters of diffusion matrices $\alpha = B_1^D - B_2^D$ and $\beta = B_2^D - B_3^D$, we obtain the gap of these lower bounds as,

$$3R(B_1^D + B_2^D) - 2R(2B_1^D + B_3^D) = R(2(B_2^D - B_3^D) - (B_1^D - B_2^D))$$
$$= R(2\beta - \alpha)$$

If $\alpha = 2\beta$, these lower bounds always coincide. If $\alpha \neq 2\beta$, different lower bounds are produced at the 18R-th rounds. In such a case, we had better choose a larger lower bound and use it to adjust lower bounds for the rounds after the 18-th rounds to get more precise estimation.

5.2   Interpolation for Skipped Rounds

The corollaries 2, 3 and 4 are not able to provide lower bounds for any number of rounds, because they are valid for only multiples of 3, 6, or 9 rounds. Besides these corollaries, we use known results for Feistel structure to interpolate guaranteed lower bounds of the rounds which are not indicated by these corollaries.

Firstly the following trivial conditions are described explicitly.

$$\text{(1-round cond.)} \quad D_i \geq 0, \quad L_i \geq 0 \ .$$

$$\text{(2-round cond.)} \quad D_i + D_{i+1} \geq 1 \ , \quad L_i + L_{i+1} \geq 1 \ .$$

Kanda has proved inequalities for 3 and 4-round for the single matrix based normal Feistel structure, which can be converted into our settings as follows [9].

$$\text{(3-round cond.)} \quad D_i + D_{i+1} + D_{i+2} \geq 2 \ .$$

$$\text{(4-round cond.)} \quad D_i + D_{i+1} + D_{i+2} + D_{i+3} \geq B_1^D \ .$$

Additionally, we use the following 5-round condition. The proof will be appeared in the appendix A.

$$\text{(5-round cond.)} \quad D_i + D_{i+1} + D_{i+2} + D_{i+3} + D_{i+4} \geq B_1^D + 1 \ .$$

We make use of these lower bounds for less than 5 consecutive rounds for differential active S-boxes and less than 2 consecutive rounds for linear active S-boxes to obtain the lower bounds for arbitrary round numbers.

5.3   Example Choice of Matrices

Here, we will demonstrate how to apply the generalized DSM theory to concrete Feistel structure to enhance the immunity against differential attack and linear attack by illustrating example matrices.

Let A_0, A_1 and A_2 be 8×8 matrices over $GF(2^8)$ with irreducible polynomial $x^8 + x^4 + x^3 + x^2 + 1 = 0$ as follows:

$$A_0 = \begin{pmatrix} 1\,9\,2\,5\,8\,1\,4\,1 \\ 1\,1\,9\,2\,5\,8\,1\,4 \\ 4\,1\,1\,9\,2\,5\,8\,1 \\ 1\,4\,1\,1\,9\,2\,5\,8 \\ 8\,1\,4\,1\,1\,9\,2\,5 \\ 5\,8\,1\,4\,1\,1\,9\,2 \\ 2\,5\,8\,1\,4\,1\,1\,9 \\ 9\,2\,5\,8\,1\,4\,1\,1 \end{pmatrix}, \quad A_1 = \begin{pmatrix} 1\,6\,8\,9\,6\,9\,5\,1 \\ 1\,1\,6\,8\,9\,6\,9\,5 \\ 5\,1\,1\,6\,8\,9\,6\,9 \\ 9\,5\,1\,1\,6\,8\,9\,6 \\ 6\,9\,5\,1\,1\,6\,8\,9 \\ 9\,6\,9\,5\,1\,1\,6\,8 \\ 8\,9\,6\,9\,5\,1\,1\,6 \\ 6\,8\,9\,6\,9\,5\,1\,1 \end{pmatrix}, \quad A_2 = \begin{pmatrix} 1\,6\,4\,8\,4\,5\,8\,9 \\ 9\,1\,6\,4\,8\,4\,5\,8 \\ 8\,9\,1\,6\,4\,8\,4\,5 \\ 5\,8\,9\,1\,6\,4\,8\,4 \\ 4\,5\,8\,9\,1\,6\,4\,8 \\ 8\,4\,5\,8\,9\,1\,6\,4 \\ 4\,8\,4\,5\,8\,9\,1\,6 \\ 6\,4\,8\,4\,5\,8\,9\,1 \end{pmatrix}.$$

Note that we chose the matrix A_0 from Whirlpool hashing function's diffusion matrix for reference [2] [8].

Let A_0', A_1' and A_2' be 64×64 matrices over $GF(2)$ which are equivalent to A_0, A_1 and A_2, respectively. These matrices have the following properties [3].

1. $\mathcal{B}r_8(A_0') = \mathcal{B}r_8(A_1') = \mathcal{B}r_8(A_2') = 9$,
2. $\mathcal{B}r_8([A_0'|A_1']) = \mathcal{B}r_8([A_1'|A_2']) = \mathcal{B}r_8([A_2'|A_0']) = 8$,
3. $\mathcal{B}r_8([A_0'|A_1'|A_2']) = 8$,
4. $\mathcal{B}r_8([^tA_0'^{-1}|^tA_1'^{-1}]) = \mathcal{B}r_8([^tA_1'^{-1}|^tA_2'^{-1}]) = \mathcal{B}r_8([^tA_2'^{-1}|^tA_0'^{-1}]) = 8$.

Property 1 indicates that each matrix is an optimal diffusion mapping itself since the branch number is column number plus 1, but the properties 2-4 indicate the combined matrices made from these matrices are not optimal. Our experiment shows that there are no set of matrices satisfying the optimality of property 2-4 within the following searching space.

- Irreducible polynomial is $x^8 + x^4 + x^3 + x^2 + 1 = 0$,
- Each element of matrices is in hex values $\{1,2,3,...,e,f\} \in GF(2^8)$. They can be represented as at most 4-bit value.

Note that the searching space applied in finding the Whirlpool's diffusion matrix by the designers is subset of our searching space, and the smallness of the matrix elements is considered to contribute efficient implementations [20, 2].

Since we could not prepare optimal matrices in this setting, now we can make use of the previously obtained generalized theorems for Feistel structures using the above non-optimal matrices.

---

[2] The matrix is transposed so as to adjust our form $y = Mx$ not $y = xM$.
[3] The notion $\mathcal{B}r_8$ is defined in section 2.

5.4   Feistel Structures Using 2 or 3 Matrices in DSM

We consider two types of Feistel structures which belong to $(8, 8, 2r)$-SPFS. One is using two matrices, the other is using three matrices. These both structures can be used for 128-bit blockciphers due to $m = n = 8$. In appendix B, we will show the cases for 64-bit block Feistel structure as well.

Let F128A be an $(8, 8, 2r)$-SPFS which employs matrices A_0 and A_1 as M_{2i+1} $= M_{2r-2i} = A_{i \bmod 2}$ for $0 \le i < r$ (see Fig 3 when $r = 6$).

Fig. 3. Allocation of Matrices A_0, A_1 in F128A

Let F128B be an $(8, 8, 2r)$-SPFS which employs matrices A_0, A_1 and A_2 as $M_{2i+1} = M_{2r-2i} = A_{i \bmod 3}$ for $0 \le i < r$ (see Fig 4 when $r = 6$).

Fig. 4. Allocation of Matrices A_0, A_1, A_2 in F128B

In the above situation, we know that in F128A, $B_1^D = 9, B_2^D = 8, B_2^L = 8$, and the corollaries 2, 4 and conditions in Sect.5.2 are effective. On the other hand, in F128B, $B_1^D = 9, B_2^D = 8, B_3^D = 8, B_2^L = 8$, and the corollaries 2, 3, 4 and conditions in Sect.5.2 are effective.

The results of the guaranteed number of active S-boxes for F128A and F128B with other additional information are shown in the Table 1. Columns labeled by 'Dif.' and 'Lin.' contain lower bounds of differential and linear active S-boxes, respectively. Additionally, we show the lower bounds for the weights of the characteristics, which is simply calculated by multiplying the lower bound and a index number of maximum differential or linear probability of S-boxes [4]. The considered S-boxes here have the maximum differential probability 2^{-6} and 2^{-5}, maximal linear probability 2^{-6} and $2^{-4.39}$ for reference[4].

These weights of characteristics can be used to practically estimate the strength of the cipher against differential attack and linear attack [4]. In this case, the weight value should be larger than 128 with reasonable margin, for example 10-round F128A and 9-round F128B using 2^{-6} S-box and 12-round F128A, F128B using the second S-box seem to hold minimum security. The difference between F128A and F128B is only the value of lower bound for 9-round

---

[4] The value 2^{-6} is known best probability, 2^{-5} and $2^{-4.39}$ are experimentally obtained values that can be achieved by randomly chosen S-boxes in reasonable trials [4].

Table 1. Lower Bounds of Number of Active S-boxes and Weights of Characteristics

round	F128A							F128B		
	Dif.	DP_{max} 2^{-6}	DP_{max} 2^{-5}	Lin.	LP_{max} 2^{-6}	LP_{max} $2^{-4.39}$	speed (cycles/byte)	Dif.	Lin.	speed (cycles/byte)
1	0	0	0	0	0	0	-	0	0	-
2	1	6	5	1	6	4.39	-	1	1	-
3	2	12	10	8	48	35.12	-	2	8	-
4	9	54	45	8	48	35.12	-	9	8	-
5	10	60	50	9	54	39.51	-	10	9	-
6	17	102	85	16	96	70.24	-	17	16	-
7	17	102	85	16	96	70.24	-	17	16	-
8	18	108	90	17	102	74.63	-	18	17	-
9	19	114	95	24	144	105.36	-	<u>26</u>	24	-
10	26	156	130	24	144	105.36	11.38	26	24	11.65
11	27	162	135	25	150	109.75	-	27	25	-
12	34	204	170	32	192	140.48	13.60	34	32	14.11
13	34	204	170	32	192	140.48	-	34	32	-
14	35	210	175	33	198	144.87	15.75	35	33	16.37
15	36	216	180	40	240	175.6	-	36	40	-
16	43	258	215	40	240	175.6	17.93	43	40	18.70
17	44	264	220	41	246	179.99	-	44	41	-
18	51	306	255	48	288	210.72	19.88	<u>52</u>	48	20.64

and 18-round of differential active S-boxes. It implies that if 9-rounds immunity against differential attack is important, usage of 3 matrices should be taken into consideration.

Additionally, we mention a software implementation aspect. Software performance (in cycles per byte) of a moderately optimized C implementation of the F128A and F128B are measured on AMD Athlon64 4000+ (2.41GHz) with Windows XP Professional x64 Edition and Visual Studio .NET 2003. To use the lookup-table based implementation suitable for a 64-bit CPU, F128A requires a 32KB table (8 × 8-bit input × 64-bit output × 2 matrices) and F128B requires a 48KB table (3 matrices) [2]. Though they need large tables, we confirm that they achieve practically enough speed in this environment.

From the above observation, it is revealed that our novel results can be used to theoretically estimate the strength of Feistel structures using *DSM*.

6   Conclusion

We provide extended theory for the guaranteed number of active S-boxes of Feistel structure with DSM, which is realized by replacing the condition of optimal mappings with general mappings. As a result, we established a simple tool to evaluate any rounds of Feistel structures using DSM which employs arbitrary types of diffusion matrices. The effects of the novel result are confirmed by evaluating certain Feistel structures with concrete example matrices.

Acknowledgments. We thank Shiho Moriai for helpful discussions and suggestions. We also thank anonymous referees for their valuable comments.

References

1. K. Aoki, T. Ichikawa, M. Kanda, M. Matsui, S. Moriai, J. Nakajima, and T. Tokita, "Camellia: A 128-bit block cipher suitable for multiple platforms." in *Proceedings of Selected Areas in Cryptography – SAC 2000* (D. R. Stinson and S. E. Tavares, eds.), no. 2012 in LNCS, pp. 41–54, Springer-Verlag, 2001.
2. P. S. L. M. Barreto and V. Rijmen, "The Whirlpool hashing function." Primitive submitted to NESSIE, Sept. 2000. Available at http://www.cryptonessie.org/.
3. E. Biham and A. Shamir, "Differential cryptanalysis of des-like cryptosystems." *Journal of Cryptology*, vol. 4, pp. 3–72, 1991.
4. J. Daemen and V. Rijmen, "Statistics of correlation and differentials in block ciphers." in *IACR ePrint archive 2005/212*, 2005.
5. J. Daemen and V. Rijmen, *The Design of Rijndael: AES - The Advanced Encryption Standard (Information Security and Cryptography)*. Springer, 2002.
6. H. Feistel, "Cryptography and computer privacy." *Scientific American*, vol. 228, pp. 15–23, May 1973.
7. Data Encryption Standard, "Federal Information Processing Standard (FIPS)." National Bureau of Standards, U.S. Department of Commerce, Washington D.C., Jan. 1977.
8. International Organization for Standardization, "ISO/IEC 10118-3: Information Technology - Security Techniques - Hash-functions - Part 3: Dedicated hash-functions." 2003.
9. M. Kanda, "Practical security evaluation against differential and linear cryptanalyses for Feistel ciphers with SPN round function." in *Proceedings of Selected Areas in Cryptography – SAC'00* (D. R. Stinson and S. E. Tavares, eds.), no. 2012 in LNCS, pp. 324–338, Springer-Verlag, 2001.
10. M. Luby and C. Rackoff, "How to construct pseudorandom permutations from pseudorandom functions." *SIAM Journal on Computing*, vol. 17, pp. 373–386, 1988.
11. M. Matsui, "Linear cryptanalysis of the data encryption standard." in *Proceedings of Eurocrypt'93* (T. Helleseth, ed.), no. 765 in LNCS, pp. 386–397, Springer-Verlag, 1994.
12. M. Matsui, "New structure of block ciphers with provable security against differential and linear cryptanalysis." in *Proceedings of Fast Software Encryption – FSE'96* (D. Gollmann, ed.), no. 1039 in LNCS, pp. 205–218, Springer-Verlag, 1996.
13. K. Nyberg and L. R. Knudsen, "Provable security against a differential cryptanalysis." in *Proceedings of Crypto'92* (E. F. Brickell, ed.), no. 740 in LNCS, pp. 566–574, Springer-Verlag, 1993.
14. V. Rijmen, J. Daemen, B. Preneel, A. Bossalaers, and E. D. Win, "The cipher SHARK." in *Proceedings of Fast Software Encryption – FSE'96* (D. Gollmann, ed.), no. 1039 in LNCS, pp. 99–111, Springer-Verlag, 1996.
15. R. L. Rivest, M. J. B. Robshaw, R. Sidney, and Y. L. Yin, "The RC6 block cipher." Primitive submitted to AES, 1998. Available at http://www.rsasecurity.com/.
16. B. Schneier and J. Kelsey, "Unbalanced Feistel networks and block cipher design." in *Proceedings of Fast Software Encryption – FSE'96* (D. Gollmann, ed.), no. 1039 in LNCS, pp. 121–144, Springer-Verlag, 1996.

17. B. Schneier, J. Kelsey, D. Whiting, D. Wagner, C. Hall, and N. Ferguson, "Twofish: A 128-bit block cipher." Primitive submitted to AES, 1998. Available at http://www.schneier.com/.
18. T. Shirai, S. Kanamaru, and G. Abe, "Improved upper bounds of differential and linear characteristic probability for Camellia." in *Proceedings of Fast Software Encryption – FSE'02* (J. Daemen and V. Rijmen, eds.), no. 2365 in LNCS, pp. 128–142, Springer-Verlag, 2002.
19. T. Shirai and B. Preneel, "On feistel ciphers using optimal diffusion mappings across multiple rounds." in *Proceedings of Asiacrypt'04* (P. J. Lee, ed.), no. 3329 in LNCS, pp. 1–15, Springer-Verlag, 2004.
20. T. Shirai and K. Shibutani, "On the diffusion matrix employed in the Whirlpool hashing function." NESSIE Public reports, 2003. Available at http://www.cryptonessie.org/.
21. T. Shirai and K. Shibutani, "Improving immunity of Feistel ciphers against differential cryptanalysis by using multiple MDS matrices." in *Proceedings of Fast Software Encryption – FSE'04* (B. Roy and W. Meier, eds.), no. 3017 in LNCS, pp. 260–278, Springer-Verlag, 2004.

Appendix A

Here, we show the proof for the condition presented in the section 5.

By simply replacing the branch number symbol of the Kanda's Corollary 1 in [9] with our symbol B_1^D, we obtain

Corollary 5. *The minimum number of differential active S-boxes in any four consecutive rounds satisfies*

(i) $D_i + D_{i+1} + D_{i+2} + D_{i+3} \geq B_1^D$ if and only if $D_i = D_{i+3} = 0$,

(ii) $D_i + D_{i+1} + D_{i+2} + D_{i+3} \geq B_1^D + 1$ in the other cases.

Using the above corollary, we show the following lemma.

Lemma 4. *The minimum number of differential active S-boxes in any 5 consecutive rounds satisfies*

$$D_i + D_{i+1} + D_{i+2} + D_{i+3} + D_{i+4} \geq B_1^D + 1 .$$

Proof. If $D_i \neq 0$ or $D_{i+3} \neq 0$, the inequality (ii) of the corollary 5 directly implies the above inequality. If $D_i = D_{i+3} = 0$, (i) implies $D_{i+1} + D_{i+2} \geq B_1^D$. By combining trivial condition $D_{i+3} + D_{i+4} \geq 1$, the desired condition is obtained immediately. □

Appendix B

Let A_0, A_1 and A_2 be 4×4 matrices over $GF(2^8)$ with irreducible polynomial $x^8 + x^4 + x^3 + x + 1 = 0$ as follows.

$$A_0 = \begin{pmatrix} 2&3&1&1 \\ 1&2&3&1 \\ 1&1&2&3 \\ 3&1&1&2 \end{pmatrix} , \quad A_1 = \begin{pmatrix} 1&6&8&4 \\ 4&1&6&8 \\ 8&4&1&6 \\ 6&8&4&1 \end{pmatrix} , \quad A_2 = \begin{pmatrix} 1&9&4&a \\ a&1&9&4 \\ 4&a&1&9 \\ 9&4&a&1 \end{pmatrix} .$$

Table 2. Lower Bounds of Number of Active S-boxes and Weights of Characteristics

	F64A								F64B		
round	Dif.	DP_{max} 2^{-6}	DP_{max} 2^{-5}	Lin.	LP_{max} 2^{-6}	LP_{max} $2^{-4.39}$	speed (cycles/byte)		Dif.	Lin.	speed (cycles/byte)
1	**0**	0	0	**0**	0	0	-		**0**	**0**	-
2	**1**	6	5	**1**	6	4.39	-		**1**	**1**	-
3	**2**	12	10	**5**	30	21.95	-		**2**	**5**	-
4	**5**	30	25	**5**	30	21.95	-		**5**	**5**	-
5	**6**	36	30	**6**	36	26.34	-		**6**	**6**	-
6	**10**	60	50	**10**	60	43.9	-		**10**	**10**	-
7	**10**	60	50	**10**	60	43.9	-		**10**	**10**	-
8	**11**	66	55	**11**	66	48.29	-		**11**	**11**	-
9	**12**	72	60	**15**	80	65.85	-		**15**	**15**	-
10	**15**	80	75	**15**	80	65.85	17.52		**15**	**15**	17.53
11	**16**	86	80	**16**	86	70.24	-		**16**	**16**	-
12	**20**	120	100	**20**	120	87.8	20.52		**20**	**20**	20.52
13	**20**	120	100	**20**	120	87.8	-		**20**	**20**	-
14	**21**	126	105	**21**	126	92.19	23.66		**21**	**21**	23.66
15	**22**	132	110	**25**	130	109.75	-		**25**	**25**	-
16	**25**	150	125	**25**	130	109.75	26.17		**25**	**25**	26.17
17	**26**	156	130	**26**	136	114.14	-		**26**	**26**	-
18	**30**	180	150	**30**	180	131.7	29.03		**30**	**30**	29.03

Let A'_0, A'_1 and A'_2 be 32×32 matrices over $GF(2)$ which are equivalent to A_0, A_1 and A_2, respectively. These matrices have the following branch number properties.

1. $\mathcal{B}r_8(A'_0) = \mathcal{B}r_8(A'_1) = \mathcal{B}r_8(A'_2) = 5$,
2. $\mathcal{B}r_8([A'_0|A'_1]) = \mathcal{B}r_8([A'_1|A'_2]) = \mathcal{B}r_8([A'_2|A'_0]) = 5$,
3. $\mathcal{B}r_8([A'_0|A'_1|A'_2]) = 5$,
4. $\mathcal{B}r_8([^tA'^{-1}_0|^tA'^{-1}_1]) = \mathcal{B}r_8([^tA'^{-1}_1|^tA'^{-1}_2]) = \mathcal{B}r_8([^tA'^{-1}_2|^tA'^{-1}_0]) = 5$.

Contrary to the 8×8 matrices cases, all of these conditions indicate that the branch numbers are optimal. Note that we chose the matrix A_0 is from AES/Rijndael's diffusion matrix for reference [5].

Let F64A and F64B be a $(4, 8, 2r)$-SPFS which employs the matrices A_0, A_1 and A_2 same as in F128A and F128B, respectively. F64A and F64B can be used for 64-bit blockciphers. The lower bounds of active S-boxes indicated by our theory and weight of the characteristics are shown in Table 2.

We evaluate software performance (in cycles per byte) of a moderately op-timized C implementation of the F64A and F64B are measured on an AMD Athlon64 4000+ (2.41GHz) with Windows XP Professional x64 Edition and Vi-sual Studio .NET 2003 (same as Table 1 environment). We confirmed that they achieve practically enough speed in this environment. Moreover we expect that F64A and F64B can be implemented efficiently on 32-bit processors, because they require smaller tables than F128A and F128B do.

Pseudorandom Permutation Families over Abelian Groups

Louis Granboulan, Éric Levieil, and Gilles Piret[*]

École Normale Supérieure
Louis.Granboulan@ens.fr,Eric.Levieil@ens.fr,
Gilles.Piret@ens.fr

Abstract. We propose a general framework for differential and linear cryptanalysis of block ciphers when the block is not a bitstring. We prove piling-up lemmas for the generalized differential probability and the linear potential, and we study their lower bounds and average value, in particular in the case of permutations of $\mathbb{F}_p$. Using this framework, we describe a toy cipher, that operates on blocks of 32 decimal digits, and study its security against common attacks.

Keywords: block cipher, arbitrary domain, differential and linear cryptanalysis.

1 Introduction

1.1 Motivations

While all well-known block ciphers are pseudo-random permutation families of some set $\{0,1\}^n$ where $n = 64$ or 128, there exists some applications where a pseudo-random permutation of an arbitrary set is needed. For example, if one wants to add an encryption layer within a system that stores its data in decimal value, this encryption layer should encrypt decimal numbers without any expansion, and this is not possible if a binary encoding of these numbers is encrypted by a standard block cipher. Another example appears in some public-key cryptography protocols, where one assumes the existence of some ideal permutation or of some ideal cipher, that permutes elements of a set of cardinality other than 2^n, for example the set of points of an elliptic curve.

Moreover, while there are many studies on the cryptographic properties of boolean functions, and some studies on the cryptographic properties of addition modulo 2^n, no published results really looks into the generalization of these binary properties to the case where the characteristic is $\ell > 2$. A general framework for differential and linear cryptanalysis for arbitrary characteristic may bring a new insight into the understanding of these attacks.

[*] This work is supported in part by the French government through X-Crypt, in part by the European Commission through ECRYPT.

M.J.B. Robshaw (Ed.): FSE 2006, LNCS 4047, pp. 57–77, 2006.

1.2   Previous Work

Black and Rogaway [3] have described how to design block ciphers that permute arbitrary domains. Hence our problem already has a solution. However, their techniques are modes of use of conventional ciphers, and we prefer to study the feasibility of ad hoc designs.

The generalization of differential cryptanalysis to any abelian group is classical, and this generalization appears in the study of ciphers using addition modulo 2^n [10,16,11] but also more exotic operations like the $\otimes$ in IDEA [10] or multiplication in Multiswap [4]. Nyberg [13] wrote one of the few papers that study S-boxes over $\mathbb{F}_p$ with respect to differential cryptanalysis[1]. The generalization of linear cryptanalysis is a new result, the only similar work being the $\mathbb{Z}_4$-linear cryptanalysis by Parker and Raddum [15].

Our toy cipher we describe is based on a straightforward adaptation of Rijndael [6], which is a typical example of key-alternating cipher [8].

1.3   Our Setting

To study the differential cryptanalysis of a function $f : G \to G'$, we need to provide both G and G' with a structure of abelian group. The number of elements in these groups will be denoted q and q'. The minimal integer ℓ such that all elements of these groups are of ℓ-torsion (i.e. $\ell.G = \ell.G' = \{0\}$) will be called the characteristic of f and be a key parameter for linear cryptanalysis. We will investigate more deeply the prime case, where $G = G' = \mathbb{F}_p$.

1.4   Outline of the Paper

Sections 2 and 3 explain how differential and linear cryptanalysis are generalized. We give definitions and basic properties, then we show that *piling-up* theorems exist, and therefore these techniques can be used to evaluate the security of a whole cipher, based on the study of its non-linear components. We also show that in the prime case, optimality is equivalent to f being a degree 2 polynomial. Finally, we give an estimation of the non-linearity of a random function, with respect to differential cryptanalysis.

In section 4 we describe our cipher TOY100. We explain the design criteria. Because our toy cipher has non-prime characteristic $\ell = 100$, some technical difficulties appear in the study of the linear part of the cipher. We solve the problem for our specific example. Section 5 is a security analysis of TOY100.

2   Differential Cryptanalysis

2.1   Definition

Introduced by Biham and Shamir in [2], differential cryptanalysis is one of the most useful techniques in modern cryptanalysis. The idea is to encrypt pairs of

---

[1] But there is a small mistake in its proposition 7.

plaintexts having a fixed difference, and to observe the differences between the pairs of ciphertexts.

We recall that in our setting f is a function from G to G', abelian groups of cardinality q and q'. Let us define a $q \times q'$ matrix Δ that describes the action of f over differences by $\Delta(f)_{a,b} = \#\{x | f(x+a) - f(x) = b\}$.

The complexity of an attack by differential cryptanalysis is of order $1/DP(f)$ where the differential probability $DP(f) = D(f)/q$ is defined by:

$$D(f) = \max_{(a,b) \in G \times G' \setminus \{(0,0)\}} \Delta(f)_{a,b}.$$

The exact value of $D(f)$ being too expensive to compute, one usually computes the exact values for the elementary functions used in f and combine these values using piling-up theorems.

2.2   Properties Valid for Any Group

Differential probability for the inverse. If $f : G \to G$ is bijective, then $\Delta(f^{-1}) = {}^t\Delta(f)$ and therefore $D(f^{-1}) = D(f)$.

Proof. If $f(x+a) - f(x) = b$, then $f^{-1}(y) + a = f^{-1}(y+b)$ with $y = f(x)$.   □

Parallel execution. If f is the parallel execution of functions f_1 and f_2, then its differential properties are easily deduced from the differential properties of f_1 and f_2. More precisely, if we define f over $G_1 \times G_2$ by $f(x,y) = (f_1(x), f_2(y))$, then $\Delta(f)_{(a_1,a_2),(b_1,b_2)} = \Delta(f_1)_{a_1,b_1} \Delta(f_2)_{a_2,b_2}$ and $D(f) = \max(q_2 D(f_1), q_1 D(f_2))$.

Sequential execution. If f is the sequential execution of two functions, the differential properties cannot be directly combined, because the image of a uniform distribution of input pairs with fixed difference does not necessarily have uniform distribution for all output pairs with given difference. The distribution can be made uniform by adding a random key, and ciphers using this design are named Markov ciphers [10]. In this setting, we compose the function $f : G \to G'$, the translation in G' that we name ADD_KEY, and the function $g : G' \to G''$ to obtain $h_K = g \circ \text{ADD\_KEY}(K) \circ f$.

Theorem 1

$$DP(h_K) \approx DP(f)DP(g)$$

if the following hypothesis hold:

- *Stochastic equivalence.* $\Delta(h_K)$ *does not depend heavily on* K;
- *Dominant characteristic for* (a, c). $\sum_b \Delta(f)_{a,b} \Delta(g)_{b,c} \approx \max_b \Delta(f)_{a,b} \Delta(g)_{b,c}$.
- *Independence.* $\max_b \Delta(f)_{a,b} \Delta(g)_{b,c} \approx \max_{b,b'} \Delta(f)_{a,b} \Delta(g)_{b',c}$.

Proof. Let $\phi_a(x, K) = (x, f(x) + K, f(x+a) - f(x))$. The restriction of ϕ_a to the set of solutions (x, K) of the equation $h_K(x+a) - h_K(x) = c$ is a one-to-one mapping to the set of solutions (x, y, b) of the pair of equations $g(y+b) - g(y) = c$ and $f(x+a) - f(x) = b$. Therefore the following formula over Δ matrices holds:

$$\sum_{K \in G'} \Delta(h_K) = \Delta(f)\Delta(g).$$

Under the first hypothesis, $D(h_K) \approx \frac{1}{q'} \max_{(a,b) \neq (0,0)} \left(\Delta(f)\Delta(g) \right)_{a,b}$. Now we apply the second hypothesis to a pair (a, b) for which $D(h_K) \approx \frac{1}{q'} \left(\Delta(f)\Delta(g) \right)_{a,b}$, and then we apply the third hypothesis.   □

Lower bound. If $D(f) = 1$ then f is not bijective.

Proof. For any non-zero a, $\sum_{b \in G} \Delta(f)_{a,b} = q$, therefore all elements of the a-th row of $\Delta(f)$ are equal to 1 and in particular $\Delta(f)_{a,0}$.   □

2.3   The Case of $\mathbb{F}_p$

All functions in $\mathbb{F}_p$ can be interpolated by a polynomial which is unique if its degree is less than p. The degree of f has some impact on $D(f)$.

Proposition 1. *(i)$D(f)$ $D(f) = p$ is equivalent to f linear or constant.*

   (ii) $D(f) = p - 1$ is impossible.
   (iii) If f has degree $d \geq 2$, then $D(f) \leq d - 1$. In particular, if f is of degree 2, then $D(f) = 1$.
   (iv) For all d between 2 and $p - 1$, there are polynomials of degree d, such as $D(f) = d - 1$.

Proof. (i): Let $a \neq 0$ and b be such that $\Delta(f)_{a,b} = p$. Then $f(x) = a^{-1}bx + f(0)$.
(ii): Let $a \neq 0$ and b be such that $\Delta(f)_{a,b} = p - 1$. There exists $b' \neq b$ such that $\Delta(f)_{a,b'} = 1$. But $0 = \sum_{x \in G} f(x + a) - f(x) = (p - 1)b + b' = b' - b$.
(iii): $f(x + a) - f(x) - b$ is a polynomial of degree $d - 1$, so it has at most $d - 1$ roots.
(iv): We want to find f such that $f(x + 1) - f(x)$ is a polynomial with $d - 1$ distinct roots. First, we choose any polynomial with $d - 1$ distinct roots then we write the equality between the coefficients. We obtain a triangular system with a non-zero diagonal, which implies it is invertible.   □

Conjecture for the lower bound

Conjecture 1. If $D(f) = 1$, then the degree of f is 2.

If we define the differential $df_a(x) = f(x + a) - f(x)$, it has the property of being a zero-sum function i.e. $\sum_{x \in G} df_a(x) = 0$. The hypothesis $D(f) = 1$ of our conjecture is equivalent to $\forall a \neq 0$, df_a is bijective.

In spite of this simple formulation, and a computer-aided verification that it is true for $p \leq 19$, we could not prove this conjecture. However, if the following lemma holds, then this conjecture is true, as shown in appendix A.5.

Lemma 1 (Key lemma). *If $\varphi : \mathbb{F}_p \to \mathbb{Z}$ satisfies $\sum_{y \in \mathbb{F}_p} \varphi(y)^2 = p - 1$, and $\forall x \neq 0$, $\sum_{y \in \mathbb{F}_p} \varphi(y)\varphi(x + y) = -1$ then $\forall x$, $\varphi(x) \in \{0, \pm 1\}$.*

Average value. To find functions with high degree but low $D(f)$, we can try random functions. The following theorem evaluates the average value of $D(f)$ and its proof (in appendix A.1) contains upper bounds on the number of functions with low or high $D(f)$.

Theorem 2. *Let us define* $z(p) = \lfloor \Gamma^{-1}(p/(6 \log p)) \rfloor - 1$ *where as usual* $\Gamma(z + 1) = z!$, *then*

$$\lim_{p \to \infty} Pr[z(p) \le D(f) \le 3z(p)] = 1$$

It is possible to decrease the constant 3 to 2 (the proof will be in the full version of the paper). There is no reason that prevents this result to be applied to $\mathbb{Z}/q\mathbb{Z}$, except perhaps the human's lack of taste for lengthy computations. However, it is impossible to have really precise results on this subject, unless one can explicit the dependence between the differentials of a function. Assuming independence is the usual way to deal with this problem (see for example [8]), but it is not true for small p.

2.4   The Case of $\mathbb{Z}/q\mathbb{Z}$

The case where G is isomorphic to $\mathbb{Z}/q\mathbb{Z}$ cannot be seen as a generalization of the prime case for two reasons:

- there exist many functions that cannot be interpolated by polynomials
- even when this interpolation exists, the form of canonical interpolations is tricky to define

The following theorem, proven in appendix A.2, shows that polynomials are a negligible fraction of the functions over $\mathbb{Z}/q\mathbb{Z}$. For example, over $\mathbb{Z}/100\mathbb{Z}$ there are $2 \cdot 10^{12}$ polynomials and 10^{200} functions.

Theorem 3. (i) *Let* $q = p^2$ *with* p *prime. Then the number of distinct polynomials over* $\mathbb{Z}/q\mathbb{Z}$ *is equal to* p^{3p}.
(ii) *Let* $q = q_1 q_2$, *with* q_1, q_2 *coprime. Then the number of distinct polymials over* $\mathbb{Z}/q\mathbb{Z}$ *is the product of this number over* $\mathbb{Z}/q_1\mathbb{Z}$ *and* $\mathbb{Z}/q_2\mathbb{Z}$.

If $q = q_1 q_2$, with q_1, q_2 coprime, and if f is a polynomial, then its differential properties need only to be studied over $\mathbb{Z}/q_1\mathbb{Z}$ and over $\mathbb{Z}/q_2\mathbb{Z}$, as proved in the following theorem. If it is not a polynomial, such a decomposition is not possible.

Theorem 4. *Let* $f \in \mathbb{Z}/q\mathbb{Z}[X]$ *and for* $i = 1, 2$ $f_i \in \mathbb{Z}/q_i\mathbb{Z}$ *defined by* $f_i(x) = f(x) \pmod{q_i}$. *Then* $D(f) = D(f_1)D(f_2)$.

Proof. $z \to (z \bmod q_1, z \bmod q_2)$ is an isomorphism.   □

3   Linear Cryptanalysis

3.1   Definition

Linear cryptanalysis is a known-plaintext attack that was discovered just after differential cryptanalysis [18,17,12]. It is based on the study of linear approximations of the cipher. Linear cryptanalysis has been defined for boolean functions: a linear approximation of a function f is described by two masks (a, b) which select respectively bits of the input and of the output. If we denote by $\langle a|x \rangle$ the dot product of a and x, then linear approximations are given by comparing $\langle a|x \rangle - \langle b|f(x) \rangle$ for random x and $\langle a|x \rangle - \langle b|y \rangle$ for random x and y.

Linear cryptanalysis can be generalized to the study of the functions $f : G \rightarrow G'$, if there is some integer q such that all elements of both groups are of ℓ-torsion. This condition implies that both G and G' are isomorphic to a product of cyclic groups of order dividing ℓ. Under this condition and using this isomorphism, we can define scalar products over G and G' with output in $\mathbb{Z}/\ell\mathbb{Z}$, denoted $\langle \cdot | \cdot \rangle$. And finally we define the scalar product on $G \times G'$ by $\langle a, b | x, y \rangle = \langle a|x \rangle - \langle b|y \rangle$. The generalization of linear cryptanalysis can be done using two approaches.

Bias from random behavior. For any pair $(a, b) \in (G, G')$, let us define the distribution vector $\Lambda_0(f)_{a,b} = (\#\{x \in G \mid \langle a, b|x, f(x) \rangle = u\})_{u \in \mathbb{Z}/\ell\mathbb{Z}}$. The random behavior is given by $S_{a,b;u} = \frac{1}{q'}\#\{(x, y) \in G \times G' \mid \langle a, b|x, y \rangle = u\}$.

Therefore, if we define the bias $\Lambda_S(f)_{a,b;u} = \Lambda_0(f)_{a,b;u} - S_{a,b;u}$, then all elements of this matrix sum up to zero $\sum_u \Lambda_S(f)_{a,b;u} = 0$ and its greatest term is a measure of non-linearity.

$$L(f) = \max_{a,b \neq 0, u} \left(\Lambda_S(f)_{a,b;u} \right)^2$$

The complexity of the attack is expected to be of order $1/LP(f)$, where the linear potential $LP(f) = L(f)/q^2$.

Dual of differential cryptanalysis. The other approach generalizes the duality between differential and linear cryptanalysis, as it has been done for example by Chabaud and Vaudenay [5]. First, we need to define the characteristic function of f, which is $\theta_f : G \times G' \rightarrow \{0, 1\}$ such that $\theta_f(x, y) = 1$ iff $y = f(x)$. We also define the convolutional product of two functions by $(f*g)(a) = \sum_x f(x)g(a+x)$. As in Chabaud-Vaudenay, we can prove that $(\theta_f * \theta_f)(a, b) = \Delta_{a,b}$.

Let us choose a ℓ-th root[2] of unity $\xi \in \mathbb{C}$ and define the transform of $\phi : X \rightarrow Y$ by $\hat{\phi}(a) = \sum_x \phi(x)\xi^{\langle a|x \rangle}$. Note that $\hat{\phi}(-a)$ and $\hat{\phi}(a)$ are complex conjugates, that $\hat{\hat{\phi}}(x) = \#Y.\phi(-x)$, and also that $\widehat{(\phi * \phi)} = |\hat{\phi}|^2$ and therefore is real-valued. By duality, we define $\lambda(f)_{a,b} = \widehat{(\theta_f * \theta_f)}(a, b)$ and $\lambda(f) = \max_{(a,b) \neq (0,0)} \lambda(f)_{a,b}$.

---

[2] Replacing -1 by a ℓ-th root of unity is not a new idea. For example, it appeared as footnote 4 of [1]. The fact that it is a different approach than computing the bias was probably not noticed.

Links between both approaches. In the binary case (i.e. $\ell = 2$) we have $\xi = -1$ and $L(f)_{a,b;1} = -L(f)_{a,b;0}$ therefore $\hat{\theta}_f(a,b) = 2L(f)_{a,b;0}$ and $\lambda(f) = 4L(f)$. When $\ell \neq 2$, no such simple relation exists. For example, in $\mathbb{Z}/7\mathbb{Z}$, let us take $f(x) = x^6 + x^3$ and $g(x) = x^6 + x^3 + x^2$. Then $L(f) = L(g) = 9$ but $\lambda(f) = 39.96 \cdots$ while $\lambda(g) = 26.19 \cdots$. The list of all possible values for $\mathbb{Z}/5\mathbb{Z}$ and $\mathbb{Z}/7\mathbb{Z}$ is in appendix A.4.

Both approaches give some insight into the security of a cipher. However, in the following, we mainly consider the measure of bias, which is easier to implement as a concrete cryptanalysis.

3.2 Properties Valid for Any Group

Main properties of $S_{a,b;u}$. When a, b are fixed, $S_{a,b;u}$ is either 0 or another fixed value denoted $S_{a,b}$. The set $T_{a,b} = \{u | S_{a,b;u} \neq 0\}$ is a subgroup of $\mathbb{Z}/\ell\mathbb{Z}$, and $q' = S_{a,b} \#T_{a,b}$.

Proof. If $\langle a, b|x_0, y_0 \rangle = u_0$ and $\langle a, b|x_1, y_1 \rangle = u_1$, then $\langle a, b|x_0 + x_1, y_0 + y_1 \rangle = u_0 + u_1$. Therefore, the sets of solutions of the equations $\langle a, b|x, y \rangle = u$ can be translated one to another. □

The inverse. If $f : G \to G$ is bijective, then $\Lambda_S(f^{-1})_{a,b;u} = \Lambda_S(f)_{a,b;-u}$ and therefore $L(f^{-1}) = L(f)$.

Parallel execution. If f is the parallel execution of functions f_1 and f_2 of same characteristic, then bias matrices are combined by convolution. More precisely, if we define f over $G_1 \times G_2$ by $f(x, y) = (f_1(x), f_2(y))$, then $\Lambda_0(f)_{(a_1,a_2),(b_1,b_2)} = \Lambda_0(f_1)_{a_1,b_1} * \Lambda_0(f_2)_{a_2,b_2}$. If the sets T_{a_1,b_1} and T_{a_2,b_2} are equal, then this formula also applies to Λ_S.

Proof. Note that the hypothesis on the sets T_{a_i,b_i} is mandatory. A simple counterexample is $G_1 = G_2 = \mathbb{Z}/100\mathbb{Z}$, $f_1(x) = 2x$, $f_2(x) = x$, $a_1 = 5$, $a_2 = 10$, $b_1 = b_2 = 0$ and $\langle .|. \rangle$ is the usual multiplication over $\mathbb{Z}/100\mathbb{Z}$.

To prove the formula for Λ_0, we decompose $\langle a, b|x, f(x) \rangle = u$ into its components $\langle a_1, b_1|x_1, f_1(x_1) \rangle = v$ and $\langle a_2, b_2|x_2, f_2(x_2) \rangle = u - v$. Then we use the following facts: $S_{a,b;u}^{G_1 \times G_2} = \sum_v S_{a_1,b_1;u-v}^{G_1} S_{a_2,b_2;v}^{G_2}$ and $\sum_{v \in T_{a,b}} \Lambda_G(f)_{a,b;v} = 0$. □

Sequential execution. As for the differential cryptanalysis, we suppose we have a Markov cipher. In this case, the following theorem, proven in appendix A.3 allows us to approximate the value of $LP(h_K)$, for $h_K = g \circ \text{ADD\_KEY}(K) \circ f$:

Theorem 5 (Piling-up for LP)

$$LP(h_K) \approx LP(f)LP(g)$$

if the following hypothesis hold:

- *Stochastic equivalence.* $\Lambda_S(h_K)_{a,c;u+\langle b|K\rangle}$ *does not depend heavily on* K;
- *Dominant trail and independence.*

$$\max_{\substack{a,b,c,u \\ T_{a,b}=T_{b,c}}} \left| \sum_v \Lambda_S(f)_{a,b;u-v}\Lambda_S(g)_{b,c;v} \right| \approx \max_{a,b_f,b_g,c,u,v} \left| \Lambda_S(f)_{a,b_f;u-v}\Lambda_S(g)_{b_g,c;v} \right|$$

Piling-up $\lambda(f)$. This other approach also has composition results. For example, we prove in appendix A.3 a piling-up lemma that shows that under some appropriate hypothesis, $\lambda(h_K)/q^2 \approx \lambda(f)/q^2\lambda(g)/q'^2$.

3.3   The Case of $\mathbb{F}_p$

Functions over $\mathbb{F}_p$ that have optimal resistance against linear cryptanalysis have degree 2.

Theorem 6. *Let G be a group of cardinality p, with p prime. If $L(f) = 1$, then f can be interpolated by a polynomial of degree 2.*

Proof. Let us work in $PG(2, p)$, the projective plane over $\mathbb{F}_p$. Let

$$E(f) = \{x, f(x), 1 | x \in \mathbb{F}_p\} \cup (0, 1, 0).$$

$E(f)$ is a $p + 1$-arc, i.e. a set of $p + 1$ points, no three of which are collinear. According to the corollary of theorem 10.4.1, p.236, of [9], a $p+1$-arc in $PG(2, p)$ with p odd, is a conic. So

$$E(f) = \{(x_0, x_1, x_2) | a_{00}x_0^2 + a_{11}x_1^2 + a_{22}x_2^2 + a_{01}x_0x_1 + a_{02}x_0x_2 + a_{12}x_1x_2 = 0\}$$

But $(0, 1, 0) \in E(f)$, therefore $a_{11} = 0$.
And $(a_{01}, -a_{00}, 0) \in E(f)$, therefore $(a_{01}, -a_{00}, 0) \equiv (0, 1, 0)$. Therefore $a_{01} = 0$.
If $a_{12} = 0$, $(0, f(0), 1) \in E(f)$ implies $\{(0, y, 1) | y \in \mathbb{F}_p\} \subset E(f)$. Therefore a_{12} is not null and f is described by a degree 2 polynomial. $\qquad\square$

3.4   Relation with the Linear Cryptanalysis over $\mathbb{Z}/4\mathbb{Z}$

Matthew Parker and Haavard Raddum have suggested a generalization of linear cryptanalysis over $\mathbb{Z}/4\mathbb{Z}$ in [15]. Their method allows better approximations of the S-boxes but the combination of those approximations is less efficient than in classical linear cryptanalysis. Their method is a very particular case of ours, where a $2n$-bit string is seen as an element of $(\mathbb{Z}/4\mathbb{Z})^n$.

4   A Toy Cipher: TOY100

4.1   High-Level Description

In this section we aim at showing that it is possible to design a secure and efficient block cipher that does not use words of n bits as a block.

The structure of the cipher is quite similar to Rijndael [6,7]. It works on blocks of 32 decimal digits, with keys of the same size. It is composed of 11 identical rounds, followed by a slightly different final round. A block A is divided in 16 subblocks, each subblock being a number between 0 and 99. A block is represented as a 4×4 matrix $A = (a_{i,j})_{i,j \in \{0,\dots,3\}}$, of which each element is a subblock. Round r $(r = 0 \dots 10)$ is made out of the application of a key addition layer $\sigma[K^r]$ which adds modulo 100 a subkey to each subblock, followed by the parallel application, denoted γ, of a certain S-box to each subblock, and finally a linear function θ that mixes the subblocks. The last round has a final key addition instead of the linear layer, so it is written as $\sigma[K^{11}] \circ \gamma \circ \sigma[K^{12}]$.

4.2 Our Choice of Components

The S-Box. The S-box was chosen to satisfy $D(f) \le 5$ and $L(f) \le 5^2$. An iteration of RC4-100 consists, being given an array of 100 numbers, and two pointers i, j, to increment i, add $t[i]$ to j (modulo 100) then exchange $t[i]$ and $t[j]$. Starting from the permutation identity, $i = 1, j = 0$, we checked the permutation every 100 iterations until we find a permutation satisfying the criteria on $D(f)$ and $L(f)$. The permutation found is the 3 409 672th, after 340 967 200 iterations of RC4-100.

This function has $D(f) = 5$, $L(f) = 5^2$ and $\lambda(f) = 734.122\cdots$

	0	1	2	3	4	5	6	7	8	9
0	0	67	12	32	30	53	34	37	71	38
10	42	94	58	95	78	35	6	22	36	81
20	61	93	43	72	25	27	15	69	90	47
30	1	91	84	86	24	79	66	40	10	33
40	59	8	11	48	28	76	73	82	39	51
50	45	13	97	74	9	7	52	88	62	96
60	23	29	3	4	75	56	5	64	17	49
70	68	77	80	55	85	92	44	21	98	50
80	20	31	65	83	19	57	41	70	18	99
90	89	60	46	26	63	14	87	16	54	2

The Diffusion Function. The diffusion function θ is composed of two similar parts, MixColumns and MixRows.
First, we define a function Mix that takes 4 subblocks as an input:

$$
Mix \begin{pmatrix} a_1 \\ a_2 \\ a_3 \\ a_4 \end{pmatrix} = \begin{pmatrix} a_4 + a_1 + a_2 \\ a_1 + a_2 + a_3 \\ a_2 + a_3 + a_4 \\ a_3 + a_4 + a_1 \end{pmatrix}
$$

Mix is bijective and its inverse is:

$$Mix^{-1} \begin{pmatrix} a_1 \\ a_2 \\ a_3 \\ a_4 \end{pmatrix} = \begin{pmatrix} S - a_3 \\ S - a_4 \\ S - a_1 \\ S - a_2 \end{pmatrix}$$

with $S = (a_1 + a_2 + a_3 + a_4)/3$.

MixColumns (resp. MixRows) consists in applying *Mix* to each column (resp. row). Note that MixColumns and MixRows commute.

We define the subblock weight of a block B as the number of non-zero subblocks, and we denote it as $SW(B)$. The *branch number* is a measure of the efficiency of a diffusion layer.

Definition 1. *The branch number of a diffusion function f, $BN(f)$ is defined as:*

$$BN(f) = \min_{B \neq 0}(SW(B) + SW(f(B))$$

Proposition 2

$$BN(\theta) = 6$$

Proof. The first step of the proof enumerates the cases where there are one or two non-zero subblocks in the input B, and show that there will be at least six non-zero subblocks in $\theta(B)$; it is the same for $\theta^{-1}(B)$. We conclude by observing that if $b_{21} = b_{22} = b_{23} = 50$ and the other subblocks of B are 0, then $C = \theta(B)$ is such that $c_{12} = c_{22} = c_{32} = 50$ and the other subblocks are 0. □

The Key Schedule. The key expansion is very similar to the one of AES. As always, additions are modulo 100. The first round key K^0 is the key itself. For the following rounds, we iterate as follows:

$$k_{0,j}^{r+1} = k_{0,j}^r + S(k_{3,(j+1) \bmod 4}^r) + 3^r \quad (j \in \{0, 1, 2, 3\})$$
$$k_{i,j}^{r+1} = k_{i,j}^r + k_{i-1,j}^{r+1} \quad (i \in \{1, 2, 3\}, j \in \{0, 1, 2, 3\})$$

5   Security Analysis of TOY100

5.1   Differential Cryptanalysis

The best differentials we found rely on the following property of the linear layer:

$$\begin{bmatrix} \delta & -\delta & 0 & 0 \\ -\delta & \delta & 0 & 0 \\ 0 & 0 & 0 & 0 \\ 0 & 0 & 0 & 0 \end{bmatrix} \xrightarrow{\theta} \begin{bmatrix} 0 & 0 & 0 & 0 \\ 0 & 0 & 0 & 0 \\ 0 & 0 & \delta & -\delta \\ 0 & 0 & -\delta & \delta \end{bmatrix} \quad (1)$$

We estimated the probability of these differentials for $n = 2, 3, 4, 5, 6$. That is to say, we computed

$$\max_{\Delta_0, \Delta_n \in \{1,...,99\}} \sum_{\Delta_1,...,\Delta_{n-1} \in \{1,...,99\}} \frac{\Pi(\Delta_0 \to \Delta_1)^2 \cdot .... \cdot \Pi(\Delta_{n-1} \to \Delta_n)^2}{10^{8n}}, \quad (2)$$

where $\Pi(\Delta_i \to \Delta_j) := \Delta(f)_{\Delta_i,\Delta_j} \cdot \Delta(f)_{-\Delta_i,-\Delta_j}$. Remark that our choice of the linear transform makes the "modified difference distribution table" Π particularly important. There is always some "interaction" between the linear transform and the S-box regarding resistance against differential (and linear) cryptanalysis, but it is rarely so explicit.

Our results are given in Table 1. Note that the probabilities given are only lower bounds, as other characteristics exist for the same differential; however they have more active S-boxes, so we expect their contribution to the overall probability to be small. Such n-round differential can be used in an attack on $n + 1$ rounds. This way we can attack up to 6 (and maybe 7) rounds. Details are given in appendix B.1.

Table 1. Estimated probability for the best n-round differential

# Rounds n	Best Probability
2	$4.05 \cdot 10^{-11}$
3	$2.83 \cdot 10^{-16}$
4	$2.61 \cdot 10^{-21}$
5	$2.72 \cdot 10^{-26}$
6	$3.47 \cdot 10^{-31}$

5.2 Linear Cryptanalysis

The best linear characteristic we found relies on the same type of observation as the one used for differential cryptanalysis.

Namely, θ transforms mask $\begin{bmatrix} \alpha & -\alpha & 0 & 0 \\ -\alpha & \alpha & 0 & 0 \\ 0 & 0 & 0 & 0 \\ 0 & 0 & 0 & 0 \end{bmatrix}$ into mask $\begin{bmatrix} 0 & 0 & 0 & 0 \\ 0 & 0 & 0 & 0 \\ 0 & 0 & \alpha & -\alpha \\ 0 & 0 & -\alpha & \alpha \end{bmatrix}$.

The piling-up lemma can be iterated, so for an $(n + 1)$-round characteristic we have

$$\sum_{K^1,...,K^n \in (\mathbb{Z}/100\mathbb{Z})^{16}} \Lambda_S(h_{K^1,...,K^n})_{a,c;u+b_1 K^1+...+b_n K^n}$$

$$(3)$$

$$= \sum_{v_1,...,v_n \in \mathbb{Z}/100\mathbb{Z}} \Lambda_S(\rho_{n+1})_{a,b_n;u-v_1-...-v_n} \cdot$$
$$\Lambda_S(\rho_n)_{b_n,b_{n-1};v_n} \cdot .... \cdot \Lambda_S(\rho_1)_{b_1,c;v_1}$$

where $\rho_i = \gamma \cdot \theta$ $(i \neq n + 1)$ and $\rho_{n+1} = \gamma$.

This equation holds under the hypothesis that

$$S_{a,b_n} = S_{b_n,b_{n-1}} = ... = S_{b_1,c}. \tag{4}$$

Informally, equation (3) gives the average bias taken over all n-tuples of round keys (the first and last round keys are not considered here; they only contribute to the linear equation by a constant, which is unknown). Note that for the equation to be useful for linear cryptanalysis, it is required that the characteristic roughly equally holds for all keys. This hypothesis is common; it is known as *hypothesis of stochastic equivalence* [10,14]. We computed the maximum of (3) over all possible $(n+3)$-uples $(a, b_1, ..., b_n, c; u)$, for various numbers of rounds. The maxima we found correspond to $a = b_1 = ... = b_n = c = 10$ and $u = 0$. We note that condition (4) is satisfied. Detailed figures are given in Appendix B.2. Taking the first and last round keys into consideration, the corresponding linear approximation for $n+1$ rounds of the cipher is

$$10 \cdot (c_{33} + c_{44} - c_{34} - c_{43}) - 10 \cdot (p_{11} + p_{22} - p_{12} - p_{21})$$

$$= \sum_{i=0}^{\lfloor \frac{n+1}{2} \rfloor} 10(k_{11}^{2i} + k_{22}^{2i} - k_{12}^{2i} - k_{21}^{2i}) + \sum_{i=1}^{\lceil \frac{n+1}{2} \rceil} 10(k_{33}^{2i-1} + k_{44}^{2i-1} - k_{34}^{2i-1} - k_{43}^{2i-1}), \tag{5}$$

if r is odd; $c_{33} + c_{44} - c_{34} - c_{43}$ must be replaced by $c_{11} + c_{22} - c_{12} - c_{21}$ if it is even.

5.3   Structural Attacks

The diffusion layer of our cipher operates on well-aligned blocks, which could make it vulnerable to structural attacks. We explored truncated differential, impossible differential, and square attacks. The best such attack we found is a square-like attack, which can be used for a practical cryptanalysis of up to 4 rounds of TOY100. Details are given in appendix B.3.

6   Conclusion

In this paper we extended usual block cipher theory over $\mathbb{Z}_2^n$ to a more general framework in which the input and output spaces are arbitrary abelian groups. We studied quite extensively how differential and linear cryptanalysis apply in this context. We observe that many concepts, such as differential and linear parameters of a function or piling-up lemmas, can be generalized. Moreover, constructing a cipher by using the classical key-alternating paradigm still seems to be appropriate.

However several problems remain unsolved. The link between the differential parameter $D(f)$ and linear parameters $L(f)$ and $\lambda(f)$ should be investigated. Constructing functions with good such parameters, without using some kind of random search, is an open problem as well. A formalization of the "special role" of elements of small characteristic is also a goal for further research. Finally, our toy cipher would deserve a more consequent cryptanalytic effort.

Acknowledgement

The idea of the proof of theorem 6 was found by Mathieu Dutour and David Madore. We also thank David Madore for the proof of theorem 3.

References

1. T. Baignères, P. Junod, and S. Vaudenay. How Far Can We Go Beyond Linear Cryptanalysis? *Advances in Cryptology - Asiacrypt 2004*, LNCS 3329, Springer-Verlag, 2004. http://lasecwww.epfl.ch/php_code/publications/search.php?ref=BJV04

2. E. Biham and A. Shamir. Differential Cryptanalysis of DES-like cryptosystems. *Advances in Cryptology, CRYPTO '90*, Springer-Verlag, pp. 2-21.

3. John Black and Phillip Rogaway. Ciphers with Arbitrary Finite Domains. *RSA Data Security Conference, Cryptographer's Track (RSA CT '02)*, LNCS, vol. 2271, pp. 114-130, Springer, 2002.

4. Nikita Borisov, Monica Chew, Rob Johnson, and David Wagner. Cryptanalysis of Multiswap. 2001. http://www.cs.berkeley.edu/~rtjohnso/multiswap/

5. F. Chabaud and S. Vaudenay. Links between differential and linear cryptalysis. *Advances in Cryptology, Proceedings Eurocrypt'94*, LNCS 950, Springer-Verlag, 1995, pp.356-365.

6. J. Daemen and V. Rijmen. The Design of Rijndael: AES- the Advanced Encryption Standard. Springer-Verlag, 2002.

7. J. Daemen and V. Rijmen. AES proposal: Rijndael. *First Advanced Encryption Standard (AES) Conference*, Ventura, Canada National Institute of Standards and Technology, 1998.

8. J. Daemen et V. Rijmen. Statistics of Correlation and Differentials in Block Ciphers. Cryptology ePrint Archive, Report 2005/212, 2005 http://eprint.iacr.org/2005/212

9. J.W.P. Hirschfeld. Projective Geometries Over Finite Fields. *Oxford University Press, Oxford.* 1979.

10. X. Lai, J.L. Massey, and S. Murphy. Markov ciphers and differential cryptanalysis. *Advances in Cryptology, Proceedings Eurocrypt'91*, LNCS 547, D.W. Davies, Ed., Springer-Verlag, 1991, pp. 17-38.

11. Helger Lipmaa, Johan Wallén and Philippe Dumas. On the Additive Differential Probability of Exclusive-Or. In Bimal Roy and Willi Meier, editors, *Fast Software Encryption 2004*, volume 3017 of Lecture Notes in Computer Science, pages 317–331, Delhi, India, February 5–7, 2004. Springer-Verlag.

12. M. Matsui. Linear cryptanalysis method for DES cipher. *Advances in Cryptology, Proceedings Eurocrypt'93*, LNCS 765, T. Helleseth, Ed., Springer-Verlag, 1994, pp. 386-397.

13. K. Nyberg. Differentially uniform mappings for cryptography. *Advances in Cryptology, Proceedings Eurocrypt'93*, LNCS 765, T. Helleseth, Ed., Springer-Verlag, 1994, pp. 55-64.
14. K. Nyberg. Linear Approximation of Block Ciphers. *Advances in Cryptology, Proceedings Eurocrypt'94*, LNCS 950, pages 439–444. Springer-Verlag, 1995.
15. M.G.Parker and H.Raddum. $\mathbb{Z}_4$-Linear Cryptanalysis. NESSIE Internal Report, 27/06/2002: NES/DOC/UIB/WP5/018/1
16. B. Schneier, J. Kelsey, D. Whiting, D. Wagner, C. Hall, and N. Ferguson. New Results on the Twofish Encryption Algorithm. *Second AES Candidate Conference*, April 1999.
17. M. Matsui, A. Yamagishi. A New Method for Known Plaintext Attack of FEAL Cipher. *Advances in Cryptology, Proceedings Eurocrypt'92*, pages 81–91.
18. A. Tardy-Corfdir, H. Gilbert. A Known Plaintext Attack of FEAL-4 and FEAL-6. *Advances in Cryptology, CRYPTO 1991*, pages 172–181.

A Proofs

A.1 Average Value of $D(f)$ in the Prime Case

We prove theorem 2.

We note $f^{<-1>}(y)$ the preimage of y and we define the function bp (biggest preimage) as $bp(f) = \max_y \#f^{<-1>}(y)$.

If $df_1 = dg_1$, then $f - g$ is a constant. Moreover, $bp(df_1) \leq D(f)$. Therefore, the function $f \to (df_1, f(0))$ is injective from the set of functions with $D(f) < k$ to the product of the set of functions with $bp < k$ by G.

We define $C_{k,p}$ as $\binom{p}{k}(p-1)^{p-k}$.

Using the precedent remark and lemma 2 just below, we deduce that:

$$Pr[D(f) < k] \leq p^{1-p}\#\{f \mid bp(f) < k\} \leq p(1 - \frac{C_{k,p}}{p^p})^p$$

$z(p)$ satisfies $z(p)! \leq \frac{p}{6 \log p} < (z(p)+1)!$

Then $C_{z(p),p} \sim \frac{p^p}{ez(p)!}$

For p big enough, we have $\frac{C_{z(p),p}}{p^p} \geq 2 \log p/p$ and therefore

$$\lim_{p \to \infty} Pr[D(f) < z(p)] = 0$$

If $D(f) > 3z(p)$ there is an x such that $bp(df_x) > 3z(p)$. But knowing x, df_x, and $f(0)$ determines uniquely f.

Therefore $\Pr[D(f) > 3z(p)] \leq p^{2-p}\#\{f \mid bp(f) > 3z(p)\}$.

Using lemma 2 just below, we obtain that $\Pr[D(f) > 3z(p)] \leq p^{3-p}C_{3z(p),p}$.

Using Stirling's formula, we deduce:

$$\lim_{p \to \infty} Pr[D(f) > 3z(p)] = 0.$$

Lemma 2. $\#\{f \mid bp(f) = k\} \le pC_{k,p}$ and $\#\{f \mid bp(f) < k\} \le p^p(1 - \frac{C_{k,p}}{p^p})^p$.

Proof. First, we remark that $\#\{f \mid$ the cardinality of the preimage of $i = k\} = C_{k,p}$

(i) If $bp(f) = k$ then it exists $i \in G$ such as $\#f^{<-1>}(i) = k$. We conclude using the above remark and the fact that the cardinality of an union is upper-bounded by the sum of cardinalities.

(ii) If $bp(f) < k$ then for all y we have $\#\{x \mid f(x) = y\} < k$, and also $\#\{f \mid \#f^{<-1>}(y) < k\} \le p^p - \#\{f \mid \#f^{<-1>}(y) = k\} \le p^p(1 - \frac{C_{k,p}}{p^p})$.

Those events are anti-correlated, i.e. if an element has a small preimage, then the probability that the other elements have also a small preimage is smaller. So we can bound the global probablity by the product of probabilities.
Therefore

$$\#\{f \mid bp(f) < k\} \le p^p \left(1 - \frac{C_{k,p}}{p^p}\right)^p.$$

$\square$

A.2   Counting Polynomial over $\mathbb{Z}/q\mathbb{Z}$

We prove theorem 3.

(i): Let P be a polynomial over $\mathbb{Z}/q\mathbb{Z}$. We can write P in the form: $P(X) = A(X)(X^p - X)^2 + B(X)p(X^p - X) + C(X)(X^p - X) + pD(X) + E(X)$ with A, B, C, D, E polynomials such as B, C, D, E have degree at most $p - 1$ and coefficients between 0 and $p - 1$.
We want to prove that

$$\forall x \in \mathbb{Z}/q\mathbb{Z} \quad P(x) = 0 \pmod{p^2} \Leftrightarrow C = D = E = 0$$

Only the direct sense is difficult. Clearly, $E = 0$ because the equation is also true modulo p. We remark that $Q(xp + y) = Q(y) \pmod{p}$.
Then, we have $P(p + y) = C(y)(y^p - y) - pC(y) + pD(y) \pmod{p^2}$. And $0 = P(P + y) - P(y) = -pC(y) \pmod{p^2}$. Therefore $C = 0$ and $D = 0$.
(ii) We define the function ϕ from $\mathbb{Z}/q\mathbb{Z}[X]$ to $\mathbb{Z}/q_1\mathbb{Z}[X] \times \mathbb{Z}/q_2\mathbb{Z}[X]$ as $\phi(P) = (P_1, P_2)$ such that $P_i(x) = P(x) \pmod{q_i}$. The function ϕ is well-defined and bijective.

A.3   Piling-Up for Linear Cryptanalysis

Piling-up for bias-based approach. We prove theorem 5.
Let $\phi_{a,c}(x, K) = (x, f(x) + K, \langle a, c|x, h_K(x)\rangle)$. For any b, $\phi_{a,c}$ is a bijection from the set of solutions (x, K) of the equation $\langle a, c|x, h_K(x)\rangle = u + \langle b|K\rangle$ onto the set of solutions (x, y, v) of the equations $\langle a, b|x, f(x)\rangle = u - v$ and

$\langle b, c | y, g(y) \rangle = v$. Therefore, for any b, we have a sort of generalized matrix product $\sum_K \Lambda_0(h_K) = \Lambda_0(f)\Lambda_0(g)$ where the elements of these matrix are multiplied by convolution with respect to u. More precisely,

$$\sum_{K \in G'} \Lambda_0(h_K)_{a,c;u+\langle b|K \rangle} = \sum_v \Lambda_0(f)_{a,b;u-v} \Lambda_0(g)_{b,c;v}.$$

We will prove now that the formula remains true translated to Λ_S, for any b such that $T_{a,b} = T_{b,c}$. Both sides are zero if $u \notin T_{a,b}$, therefore we suppose that $u \in T_{a,b}$.

First, we recall that $\sum_{v \in T_{a,b}} \Lambda_S(f)_{a,b;v} = 0$. On one hand, we compute $\sum_v \Lambda_0(f)_{a,b;u-v} S_{b,c;v} = \sum_{v \in T_{b,c}} \Lambda_0(f)_{a,b;u-v} S_{b,c} = q' S_{b,c}$, and also, because $T_{a,b}$ is a group and $\langle c|y \rangle \in T_{a,b}$, we compute $\sum_K S_{a,c;u+\langle b|K \rangle} = \sum_v S_{a,b;v} \#\{y \mid \langle c|y \rangle = u - v\} = S_{a,b} \sum_{v \in T_{a,b}} \#\{y \mid \langle c|y \rangle = u - v\} = q' S_{a,b}$.

Therefore

$$\sum_{K \in G'} \Lambda_S(h_K)_{a,c;u+\langle b|K \rangle} = \sum_v \Lambda_S(f)_{a,b;u-v} \Lambda_S(g)_{b,c;v}.$$

We need the additional hypothesis that there exists some b such that $T_{a,b} = T_{b,c}$. This is true if $G = G' = G''$, because $\langle a, a+c | x, y \rangle = \langle a+c, c | x - y, x \rangle$ and therefore $T_{a,a+c} = T_{a+c,c}$. It follows that:

$$LP(h_K) = \max_{a,c,u} \left(\frac{\Lambda_S(h_K)_{a,c;u}}{q} \right)^2 \approx \max_{\substack{a,b,c,u \\ T_{a,b}=T_{b,c}}} \left(\frac{\sum_v \Lambda_S(f)_{a,b;u-v} \Lambda_S(g)_{b,c;v}}{qq'} \right)^2$$

$$\approx \max_{a,b_f,b_g,c,u,v} \left(\frac{\Lambda_S(f)_{a,b_f;u-v}}{q} \frac{\Lambda_S(g)_{b_g,c;v}}{q'} \right)^2 = LP(f)LP(g)$$

Piling-up for duality-based approach. $\lambda(h_K)_{a,c} = \sum_{x,z} \Delta(h_K)_{x,z} \xi^{\langle a,c|x,z \rangle}$. If $\Delta(h_K)$ does not depend heavily on K, then $\Delta(h_K)_{x,z} \approx \frac{1}{q'} \sum_y \Delta(f)_{x,y} \Delta(g)_{y,z}$. If $\sum_y \Delta(f)_{x,y} \Delta(g)_{y,z} \xi^{\langle a,b|x,y \rangle} \xi^{\langle b,c|y,z \rangle} \approx \frac{1}{q'} \sum_{y_f,y_g} \Delta(f)_{x,y_f} \Delta(g)_{y_g,z} \xi^{\langle a,b|x,y_f \rangle} \xi^{\langle b,c|y_g,z \rangle}$, then $\lambda(h_K)_{a,c} \approx \sum_{x,z} \frac{1}{q'} \frac{1}{q'} \sum_{y_f,y_g} \Delta(f)_{x,y_f} \Delta(g)_{y_g,z} \xi^{\langle a,b|x,y_f \rangle} \xi^{\langle b,c|y_g,z \rangle}$ which means that $\lambda(h_K)_{a,c} \approx \frac{1}{q'^2} \lambda(f)_{a,b} \lambda(g)_{b,c}$ and therefore

$$\frac{\lambda(h_K)}{q^2} \approx \frac{\lambda(f)}{q'^2} \frac{\lambda(g)}{q^2}$$

A.4   A List of all Triples $D(f), L(f), \lambda(f)$ for Small Values of p.

This is a table of all possible values for non affine functions and for $p = 5$ and 7:

p	D	L	λ	example
5	1	1	5	x^2
	2	4	$9.472\cdots$	x^4+2x^2
	2	4	$13.090\cdots$	x^3
	3	4	$16.708\cdots$	x^4+x^2
	3	9	$19.472\cdots$	x^4
7	1	1	7	x^2
	2	4	$13.097\cdots$	$x^6+x^4+6x^2$
	2	9	14	x^4
	2	4	$14.185\cdots$	$x^6+6x^4+x^2$
	2	9	$14.454\cdots$	x^5+x^2
	2	4	$14.603\cdots$	x^6+x^4
	3	4	$14.603\cdots$	x^6+3x^4
	2	4	$15.207\cdots$	$x^6+x^5+4x^2$
	2	4	$16.899\cdots$	x^6+x^2
	3	4	$16.899\cdots$	$x^6+5x^5+x^4+x^3$
	2	9	$17.048\cdots$	$x^6+x^5+x^3+5x^2$
	3	9	$17.048\cdots$	$x^6+3x^4+2x^3$
	2	9	$17.234\cdots$	$x^6+x^4+x^3+3x^2$
	3	4	$17.234\cdots$	x^6+5x^2
	3	4	$18.256\cdots$	$x^6+3x^3+x^2$
	3	9	$18.256\cdots$	$x^6+x^4+3x^3$
	2	4	$18.591\cdots$	x^5+x^4
	3	4	$18.591\cdots$	x^5+2x^2
	3	9	$18.591\cdots$	$x^5+x^4+2x^3$
	2	9	$19.076\cdots$	$x^6+5x^3+5x^2$
	2	4	$19.195\cdots$	$x^6+2x^4+5x^2$
	3	4	$19.195\cdots$	$x^6+2x^5+x^4+x^3$
	3	9	$19.195\cdots$	$x^6+2x^5+4x^3$
	2	4	$21.640\cdots$	$x^6+2x^5+x^3+x^2$
	3	4	$21.640\cdots$	$x^6+2x^3+x^2$

p	D	L	λ	example
7	2	4	$22.476\cdots$	x^3
	3	9	$22.878\cdots$	$x^6+4x^5+x^3+x^2$
	4	9	$22.878\cdots$	$x^6++2x^4+2x^3$
	2	4	$23.481\cdots$	x^6+3x^3
	2	4	$23.481\cdots$	$x^6+2x^5+x^4+2x^2$
	2	4	$24.689\cdots$	$x^6+2x^5+3x^4$
	3	4	$24.689\cdots$	$x^6+2x^4+x^3+x^2$
	4	4	$24.689\cdots$	$x^6+3x^5+x^3$
	3	9	$24.921\cdots$	$x^6+3x^4+6x^2$
	4	9	$24.921\cdots$	$x^6+3x^3+3x^2$
	3	9	$25.591\cdots$	x^4+x^3
	4	9	$25.591\cdots$	$x^5+3x^3+x^2$
	3	9	$26.195\cdots$	$x^6+x^4+x^3$
	3	4	$26.799\cdots$	x^6+2x^5
	4	16	$29.207\cdots$	$x^6+3x^4+x^3$
	3	4	$30.183\cdots$	x^4+3x^2
	4	4	$30.183\cdots$	x^5
	3	9	$31.689\cdots$	$x^6+2x^4+x^2$
	4	9	$31.689\cdots$	$x^6+2x^5+x^4+4x^2$
	4	16	$32.256\cdots$	$x^6+x^4+2x^3+5x^2$
	3	4	$32.628\cdots$	x^6+2x^3
	4	4	$32.628\cdots$	$x^6+3x^5+x^2$
	3	9	$35.073\cdots$	$x^6+x^4+x^3+x^2$
	4	9	$35.073\cdots$	$x^6+2x^5+2x^3$
	4	16	$39.024\cdots$	x^5+x^3
	3	9	$39.963\cdots$	x^6+x^3
	5	9	$39.963\cdots$	$x^6+x^5+5x^3$
	5	16	$41.169\cdots$	$x^6+x^4+x^2$
	5	25	$44.481\cdots$	x^6

A.5   In the Prime Case, $D(f)=1 \Rightarrow L(f)=1$

We will use the following lemma, for which we did not find a proof.

Lemma 1 (Key lemma). *If $\varphi : \mathbb{F}_p \to \mathbb{Z}$ satisfies $\forall x \neq 0$, $(\varphi * \varphi)(x) = -1$, and $(\varphi * \varphi)(0) = p-1$ then $\forall x$, $\varphi(x) \in \{0, \pm 1\}$.*

Let us fix f, a, and b. We denote $\eta(u) = \Lambda_S(f)_{a,b;u}$ and $\sigma = \eta * \eta$. Note that $\sum_u S_{a,b;u}\xi^u = 0$ and therefore $\hat{\theta}_f(a,b) = \sum_u \eta(u)\xi^u$ and also $\lambda(f)_{a,b} = \sum_{u,v} \eta(u)\eta(v)\xi^{u-v} = \sum_v \eta(v)^2 + \sum_{u=1}^{(\ell+1)/2} (\xi^u + \xi^{-u})\sigma(u)$ which is a real number.

In general, the lower bound for $D(f)$ is q/q'; if this lower bound is reached, then the matrix $\Delta(f)$ if fully known: $\Delta(f)_{a\neq 0,b} = q/q'$, $\Delta(f)_{0,b\neq 0} = 0$, and $\Delta(f)_{0,0} = q$, and therefore we can completely compute its transform: $\lambda(f)_{a,b\neq 0} = q$, $\lambda(f)_{a\neq 0,0} = 0$, and $\lambda(f)_{0,0} = q^2$.

Now, let us look at the case where $G = G' = \mathbb{F}_p$. If f is a polynomial of degree 2, we can check that $D(f) = L(f) = 1$. We want to prove that $D(f) = 1 \Rightarrow L(f) = 1$.

Let us suppose that $D(f) = 1$, then the duality implies that $\lambda(f)_{a,b\neq 0} = p$. However, $\lambda(f)_{a,b} = \sum_v \eta(v)^2 + \sum_{u=1}^{\ell-1} \sigma(u)\xi^u$. Since $\eta(v)$ is an integer, the second sum is also an integer. Because the $(\xi^u)_{u=1...\ell-2}$ are linearly independent over $\mathbb{Q}$, the fact that $\sum_{u\neq 0} \sigma(u)\xi^u$ is an integer implies that all $\sigma(u)$ are equal to some common value σ. Therefore $\sum_v \eta(v)^2 = p + \sigma$.

We also know that $\sum_v \eta(v) = 0$ and therefore $0 = (\sum_v \eta(v))^2 = \sum_v \eta(v)^2 + 2\sum_u \sigma(u) = p(\sigma + 1)$ and we proved that $\sigma = -1$ and $\sigma(0) = p - 1$.

We apply the key lemma to the function η, which means that $\Lambda_S(f)_{a,b;u} = \{0, \pm 1\}$, and therefore $L(f) = 1$.

B   Security Analysis of TOY100

B.1   Differential Cryptanalysis

A Key Recovery Attack. The attack uses the differential described in Section 5.1, followed by one round of key guess. More precisely, the differential is followed by $\sigma[K^n] \cdot \gamma \cdot \sigma[K^{n+1}]$. The attack goes as follows:

(i) Encrypt N plaintext pairs $(P, P + \Delta_0)$.
(ii) The corresponding ciphertext pairs that actually follow the differential are equal on 12 words (of which the position is fixed). Consider only the pairs satisfying this condition.
(iii) The key guess is performed on the 4 words of the last round key for which the difference is non zero. A counter is set for each candidate. It is incremented when the difference before the last S-box layer corresponding to the candidate is $\theta(\Delta_n)$.
(iv) After enough pairs have been considered, the most counted candidate is selected. The remaining key material is retrieved using a similar attack or by exhaustive key search.

Let $\mathbf{T}_0$ denote the event that 12 words of the output difference are 0, as specified in step 2 of the attack. Let $\mathbf{D}$ be the event that the differential is followed. We consider the 5-round differential, with $D := Pr[\mathbf{D}] \simeq 3 \cdot 10^{-26}$. Then we have

$$
\begin{aligned}
Pr[\mathbf{T}_0] &= Pr[\mathbf{T}_0|\mathbf{D}] \cdot Pr[\mathbf{D}] + Pr[\mathbf{T}_0|\neg\mathbf{D}] \cdot Pr[\neg\mathbf{D}] \\
&\simeq 1 \cdot 3 \cdot 10^{-26} + 10^{-24} \cdot (1 - 3 \cdot 10^{-26}) \\
&\simeq 10^{-24}
\end{aligned}
\tag{6}
$$

The right 4-subblock subkey will be counted $N \cdot D \simeq 3 \cdot 10^{-26} \cdot N$ times. A wrong 4-subblock subkey will be counted $N \cdot P[\mathbf{T}_0] \cdot 100^{-4} \simeq 10^{-32} \cdot N$ times. Hence the SNR of the attack is $3 \cdot 10^6$, and the subkey can be recovered using less than $2/D = 2/3 \cdot 10^{26}$ pairs. The best way to retrieve the remaining part of the key is exhaustive search.

Applying the same attack for one more round is probably possible, but almost requires the whole codebook. To the best of our investigations, more complex variants of the attack do not significantly improve its efficiency.

Another Property of the Linear Layer. The following property of the linear layer, which corresponds to the branch number bound, seems promising:

$$\begin{bmatrix} 0 & 0 & 0 & 0 \\ 50 & 50 & 50 & 0 \\ 0 & 0 & 0 & 0 \\ 0 & 0 & 0 & 0 \end{bmatrix} \xrightarrow{\theta} \begin{bmatrix} 0 & 50 & 0 & 0 \\ 0 & 50 & 0 & 0 \\ 0 & 50 & 0 & 0 \\ 0 & 0 & 0 & 0 \end{bmatrix} \tag{7}$$

However this pattern can be used only if $\Delta(f)_{50,50}$ is big enough. For the function we selected it is 0. We note here the particular role played by $\Delta(f)_{50,50}$. The existence of such "specially important" elements in the matrix $\Delta(f)$ is related to the fact that we are working over a ring. Other elements of small characteristic can be important as well for the same kind of reason. In Table 2 we give elements of $\Delta(f)$ corresponding to input and output differences which are multiple of 25; we observe that all of them are small.

Table 2. Values of $\Delta(f)_{a,b}$ when a and b have small characteristic

(a,b)	$\Delta(f)_{a,b}$
(25,25)	0
(25,50)	0
(25,75)	0
(50,25)	1
(50,50)	0
(50,75)	1
(75,25)	0
(75,50)	0
(75,75)	0

B.2 Linear Cryptanalysis

The following linear equation (equation 5 in section 5.2)

$$10 \cdot (c_{33} + c_{44} - c_{34} - c_{43}) - 10 \cdot (p_{11} + p_{22} - p_{12} - p_{21})$$
$$= \sum_{i=0}^{\lfloor \frac{n+1}{2} \rfloor} 10(k_{11}^{2i} + k_{22}^{2i} - k_{12}^{2i} - k_{21}^{2i}) + \sum_{i=1}^{\lceil \frac{n+1}{2} \rceil} 10(k_{33}^{2i-1} + k_{44}^{2i-1} - k_{34}^{2i-1} - k_{43}^{2i-1}),$$

holds with probability $1/10$ for a random permutation, and with probability $1/10 + \epsilon$ for TOY100 parameterized by a random key, where $|\epsilon|$ is given in Table 3. Therefore it can be used to build a distinguisher, by identifying the value of $10 \cdot (c_{33} + c_{44} - c_{34} - c_{43}) - 10 \cdot (p_{11} + p_{22} - p_{12} - p_{21})$ occurring the most often, and comparing its frequency of apparition to a certain threshold, in

Table 3. Estimated bias for the best $(n+1)$-round linear characteristic

# Rounds $n+1$	Best Bias
2	$2.49 \cdot 10^{-6}$
3	$8.78 \cdot 10^{-9}$
4	$3.10 \cdot 10^{-11}$
5	$1.09 \cdot 10^{-13}$
6	$3.86 \cdot 10^{-16}$

order to distinguish both probability distributions. The data complexity of the attack is $O(\epsilon^{-2})$. This distinguisher can be used in a key-recovery attack, by performing key guesses on the first and/or last round key. Up to 7 rounds of the cipher can be attacked this way, and we are close to an attack on 8 rounds. The data and time complexity are $O(\epsilon^{-2})$.

Finally, we note that relying on property (7) to build a characteristic is not possible, as our S-box satisfies $\Lambda_S(f)_{50,50;0} = \Lambda_S(f)_{50,50;50} = 0$.

B.3   A Square-Like Attack

Our square-like attack aims at the cipher

$$P^{(i)} \xrightarrow{\sigma[K^1] \cdot \gamma \cdot \theta} A^{(i)} \xrightarrow{\sigma[K^2] \cdot \gamma \cdot \theta} B^{(i)} \xrightarrow{\sigma[K^3] \cdot \gamma} C^{(i)} \xrightarrow{\theta} D^{(i)} \xrightarrow{\sigma[K^4]} E^{(i)} \xrightarrow{\gamma \cdot \sigma[K^5]} F^{(i)}$$

It exploits batches of 100^4 plaintexts with the following structure:

$$P^{(i)} = \begin{bmatrix} p_{11}^{(i)} & p_{12}^{(i)} & p_{13}^{(i)} & p_{14}^{(i)} \\ p_{21}^{(i)} & p_{22}^{(i)} & p_{23}^{(i)} & p_{24}^{(i)} \\ p_{31}^{(i)} & p_{32}^{(i)} & p_{33}^{(i)} & p_{34}^{(i)} \\ p_{41}^{(i)} & p_{42}^{(i)} & p_{43}^{(i)} & p_{44}^{(i)} \end{bmatrix} = \begin{bmatrix} a^{(i)} & b^{(i)} & \kappa & \kappa \\ c^{(i)} & d^{(i)} & \kappa & \kappa \\ \kappa & \kappa & \kappa & \kappa \\ \kappa & \kappa & \kappa & \kappa \end{bmatrix},$$

where $(a^{(i)}, b^{(i)}, c^{(i)}, d^{(i)})$ takes every possible value. As the value of constants does not matter for our attack, all κ's denote constants that are *not* necessarily equal.

Let us define:

$$S_{rs}(x) := S(x + k_{rs}^1)$$
$$m^{(i)} := S_{11}(a^{(i)}) + S_{12}(b^{(i)})$$
$$n^{(i)} := S_{21}(c^{(i)}) + S_{22}(d^{(i)})$$
$$o^{(i)} := S_{11}(a^{(i)}) + S_{21}(c^{(i)})$$
$$p^{(i)} := S_{12}(b^{(i)}) + S_{22}(d^{(i)})$$
$$x^{(i)} := m^{(i)} + n^{(i)} = o^{(i)} + p^{(i)}$$

After the first round $\sigma[k^1] \cdot \gamma \cdot \theta$ the data become:

$$\begin{bmatrix} x^{(i)} & x^{(i)} & p^{(i)} & o^{(i)} \\ x^{(i)} & x^{(i)} & p^{(i)} & o^{(i)} \\ n^{(i)} & n^{(i)} & S_{22}(d^{(i)}) & S_{21}(c^{(i)}) \\ m^{(i)} & m^{(i)} & S_{12}(b^{(i)}) & S_{11}(a^{(i)}) \end{bmatrix} + \begin{bmatrix} \kappa & \kappa & \kappa & \kappa \\ \kappa & \kappa & \kappa & \kappa \\ \kappa & \kappa & \kappa & \kappa \\ \kappa & \kappa & \kappa & \kappa \end{bmatrix}$$

It is then easy to see that the state $B^{(i)}$ after the second round $\sigma[k^2] \cdot \gamma \cdot \theta$ is such that $b_{11}^{(i)}, b_{12}^{(i)}, b_{21}^{(i)}, b_{22}^{(i)}$ are still active (i.e. take every value equally often). This property is preserved after passing through $\sigma[k^3] \cdot \gamma$. In order to push the distinguisher further, we use the following property of θ again:

$$D^{(i)} = \theta(C^{(i)}) \Rightarrow d_{33}^{(i)} + d_{44}^{(i)} - d_{34}^{(i)} - d_{43}^{(i)} = c_{11}^{(i)} + c_{22}^{(i)} - c_{12}^{(i)} - c_{21}^{(i)} \tag{8}$$

So we have

$$\sum_{1 \le i \le 100^4} e_{33}^{(i)} + e_{44}^{(i)} - e_{34}^{(i)} - e_{43}^{(i)}$$

$$= \sum_{1 \le i \le 100^4} (d_{33}^{(i)} + k_{33}^4) + (d_{44}^{(i)} + k_{44}^4) - (d_{34}^{(i)} + k_{34}^4) - (d_{43}^{(i)} + k_{43}^4)$$

$$= \sum_{1 \le i \le 100^4} d_{33}^{(i)} + d_{44}^{(i)} - d_{34}^{(i)} - d_{43}^{(i)}$$

$$= \sum_{1 \le i \le 100^4} c_{11}^{(i)} + c_{22}^{(i)} - c_{12}^{(i)} - c_{21}^{(i)}$$

$$= \sum_{1 \le i \le 100^4} c_{11}^{(i)} + \sum_{1 \le i \le 100^4} c_{22}^{(i)} - \sum_{1 \le i \le 100^4} c_{12}^{(i)} - \sum_{1 \le i \le 100^4} c_{21}^{(i)} = 0,$$

where the last equality results from the fact that $c_{11}^{(i)}, c_{12}^{(i)}, c_{21}^{(i)}$ and $c_{22}^{(i)}$ are active.

By guessing 4 words $k_{33}^5, k_{34}^5, k_{43}^5, k_{44}^5$ of the last round key we can check this property. The probability of a false alarm is $1/100$, so about 4 batches of 100^4 plaintexts are necessary to retrieve this part of the key. Besides it is clear that our analysis holds for any "square of four words" of the plaintext. Hence we can retrieve the remaining 12 subblocks using the same method. The global complexity is about $16 \cdot 100^4$ chosen plaintexts. The offline work is of the same order of magnitude.

A Zero-Dimensional Gröbner Basis for AES-128

Johannes Buchmann, Andrei Pyshkin*, and Ralf-Philipp Weinmann

Technische Universität Darmstadt, Fachbereich Informatik,
Hochschulstr. 10, D-64289 Darmstadt, Germany
{buchmann, pyshkin, weinmann}@cdc.informatik.tu-darmstadt.de

Abstract. We demonstrate an efficient method for computing a Gröbner basis of a zero-dimensional ideal describing the key-recovery problem from a single plaintext/ciphertext pair for the full AES-128. This Gröbner basis is relative to a degree-lexicographical order. We investigate whether the existence of this Gröbner basis has any security implications for the AES.

Keywords: block ciphers, Gröbner bases, AES, Rijndael.

1 Introduction

Gröbner bases are standard representations of polynomial ideals that possess several useful properties:

- given a Gröbner basis of an ideal $I \subset R$, we can efficiently decide whether a polynomial $f \in R$ lies in I
- for suitable term orders (e.g. lexicographical orders), the variety of the ideal can be efficiently computed; this yields solutions for the polynomial system induced by the ideal.

Usually, the Gröbner basis of a set of polynomials is computed using either a variant of Buchberger's algorithm [3] or using Faugere's F_4 [10] or F_5 [11] algorithm. These algorithms involve polynomials reductions which are costly. In general the time and the space complexity of these algorithms is difficult to predict. For polynomials in a large number of variables, these algorithms quickly become infeasible.

Rijndael, the block cipher that has been selected as the Advanced Encryption Standard (AES) in 2001, has become the industry-wide standard block cipher by now. Its design, the wide-trail strategy, is considered state of the art. However, Rijndael has from the beginning been critized for its mathematical simplicity and rich algebraic structure [15,13,8]. On the other hand this criticism has not yet substantiated into an attack; quite to the contrary, claims of an algebraic attack using XSL [8] have recently been debunked [6].

For the Rijndael block cipher, two algebraic representations in the form of multivariate polynomial systems of equations have been proposed so far. Courtois and Pieprzyk have demonstrated how to obtain overdefined systems of quadratic

---

* Supported by a stipend of the Marga und Kurt-Möllgaard-Stiftung.

equations over $GF(2)$, while Murphy and Robshaw have constructed an embedding for the AES called Big Encryption System (BES) for which a system of overdefined quadratic equations over $GF(2^8)$ exists [16].

A representation considering the output of the S-Box as a polynomial expression of the input over $GF(2^8)$ has thus far been neglected because the polynomials in this case are of relatively high degree. Using this representation we can describe the key recovery problem for the AES cipher with a key length of 128 bits as a system of 200 polynomial equations of degree 254 and 152 linear equations. In this paper we will show that by choosing an appropriate term order and by applying linear operations only, we can generate a Gröbner basis for AES-128 from this system without a single polynomial reduction.

The structure of this paper is as follows: in Section 2 we establish the notation used in this paper, in Section 3 we explain how to construct the Gröbner basis for Rijndael, in Section 4 we study the cryptanalytic importance of our result. Finally we summarize the impact of our result in Section 5 and conclude.

2   Notation

We assume the reader to be familiar with the description of AES as given in [17]. In the following we restrict ourselves to AES-128, i.e. Rijndael with a block and key size of 128 bits.

We will deviate from the standard representation by using a column vector instead of a matrix for the internal state and the round keys. The elements in the column vector are identified with the elements of the matrix in a column-wise fashion by the following map:

$$\varphi : F^{4\times 4} \to F^{16}, \begin{pmatrix} s_{0,0} & s_{0,1} & s_{0,2} & s_{0,3} \\ s_{1,0} & s_{1,1} & s_{1,2} & s_{1,3} \\ s_{2,0} & s_{2,1} & s_{2,2} & s_{2,3} \\ s_{3,0} & s_{3,1} & s_{3,2} & s_{3,3} \end{pmatrix} \mapsto (s_{0,0}, s_{1,0}, \ldots, s_{0,1}, s_{1,1}, \ldots)^T \quad (1)$$

Furthermore we define the 16×16 matrix P to be the permutation matrix that achieves the exchange of elements in the column vector that is equivalent to transposing the state matrix.

The above notation allows us to express the diffusion performed by the MixColumns and ShiftRows operations as a single matrix multiplication.

Let $x_{i,j}$ denote the variable referring to the ith component of the state vector after the jth round execution. By this definition the variables $x_{i,0}$ are called *plaintext variables*, correspondingly $x_{i,10}$ are called *ciphertext variables*. All other variables $x_{i,j}$ are called *intermediate state variables*; variables $k_{i,j}$ are called *key variables*. We will also refer to $k_{i,0}$ as *cipher key variables*.

The field F is the finite field $GF(2^8)$ as defined for Rijndael. The polynomial ring R is defined as

$$R := F[x_{i,j}, k_{i,j} : \{0 \le i \le 15, 0 \le j \le 10\}]$$

3   Construction of the Gröbner Basis

In this section we will explain how to construct a degree lexicographical Gröbner basis describing the AES key recovery problem step by step. To accomplish this task we will first give a very minimal introduction to Gröbner bases; just enough to follow this paper. We kindly refer the inclined reader to [9] and [2] for a more gentle introduction to the topic.

3.1   Gröbner Bases

Some confusion regularly arises out of the expressions *term* and *monomial*. One school calls a product of variables a *term* and the product of said term and a coefficient a *monomial*; notably this is done in [2]. The other camp, e.g. the authors of [9], uses *term* and *monomial* in an interchanged fashion. We adopt the conventions of [2].

For a given ideal there usually exists more than one Gröbner basis. These are relative to a so called term order, which we shall now define:

Definition 1 (Term order). *A term order $\leq$ is a linear order on the set of terms $\mathcal{T}(R)$ such that*

1. *$1 \leq t$ for all terms $t \in \mathcal{T}(R)$*
2. *for all terms $s, t_1, t_2 \in \mathcal{T}(R)$ whenever $t_1 \leq t_2$ then $st_1 \leq st_2$*

The maximum element of the set of terms of a polynomial p under a fixed term order $\leq$ shall be referred to as the head term of p, short $HT(p)$.

We will now introduce two useful and widely used term orders. First, however, we define two technicalities: For a term $t = \mathfrak{v}_1^{e_1} \mathfrak{v}_2^{e_2} \cdots \mathfrak{v}_k^{e_k} \in \mathcal{T}(R)$ we define the *exponent vector* of t to be $\epsilon(t) = (e_1, e_2, \ldots, e_k) \in \mathbb{N}_0^k$. The total degree of the term t then is $\deg(t) = \sum_{i=1}^{k} e_i$.

Example 1 (lexicographical term order). For terms s, t we define $s <_{lex} t$ iff there exists an i with $1 \leq i \leq k$ such that the first $i - 1$ components of $\epsilon(s)$ and $\epsilon(t)$ are equal but the ith component of $\epsilon(s)$ is smaller than the ith component of $\epsilon(t)$.

Example 2 (degree lexicographical term order). For terms s, t we define $s <_{dlex} t$ iff either $\deg(s) < \deg(t)$ or if $\deg(s) = \deg(t)$ and $s <_{lex} t$.

Remark 1. Note that there is more than one lexicographical order and more than one degree lexicographical term order. Different orderings on the variables induce different term orders!

The formal definition of a Gröbner basis does not give much insight about how to construct one:

Definition 2 (Gröbner basis). *Let $\mathcal{I}$ be an ideal of R. A set of polynomials $\{g_1, \ldots, g_m\} \subset \mathcal{I}$ is a Gröbner basis if the following holds:*

$$\langle HT(g_1), \ldots, HT(g_m) \rangle = \langle \{HT(p) : p \in I\} \rangle$$

The first Buchberger criterion [4] is a basic test that is used in most implementations of Buchberger's algorithm to avoid "useless" polynomial reductions. The following theorem follows almost instantaneously from this criterion and gives an important hint how a Gröbner basis can be attained without knowing anything about polynomial reductions.

Theorem 1. *Let G be a set of polynomials and $H = \{HT(f) : f \in G\}$. If all elements in H are pairwise prime, then G is a Gröbner basis.*

Proof. See [5].

A zero-dimensional ideal is an ideal that has a finite number of solutions over the closure of the field. It usually is advantageous to have this property for Gröbner basis computations. By using Corollary 6.56 of [2] we can determine whether an ideal I is zero-dimensional. Below we state a reduced version of this corollary:

Lemma 1. *Let I be a proper ideal of $F[x_1, \ldots, x_n]$. Then the following assertions are equivalent:*

- $\dim(I) = 0$
- *There exists a term order $\leq$ such that for each $1 \leq i \leq n$ there is $g_i \in I$ with $HT(g_i) = x_i^{\nu_i}$ for some $0 \leq \nu_i \in \mathbb{N}$.*

3.2   The S-Box

The S-Box used in Rijndael can be interpolated as a sparse polynomial over F:

$$\sigma : F \to F, \quad x \mapsto \begin{aligned}&05x^{254} + 09x^{253} + \text{F9}x^{251} + 25x^{247} + \text{F4}x^{239} + \\ &\text{B5}x^{223} + \text{B9}x^{191} + \text{8F}x^{127} + 63\end{aligned} \tag{2}$$

whilst the interpolation polynomial of the inverse S-Box

$$\sigma^{-1} : F \to F, \quad x \mapsto \sum_{i=0}^{254} c_i x^i \tag{3}$$

is dense. This polynomial is given in Appendix A.

3.3   The Linear Transformation

The linear transformation of AES consists of two operations, ShiftRows and MixColumns. We can perform the linear transform by multiplying the state column vector with a 16×16-matrix D from the left. In the following, we calculate D; however at the start of each round we apply the transposition matrix P since it makes expressing the operations as matrices easier. At the end we multiply with the matrix P to undo the initial transposition.

A matrix that shifts the elements of a 1×4 row vector cyclically by an offset t is of the following form:

$$D_{\text{SR}_t} = \left(\Delta_{i,(j-t) \bmod 4}\right) \in F^{4 \times 4} \tag{4}$$

where $\Delta_{i,j}$ is the Kronecker delta. The ShiftRows operation is equivalent to multiplying by the matrix D_{SR}:

$$D_{SR} = \begin{pmatrix} D_{SR_0} & 0 & 0 & 0 \\ 0 & D_{SR_1} & 0 & 0 \\ 0 & 0 & D_{SR_2} & 0 \\ 0 & 0 & 0 & D_{SR_3} \end{pmatrix} \in F^{16 \times 16} \tag{5}$$

The MixColumns operation is applied to each row of the internal state. We use the matrix D_{MC} to transform the column vector equivalently:

$$D_{MC} = \begin{pmatrix} 02\ 03\ 01\ 01 \\ 01\ 02\ 03\ 01 \\ 01\ 01\ 02\ 03 \\ 03\ 01\ 01\ 02 \end{pmatrix} \otimes I_4 \in F^{16 \times 16} \tag{6}$$

where $\otimes$ denotes the tensor product. Concatenation of the two operations in the diffusion layer is achieved by multiplying the above matrices, yielding the matrix D:

$$D = P \cdot D_{MC} \cdot D_{SR} \cdot P \tag{7}$$

The diffusion layer of the last round is missing the MixColumns transformation; it will be described by the matrix $\tilde{D}$:

$$\tilde{D} = P \cdot D_{SR} \cdot P \tag{8}$$

This enables us to obtain the following vectorial representation of a system of 16 polynomial equations that holds for rounds $1 \leq j \leq 9$ of the cipher:

$$\begin{pmatrix} \sigma(x_{0,(j-1)} + k_{0,(j-1)}) \\ \vdots \\ \sigma(x_{15,(j-1)} + k_{15,(j-1)}) \end{pmatrix} + D^{-1} \begin{pmatrix} x_{0,j} \\ \vdots \\ x_{15,j} \end{pmatrix} = 0 \tag{9}$$

For the last round we need to take the simplified diffusion layer and the final key addition into account:

$$\begin{pmatrix} \sigma(x_{0,9} + k_{0,9}) \\ \vdots \\ \sigma(x_{15,9} + k_{15,9}) \end{pmatrix} + \tilde{D}^{-1} \begin{pmatrix} x_{0,10} + k_{0,10} \\ \vdots \\ x_{15,10} + k_{15,10} \end{pmatrix} = 0 \tag{10}$$

Choosing any degree lexicographical term order, either a term $x_{i,j}^{254}$ or a term $k_{i,j}^{254}$ occurs as head term of each polynomial. We take note that none of the head terms is a power of a plaintext nor of a ciphertext variable. Moreover all of the head terms are pairwise prime. The variable order chosen will influence whether the head term is a power of a key variable or of an intermediate state variable.

3.4   The Key Schedule

In order to obtain a Gröbner basis of both the cipher and the key scheduling polynomials, we need to set up the key scheduling in a slightly different way. Usually, the key scheduling expresses the elements of the round subkey of round $1 \leq j \leq 10$ as a vector of polynomials in the key variables of the previous round as follows:

$$
\begin{pmatrix} k_{0,j} \\ k_{1,j} \\ k_{2,j} \\ k_{3,j} \\ k_{4,j} \\ \vdots \\ k_{15,j} \end{pmatrix} = \begin{pmatrix} k_{0,j-1} \\ k_{1,j-1} \\ k_{2,j-1} \\ k_{3,j-1} \\ k_{4,j-1} \\ \vdots \\ k_{15,j-1} \end{pmatrix} + \begin{pmatrix} \sigma(k_{15,j-1}) \\ \sigma(k_{12,j-1}) \\ \sigma(k_{13,j-1}) \\ \sigma(k_{14,j-1}) \\ k_{0,j} \\ \vdots \\ k_{11,j} \end{pmatrix} + \begin{pmatrix} \gamma_{j-1} \\ 0 \\ 0 \\ 0 \\ 0 \\ \vdots \\ 0 \end{pmatrix} \tag{11}
$$

where the $\gamma_0, \ldots, \gamma_9$ are the round constants. To make all head terms pairwise prime (see also Section 3.5 on the term order chosen), we we have to proceed in reverse order:

$$
\begin{pmatrix} \sigma^{-1}(k_{0,j} + k_{0,j-1} + \gamma_{j-1}) \\ \sigma^{-1}(k_{1,j} + k_{1,j-1}) \\ \sigma^{-1}(k_{2,j} + k_{2,j-1}) \\ \sigma^{-1}(k_{3,j} + k_{3,j-1}) \\ k_{4,j} + k_{4,j-1} \\ \vdots \\ k_{15,j} + k_{15,j-1} \end{pmatrix} + \begin{pmatrix} k_{15,j-1} \\ k_{12,j-1} \\ k_{13,j-1} \\ k_{14,j-1} \\ k_{0,j} \\ \vdots \\ k_{11,j} \end{pmatrix} = 0 \tag{12}
$$

3.5   Choosing a Suitable Variable Order

The plaintext and ciphertext polynomials simply are of the form

$$
x_{i,0} + p_i \qquad p_i \in F, 0 \leq i \leq 15 \tag{13}
$$

respectively

$$
x_{i,0} + c_i \qquad c_i \in F, 0 \leq i \leq 15. \tag{14}
$$

Let $\mathcal{A}$ be the union of the left-hand side of equations (9), (10) and (12) for all rounds $1 \leq j \leq 10$ as well as the plaintext and ciphertext polynomials. Ordering the variables as follows makes all head terms pairwise prime:

1. plaintext variables: $x_{0,0} < \ldots < x_{15,0}$
2. ciphertext variables: $x_{0,10} < \ldots < x_{15,10}$
3. key variables of all rounds in natural order: $k_{0,0} < k_{1,0} < \ldots < k_{15,10}$
4. intermediate state variables in their natural order

The degree lexicographical term order with the above variable order will be in the following be referred to as $<_{\mathcal{A}}$. By Theorem 1, the set of polynomials $\mathcal{A}$ is a Gröbner basis relative to this term order! Moreover, checking Lemma 1 we verify that this ideal is zero-dimensional.

4   Exploiting the Gröbner Basis

In the previous section we have shown how to obtain a zero-dimensional Gröbner basis $\mathcal{A}$ for AES-128. In this section we explore the cryptanalytic impact of this finding. To this end, we investigate the complexity of a Gröbner basis conversion algorithm, find an invariant under the elimination of variables and explain why the naïve way of applying the ideal membership test does not work for guessing parts of the round key.

4.1   Complexity of Gröbner Basis Conversions

An obvious question is whether the Gröbner basis we have computed in the previous section can be efficiently converted to a different, more suitable order, i.e. a lexicographical order or an elimination order [1].

Two algorithms and variations of them are known for performing Gröbner basis conversions, the FGLM algorithm [12] and the Gröbner Walk [7]. While the FGLM algorithm as described in [12] only works for zero-dimensional ideals, the Gröbner Walk naturally also works for ideals of positive dimension. Since we have established that $\mathcal{A}$ is zero-dimensional, we are in a position to use FGLM and give an estimate for its time complexity below.

An important characteristic of the ideal is the vector space dimension of the residue class ring obtained when factoring the polynomial ring R by the ideal I:

Definition 3. *Let $R := F[x_1, \ldots, x_n]$. Then the F-space dimension of the ideal $I \subset R$ shall be denoted by $\dim(R/I)$.*

From Lemma 6.51 and Proposition 6.52 in [2] it is straightforward to deduce the following lemma:

Lemma 2. *Let $\leq$ be a term order on $\mathcal{T}(R)$ and G a Gröbner basis of I w.r.t. $\leq$. Then*

$$\dim(R/I) = \# \{t \in \mathcal{T}(R) : s \nmid t \text{ for all } s \in HT(I)\}$$
$$= \# \{t \in \mathcal{T}(R) : s \nmid t \text{ for all } s \in HT(G)\}$$

Applying the lemma to a Gröbner basis with univariate head terms yields the following corollary:

Corollary 1. *Let $G = \{g_1, \ldots, g_n\}$ be a Gröbner basis for the ideal $I \subset F[x_1, \ldots, x_n]$ with head terms $x_1^{d_1}, \ldots, x_n^{d_n}$. Then $\dim(R/I) = d_1 \cdots d_n$.*

This result is sufficient to give a bound on the complexity of the Gröbner basis conversion using FGLM. The following theorem is a slightly rephrased version of Theorem 5.1 in [12]:

Theorem 2. *Let F be a finite field and $R = F[x_1, \ldots, x_n]$. Furthermore $G_1 \subset R$ is the Gröbner basis relative to a term order $<_1$ of an ideal I, and $D = \dim(R/I)$. We can then convert G_1 into a Gröbner basis G_2 relative to a term order $<_2$ in $O(nD^3)$ field operations.*

From Corollary 1 we conclude that the vector space dimension of the ideal generated by the Gröbner basis $\mathcal{A}$ is way too big for the FGLM algorithm be useful for cryptanalytic purposes in this case:

$$dim(R/\mathcal{A}) = 254^{200} \approx 2^{1598} \tag{15}$$

For the Gröbner Walk, the running time strongly depends on the source and the target term order. It is an open problem to give bounds on the time and space complexity for this algorithm. The only bounds known are local bounds, namely for adjacent term orders, due to Kalkbrener [14].

4.2   Elimination of Variables

In this section we establish that the dimension of the vector space of the ideal remains invariant when eliminating certain variables. We first prove the following more general statement:

Proposition 1. *Let I' be a zero-dimensional ideal of $R' := F[x_1, \ldots, x_n]$, I an ideal of $R := R'[x_{n+1}]$ and $I' = I \cap R'$. Then $\dim R/I = \dim R'/I'$ iff there exists a polynomial $g \in R'$ such that $x_{n+1} + g \in I$.*

Proof. W.l.o.g. we fix a lexicographical term ordering such that x_{n+1} is the greatest variable. Let $\mathrm{RT}(I)$ and $\mathrm{RT}(I')$ be defined as follows:

$$\mathrm{RT}(I) = \{t \in \mathcal{T}(R) : s \nmid t \text{ for all } s \in \mathrm{HT}(I)\}$$
$$\mathrm{RT}(I') = \{t \in \mathcal{T}(R') : s \nmid t \text{ for all } s \in \mathrm{HT}(I')\} \subset \mathrm{RT}(I)$$

By Lemma 2, $\dim_K(R/I) = \#\mathrm{RT}(I)$ holds. Thus it is sufficient to prove that $\#\mathrm{RT}(I) = \#\mathrm{RT}(I')$. Since $x_{n+1} \nmid t$ for $t \in \mathcal{T}(R')$, the equality $\mathrm{RT}(I) = \mathrm{RT}(I')$ holds iff $x_{n+1} \in \mathrm{HT}(I)$, i.e. exists a $g \in R'$ for which $x_{n+1} + g \in I$. $\qquad\square$

Corollary 2. *For the set of polynomials $\mathcal{A}$ the dimension $\dim(R/I)$ is invariant under the elimination of all variables except the round key variables $k_{i,0}$ with $0 \le i \le 15$ and $k_{i,j}$ with $0 \le i \le 3$, $1 \le j \le 9$.*

Proof. By induction using Proposition 1.

So even eliminating a significant amount of variables does not reduce the complexity of converting the Gröbner basis to a term order suitable for key recovery.

4.3   Taking the Field Equations into Account

Let $R = F[x_1, \ldots, x_n]$ be a polynomial ring over finite field $F = GF(2^m)$ with $q = 2^m$ elements. For every element $\tau \in F$ the relation $\tau^q = \tau$ holds; the equations

$$x_i^q + x_i = 0 \tag{16}$$

are commonly called *field equations*. The set of roots of each of these equations is the set of all elements of the field F. By adjoining the set of all field polynomials $\mathcal{F}$ — the left-hand side of Equation 16 — to the set of polynomials $\mathcal{A}$, we eliminate all points of the variety that only exist in the closure but not in the ground field. The resulting set does not form a Gröbner basis, however.

What we have to do is to compute the intersection of two varieties; this is usually achieved by computing the Gröbner basis of the sum of the corresponding ideals. We have a set of polynomials $\mathcal{A}$, describing AES which is a Gröbner basis relative to the order $<_\mathcal{A}$, and a second set of polynomials $\mathcal{F}$, which also forms a Gröbner basis relative to the same order. It is however unclear how to exploit the Gröbner basis property of the input.

4.4   Testing Keys

Gröbner bases were invented to solve the ideal membership problem. So why are we not able to simply test whether a linear polynomial of the form

$$k_i + C, \qquad C \in F \tag{17}$$

— with C being a key variable guess — lies in the ideal? After all, this would allow us to determine the key piecementally by guessing each byte.

Several problems present themselves here. First of all, the polynomial system has solutions over the closure of the ground field, which means that we have to test for a polynomial

$$g = p \cdot \prod (k_i + C_j)^{t_j}, \quad t_j \in \mathbb{N}_0,\ C_j \in F$$

instead, where the C_j denote candidate values for the key variable and p is a product of irreducible non-linear polynomials. Moreover the dimension of the ideal again plays an important role here: it is an upper bound on the number of solutions of the corresponding polynomial system in the closure of the field. Hence the degree of g is expected to be very large.

5   Implications

As far as the authors are aware at the time of writing this paper, the existence of the above Gröbner basis has no security implications for AES. We conjecture that methods similar to the one presented in this paper can be used to produce total-degree Gröbner bases for many other iterated block ciphers – however we like to point out that because of the high algebraic structure of Rijndael, it makes for an excellent example.

6   Conclusion

We have demonstrated that by choosing a particular variable order, degree lexicographical Gröbner bases for AES-128 can be constructed without polynomial reductions. We have analyzed the implications of this finding and have shown that several obvious approaches do not translate into a successful cryptanalysis. It is an open problem whether the results contained in this paper can be leveraged into an attack.

Acknowledgments

The third author acknowledges several insightful discussions with Frederik Armknecht.

References

1. David Bayer and Michael Stillman. On the complexity of computing syzygies. *Journal of Symbolic Computation*, 6(2/3):135–147, 1988.
2. Thomas Becker and Volker Weispfenning. *Gröbner Bases – A Computational Approach to Commutative Algebra*. Springer-Verlag, 1991.
3. Bruno Buchberger. *Ein Algorithmus zum Auffinden der Basiselemente des Restklassenringes nach einem nulldimensionalen Polynomideal*. PhD thesis, University of Innsbruck, Austria, 1965.
4. Bruno Buchberger. A criterion for Detecting Unnecessary Reductions in the Construction of Groebner Bases. volume 72, pages 3–21, London, UK, 1979. Johannes Kepler University Linz, Springer, Berlin - Heidelberg - New York.
5. Johannes Buchmann, Andrei Pyshkin, and Ralf-Philipp Weinmann. Block Ciphers Sensitive to Gröbner Basis Attacks. In *Topics in Cryptology - CT-RSA 2006*, volume 3860 of *Lecture Notes in Computer Science*, pages 313–331. Springer, 2006.
6. Carlos Cid and Gaëtan Leurent. An analysis of the XSL Algorithm. In Roy Bimal, editor, *Advances in Cryptology – ASIACRYPT 2005*, volume 3788 of *Lecture Notes in Computer Science*, pages 333–345. Springer, 2005.
7. Stéphane Collart, Michael Kalkbrener, and Daniel Mall. Converting Bases with the Gröbner Walk. *Journal of Symbolic Computation*, 24(3/4):465–469, 1997.
8. Nicolas Courtois and Josef Pieprzyk. Cryptanalysis of Block Ciphers with Overdefined Systems of Equations. In Yuliang Zheng, editor, *Advances in Cryptology – ASIACRYPT 2002*, volume 2501 of *Lecture Notes in Computer Science*, pages 267–287. Springer, 2002.
9. David A. Cox, John B. Little, and Don O'Shea. *Ideals, Varieties, and Algorithms*. Springer-Verlag, NY, 2nd edition, 1996. 536 pages.
10. Jean-Charles Faugère. A New Efficient Algorithm for Computing Gröbner bases (F4). *Journal of Pure and Applied Algebra*, 139(1-3):61–88, June 1999.
11. Jean-Charles Faugère. A new efficient algorithm for computing Gröbner bases without reduction to zero (F_5). In *Symbolic and Algebraic Computation – ISSAC 2002*, pages 75–83. ACM, 2002.
12. Jean-Charles Faugère, P. Gianni, Daniel Lazard, and Teo Mora. Efficient Computation of Zero-Dimensional Gröbner Bases by Change of Ordering. *Journal of Symbolic Computation*, 16(4):329–344, 1993.
13. Niels Ferguson, Richard Schroeppel, and Doug Whiting. A Simple Algebraic Representation of Rijndael. In Serge Vaudenay and Amr M. Youssef, editors, *Selected Areas in Cryptography*, volume 2259 of *Lecture Notes in Computer Science*, pages 103–111. Springer, 2001.
14. Michael Kalkbrener. On the Complexity of Gröbner Bases Conversion. *Journal of Symbolic Computation*, 28(1-2):265–273, 1999.
15. Sean Murphy and Matt Robshaw. Further Comments on the Structure of Rijndael. AES Comment to NIST, August 2000.

16. Sean Murphy and Matthew J.B. Robshaw. Essential Algebraic Structure within the AES. In Moti Yung, editor, *Advances in Cryptology – CRYPTO 2002*, volume 2442 of *Lecture Notes in Computer Science*, pages 1–16. Springer, 2002.
17. National Institute of Standards and Technology. FIPS-197: Advanced Encryption Standard, November 2001. Available at http://csrc.nist.gov/publications/fips/.

A Polynomial Interpolation of the Inverse S-Box of Rijndael

$$\sigma^{-1} : F \to F$$

$$x \mapsto 05x^{254} + \text{CF}x^{253} + \text{B3}x^{252} + 16x^{251} + 55x^{250} + \text{C0}x^{249} + 7\text{A}x^{248} + 01x^{247} +$$
$$22x^{246} + \text{D8}x^{245} + 6\text{B}x^{244} + \text{A6}x^{243} + 1\text{F}x^{242} + 0\text{D}x^{241} + \text{BC}x^{240} + 49x^{239} +$$
$$85x^{238} + \text{B4}x^{237} + 1\text{B}x^{236} + 5\text{E}x^{235} + \text{BD}x^{234} + 18x^{233} + 1\text{D}x^{232} + 6\text{D}x^{231} +$$
$$\text{C5}x^{230} + 23x^{229} + 09x^{228} + 43x^{227} + 68x^{226} + 80x^{225} + 6\text{C}x^{224} + \text{CC}x^{223} +$$
$$42x^{222} + 9\text{F}x^{221} + 0\text{F}x^{220} + \text{D2}x^{219} + 3\text{B}x^{218} + 2\text{C}x^{217} + 5\text{F}x^{216} + \text{BE}x^{215} +$$
$$\text{AE}x^{214} + \text{E4}x^{213} + 93x^{212} + 8\text{B}x^{211} + \text{CB}x^{210} + 65x^{209} + \text{C0}x^{208} + 1\text{E}x^{207} +$$
$$8\text{E}x^{206} + 32x^{205} + 1\text{D}x^{204} + \text{A5}x^{203} + 76x^{202} + \text{A9}x^{201} + 2\text{C}x^{200} + 13x^{199} +$$
$$05x^{198} + 60x^{197} + \text{FD}x^{196} + 1\text{B}x^{195} + \text{AB}x^{194} + 64x^{193} + \text{C1}x^{192} + \text{A8}x^{191} +$$
$$7\text{F}x^{190} + 55x^{189} + \text{DB}x^{188} + \text{EC}x^{187} + 20x^{186} + \text{C4}x^{185} + \text{DB}x^{184} + 7\text{E}x^{183} +$$
$$92x^{182} + 80x^{181} + \text{A3}x^{180} + 59x^{179} + 91x^{178} + 91x^{177} + 81x^{176} + 4\text{E}x^{175} +$$
$$11x^{174} + \text{DD}x^{173} + 4\text{E}x^{172} + \text{D3}x^{171} + \text{E3}x^{170} + 19x^{169} + \text{E7}x^{168} + 03x^{167} +$$
$$24x^{166} + 45x^{165} + \text{DA}x^{164} + \text{EA}x^{163} + 87x^{162} + 2\text{D}x^{161} + 23x^{160} + 82x^{159} +$$
$$38x^{158} + \text{B7}x^{157} + 9\text{E}x^{156} + \text{B3}x^{155} + 2\text{A}x^{154} + 3\text{E}x^{153} + 1\text{C}x^{152} + \text{EC}x^{151} +$$
$$\text{C3}x^{150} + 45x^{149} + \text{ED}x^{148} + \text{D5}x^{147} + 2\text{A}x^{146} + 8\text{D}x^{145} + \text{ED}x^{144} + 37x^{143} +$$
$$26x^{142} + \text{E0}x^{141} + \text{BC}x^{140} + 58x^{139} + \text{E2}x^{138} + 6\text{C}x^{137} + 24x^{136} + 55x^{135} +$$
$$\text{C7}x^{134} + \text{AA}x^{133} + 09x^{132} + 4\text{F}x^{131} + 82x^{130} + \text{CA}x^{129} + 10x^{128} + \text{EE}x^{127} +$$
$$1\text{A}x^{126} + 2\text{E}x^{125} + 40x^{124} + 27x^{123} + 81x^{122} + 92x^{121} + \text{B1}x^{120} + 02x^{119} +$$
$$8\text{B}x^{118} + 87x^{117} + 7\text{F}x^{116} + \text{B0}x^{115} + 6\text{F}x^{114} + 53x^{113} + 08x^{112} + \text{CB}x^{111} +$$
$$03x^{110} + \text{B0}x^{109} + \text{DF}x^{108} + 1\text{F}x^{107} + \text{A7}x^{106} + \text{A2}x^{105} + \text{FE}x^{104} + 8\text{E}x^{103} +$$
$$\text{A8}x^{102} + \text{E1}x^{101} + 71x^{100} + \text{FF}x^{99} + 55x^{98} + 5\text{A}x^{97} + 1\text{D}x^{96} + 9\text{D}x^{95} +$$
$$\text{BF}x^{94} + \text{E8}x^{93} + \text{BA}x^{92} + 6\text{B}x^{91} + 72x^{90} + \text{E3}x^{89} + 04x^{88} + \text{D9}x^{87} +$$
$$38x^{86} + \text{D3}x^{85} + \text{B9}x^{84} + 16x^{83} + 52x^{82} + 18x^{81} + 19x^{80} + 3\text{E}x^{79} +$$
$$9\text{E}x^{78} + 03x^{77} + 56x^{76} + \text{A6}x^{75} + 71x^{74} + 03x^{73} + \text{E4}x^{72} + 86x^{71} +$$
$$\text{F5}x^{70} + \text{B0}x^{69} + 05x^{68} + \text{D1}x^{67} + 10x^{66} + \text{E2}x^{65} + \text{E5}x^{64} + \text{CB}x^{63} +$$
$$\text{B1}x^{62} + \text{F2}x^{61} + 8\text{E}x^{60} + \text{C7}x^{59} + 0\text{C}x^{58} + \text{A7}x^{57} + \text{BF}x^{56} + 46x^{55} +$$
$$0\text{B}x^{54} + 01x^{53} + \text{C5}x^{52} + \text{A3}x^{51} + 50x^{50} + 77x^{49} + \text{EA}x^{48} + 05x^{47} +$$
$$65x^{46} + 8\text{E}x^{45} + 89x^{44} + \text{D4}x^{43} + 6\text{D}x^{42} + \text{D3}x^{41} + 75x^{40} + 65x^{39} +$$
$$13x^{38} + 2\text{F}x^{37} + 86x^{36} + \text{AF}x^{35} + 7\text{C}x^{34} + 7\text{B}x^{33} + 85x^{32} + \text{C8}x^{31} +$$
$$\text{E8}x^{30} + 04x^{29} + 7\text{B}x^{28} + \text{CF}x^{27} + 2\text{F}x^{26} + 8\text{A}x^{25} + 9\text{A}x^{24} + 3\text{D}x^{23} +$$
$$\text{CF}x^{22} + 21x^{21} + 39x^{20} + \text{D9}x^{19} + 29x^{18} + 73x^{17} + \text{F6}x^{16} + 23x^{15} +$$
$$40x^{14} + 1\text{B}x^{13} + \text{B2}x^{12} + \text{C0}x^{11} + 6\text{D}x^{10} + 85x^{9} + 1\text{C}x^{8} + 8\text{A}x^{7} +$$
$$2\text{C}x^{6} + \text{BB}x^{5} + 90x^{4} + 1\text{E}x^{3} + 7\text{E}x^{2} + \text{F3}x^{1} + 52$$

Cryptanalysis of the Full HAVAL with 4 and 5 Passes

Hongbo Yu[1], Xiaoyun Wang[2,*], Aaram Yun[3], and Sangwoo Park[3]

[1] Shandong University, Jinan 250100, China
yhb@mail.sdu.edu.cn
[2] Shandong University and Tsinghua University, China
xywang@sdu.edu.cn
[3] National Security Research Institute,
161 Gajeong-dong, Yuseong-gu, Daejeon 305-350, Korea
{aaram, psw}@etri.re.kr

Abstract. HAVAL is a cryptographic hash function with variable digest size proposed by Zheng, Pieprzyk and Seberry in 1992. It has three variants, 3-, 4-, and 5-pass HAVAL. Previous results on HAVAL suggested only practical collision attacks for 3-pass HAVAL. In this paper, we present collision attacks for 4 and 5 pass HAVAL. For 4-pass HAVAL, we describe two practical attacks for finding 2-block collisions, one with 2^{43} computations and the other with 2^{36} computations. In addition, we show that collisions for 5-pass HAVAL can be found with about 2^{123} computations, which is the first attack more efficient than the birthday attack.

Keywords: Hash function, collision, differential path, message modification.

1 Introduction

The hash function HAVAL was proposed by Zheng, Pieprzyk and Seberry at Auscrypt '92 [11]. It has a similar structure as the well-known hash functions such as MD4 [3] and MD5 [4]. In Asiacrypt '03, Rompay et al. gave a collision attack for 3-pass HAVAL with complexity 2^{29} computations [1]. The fastest attack on 3-pass HAVAL was presented by X.Y.Wang et al. [5], and it can find a collision with time complexity less than 2^7 computations. In SCN 2004, Y.Yoshida et al. showed that the compression functions of full 4-pass and 5-pass HAVAL are not random and can be distinguished from a truly random function [2].

In this paper, we use the method of *modular differential* to analyze the full 4-pass and 5-pass HAVAL. This method was presented early in 1997 by X.Y.Wang [10], and formalized in Eurocrypt '05 [6,7]. This type of cryptanalysis is powerful to break the most prevailing hash functions such as MD4 [6], MD5 [7], SHA-0 [8] and SHA-1 [9].

---

* Supported by the National Natural Science Foundation of China (NSFC Grant No.90604036 and No.60525201) and 973 Project (No.2004CB318000).

M.J.B. Robshaw (Ed.): FSE 2006, LNCS 4047, pp. 89–110, 2006.

In this paper, we provide two practical attacks for 4-pass HAVAL, with 2^{43} and 2^{36} HAVAL computations, respectively. In addition, we give the first theoretical attack for 5-pass HAVAL with a complexity less than 2^{123} computations.

The rest of the paper is organized as follows: in Section 2, we give a brief description of HAVAL algorithm. In section 3, we introduce some basic conclusions and notations used in our paper. The attack details are described in Sections 4, 5, and 6. Section 7 concludes the paper.

2   Description of HAVAL

In this section we provide a brief description of HAVAL. Since the structure of 4-pass and 5-pass version of HAVAL are essentially the same, here we only give the description of 4-pass HAVAL. We use modified and simplified notations than those in the original paper [11], and omit all non-relevant parts.

Although HAVAL supports digest sizes of 128, 160, 192, 224, and 256 bits, the main algorithm computes 256-bit digests and the other sizes are supported by post-processing the 256-bit hash value. Therefore for our purposes we may consider HAVAL as a hash function with output size of 256 bits.

HAVAL is a Merkle-Damgård hash function, which uses a compression function to digest messages. The compression function H of HAVAL takes a 1024-bit message and a 256-bit initial value as input, and produces 256-bit hash value as output. The message is represented as 32 message words, $m_0, m_1, \ldots, m_{31}$, each consisting of 32 bits. The 256-bit initial value (or chaining value) is represented as the following 8 words, $a_0, b_0, \ldots,$ and h_0:

$$a_0 = 0x243f6a88, \; b_0 = 0x85a308d3, \; c_0 = 0x13198a2e, \; d_0 = 0x03707344,$$
$$e_0 = 0xa4093822, \; f_0 = 0x299f31d0, \; g_0 = 0x082efa98, \; h_0 = 0xec4e6c89.$$

4-pass HAVAL uses the following four boolean functions:

Pass	Function
1	$f_1(x_6, x_5, x_4, x_3, x_2, x_1, x_0) = x_0 \oplus x_0 x_3 \oplus x_1 x_3 \oplus x_2 x_4 \oplus x_5 x_6$
2	$f_2(x_6, x_5, x_4, x_3, x_2, x_1, x_0) = x_1 x_3 \oplus x_4 \oplus x_1 x_4 \oplus x_0 x_5 \oplus x_2 x_5 \oplus x_1 x_2 x_5 \oplus$ $x_1 x_6 \oplus x_0 x_1 x_6 \oplus x_2 x_6$
3	$f_3(x_6, x_5, x_4, x_3, x_2, x_1, x_0) = x_2 x_3 \oplus x_0 x_4 \oplus x_5 \oplus x_1 x_6 \oplus x_0 x_2 x_6 \oplus x_5 x_6$
4	$f_4(x_6, x_5, x_4, x_3, x_2, x_1, x_0) = x_0 x_1 \oplus x_3 \oplus x_0 x_3 \oplus x_0 x_4 \oplus x_0 x_2 x_4 \oplus$ $x_0 x_5 \oplus x_1 x_2 x_5 \oplus x_4 x_5 \oplus x_0 x_6 \oplus x_2 x_6 \oplus x_5 x_6 \oplus x_0 x_5 x_6$

In HAVAL, the boolean functions are applied bitwisely to 32-bit input variables to produce 32-bit output values.

4-Pass HAVAL Compression Function. Given a 1024-bit message block $M = (m_0, m_1, ..., m_{31})$, the compressing process is as follows:

1. Let $(aa, bb, cc, dd, ee, ff, gg, hh)$ be the input of compressing process for M. Initialize chaining variables (a, b, c, d, e, f, g, h) as $(aa, bb, cc, dd, ee, ff, gg, hh)$.

2. Perform the following 128 steps:

   For $j=1, 2, 3,$ and 4

        For $i = 0$ to 31

            $p := f_j(g, f, e, d, c, b, a)$

            $r := (p \gg 7) + (h \gg 11) + m_{ord(j,i)} + k_{j,i}$

            $h := g, g := f, f := e, e := d, d := c, c := b, b := a, a := r$

   The operation in each step employs a constant $k_{j,i}$ (See ref.[11]). $\gg s$ represents the s bit rotation to the right. $+$ denotes addition modulo 2^{32}. The orders of message words in each pass can refer to [11].

3. Add a, b, c, d, e, f, g, h respectively to the input value, i.e.,

   $aa := a + aa, \ bb := b + bb, \ ... \ ..., \ hh := h + hh$

4. $H(M) = hh\|gg\|ff\|ee\|dd\|cc\|bb\|aa$, where $\|$ denotes the bit concatenation.

3 Some Basic Conclusions and Notations

In this section, we give several properties of the four boolean functions f_1, f_2, f_3, f_4.

Proposition. Let $y_1 = f_1(x_6, x_5, x_4, x_3, x_2, x_1, x_0)$, and $y_{1,x_i} = f_1(x_6, ..., x_{i+1}, \neg x_i, ..., x_0)$, where $\neg x_i$ is the complement of the bit x_i. Then

1. $y_1 = y_{1,x_0} \iff x_3 = 1$.

   $y_1 = x_0$ and $y_{1,x_0} = \neg x_0 \iff x_3 = 0$ and $x_1 x_3 \oplus x_5 x_6 \oplus x_2 x_4 = 0$.

   $y_1 = \neg x_0$ and $y_{1,x_0} = x_0 \iff x_3 = 0$ and $x_1 x_3 \oplus x_5 x_6 \oplus x_2 x_4 = 1$.

2. $y_1 = y_{1,x_1} \iff x_3 = 0$.

   $y_1 = x_1$ and $y_{1,x_1} = \neg x_1 \iff x_3 = 1$ and $x_5 x_6 \oplus x_2 x_4 \oplus x_0 x_3 \oplus x_0 = 0$.

   $y_1 = \neg x_1$ and $y_{1,x_1} = x_1 \iff x_3 = 1$ and $x_5 x_6 \oplus x_2 x_4 \oplus x_0 x_3 \oplus x_0 = 1$.

3. $y_1 = y_{1,x_2} \iff x_4 = 0$.

   $y_1 = x_2$ and $y_{1,x_2} = \neg x_2 \iff x_4 = 1$ and $x_1 x_3 \oplus x_5 x_6 \oplus x_0 x_3 \oplus x_0 = 0$.

   $y_1 = \neg x_2$ and $y_{1,x_2} = x_2 \iff x_4 = 1$ and $x_1 x_3 \oplus x_5 x_6 \oplus x_0 x_3 \oplus x_0 = 1$.

4. $y_1 = y_{1,x_3} \iff x_0 \oplus x_1 = 0$.

   $y_1 = x_3$ and $y_{1,x_3} = \neg x_3 \iff x_0 \oplus x_1 = 1$ and $x_5 x_6 \oplus x_2 x_4 \oplus x_0 = 0$.

   $y_1 = \neg x_3$ and $y_{1,x_3} = x_3 \iff x_0 \oplus x_1 = 1$ and $x_5 x_6 \oplus x_2 x_4 \oplus x_0 = 1$.

5. $y_1 = y_{1,x_4} \iff x_2 = 0$.

   $y_1 = x_4$ and $y_{1,x_4} = \neg x_4 \iff x_2 = 1$ and $x_1 x_3 \oplus x_5 x_6 \oplus x_0 x_3 \oplus x_0 = 0$.

   $y_1 = \neg x_4$ and $y_{1,x_4} = x_4 \iff x_2 = 1$ and $x_1 x_3 \oplus x_5 x_6 \oplus x_0 x_3 \oplus x_0 = 1$.

6. $y_1 = y_{1,x_5} \iff x_6 = 0$.

   $y_1 = x_5$ and $y_1 = \neg y_{1,x_5} \iff x_6 = 1$ and $x_1 x_3 \oplus x_2 x_4 \oplus x_0 x_3 \oplus x_0 = 0$.

   $y_1 = \neg x_5$ and $y_1 = y_{1,x_5} \iff x_6 = 1$ and $x_1 x_3 \oplus x_2 x_4 \oplus x_0 x_3 \oplus x_0 = 1$.

7. $y_1 = y_{1,x_6} \iff x_5 = 0$.

   $y_1 = x_6$ and $y_1 = \neg y_{1,x_6} \iff x_5 = 1$ and $x_3 x_1 \oplus x_2 x_4 \oplus x_0 x_3 \oplus x_0 = 0$.

   $y_1 = \neg x_6$ and $y_1 = y_{1,x_6} \iff x_5 = 1$ and $x_3 x_1 \oplus x_2 x_4 \oplus x_0 x_3 \oplus x_0 = 1$.

Here, $x_i \in \{0, 1\}$ $(0 \leq i \leq 6)$.

It is easy to deduce the similar properties of the other three functions f_2, f_3 and f_4. We omit them because of the limited pages.

Notations. In order to describe our attack conveniently, we define some notations.

1. $M = (m_i)_{i<32}$ and $M' = (m'_i)_{i<32}$ denote a collection of 32 words respectively.
2. $\Delta m_i = m'_i - m_i$, $\Delta a_i = a'_i - a_i$, ..., $\Delta h_i = h'_i - h_i$, $\Delta p_i = p'_i - p_i$ denote the modular differences of two variables.
3. a_i, b_i, c_i, d_i, e_i, f_i, g_i, h_i and a'_i, b'_i, c'_i, d'_i, e'_i, f'_i, g'_i, h'_i denote the chaining variables after the i-th step corresponding to the message blocks M and M' respectively. According to the HAVAL algorithm, we know that $b_i = a_{i-1}$, $c_i = a_{i-2}$, $d_i = a_{i-3}$, $e_i = a_{i-4}$, $f_i = a_{i-5}$, $g_i = a_{i-6}$, $h_i = a_{i-7}$.
4. $x_{i,j}$ denotes the j-th bit of 32-bit word x_i. For example, $a_{i,j}$ is the j-th bit of a_i.
5. $x_i[j]$ is the value obtained by modifying the jth bit of x_i from 0 to 1 (hence this notation implicitly states that $x_{i,j} = 0$. Similarly, $x_i[-j]$ is the value obtained by modifying the jth bit of x_i from 1 to 0.
6. $x_i[\pm j_1, \pm j_2, \ldots, \pm j_k]$ is shorthand for $x_i[\pm j_1][\pm j_2]\ldots[\pm j_k]$, i.e., modifying x_i at bit positions $j_1, \ldots, j_k$ according to the $\pm$ signs.

4   The Attack Against 4-Pass HAVAL with One Message Word Difference

Our collision attack can be divided into three phases: 1. Choose a appropriate message difference and deduce the differential path according to the specified message difference. 2. Determine the corresponding chaining variable conditions. 3. Fulfill the message modification to guarantee that a portion of the conditions hold.

We have obtained two collision attacks for 4-pass HAVAL. Both methods find two-block collisions, i.e., collision pairs consisting of two 2048-bit messages $M_0 \| M_1$ and $M'_0 \| M'_1$. Since both attacks use essentially the same methodology, we will briefly give an outline for the first attack in this section, and then give more detailed exposition for the second attack in the next section.

In the first method, message differences are given only on the message word m_5 with difference 2^{31}. That is, for both blocks M_0 and M_1, we have

$$\Delta m_i = m'_i - m_i = \begin{cases} 2^{31} & \text{if } i = 5, \\ 0 & \text{otherwise.} \end{cases}$$

In the first block, the difference introduced at step 6 by m_5 is propagated until step 33, where m_5 is again used and the first inner collision is produced. The message word m_5 is again used at step 95, near the end of the pass 3. From step 95 to step 122, the differences are propagated so that at each step only one chaining variable difference is active. At step 123, the message word m_5 is again used and from then two chaining variables are active at each step, ending up as a near-collision with two active variables.

In the second block, the initial differences produced by the first block, as well as the one introduced by m_5 at step 6 is again eliminated at step 33. From step 95 to the end of the second block, the differences propagate in a similar fashion as in the first block, except that all the signs are reversed. Therefore at the end of the second block the output differences cancel the input difference of the second block, producing a two-block collision. The differential paths are given in Table 4 and Table 5. Due to space constraint, we will omit the tables for sufficient conditions for the differential paths.

Using the message modification technique, explained in Section 5.3, we may satisfy all the conditions in the first pass with probability 1. Therefore the probability for the third and fourth passes is the success probability of the whole algorithm, which can be estimated to be greater than 2^{-43}. In Table 1 we provide an example of a collision pair we found.

Note that the message word m_5 appears at step 33, the beginning of the second pass, and it again appears at step 95, almost at the end of the third pass, which gives a long stretch of steps without differences.

Table 1. A collision pair for 4-pass HAVAL. H is the common hash value with little-endian and no message padding.

M_0	00000000 00000000 00000000 00000080 00000000 00000080 00000080 00000000
	00000000 00000000 00002000 0000e0ff 0000e0ff 00000000 0080f3ff 00c0ecff
	0040ecff 0040ffff 0080feff 0080feff 0080ffff 00fcffff 00000000 00fcffff
	00fcffff 00fcffff 00000000 00fcffff 00000000 00000000 40070000 d9dc1fdc
M_1	0000e87f 0000f8ff 0020f0ff 000100ff 00ff0174 000ff0f2 c001e484 00daf1fb
	c01706fa 80eff3f9 00d6f1ff 80f7ff1f f7fffd40 00000000 00200028 0000003e
	00002088 000020a0 00007ef9 00000008 00c045ba 00003bc0 003cfcfc 007c1f03
	00bc81fe 00c4ddfb 003cfeff 00000000 00000000 00000200 3f000000 a095d965
M_0	00000000 00000000 00000000 00000080 00000000 00000000 00000080 00000000
	00000000 00000000 00002000 0000e0ff 0000e0ff 00000000 0080f3ff 00c0ecff
	0040ecff 0040ffff 0080feff 0080feff 0080ffff 00fcffff 00000000 00fcffff
	00fcffff 00fcffff 00000000 00fcffff 00000000 00000000 40070000 d9dc1fdc
M_1	0000e87f 0000f8ff 0020f0ff 000100ff 00ff0174 000ff072 c001e484 00daf1fb
	c01706fa 80eff3f9 00d6f1ff 80f7ff1f f7fffd40 00000000 00200028 0000003e
	00002088 000020a0 00007ef9 00000008 00c045ba 00003bc0 003cfcfc 007c1f03
	00bc81fe 00c4ddfb 003cfeff 00000000 00000000 00000200 3f000000 a095d965
H	481a1bf8 04defc01 a62b7444 63979a59 93e9b12d b20d82bd 7e626c25 22db74ca

5   The Attack against 4-Pass HAVAL with Two Message Word Differences

5.1   Choosing the Differential Path

For this second attack, we have found another differential path using differences at message words m_8 and m_{16}. In the first block, we will use $\Delta m_8 = 2^{13}$ and $\Delta m_{16} = -2^2$, and in the second block $\Delta m_8 = -2^{13}$ and $\Delta m_{16} = 2^2$.

The differential path for the first block consists of two inner collisions in steps 9–48 and steps 71–79, and a near-collision (steps 117–128). The path for the second block has similar structure.

The main difference between this attack and the attack described in Section 4 is that, in the current attack we also use the advanced message modification technique, which will be explained in Subsection 5.3. This enables us to correct more conditions in the second pass. Therefore here we can afford to have our first inner collision to stretch further into the second pass by using two message word differences. Hence we select ΔM_0 and ΔM_1 to ensure that in this path the differences of 3–4 rounds happen with high probability.

5.2   Deriving the Sufficient Conditions for Collision Path

In this section, we derive a set of sufficient conditions, summarized in Table 8, which ensures the collision path to hold. We give an example explaining how to deduce the set of sufficient conditions.

In step 9 of the differential path presented in Table 6, the message difference $\Delta m_8 = 2^{13}$ produces the changed variable $a_9[-14, 15]$. The difference $a_9[-14]$ doesn't produce any more bit differences between step 10 and step 16, and the difference $a_9[15]$ is used to produce the difference $a_{15}[-8, -9, -10, 11]$ in step 15.

1. In step 9, $a_9' = a_9[-14, 15]$ iff $a_{9,14} = 1$ and $a_{9,15} = 0$.
2. In step 10, $(a_9[-14, 15], a_8, a_7, a_6, a_5, a_4, a_3, a_2)$
    $\rightarrow (a_{10}, a_9[-14, 15], a_8, a_7, a_6, a_5, a_4, a_3)$.
    From 1 of Proposition , $a_{10}' = a_{10}$ iff $a_{6,14} = 1$ and $a_{6,15} = 1$.
3. In step 11, $(a_{10}, a_9[-14, 15], a_8, a_7, a_6, a_5, a_4, a_3)$
    $\rightarrow (a_{11}, a_{10}, a_9[-14, 15], a_8, a_7, a_6, a_5, a_4)$.
    From 2 of Proposition , $a_{11}' = a_{11}$ iff $a_{7,14} = 0$ and $a_{7,15} = 0$.
4. In step 12, $(a_{11}, a_{10}, a_9[-14, 15], a_8, a_7, a_6, a_5, a_4)$
    $\rightarrow (a_{12}, a_{11}, a_{10}, a_9[-14, 15], a_8, a_7, a_6, a_5)$.
    From 3 of Proposition , $a_{12}' = a_{12}$ iff $a_{7,14} = 0$ and $a_{7,15} = 0$.
5. In step 13, $(a_{12}, a_{11}, a_{10}, a_9[-14, 15], a_8, a_7, a_6, a_5)$
    $\rightarrow (a_{13}, a_{12}, a_{11}, a_{10}, a_9[-14, 15], a_8, a_7, a_6)$.
    From 4 of Proposition , $a_{13}' = a_{13}$ iff $a_{12,14} \oplus a_{11,14} = 0$ and $a_{12,15} \oplus a_{11,15} = 0$.
6. In step 14, $(a_{13}, a_{12}, a_{11}, a_{10}, a_9[-14, 15], a_8, a_7, a_6)$
    $\rightarrow (a_{14}, a_{13}, a_{12}, a_{11}, a_{10}, a_9[-14, 15], a_8, a_7)$.
    From 5 of Proposition , $a_{14}' = a_{14}$ iff $a_{11,14} = 0$ and $a_{11,15} = 0$.

7. In step 15, $(a_{14}, a_{13}, a_{12}, a_{11}, a_{10}, a_9[-14, 15], a_8, a_7)$
   $\rightarrow (a_{15}[-8, -9, -10, 11], a_{14}, a_{13}, a_{12}, a_{11}, a_{10}, a_9[-14, 15], a_8)$.
   From 6 of Proposition , $a'_{15} = a_{15}[-8, -9, -10, 11]$ iff
   $a_{15,8} = 1$, $a_{15,9} = 1$, $a_{15,10} = 1$, $a_{15,11} = 0$, $a_{8,14} = 0$, $a_{8,15} = 1$ and
   $a_{13,15}a_{11,15} \oplus a_{12,15}a_{10,15} \oplus a_{14,15}a_{11,15} \oplus a_{14,15} = 0$.
8. In step 16, $(a_{15}[-8, -9, -10, 11], a_{14}, a_{13}, a_{12}, a_{11}, a_{10}, a_9[-14, 15], a_8)$
   $\rightarrow (a_{16}[4], a_{15}[-8, -9, -10, 11], a_{14}, a_{13}, a_{12}, a_{11}, a_{10}, a_9[-14, 15])$.
   From 6 of Proposition , $a'_{16} = a_{16}[4]$ iff $a_{16,4} = 0$, $a_{10,14} = 0$, $a_{10,15} = 0$,
   $a_{12,8} = 1$, $a_{12,9} = 1$, $a_{12,10} = 1$, $a_{12,11} = 0$ and
   $a_{14,11}a_{12,11} \oplus a_{10,11}a_{9,11} \oplus a_{13,11}a_{11,11} = 0$.

There are 25 equations in steps 9–16. We can simplify the above 25 conditions and classify them into two types:

- Conditions with exact form

  After a little simplification, there are 24 conditions with exact form, i.e., of form $a_i = 0$ or $a_i = 1$:

  $a_{16,4} = 0$, $a_{6,14} = 1$, $a_{6,15} = 1$, $a_{7,14} = 0$, $a_{7,15} = 0$, $a_{8,14} = 0$, $a_{8,15} = 1$,
  $a_{9,14} = 1$, $a_{9,15} = 0$, $a_{10,14} = 0$, $a_{10,15} = 0$, $a_{11,14} = 0$, $a_{11,15} = 0$, $a_{12,8} = 1$,
  $a_{12,9} = 1$, $a_{12,10} = 1$, $a_{12,11} = 0$, $a_{12,14} = 0$, $a_{12,15} = 0$, $a_{14,15} = 0$, $a_{15,8} = 1$,
  $a_{15,9} = 1$, $a_{15,10} = 1$, $a_{15,11} = 0$

  $a_{12,14} = 0$ is derived from two equations: $a_{11,14} \oplus a_{12,14} = 0$ and $a_{11,14} = 0$.
  $a_{12,15} = 0$ is derived from two equations: $a_{11,15} \oplus a_{12,15} = 0$ and $a_{11,15} = 0$.
  $a_{14,15} = 0$ is deduced by the following three equations: $a_{13,15}a_{11,15} \oplus a_{12,15}$
  $a_{10,15} \oplus a_{14,15}a_{11,15} \oplus a_{14,15} = 0$, $a_{11,15} = 0$, and $a_{12,15} = 0$.

- Conditions expressed as multi-variable equations:

  There is only one condition which is expressed as a multi-variable equation:

$$a_{10,11}a_{9,11} \oplus a_{13,11}a_{11,11} = 0 \qquad (1)$$

Each equation with the first form holds with probability $\frac{1}{2}$, and the equation (1) holds with probability $\frac{5}{8}$. So the total probability for the 9-16 step differential is $\frac{5}{2^{27}}$.

Similarly, we can determine all the other conditions which result in the differential paths in Table 6 and Table 7. Summing up all these sufficient conditions, we obtain Table 8 and Table 9.

5.3   Message Modification

We modify M_0 and M_1 so that almost all conditions in Table 8 and 9 hold. The modification include the basic modification and advanced modification techniques.

Basic Modification. The basic modification is a simple message modification used to ensure all the conditions in the first round (step 1-32) hold. For example, given a message $M_0 = (m_i)_{i<32}$, we compute a_6 and correct a_6 to satisfy the two conditions in Table 8 by setting $a_6 = a_6 \vee 0x6000$, then update m_5 as:

$$m_5 = a_6 - (f(b_0, a_0, a_1, a_2, a_3, a_4, a_5) \ggg 7) - (c_0 \ggg 11)$$

It is easy to correct all the conditions from step 1 to step 32 of the differential paths in Table 8 and Table 9.

Advanced Message Modification. We correct some more conditions in round 2 by the advanced modification. If the condition on $a_{i,j}$ is wrong, we change the j-th bit of the corresponding message m and some other message words which produce a partial collision in the first round. A sample for correcting $a_{34,4}$ is given in Table 2. We define this kind of corrected condition as *rectifiable condition*.

Table 2. The message modification for correcting $a_{34,4}$

step	m_i	the modified m_i	new variable value	conditions
8	m_7	$m_7 \leftarrow m_7 + 2^{10}$	$a_8[11], a_7, a_6, a_5, a_4, a_3, a_2, a_1$	$a_{8,11} = 0$
9	m_8		$a_9, a_8[11], a_7, a_6, a_5, a_4, a_3, a_2$	$a_{5,11} = 1$
10	m_9		$a_{10}, a_9, a_8[11], a_7, a_6, a_5, a_4, a_3$	$a_{6,11} = 0$
11	m_{10}		$a_{11}, a_{10}, a_9, a_8[11], a_7, a_6, a_5, a_4$	$a_{6,11} = 0$
12	m_{11}		$a_{12}, a_{11}, a_{10}, a_9, a_8[11], a_7, a_6, a_5$	$a_{11,11} \oplus a_{10,11} = 0$
13	m_{12}		$a_{13}, a_{12}, a_{11}, a_{10}, a_9, a_8[11], a_7, a_6$	$a_{10,11} = 0$
14	m_{13}		$a_{14}, a_{13}, a_{12}, a_{11}, a_{10}, a_9, a_8[11], a_7$	$a_{7,11} = 0$
15	m_{14}	$m_{14} \leftarrow m_{14} - 2^3$	$a_{15}, a_{14}, a_{13}, a_{12}, a_{11}, a_{10}, a_9, a_8[11]$	$a_{9,11} = 1,$ $a_{13,11}a_{11,11} \oplus a_{12,11}a_{10,11}$ $\oplus a_{14,11}a_{11,11} \oplus a_{14,11} = 0$
16	m_{15}	$m_{15} \leftarrow m_{15} + 2^{31}$	$a_{16}, a_{15}, a_{14}, a_{13}, a_{12}, a_{11}, a_{10}, a_9$	

In the first block, the rectifiable conditions are as follows:

$a_{34,4}, a_{34,14}, a_{35,4}, a_{35,14}, a_{35,25}, a_{36,14}, a_{36,25}, a_{37,25}, a_{37,14}, a_{37,4}, a_{38,25}, a_{39,25}, a_{40,25}.$

In the second block, the rectifiable conditions are as follows:

$a_{34,4}, a_{34,14}, a_{35,4}, a_{35,14}, a_{35,25}, a_{36,14}, a_{36,25}, a_{37,4}, a_{37,14}, a_{37,25}, a_{38,25}, a_{40,25}.$

By the two types of modification, there are 8 remaining conditions in Table 8, and 9 conditions in Table 9 that need to be satisfied.

5.4   Complexity Evaluation

In order to calculate the attack complexity, we need to estimate the probabilities of two truncated differentials, one is from step 1 to step 64, the other is from step 65 to step 128. After message modification, we know that the 1-64 step differential of the first block holds with probability 2^{-8} and that of the second block with probability 2^{-9}. What is left is to calculate the probability that all the equations in rounds 3-4 hold concurrently for both blocks.

Complexity Evaluation for the First Block

There are total 22 equations in rounds 3-4 for the first block. In order to deduce their probability, we divide these equations into three equation systems.

Equation System 1

$$
\begin{cases}
0 = a_{71,14} \\
0 = a_{69,14}a_{65,14} \oplus a_{67,14} \\
0 = a_{66,14} \\
0 = a_{73,14}a_{67,14} \oplus a_{70,14} \\
0 = a_{72,14} \\
0 = a_{75,14} \\
1 = a_{70,14} \\
0 = a_{77,14}a_{75,14} \oplus a_{76,14} \oplus a_{72,14}
\end{cases}
$$

The equation system 1 ensures the differential characteristics from step 71 to 79 in Table 6 hold.

There are two solutions for the 11 variables, so the equation system 1 holds with probability 2^{-10}.

Equation System 2

$$
\begin{cases}
0 = a_{117,14} \\
0 = a_{115,14}a_{113,14} \oplus a_{112,14}a_{111,14} \oplus a_{116,14} \oplus a_{112,14} \oplus a_{113,14} \\
\qquad \oplus a_{111,14} \oplus a_{114,14} \\
0 = a_{116,14}a_{113,14} \oplus a_{118,14} \\
0 = a_{118,14}a_{114,14} \oplus a_{119,14}a_{115,14} \oplus a_{113,14} \\
1 = a_{120,14} \\
0 = a_{121,14}a_{119,14} \oplus a_{116,14} \oplus a_{121,14} \\
0 = a_{121,14}a_{120,14} \oplus a_{122,14}a_{116,14} \oplus a_{122,14} \oplus a_{118,14} \oplus a_{116,14} \\
\qquad a_{123,14}a_{118,14} \oplus a_{118,14} \oplus a_{123,14} \oplus a_{121,14}
\end{cases}
$$

The equation system 2 guarantees the differential characteristics from step 117 to 124 hold.

It is easy to show that there are 32 solutions for 13 variables which imply that the equation system 2 holds with probability 2^{-8}.

Equation System 3

$$
\begin{cases}
0 = a_{125,3} \\
0 = a_{123,3}a_{121,3} \oplus a_{120,3}a_{119,3} \oplus a_{124,3} \oplus a_{120,3} \oplus a_{121,3} \oplus a_{119,3} \oplus a_{122,3} \\
0 = a_{124,3}a_{121,3} \oplus a_{126,3} \\
1 = a_{127,3} \\
0 = a_{126,3}a_{122,3} \oplus a_{121,3} \\
0 = a_{122,3}a_{121,3} \oplus a_{126,3} \oplus a_{122,3} \oplus a_{123,3} \oplus a_{121,3} \oplus a_{124,3}
\end{cases}
$$

The equation system 3 ensures the differential characteristics in steps 125-128 hold.

Similarly, the equation system 3 has 7 solutions with 9 variables which the probability is $\frac{7}{2^9}$.

Additional Conditions for Near Collision

Our attack is to find collisions with two blocks, so the output difference for the first block should be $(0, -2^2, 0, 2^2, 0, 0, 0, 0)$ with no bit carries which results in two other conditions on outputs bb_0 and dd_0. For the second block, the differential path needs two conditions in IVs which come from two conditions on aa_0 and cc_0 in the first block. So the additional four conditions for the first block are as follows:

$$aa_{0,3} = 0, \quad bb_{0,3} = 1, \quad cc_{0,3} = 0, \quad dd_{0,3} = 0$$

where

$$aa_0 = a_0 + a_{128}, \quad bb_0 = b_0 + a_{127}, \quad cc_0 = c_0 + a_{126}, \quad dd_0 = d_0 + a_{125}$$

It is noted that the four input words in the second block aa_0, bb_0, cc_0, dd_0 are also the output words in the first block.

Considering 8 conditions left after the message modification, the probability for the differential path in the first block is about

$$\frac{1}{2^8} \cdot \frac{1}{2^{10}} \cdot \frac{1}{2^8} \cdot \frac{7}{2^9} \cdot \frac{1}{2^4} \approx \frac{1}{2^{36}}.$$

Complexity Evaluation for the Second Block

For the second block, given a message M_1, after the modifications, M_1 and M_1' generate the differential in Table 7 with the probability

$$\frac{1}{2^9} \cdot \frac{1}{2^{10}} \cdot \frac{1}{2^8} \cdot \frac{7}{2^9} \approx \frac{1}{2^{33}}$$

The two differential paths corresponding to two blocks consist of a collision for 4 pass HAVAL, and the time complexity for the attack is about 2^{36} HAVAL computations.

5.5   Collision Search Algorithm

Summarizing the above technique details, we give an overview of the collision search algorithm.

1. Searching the first block M_0.

   (a) Choose a 1024-bit message $M_0 = (x_i)_{i<31}$ randomly and modify its first 30 words by the basic modification technique such that the conditions in steps 6-30 of Table 8 are satisfied.

   (b) Modify x_{30} and x_{31} to correct the conditions in steps 31-32 of Table 8 by the basic modification technique.

   (c) Apply the advanced modification to make the 13 rectifiable conditions to hold in round 2.

   (d) Compute $H(M_0) = (aa_0, bb_0, cc_0, dd_0, ee_0, ff_0, gg_0, hh_0)$ and
   $$H(M_0') = (aa_0', bb_0', cc_0', dd_0', ee_0', ff_0', gg_0', hh_0').$$
   If $H(M_0') - H(M_0) = \Delta H_1$, $aa_{0,3} = 0$, $bb_{0,3} = 1$, $cc_{0,3} = 0$ and $dd_{0,3} = 0$ hold , output M_0 and M_0'. Otherwise, select another x_{30} and x_{31} randomly and go to step (b).

2. Searching the second block M_1 by the similar method as M_0.

Using our search algorithm, it takes roughly 8 hours to find a 4-pass collision on a standard notebook PC, and we give a collision example in Table 3.

Table 3. A collision for 4-pass HAVAL. H is the common hash value with little-endian and no message padding.

M_0	1c6574fd	b56fff65	0feff335	d7404793	095e0c30	dcc386ab	86e85ecd	eb730b21
	0ba01f27	8e3e84e2	39e35d80	afdf0ea8	23a57ffb	903fbb44	24e03671	d63ffe68
	375e43b1	2dd81090	f408a2c5	ecc32b28	43f17d20	062e68d3	b9d1bd80	f0572c76
	e3d648b1	184ebe01	92def272	f43fe3d4	6bde4810	fc5666f3	17eec0a9	24b1dda8
M_1	7a329389	28a58673	3b7f4890	6cbb79b7	c33fac13	65ad0193	60d345c4	fa126a11
	476dcbe0	5d582432	6f782165	e8875939	dc262382	ea5d1608	23893c79	d396a5c5
	ff8d6cfb	73d43ab1	ac0b2882	a4642004	69ac7042	1cec975e	a0c5a43a	f7fa309a
	661e6061	aad0c8f0	684e80da	d8540f60	960f8720	257a61c5	87eb3f8c	98c490a3
M_0'	1c6574fd	b56fff65	0feff335	d7404793	095e0c30	dcc386ab	86e85ecd	eb730b21
	0ba03f27	8e3e84e2	39e35d80	afdf0ea8	23a57ffb	903fbb44	24e03671	d63ffe68
	375e43ad	2dd81090	f408a2c5	ecc32b28	43f17d20	062e68d3	b9d1bd80	f0572c76
	e3d648b1	184ebe01	92def272	f43fe3d4	6bde4810	fc5666f3	17eec0a9	24b1dda8
M_1'	7a329389	28a58673	3b7f4890	6cbb79b7	c33fac13	65ad0193	60d345c4	fa126a11
	476dabe0	5d582432	6f782165	e8875939	dc262382	ea5d1608	23893c79	d396a5c5
	ff8d6cff	73d43ab1	ac0b2882	a4642004	69ac7042	1cec975e	a0c5a43a	f7fa309a
	661e6061	aad0c8f0	684e80da	d8540f60	960f8720	257a61c5	87eb3f8c	98c490a3
H	9dcc0bd8	009a1246	4e0b128c	1193ec10	86ddc85e	a90ea714	8c95871c	946cabf1

6   The Attack Against 5-Pass HAVAL

We adopt the similar notations for the description of 5-pass HAVAL and its details can refer to [11].

A one-block collision for 5-pass HAVAL is found with probability higher than the birthday attack. Similar to section 3, it's easy to deduce the properties of the five round functions. We choose a message difference $\Delta M = (\Delta m_i)_{i<32}$ with $\Delta m_i = 0, i \neq 8$ and $\Delta m_8 = -1$.

The collision differential path is given in Table 10 and 11. A set of sufficient conditions for the collision path are listed in Table 12 and 13. Given any 1024-bit message M, after the message modification, M and M' produce a partial collision from step 9 to step 71 with probability higher than 2^{-40}. Utilizing the same method as in Section 5.4, it is easy to prove that the second partial collision from step 117 to step 142 holds with probability 2^{-83}. So M and M' consist of a collision with probability about 2^{-123}, and the resulting attack is faster than the birthday attack.

7   Conclusion

In this paper, we describe two practical attacks on 4-pass HAVAL with probability 2^{-43} and 2^{-36} respectively, and also give a theoretical attack on 5-pass HAVAL which is faster than birthday attack.

Acknowledgement. We would like to thank Orr Dunkelman for his valuable comments and suggestions for this paper.

References

1. B. V. Rompay, A. Biryukov, B. Preneel, and J. Vandewalle.   Cryptanalysis of 3-Pass HAVAL, Asiacrypt 2003, LNCS 2894, pp. 228–245, 2003.
2. H. Yoshida, A. Biryukov, C. D. Canniere, J. Lano, and B. Preneel. Non-randomness of the Full 4 and 5-Pass HAVAL, SCN 2004, LNCS 3352, pp. 324–336, 2005.
3. R. L. Rivest. The MD4 Message Digest Algorithm, Crypto '90, LNCS 537, pp. 303–311, 1991.
4. R. L. Rivest. The MD5 Message-Digest Algorithm, Request for Comments(RFC 1320), Internet Activities Board, Internet Privacy Task Force, 1992.
5. X. Y. Wang, D. Feng, and X. Yu. An attack on Hash Function HAVAL-128. Science in China Ser. F Information Sciences, Vol. 48, No. 5, pp. 545–556, 2005.
6. X. Y. Wang, X. J. Lai, D. Feng, H. Chen, and X. Yu. Cryptanalysis for Hash Functions MD4 and RIPEMD, Eurocrypt '05, LNCS 3494, pp. 1–18, 2005.
7. X. Y. Wang and H. B. Yu. How to Break MD5 and Other Hash Functions, Eurocrypt '05, LNCS 3494, pp. 19–35, 2005.
8. X. Y. Wang, H. B. Yu, and Y. L. Yin. Efficient Collision Search Attacks on SHA-0, Crypto '05, LNCS 3621, pp. 1–16, 2005.
9. X. Y. Wang, Y. L. Yin, and H. B. Yu. Finding collisions on the Full SHA-1, Crypto '05, LNCS 3621, pp. 17–36, 2005.

10. X. Y. Wang. The Collision attack on SHA-0, in Chinese, to appear on www. infosec.sdu.edu.cn, 1997.

11. Y. Zheng, J. Pieprzyk and J. Seberry. HAVAL — A One-way Hashing Algorithm with Variable Length of Output, Auscrypt '92, LNCS 718, pp. 83–104, 1993.

Appendix: Tables

Table 4. A differential path for the first block of 4-pass HAVAL, for 2^{43} attack. Here $m'_5 = m_5 + 2^{31}$.

Step i	m'_{i-1}	Δa_i	Outputs for M'_0
6	m'_5	2^{31}	$a_6[32], a_5, a_4, a_3, a_2, a_1, a_0, a_{-1}$
7	m_6		$a_7, a_6[32], a_5, a_4, a_3, a_2, a_1, a_0$
...	...	...	...
13	m_{12}		$a_{13}, a_{12}, a_{11}, a_{10}, a_9, a_8, a_7, a_6[32]$
14	m_{13}	2^{20}	$a_{14}[-21, 22], a_{13}, a_{12}, a_{11}, a_{10}, a_9, a_8, a_7$
15	m_{14}		$a_{15}, a_{14}[-21, 22], a_{13}, a_{12}, a_{11}, a_{10}, a_9, a_8$
16	m_{15}		$a_{16}, a_{15}, a_{14}[-21, 22], a_{13}, a_{12}, a_{11}, a_{10}, a_9$
17	m_{16}		$a_{17}, a_{16}, a_{15}, a_{14}[-21, 22], a_{13}, a_{12}, a_{11}, a_{10}$
18	m_{17}	-2^{14}	$a_{18}[15, 16, 17, -18], a_{17}, a_{16}, a_{15}, a_{14}[-21, 22], a_{13}, a_{12}, a_{11}$
19	m_{18}		$a_{19}, a_{18}[15, 16, 17, -18], a_{17}, a_{16}, a_{15}, a_{14}[-21, 22], a_{13}, a_{12}$
...	...	...	...
24	m_{23}		$a_{24}, a_{23}, a_{22}, a_{21}, a_{20}, a_{19}, a_{18}[15, 16, 17, -18], a_{17}$
25	m_{24}	2^{10}	$a_{25}[11], a_{24}, a_{23}, a_{22}, a_{21}, a_{20}, a_{19}, a_{18}[15, 16, 17, -18]$
...	...	...	...
32	m_{31}		$a_{32}, a_{31}, a_{30}, a_{29}, a_{28}, a_{27}, a_{26}, a_{25}[11]$
33	m'_5		$a_{33}, a_{32}, a_{31}, a_{30}, a_{29}, a_{28}, a_{27}, a_{26}$
...	...	...	...
95	m'_5	2^{31}	$a_{95}[-32], a_{94}, a_{93}, a_{92}, a_{91}, a_{90}, a_{89}, a_{88}$
...	...	...	...
102	m_7		$a_{102}, a_{101}, a_{100}, a_{99}, a_{98}, a_{97}, a_{96}, a_{95}[-32]$
103	m_{28}	-2^{20}	$a_{103}[-21], a_{102}, a_{101}, a_{100}, a_{99}, a_{98}, a_{97}, a_{96}$
...	...	...	...
110	m_{25}		$a_{110}, a_{109}, a_{108}, a_{107}, a_{106}, a_{105}, a_{104}, a_{103}[-21]$
111	m_{19}	-2^9	$a_{111}[-10], a_{110}, a_{109}, a_{108}, a_{107}, a_{106}, a_{105}, a_{104}$
...	...	...	...
118	m_{27}		$a_{118}, a_{117}, a_{116}, a_{115}, a_{114}, a_{113}, a_{112}, a_{111}[-10]$
119	m_{12}	-2^{30}	$a_{119}[-31], a_{118}, a_{117}, a_{116}, a_{115}, a_{114}, a_{113}, a_{112}$
120	m_9		$a_{120}, a_{119}[-31], a_{118}, a_{117}, a_{116}, a_{115}, a_{114}, a_{113}$
121	m_1		$a_{121}, a_{120}, a_{119}[-31], a_{118}, a_{117}, a_{116}, a_{115}, a_{114}$
122	m_{29}		$a_{122}, a_{121}, a_{120}, a_{119}[-31], a_{118}, a_{117}, a_{116}, a_{115}$
123	m'_5	2^{31}	$a_{123}[32], a_{122}, a_{121}, a_{120}, a_{119}[-31], a_{118}, a_{117}, a_{116}$
124	m_{15}		$a_{124}, a_{123}[32], a_{122}, a_{121}, a_{120}, a_{119}[-31], a_{118}, a_{117}$
125	m_{17}		$a_{125}, a_{124}, a_{123}[32], a_{122}, a_{121}, a_{120}, a_{119}[-31], a_{118}$
126	m_{10}		$a_{126}, a_{125}, a_{124}, a_{123}[32], a_{122}, a_{121}, a_{120}, a_{119}[-31]$
127	m_{16}	-2^{19}	$a_{127}[-20], a_{126}, a_{125}, a_{124}, a_{123}[32], a_{122}, a_{121}, a_{120}$
128	m_{13}		$a_{128}, a_{127}[-20], a_{126}, a_{125}, a_{124}, a_{123}[32], a_{122}, a_{121}$

Table 5. A differential path for the second block of 4-pass HAVAL, up to step 95, for 2^{43} attack. From step 95, the path is the same as in the Table 4 except the signs.

Step	m'_{i-1}	Δa_i	Outputs for M'_1
0			$aa_0,\ bb_0[-20],\ cc_0,\ dd_0,\ ee_0,\ ff_0[32],\ gg_0,\ hh_0$
1	m_0		$a_1,\ aa_0,\ bb_0[-20],\ cc_0,\ dd_0,\ ee_0,\ ff_0[32],\ gg_0$
2	m_1		$a_2,\ a_1,\ aa_0,\ bb_0[-20],\ cc_0,\ dd_0,\ ee_0,\ ff_0[32]$
3	m_2	2^{20}	$a_3[21],\ a_2,\ a_1,\ aa_0,\ bb_0[-20],\ cc_0,\ dd_0,\ ee_0$
4	m_3		$a_4,\ a_3[21],\ a_2,\ a_1,\ aa_0,\ bb_0[-20],\ cc_0,\ dd_0$
5	m_4		$a_5,\ a_4,\ a_3[21],\ a_2,\ a_1,\ aa_0,\ bb_0[-20],\ cc_0$
6	m'_5	2^{31}	$a_6[32],\ a_5,\ a_4,\ a_3[21],\ a_2,\ a_1,\ aa_0,\ bb_0[-20]$
7	m_6	-2^8-2^{24}	$a_7[-9,25,26,27,-28],\ a_6[32],\ a_5,\ a_4,\ a_3[21],\ a_2,\ a_1,\ aa_0$
8	m_7	-2^{17}	$a_8[18,19,20,-21],\ a_7[-9,25,...,-28],\ a_6[32],\ a_5,\ a_4,\ a_3[21],\ a_2,\ a_1$
9	m_8	-2^{11}	$a_9[12,13,-14],\ a_8[18,...,-21],\ a_7[-9,25,...,-28],\ a_6[32],\ a_5,\ a_4,$ $a_3[21],\ a_2$
10	m_9	2^6	$a_{10}[-7,8],\quad a_9[12,13,-14],\quad a_8[18,...,-21],\quad a_7[-9,25,...,-28],$ $a_6[32],\ a_5,\ a_4,\ a_3[21]$
11	m_{10}	2^9	$a_{11}[10],\quad a_{10}[-7,8],\quad a_9[12,13,-14],\quad a_8[18,...,-21],$ $a_7[-9,25,...,-28],\ a_6[32],\ a_5,\ a_4$
12	m_{11}		$a_{12},\quad a_{11}[10],\quad a_{10}[-7,8],\quad a_9[12,13,-14],\quad a_8[18,...,-21],$ $a_7[-9,25,...,-28],\ a_6[32],\ a_5$
13	m_{12}		$a_{13},\quad a_{12},\quad a_{11}[10],\quad a_{10}[-7,8],\quad a_9[12,13,-14],\quad a_8[18,...,-21],$ $a_7[-9,25,...,-28],\ a_6[32]$
14	m_{13}		$a_{14},\ a_{13},\ a_{12},\ a_{11}[10],\ a_{10}[-7,8],\ a_9[12,13,-14],\ a_8[18,...,-21],$ $a_7[-9,25,...,-28]$
15	m_{14}	-2^{29}	$a_{15}[-30],\quad a_{14},\quad a_{13},\quad a_{12},\quad a_{11}[10],\quad a_{10}[-7,8],\quad a_9[12,13,-14],$ $a_8[18,...,-21]$
16	m_{15}		$a_{16},\ a_{15}[-30],\ a_{14},\ a_{13},\ a_{12},\ a_{11}[10],\ a_{10}[-7,8],\ a_9[12,13,-14]$
17	m_{16}		$a_{17},\ a_{16},\ a_{15}[-30],\ a_{14},\ a_{13},\ a_{12},\ a_{11}[10],\ a_{10}[-7,8]$
18	m_{17}	2^{27}	$a_{18}[-28,29],\ a_{17},\ a_{16},\ a_{15}[-30],\ a_{14},\ a_{13},\ a_{12},\ a_{11}[10]$
19	m_{18}	2^{30}	$a_{19}[31],\ a_{18}[-28,29],\ a_{17},\ a_{16},\ a_{15}[-30],\ a_{14},\ a_{13},\ a_{12}$
20	m_{19}		$a_{20},\ a_{19}[31],\ a_{18}[-28,29],\ a_{17},\ a_{16},\ a_{15}[-30],\ a_{14},\ a_{13}$
21	m_{20}		$a_{21},\ a_{20},\ a_{19}[31],\ a_{18}[-28,29],\ a_{17},\ a_{16},\ a_{15}[-30],\ a_{14}$
22	m_{21}	-2^{22}	$a_{22}[23,...,26,-27],\ a_{21},\ a_{20},\ a_{19}[31],\ a_{18}[-28,29],\ a_{17},\ a_{16},$ $a_{15}[-30]$
23	m_{22}	2^{21}	$a_{23}[22],\ a_{22}[23,...,-27],\ a_{21},\ a_{20},\ a_{19}[31],\ a_{18}[-28,29],\ a_{17},\ a_{16}$
24	m_{23}	-2^{14}	$a_{24}[15,...,18,-19],\ a_{23}[22],\ a_{22}[23,...,-27],\ a_{21},\ a_{20},\ a_{19}[31],$ $a_{18}[-28,29],\ a_{17}$
25	m_{24}	2^{10}	$a_{25}[11],\ a_{24}[15,...,-19],\ a_{23}[22],\ a_{22}[23,...,-27],\ a_{21},\ a_{20},\ a_{19}[31],$ $a_{18}[-28,29]$
26	m_{25}		$a_{26},\ a_{25}[11],\ a_{24}[15,...,-19],\ a_{23}[22],\ a_{22}[23,...,-27],\ a_{21},\ a_{20},$ $a_{19}[31]$
...	...	...	...
31	m_{30}		$a_{31},\ a_{30},\ a_{29},\ a_{28},\ a_{27},\ a_{26},\ a_{25}[11],\ a_{24}[15,...,-19]$
32	m_{31}		$a_{32},\ a_{31},\ a_{30},\ a_{29},\ a_{28},\ a_{27},\ a_{26},\ a_{25}[11]$
33	m'_5		$a_{33},\ a_{32},\ a_{31},\ a_{30},\ a_{29},\ a_{28},\ a_{27},\ a_{26}$
...	...	...	...
95	m'_5	2^{31}	$a_{95}[32],\ a_{94},\ a_{93},\ a_{92},\ a_{91},\ a_{90},\ a_{89},\ a_{88}$

Table 6. A differential path for the first block of 4-pass HAVAL, for 2^{36} attack. Here $m'_8 = m_8 + 2^{13}$, $m'_{16} = m_{16} - 2^2$.

step i	m'_{i-1}	Δa_i	Outputs for M'_0
9	m'_8	2^{13}	$a_9[-14, 15]$, a_8, a_7, a_6, a_5, a_4, a_3, a_2
10	m_9		a_{10}, $a_9[-14, 15]$, a_8, a_7, a_6, a_5, a_4, a_3
11	m_{10}		a_{11}, a_{10}, $a_9[-14, 15]$, a_8, a_7, a_6, a_5, a_4
12	m_{11}		a_{12}, a_{11}, a_{10}, $a_9[-14, 15]$, a_8, a_7, a_6, a_5
13	m_{12}		a_{13}, a_{12}, a_{11}, a_{10}, $a_9[-14, 15]$, a_8, a_7, a_6
14	m_{13}		a_{14}, a_{13}, a_{12}, a_{11}, a_{10}, $a_9[-14, 15]$, a_8, a_7
15	m_{14}	2^7	$a_{15}[-8, -9, -10, 11]$, a_{14}, a_{13}, a_{12}, a_{11}, a_{10}, $a_9[-14, 15]$, a_8
16	m_{15}	2^3	$a_{16}[4]$, $a_{15}[-8, -9, -10, 11]$, a_{14}, a_{13}, a_{12}, a_{11}, a_{10}, $a_9[-14, 15]$
17	m'_{16}	2^{28}	$a_{17}[29]$, $a_{16}[4]$, $a_{15}[-8, -9, -10, 11]$, a_{14}, a_{13}, a_{12}, a_{11}, a_{10}
18	m_{17}		a_{18}, $a_{17}[29]$, $a_{16}[4]$, $a_{15}[-8, -9, -10, 11]$, a_{14}, a_{13}, a_{12}, a_{11}
19	m_{18}		a_{19}, a_{18}, $a_{17}[29]$, $a_{16}[4]$, $a_{15}[-8, -9, -10, 11]$, a_{14}, a_{13}, a_{12}
20	m_{19}		a_{20}, a_{19}, a_{18}, $a_{17}[29]$, $a_{16}[4]$, $a_{15}[-8, -9, -10, 11]$, a_{14}, a_{13}
21	m_{20}		a_{21}, a_{20}, a_{19}, a_{18}, $a_{17}[29]$, $a_{16}[4]$, $a_{15}[-8, -9, -10, 11]$, a_{14}
22	m_{21}		a_{22}, a_{21}, a_{20}, a_{19}, a_{18}, $a_{17}[29]$, $a_{16}[4]$, $a_{15}[-8, -9, -10, 11]$
23	m_{22}	2^{21}	$a_{23}[22]$, a_{22}, a_{21}, a_{20}, a_{19}, a_{18}, $a_{17}[29]$, $a_{16}[4]$
24	m_{23}	$-2^{14} + 2^{24}$	$a_{24}[15, 16, 17, -18, 25]$, $a_{23}[22]$, a_{22}, a_{21}, a_{20}, a_{19}, a_{18}, $a_{17}[29]$
25	m_{24}		a_{25}, $a_{24}[15, 16, 17, -18, 25]$, $a_{23}[22]$, a_{22}, a_{21}, a_{20}, a_{19}, a_{18}
26	m_{25}		a_{26}, a_{25}, $a_{24}[15, 16, 17, -18, 25]$, $a_{23}[22]$, a_{22}, a_{21}, a_{20}, a_{19}
27	m_{26}		a_{27}, a_{26}, a_{25}, $a_{24}[15, 16, 17, -18, 25]$, $a_{23}[22]$, a_{22}, a_{21}, a_{20}
28	m_{27}		a_{28}, a_{27}, a_{26}, a_{25}, $a_{24}[15, 16, 17, -18, 25]$, $a_{23}[22]$, a_{22}, a_{21}
29	m_{28}		a_{29}, a_{28}, a_{27}, a_{26}, a_{25}, $a_{24}[15, 16, 17, -18, 25]$, $a_{23}[22]$, a_{22}
30	m_{29}		a_{30}, a_{29}, a_{28}, a_{27}, a_{26}, a_{25}, $a_{24}[15, 16, 17, -18, 25]$, $a_{23}[22]$
31	m_{30}		a_{31}, a_{30}, a_{29}, a_{28}, a_{27}, a_{26}, a_{25}, $a_{24}[15, 16, 17, -18, 25]$
32	m_{31}	$-2^3 + 2^{13}$	$a_{32}[-4, 14]$, a_{31}, a_{30}, a_{29}, a_{28}, a_{27}, a_{26}, a_{25}
...	...	...	...
39	m_7		a_{39}, a_{38}, a_{37}, a_{36}, a_{35}, a_{34}, a_{33}, $a_{32}[-4, 14]$
40	m'_{16}	-2^{24}	$a_{40}[-25]$, a_{39}, a_{38}, a_{37}, a_{36}, a_{35}, a_{34}, a_{33}
...	...	...	...
47	m_4		a_{47}, a_{46}, a_{45}, a_{44}, a_{43}, a_{42}, a_{41}, $a_{40}[-25]$
48	m'_8		a_{48}, a_{47}, a_{46}, a_{45}, a_{44}, a_{43}, a_{42}, a_{41}
...	...	...	...
71	m'_8	2^{13}	$a_{71}[14]$, a_{70}, a_{69}, a_{68}, a_{67}, a_{66}, a_{65}, a_{64}
...	...	...	...
78	m_{30}		a_{78}, a_{77}, a_{76}, a_{75}, a_{74}, a_{73}, a_{72}, $a_{71}[14]$
79	m'_{16}		a_{79}, a_{78}, a_{77}, a_{76}, a_{75}, a_{74}, a_{73}, a_{72}
...	...	...	...
117	m'_8	2^{13}	$a_{117}[14]$, a_{116}, a_{115}, a_{114}, a_{113}, a_{112}, a_{111}, a_{110}
...	...	...	...
124	m_{15}		a_{124}, a_{123}, a_{122}, a_{121}, a_{120}, a_{119}, a_{118}, $a_{117}[14]$
125	m_{17}	2^2	$a_{125}[3]$, a_{124}, a_{123}, a_{122}, a_{121}, a_{120}, a_{119}, a_{118}
126	m_{10}		a_{126}, $a_{125}[3]$, a_{124}, a_{123}, a_{122}, a_{121}, a_{120}, a_{119}
127	m'_{16}	-2^2	$a_{127}[-3]$, a_{126}, $a_{125}[3]$, a_{124}, a_{123}, a_{122}, a_{121}, a_{120}
128	m_{13}		a_{128}, $a_{127}[-3]$, a_{126}, $a_{125}[3]$, a_{124}, a_{123}, a_{122}, a_{121}

Table 7. A differential path for the second block of 4-pass HAVAL, for 2^{36} attack. Here, $m'_8 = m_8 - 2^{13}$, $m'_{16} = m_{16} + 2^2$.

Step	m'_{i-1}	Δa_i	Output for M'_1
0			$aa_0, bb_0[-3], cc_0, dd_0[3], ee_0, ff_0, gg_0, hh_0$
1	m_0		$a_1, aa_0, bb_0[-3], cc_0, dd_0[3], ee_0, ff_0, gg_0$
2	m_1		$a_2, a_1, aa_0, bb_0[-3], cc_0, dd_0[3], ee_0, ff_0$
3	m_2		$a_3, a_2, a_1, aa_0, bb_0[-3], cc_0, dd_0[3], ee_0$
4	m_3		$a_4, a_3, a_2, a_1, aa_0, bb_0[-3], cc_0, dd_0[3]$
5	m_4	2^{23}	$a_5[24], a_4, a_3, a_2, a_1, aa_0, bb_0[-3], cc_0$
6	m_5		$a_6, a_5[24], a_4, a_3, a_2, a_1, aa_0, bb_0[-3]$
7	m_6	-2^{23}	$a_7[-24], a_6, a_5[24], a_4, a_3, a_2, a_1, aa_0$
8	m_7		$a_8, a_7[-24], a_6, a_5[24], a_4, a_3, a_2, a_1$
9	m'_8	-2^{13}	$a_9[14, 15, 16, 17, 18, 19, 20, -21], a_8, a_7[-24], a_6, a_5[24], a_4, a_3, a_2$
10	m_9		$a_{10}, a_9[14, 15, 16, 17, 18, 19, 20, -21], a_8, a_7[-24], a_6, a_5[24], a_4, a_3$
11	m_{10}		$a_{11}, a_{10}, a_9[14, 15, 16, 17, 18, 19, 20, -21], a_8, a_7[-24], a_6, a_5[24], a_4$
12	m_{11}		$a_{12}, a_{11}, a_{10}, a_9[14, 15, 16, 17, 18, 19, 20, -21], a_8, a_7[-24], a_6, a_5[24]$
13	m_{12}		$a_{13}, a_{12}, a_{11}, a_{10}, a_9[14, 15, 16, 17, 18, 19, 20, -21], a_8, a_7[-24], a_6$
14	m_{13}		$a_{14}, a_{13}, a_{12}, a_{11}, a_{10}, a_9[14, 15, 16, 17, 18, 19, 20, -21], a_8, a_7[-24]$
15	m_{14}	2^7	$a_{15}[-8, -9, -10, 11], a_{14},..., a_9[14, 15, 16, 17, 18, 19, 20, -21], a_8$
16	m_{15}	-2^3	$a_{16}[-4], a_{15}[-8, -9, -10, 11],..., a_{10}, a_9[14, 15, 16, 17, 18, 19, 20, -21]$
17	m'_{16}	2^{28}	$a_{17}[29], a_{16}[-4], a_{15}[-8, -9, -10, 11], a_{14}, a_{13}, a_{12}, a_{11}, a_{10}$
...	...	...	...
22	m_{21}		$a_{22}, a_{21}, a_{20}, a_{19}, a_{18}, a_{17}[29], a_{16}[-4], a_{15}[-8, -9, -10, 11]$
23	m_{22}	2^{21}	$a_{23}[22], a_{22}, a_{21}, a_{20}, a_{19}, a_{18}, a_{17}[29], a_{16}[-4]$
24	m_{23}	2^{14}-2^{24}	$a_{24}[-15, -16, -17, 18, -25], a_{23}[22], a_{22}, a_{21}, a_{20}, a_{19}, a_{18}, a_{17}[29]$
...	...	...	...
31	m_{30}		$a_{31}, a_{30}, a_{29}, a_{28}, a_{27}, a_{26}, a_{25}, a_{24}[-15, -16, -17, 18, -25]$
32	m_{31}	2^3-2^{13}	$a_{32}[4, -14], a_{31}, a_{30}, a_{29}, a_{28}, a_{27}, a_{26}, a_{25}$
...	...	...	...
39	m_7		$a_{39}, a_{38}, a_{37}, a_{36}, a_{35}, a_{34}, a_{33}, a_{32}[4, -14]$
40	m'_{16}	2^{24}	$a_{40}[25], a_{39}, a_{38}, a_{37}, a_{36}, a_{35}, a_{34}, a_{33}$
...	...	...	...
47	m_4		$a_{47}, a_{46}, a_{45}, a_{44}, a_{43}, a_{42}, a_{41}, a_{40}[25]$
48	m'_8		$a_{48}, a_{47}, a_{46}, a_{45}, a_{44}, a_{43}, a_{42}, a_{41}$
...	...	...	...
71	m'_8	-2^{13}	$a_{71}[-14], a_{70}, a_{69}, a_{68}, a_{67}, a_{66}, a_{65}, a_{64}$
...	...	...	...
78	m_{30}		$a_{78}, a_{77}, a_{76}, a_{75}, a_{74}, a_{73}, a_{72}, a_{71}[-14]$
79	m'_{16}		$a_{79}, a_{78}, a_{77}, a_{76}, a_{75}, a_{74}, a_{73}, a_{72}$
...	...	...	...
117	m'_8	-2^{13}	$a_{117}[-14], a_{116}, a_{115}, a_{114}, a_{113}, a_{112}, a_{111}, a_{110}$
...	...	...	...
124	m_{15}		$a_{124}, a_{123}, a_{122}, a_{121}, a_{120}, a_{119}, a_{118}, a_{117}[-14]$
125	m_{17}	-2^2	$a_{125}[-3], a_{124}, a_{123}, a_{122}, a_{121}, a_{120}, a_{119}, a_{118}$
126	m_{10}		$a_{126}, a_{125}[-3], a_{124}, a_{123}, a_{122}, a_{121}, a_{120}, a_{119}$
127	m'_{16}	2^2	$a_{127}[3], a_{126}, a_{125}[-3], a_{124}, a_{123}, a_{122}, a_{121}, a_{120}$
128	m_{13}		$a_{128}, a_{127}[3], a_{126}, a_{125}[-3], a_{124}, a_{123}, a_{122}, a_{121}$

Table 8. A set of sufficient conditions on a_i for the differential path given in Table 6

Step i	a_i	Conditions of the chaining variable in each step
6	a_6	$a_{6,14} = 1$, $a_{6,15} = 1$
7	a_7	$a_{7,14} = 0$, $a_{7,15} = 0$
8	a_8	$a_{8,14} = 0$, $a_{8,15} = 1$
9	a_9	$a_{9,14} = 1$, $a_{9,15} = 0$
10	a_{10}	$a_{10,11} = 0$, $a_{10,14} = 0$, $a_{10,15} = 0$
11	a_{11}	$a_{11,4} = 0$, $a_{11,14} = 0$, $a_{11,15} = 0$
12	a_{12}	$a_{12,8} = 1$, $a_{12,9} = 1$, $a_{12,10} = 1$, $a_{12,11} = 0$, $a_{12,14} = 0$, $a_{12,15} = 0$,
13	a_{13}	$a_{13,4} = 0$, $a_{13,8} = 0$, $a_{13,9} = 0$, $a_{13,10} = 0$, $a_{13,11} = 0$
14	a_{14}	$a_{14,4} = 0$, $a_{14,8} = 0$, $a_{14,9} = 0$, $a_{14,10} = 0$, $a_{14,11} = 0$, $a_{14,15} = 0$, $a_{14,29} = 1$
15	a_{15}	$a_{15,4} = 0$, $a_{15,8} = 1$, $a_{15,9} = 1$, $a_{15,10} = 1$, $a_{15,11} = 0$, $a_{15,29} = 0$
16	a_{16}	$a_{16,4} = 0$, $a_{16,8} = 0$, $a_{16,9} = 0$, $a_{16,10} = 0$, $a_{16,11} = 0$, $a_{16,29} = 1$
17	a_{17}	$a_{17,4} = 1$, $a_{17,8} = 0$, $a_{17,9} = 0$, $a_{17,10} = 0$, $a_{17,11} = 0$, $a_{17,22} = 1$, $a_{17,29} = 0$
18	a_{18}	$a_{18,4} = 0$, $a_{18,8} = 0$, $a_{18,9} = 0$, $a_{18,10} = 0$, $a_{18,11} = 0$, $a_{18,22} = 1$, $a_{18,25} = 1$, $a_{18,29} = 0$
19	a_{19}	$a_{19,4} = 0$, $a_{19,25} = 1$, $a_{19,29} = 0$
20	a_{20}	$a_{20,22} = 0$, $a_{20,29} = 0$
21	a_{21}	$a_{21,15} = 1$, $a_{21,16} = 1$, $a_{21,17} = 1$, $a_{21,18} = 1$, $a_{21,22} = 0$, $a_{21,25} = 0$
22	a_{22}	$a_{22,4} = 1$, $a_{22,15} = 0$, $a_{22,16} = 0$, $a_{22,17} = 0$, $a_{22,18} = 0$, $a_{22,22} = 0$, $a_{22,25} = 0$, $a_{22,29} = 0$
23	a_{23}	$a_{23,15} = 0$, $a_{23,16} = 0$, $a_{23,17} = 0$, $a_{23,18} = 0$, $a_{23,22} = 0$, $a_{23,25} = 0$
24	a_{24}	$a_{24,15} = 0$, $a_{24,16} = 0$, $a_{24,17} = 0$, $a_{24,18} = 1$, $a_{24,22} = 0$, $a_{24,25} = 0$
25	a_{25}	$a_{25,15} = 0$, $a_{25,16} = 0$, $a_{25,17} = 0$, $a_{25,18} = 1$, $a_{25,22} = 0$, $a_{25,25} = 0$
26	a_{26}	$a_{26,15} = 0$, $a_{26,16} = 0$, $a_{26,17} = 0$, $a_{26,18} = 0$, $a_{26,22} = 0$, $a_{26,25} = 0$
27	a_{27}	$a_{27,4} = 0$, $a_{27,14} = 0$, $a_{27,15} = 0$, $a_{27,16} = 0$, $a_{27,17} = 0$, $a_{27,18} = 0$, $a_{27,25} = 0$
28	a_{28}	$a_{28,4} = 0$, $a_{28,14} = 0$
29	a_{29}	$a_{29,4} = 0$, $a_{29,14} = 0$
30	a_{30}	$a_{30,4} = 0$, $a_{30,14} = 0$, $a_{30,18} = 0$
31	a_{31}	$a_{31,4} = 0$, $a_{31,14} = 0$
32	a_{32}	$a_{32,4} = 1$, $a_{32,14} = 0$
34	a_{34}	$a_{34,4} = 0$, $a_{34,14} = 0$
35	a_{35}	$a_{35,4} = 1$, $a_{35,14} = 1$ $a_{35,25} = 0$
36	a_{36}	$a_{36,4} = 0$, $a_{36,14} = 0$, $a_{36,25} = 0$
37	a_{37}	$a_{37,4} = 1$, $a_{37,14} = 1$, $a_{37,25} = 0$
38	a_{38}	$a_{38,4} = 1$, $a_{38,14} = 1$, $a_{38,25} = 0$
39	a_{39}	$a_{39,25} = 0$
40	a_{40}	$a_{40,25} = 1$
42	a_{42}	$a_{42,25} = 0$
43	a_{43}	$a_{43,25} = 1$
44	a_{44}	$a_{44,25} = 0$
45	a_{45}	$a_{45,25} = 1$
46	a_{46}	$a_{46,25} = 1$

Table 9. A set of sufficient conditions on a_i for the differential path given in Table 7

Step		Conditions of the chaining variable in each step
0	IVs	$aa_{0,3} = 0$, $bb_{0,3} = 1$, $cc_{0,3} = 0$, $dd_{0,3} = 0$
1	a_1	$a_{1,3} = 0$
2	a_2	$a_{2,3} = ee_{0,3}$, $a_{2,24} = 1$
3-5	a_3	$a_{3,24} = 0$, $a_{4,24} = 1$, $a_{5,24} = 0$
6	a_6	$a_{6,14} = 1$, $a_{6,15} = 1$, $a_{6,16} = 1$, $a_{6,17} = 1$, $a_{6,18} = 1$, $a_{6,19} = 1$, $a_{6,20} = 1$, $a_{6,21} = 1$, $a_{6,24} = 0$
7	a_7	$a_{7,14} = 0$, $a_{7,15} = 0$, $a_{7,16} = 0$, $a_{7,17} = 0$, $a_{7,18} = 0$, $a_{7,19} = 0$, $a_{7,20} = 0$, $a_{7,21} = 0$, $a_{7,24} = 1$
8	a_8	$a_{8,14} = 0$, $a_{8,15} = 1$, $a_{8,16} = 0$, $a_{8,17} = 0$, $a_{8,18} = 0$, $a_{8,19} = 0$, $a_{8,20} = 1$, $a_{8,21} = 0$, $a_{8,24} = 0$
9	a_9	$a_{9,11} = 1$, $a_{9,14} = 0$, $a_{9,15} = 0$, $a_{9,16} = 0$, $a_{9,17} = 0$, $a_{9,18} = 0$, $a_{9,19} = 0$, $a_{9,20} = 0$, $a_{9,21} = 1$, $a_{9,24} = 0$
10	a_{10}	$a_{10,4} = 1$, $a_{10,11} = 1$, $a_{10,14} = 0$, $a_{10,15} = 0$, $a_{10,16} = 0$, $a_{10,17} = 0$, $a_{10,18} = 0$, $a_{10,19} = 0$, $a_{10,20} = 0$, $a_{10,21} = 0$, $a_{10,24} = 1$
11	a_{11}	$a_{11,4} = 1$, $a_{11,14} = 0$, $a_{11,15} = 0$, $a_{11,16} = 0$, $a_{11,17} = 0$, $a_{11,18} = 0$, $a_{11,19} = 0$, $a_{11,20} = 0$, $a_{11,21} = 0$
12	a_{12}	$a_{12,8} = 1$, $a_{12,9} = 1$, $a_{12,10} = 1$, $a_{12,14} = 0$, $a_{12,15} = 0$, $a_{12,16} = 0$, $a_{12,17} = 0$, $a_{12,18} = 0$, $a_{12,19} = 0$, $a_{12,20} = 1$, $a_{12,21} = 0$
13	a_{13}	$a_{13,4} = 0$, $a_{13,8} = 0$, $a_{13,9} = 0$, $a_{13,10} = 0$, $a_{13,11} = 0$
14	a_{14}	$a_{14,4} = 0$, $a_{14,8} = 0$, $a_{14,9} = 0$, $a_{14,10} = 0$, $a_{14,11} = 0$, $a_{14,15} = 0$, $a_{14,20} = 0$, $a_{14,29} = 1$
15	a_{15}	$a_{15,4} = 0$, $a_{15,8} = 1$, $a_{15,9} = 1$, $a_{15,10} = 1$, $a_{15,11} = 0$, $a_{15,29} = 0$
16	a_{16}	$a_{16,4} = 1$, $a_{16,8} = 0$, $a_{16,9} = 0$, $a_{16,10} = 0$, $a_{16,11} = 0$, $a_{16,29} = 1$
17	a_{17}	$a_{17,4} = 1$, $a_{17,8} = 0$, $a_{17,9} = 0$, $a_{17,10} = 0$, $a_{17,11} = 0$, $a_{17,29} = 0$
18	a_{18}	$a_{18,4} = 0$, $a_{18,8} = 0$, $a_{18,9} = 0$, $a_{18,10} = 0$, $a_{18,11} = 0$, $a_{18,22} = 0$, $a_{18,29} = 0$
19	a_{19}	$a_{19,4} = 0$, $a_{19,25} = 0$, $a_{19,29} = 0$
20	a_{20}	$a_{20,22} = 0$, $a_{20,29} = 0$
21	a_{21}	$a_{21,15} = 1$, $a_{21,16} = 1$, $a_{21,17} = 1$, $a_{21,18} = 1$, $a_{21,22} = 0$, $a_{21,25} = 0$
22	a_{22}	$a_{22,4} = 0$, $a_{22,15} = 0$, $a_{22,16} = 0$, $a_{22,17} = 0$, $a_{22,18} = 0$, $a_{22,22} = 0$, $a_{22,25} = 0$, $a_{22,29} = 0$
23	a_{23}	$a_{23,15} = 0$, $a_{23,16} = 0$, $a_{23,17} = 0$, $a_{23,18} = 0$, $a_{23,22} = 0$, $a_{23,25} = 0$
24	a_{24}	$a_{24,15} = 1$, $a_{24,16} = 1$, $a_{24,17} = 1$, $a_{24,18} = 0$, $a_{24,22} = 0$, $a_{24,25} = 1$
25	a_{25}	$a_{25,15} = 0$, $a_{25,16} = 0$, $a_{25,17} = 0$, $a_{25,18} = 1$, $a_{25,22} = 0$, $a_{25,25} = 0$
26	a_{26}	$a_{26,15} = 0$, $a_{26,16} = 0$, $a_{26,17} = 0$, $a_{26,18} = 0$, $a_{26,22} = 0$, $a_{26,25} = 0$
27	a_{27}	$a_{27,4} = 0$, $a_{27,14} = 0$, $a_{27,15} = 0$, $a_{27,16} = 0$, $a_{27,17} = 0$, $a_{27,18} = 0$, $a_{27,25} = 0$
28-29	a_{28}	$a_{28,4} = 0$, $a_{28,14} = 0$, $a_{29,4} = 0$, $a_{29,14} = 0$
30	a_{30}	$a_{30,4} = 0$, $a_{30,14} = 0$, $a_{30,18} = 1$
31-32	a_{31}	$a_{31,4} = 0$, $a_{31,14} = 0$, $a_{32,4} = 0$, $a_{32,14} = 1$
34-35	a_{34}	$a_{34,4} = 0$, $a_{34,14} = 0$, $a_{35,4} = 1$, $a_{35,14} = 1$, $a_{35,25} = 0$
36	a_{36}	$a_{36,4} = 0$, $a_{36,14} = 0$, $a_{36,25} = 0$
37	a_{37}	$a_{37,4} = 1$, $a_{37,14} = 1$, $a_{37,25} = 0$
38-39	a_{38}	$a_{38,4} = 1$, $a_{38,14} = 1$, $a_{38,25} = 0$, $a_{39,25} = 0$
40-46	a_{40}	$a_{40,25} = 0$, $a_{42,25} = 0$, $a_{43,25} = 1$, $a_{44,25} = 0$, $a_{45,25} = 1$, $a_{46,25} = 1$

Table 10. A differential path for the 5-pass HAVAL. Here $m_8' = m_8 - 1$.

Step	m_{i-1}'	Δa_i	Outputs for M'
9	m_8'	-1	$a_9[1,2,3,4,-5], a_8, a_7, a_6, a_5, a_4, a_3, a_2$
10	m_9		$a_{10}, a_9[1,2,3,4,-5], a_8, a_7, a_6, a_5, a_4, a_3$
11	m_{10}	-2^{28}	$a_{11}[29,-30], a_{10}, a_9[1,2,3,4,-5], a_8, a_7, a_6, a_5, a_4$
12	m_{11}		$a_{12}, a_{11}[29,-30], a_{10}, a_9[1,2,3,4,-5], a_8, a_7, a_6, a_5$
13	m_{12}		$a_{13}, a_{12}, a_{11}[29,-30], a_{10}, a_9[1,2,3,4,-5], a_8, a_7, a_6$
14	m_{13}	2^{21}	$a_{14}[22], a_{13}, a_{12}, a_{11}[29,-30], a_{10}, a_9[1,2,3,4,-5], a_8, a_7$
15	m_{14}		$a_{15}, a_{14}[22], a_{13}, a_{12}, a_{11}[29,-30], a_{10}, a_9[1,2,3,4,-5], a_8$
16	m_{15}		$a_{16}, a_{15}, a_{14}[22], a_{13}, a_{12}, a_{11}[29,-30], a_{10}, a_9[1,2,3,4,-5]$
17	m_{16}	-2^{14}	$a_{17}[15,-16], a_{16}, a_{15}, a_{14}[22], a_{13}, a_{12}, a_{11}[29,-30], a_{10}$
18	m_{17}		$a_{18}, a_{17}[15,-16], a_{16}, a_{15}, a_{14}[22], a_{13}, a_{12}, a_{11}[29,-30]$
19	m_{18}		$a_{19}, a_{18}, a_{17}[15,-16], a_{16}, a_{15}, a_{14}[22], a_{13}, a_{12}$
20	m_{19}	$-2^7 - 2^{17}$	$a_{20}[8,9,-10,-18], a_{19}, a_{18}, a_{17}[15,-16], a_{16}, a_{15}, a_{14}[22], a_{13}$
21	m_{20}		$a_{21}, a_{20}[8,9,-10,-18], a_{19}, a_{18}, a_{17}[15,-16], a_{16}, a_{15}, a_{14}[22]$
22	m_{21}	2^{10}	$a_{22}[11], a_{21}, a_{20}[8,9,-10,-18], a_{19}, a_{18}, a_{17}[15,-16], a_{16}, a_{15}$
23	m_{22}	2^2	$a_{23}[3], a_{22}[11], a_{21}, a_{20}[8,9,-10,-18], a_{19}, a_{18}, a_{17}[15,-16], a_{16}$
24	m_{23}	2	$a_{24}[2], a_{23}[3], a_{22}[11], a_{21}, a_{20}[8,9,-10,-18], a_{19}, a_{18}, a_{17}[15,-16]$
25	m_{24}	$-2^3 - 2^{26}$	$a_{25}[-4,-27], a_{24}[2], a_{23}[3], a_{22}[11], a_{21}, a_{20}[8,9,-10,-18], a_{19}, a_{18}$
26	m_{25}		$a_{26}, a_{25}[-4,-27], a_{24}[2], a_{23}[3], a_{22}[11], a_{21}, a_{20}[8,9,-10,-18], a_{19}$
27	m_{26}		$a_{27}, a_{26}, a_{25}[-4,-27], a_{24}[2], a_{23}[3], a_{22}[11], a_{21}, a_{20}[8,9,-10,-18]$
28	m_{27}	$-2^6 + 2^{19}$	$a_{28}[-7,20], a_{27}, a_{26}, a_{25}[-4,-27], a_{24}[2], a_{23}[3], a_{22}[11], a_{21}$
29	m_{28}		$a_{29}, a_{28}[-7,20], a_{27}, a_{26}, a_{25}[-4,-27], a_{24}[2], a_{23}[3], a_{22}[11]$
30	m_{29}	2^{31}	$a_{30}[32], a_{29}, a_{28}[-7,20], a_{27}, a_{26}, a_{25}[-4,-27], a_{24}[2], a_{23}[3]$
31	m_{30}	2^{23}	$a_{31}[24], a_{30}[32], a_{29}, a_{28}[-7,20], a_{27}, a_{26}, a_{25}[-4,-27], a_{24}[2]$
32	m_{31}	2^{22}	$a_{32}[23], a_{31}[24], a_{30}[32], a_{29}, a_{28}[-7,20], a_{27}, a_{26}, a_{25}[-4,-27]$
33	m_5		$a_{33}, a_{32}[23], a_{31}[24], a_{30}[32], a_{29}, a_{28}[-7,20], a_{27}, a_{26}$
34	m_{14}	-2^{15}	$a_{34}[-16], a_{33}, a_{32}[23], a_{31}[24], a_{30}[32], a_{29}, a_{28}[-7,20], a_{27}$
35	m_{26}		$a_{35}, a_{34}[-16], a_{33}, a_{32}[23], a_{31}[24], a_{30}[32], a_{29}, a_{28}[-7,20]$
36	m_{18}	-2^{27}	$a_{36}[-28], a_{35}, a_{34}[-16], a_{33}, a_{32}[23], a_{31}[24], a_{30}[32], a_{29}$
37	m_{11}		$a_{37}, a_{36}[-28], a_{35}, a_{34}[-16], a_{33}, a_{32}[23], a_{31}[24], a_{30}[32]$
38	m_{28}		$a_{38}, a_{37}, a_{36}[-28], a_{35}, a_{34}[-16], a_{33}, a_{32}[23], a_{31}[24]$
39	m_7	2^{12}	$a_{39}[13], a_{38}, a_{37}, a_{36}[-28], a_{35}, a_{34}[-16], a_{33}, a_{32}[23]$
40	m_{16}	$2^5 + 2^{11}$	$a_{40}[6,12], a_{39}[13], a_{38}, a_{37}, a_{36}[-28], a_{35}, a_{34}[-16], a_{33}$
41	m_0		$a_{41}, a_{40}[6,12], a_{39}[13], a_{38}, a_{37}, a_{36}[-28], a_{35}, a_{34}[-16]$
42	m_{23}	2^{30}	$a_{42}[31], a_{41}, a_{40}[6,12], a_{39}[13], a_{38}, a_{37}, a_{36}[-28], a_{35}$
43	m_{20}	2^{23}	$a_{43}[24], a_{42}[31], a_{41}, a_{40}[6,12], a_{39}[13], a_{38}, a_{37}, a_{36}[-28]$
44	m_{22}		$a_{44}, a_{43}[24], a_{42}[31], a_{41}, a_{40}[6,12], a_{39}[13], a_{38}, a_{37}$
45	m_1		$a_{45}, a_{44}, a_{43}[24], a_{42}[31], a_{41}, a_{40}[6,12], a_{39}[13], a_{38}$
46	m_{10}		$a_{46}, a_{45}, a_{44}, a_{43}[24], a_{42}[31], a_{41}, a_{40}[6,12], a_{39}[13]$
47	m_4	2	$a_{47}[2], a_{46}, a_{45}, a_{44}, a_{43}[24], a_{42}[31], a_{41}, a_{40}[6,12]$
48	m_8'		$a_{48}, a_{47}[2], a_{46}, a_{45}, a_{44}, a_{43}[24], a_{42}[31], a_{41}$
49	m_{30}		$a_{49}, a_{48}, a_{47}[2], a_{46}, a_{45}, a_{44}, a_{43}[24], a_{42}[31]$
50	m_3	2^{19}	$a_{50}[20], a_{49}, a_{48}, a_{47}[2], a_{46}, a_{45}, a_{44}, a_{43}[24]$
51	m_{21}		$a_{51}, a_{50}[20], a_{49}, a_{48}, a_{47}[2], a_{46}, a_{45}, a_{44}$
52	m_9		$a_{52}, a_{51}, a_{50}[20], a_{49}, a_{48}, a_{47}[2], a_{46}, a_{45}$

Table 11. A differential path for the 5-pass HAVAL(continued from Table 10)

53	m_{17}		$a_{53}, a_{52}, a_{51}, a_{50}[20], a_{49}, a_{48}, a_{47}[2], a_{46}$
54	m_{24}		$a_{54}, a_{53}, a_{52}, a_{51}, a_{50}[20], a_{49}, a_{48}, a_{47}[2]$
55	m_{29}	2^{22}	$a_{55}[23], a_{54}, a_{53}, a_{52}, a_{51}, a_{50}[20], a_{49}, a_{48}$
56	m_6		$a_{56}, a_{55}[23], a_{54}, a_{53}, a_{52}, a_{51}, a_{50}[20], a_{49}$
57	m_{19}	2^{15}	$a_{57}[16], a_{56}, a_{55}[23], a_{54}, a_{53}, a_{52}, a_{51}, a_{50}[20]$
58	m_{12}		$a_{58}, a_{57}[16], a_{56}, a_{55}[23], a_{54}, a_{53}, a_{52}, a_{51}$
59	m_{15}		$a_{59}, a_{58}, a_{57}[16], a_{56}, a_{55}[23], a_{54}, a_{53}, a_{52}$
60	m_{13}		$a_{60}, a_{59}, a_{58}, a_{57}[16], a_{56}, a_{55}[23], a_{54}, a_{53}$
61	m_2		$a_{61}, a_{60}, a_{59}, a_{58}, a_{57}[16], a_{56}, a_{55}[23], a_{54}$
62	m_{25}		$a_{62}, a_{61}, a_{60}, a_{59}, a_{58}, a_{57}[16], a_{56}, a_{55}[23]$
63	m_{31}	2^{11}	$a_{63}[12], a_{62}, a_{61}, a_{60}, a_{59}, a_{58}, a_{57}[16], a_{56}$
64	m_{27}		$a_{64}, a_{63}[12], a_{62}, a_{61}, a_{60}, a_{59}, a_{58}, a_{57}[16]$
65	m_{19}		$a_{65}, a_{64}, a_{63}[12], a_{62}, a_{61}, a_{60}, a_{59}, a_{58}$
66	m_9		$a_{66}, a_{65}, a_{64}, a_{63}[12], a_{62}, a_{61}, a_{60}, a_{59}$
67	m_4		$a_{67}, a_{66}, a_{65}, a_{64}, a_{63}[12], a_{62}, a_{61}, a_{60}$
68	m_{20}		$a_{68}, a_{67}, a_{66}, a_{65}, a_{64}, a_{63}[12], a_{62}, a_{61}$
69	m_{28}		$a_{69}, a_{68}, a_{67}, a_{66}, a_{65}, a_{64}, a_{63}[12], a_{62}$
70	m_{17}		$a_{70}, a_{69}, a_{68}, a_{67}, a_{66}, a_{65}, a_{64}, a_{63}[12]$
71	m_8'		$a_{71}, a_{70}, a_{69}, a_{68}, a_{67}, a_{66}, a_{65}, a_{64}$
...	...		...
117	m_8'	-1	$a_{117}[1, -2], a_{116}, a_{115}, a_{114}, a_{113}, a_{112}, a_{111}, a_{110}$
118	m_{27}		$a_{118}, a_{117}[1, -2], a_{116}, a_{115}, a_{114}, a_{113}, a_{112}, a_{111}$
119	m_{12}		$a_{119}, a_{118}, a_{117}[1, -2], a_{116}, a_{115}, a_{114}, a_{113}, a_{112}$
120	m_9		$a_{120}, a_{119}, a_{118}, a_{117}[1, -2], a_{116}, a_{115}, a_{114}, a_{113}$
121	m_1		$a_{121}, a_{120}, a_{119}, a_{118}, a_{117}[1, -2], a_{116}, a_{115}, a_{114}$
122	m_{29}		$a_{122}, a_{121}, a_{120}, a_{119}, a_{118}, a_{117}[1, -2], a_{116}, a_{115}$
123	m_5		$a_{123}, a_{122}, a_{121}, a_{120}, a_{119}, a_{118}, a_{117}[1, -2], a_{116}$
124	m_{15}	-2^{26}	$a_{124}[27, 28, 29, -30], a_{123}, a_{122}, a_{121}, a_{120}, a_{119}, a_{118}, a_{117}[1, -2]$
125	m_{17}		$a_{125}, a_{124}[27, 28, 29, -30], a_{123}, a_{122}, a_{121}, a_{120}, a_{119}, a_{118}$
126	m_{10}	2^{22}	$a_{126}[23], a_{125}, a_{124}[27, 28, 29, -30], a_{123}, a_{122}, a_{121}, a_{120}, a_{119}$
127	m_{16}	-2^{15}	$a_{127}[-16], a_{126}[23], a_{125}, a_{124}[27, 28, 29, -30], a_{123}, a_{122}, a_{121}, a_{120}$
128	m_{13}		$a_{128}, a_{127}[-16], a_{126}[23], a_{125}, a_{124}[27, 28, 29, -30], a_{123}, a_{122}, a_{121}$
129	m_{27}		$a_{129}, a_{128}, a_{127}[-16], a_{126}[23], a_{125}, a_{124}[27, 28, 29, -30], a_{123}, a_{122}$
130	m_3		$a_{130}, a_{129}, a_{128}, a_{127}[-16], a_{126}[23], a_{125}, a_{124}[27, 28, 29, -30], a_{123}$
131	m_{21}		$a_{131}, a_{130}, a_{129}, a_{128}, a_{127}[-16], a_{126}[23], a_{125}, a_{124}[27, 28, 29, -30]$
132	m_{26}		$a_{132}, a_{131}, a_{130}, a_{129}, a_{128}, a_{127}[-16], a_{126}[23], a_{125}$
133	m_{17}		$a_{133}, a_{132}, a_{131}, a_{130}, a_{129}, a_{128}, a_{127}[-16], a_{126}[23]$
134	m_{11}	2^{11}	$a_{134}[12], a_{133}, a_{132}, a_{131}, a_{130}, a_{129}, a_{128}, a_{127}[-16]$
135	m_{20}		$a_{135}, a_{134}[12], a_{133}, a_{132}, a_{131}, a_{130}, a_{129}, a_{128}$
136	m_{29}		$a_{136}, a_{135}, a_{134}[12], a_{133}, a_{132}, a_{131}, a_{130}, a_{129}$
137	m_{19}		$a_{137}, a_{136}, a_{135}, a_{134}[12], a_{133}, a_{132}, a_{131}, a_{130}$
138	m_0		$a_{138}, a_{137}, a_{136}, a_{135}, a_{134}[12], a_{133}, a_{132}, a_{131}$
139	m_{12}		$a_{139}, a_{138}, a_{137}, a_{136}, a_{135}, a_{134}[12], a_{133}, a_{132}$
140	m_7		$a_{140}, a_{139}, a_{138}, a_{137}, a_{136}, a_{135}, a_{134}[12], a_{133}$
141	m_{13}		$a_{141}, a_{140}, a_{139}, a_{138}, a_{137}, a_{136}, a_{135}, a_{134}[12]$
142	m_8'		$a_{142}, a_{141}, a_{140}, a_{139}, a_{138}, a_{137}, a_{136}, a_{135}$

Table 12. A set of sufficient conditions on a_i for the differential path given in Table 10 and 11, up to the first inner collision

Step	Conditions of the chaining variable in each step
5	$a_{5,1} = 0$, $a_{5,2} = 0$, $a_{5,3} = 0$, $a_{5,4} = 1$, $a_{5,5} = 0$
6	$a_{6,1} = 0$, $a_{6,2} = 0$, $a_{6,3} = 0$, $a_{6,4} = 0$, $a_{6,5} = 0$
7	
8	$a_{8,1} = 0$, $a_{8,2} = 0$, $a_{8,3} = 0$, $a_{8,4} = 0$, $a_{8,5} = 0$, $a_{8,29} = 0$, $a_{8,30} = 0$
9	$a_{9,1} = 0$, $a_{9,2} = 0$, $a_{9,3} = 0$, $a_{9,4} = 1$, $a_{9,5} = 0$, $a_{9,29} = 0$, $a_{9,30} = 0$
10	$a_{10,1} = 0$, $a_{10,2} = 0$, $a_{10,3} = 0$, $a_{10,4} = 0$, $a_{10,5} = 0$, $a_{10,22} = 0$, $a_{10,29} = 0$
11	$a_{11,22} = 0$, $a_{11,29} = 1$, $a_{11,30} = 0$
12	$a_{12,1} = 0$, $a_{12,2} = 0$, $a_{12,3} = 0$, $a_{12,4} = 0$, $a_{12,5} = 0$, $a_{12,29} = 0$, $a_{12,30} = 1$
13	$a_{13,1} = 1$, $a_{13,2} = 1$, $a_{13,3} = 1$, $a_{13,4} = 1$, $a_{13,5} = 1$, $a_{13,15} = 1$, $a_{13,16} = 0$, $a_{13,22} = 0$, $a_{13,29} = 0$, $a_{13,30} = 0$
14	$a_{14,15} = 0$, $a_{14,16} = 0$, $a_{14,22} = 0$
15	$a_{15,22} = 0$, $a_{15,29} = 0$, $a_{15,30} = 0$
16	$a_{16,8} = 0$, $a_{16,9} = 0$, $a_{16,10} = 1$, $a_{16,15} = 0$, $a_{16,16} = 0$, $a_{16,18} = 0$, $a_{16,29} = 1$, $a_{16,30} = 1$
17	$a_{17,8} = 0$, $a_{17,9} = 0$, $a_{17,10} = 0$, $a_{17,15} = 0$, $a_{17,16} = 1$, $a_{17,18} = 0$, $a_{17,22} = 0$
18	$a_{18,2} = 1$, $a_{18,11} = 0$, $a_{18,15} = 0$, $a_{18,16} = 0$, $a_{18,22} = 1$
19	$a_{19,3} = 0$, $a_{19,8} = 0$, $a_{19,9} = 0$, $a_{19,10} = 0$, $a_{19,11} = 0$, $a_{19,18} = 0$
20	$a_{20,2} = 0$, $a_{20,3} = 0$, $a_{20,8} = 0$, $a_{20,9} = 0$, $a_{20,10} = 0$, $a_{20,15} = 0$, $a_{20,16} = 0$, $a_{20,18} = 0$
21	$a_{21,2} = 1$, $a_{21,8} = 0$, $a_{21,9} = 0$, $a_{21,10} = 0$, $a_{21,11} = 0$, $a_{21,15} = 1$, $a_{21,16} = 1$, $a_{21,18} = 0$
22	$a_{22,2} = 0$, $a_{22,3} = 0$, $a_{22,4} = 0$, $a_{22,11} = 0$, $a_{22,27} = 0$
23	$a_{23,2} = 0$, $a_{23,3} = 0$, $a_{23,4} = 1$, $a_{23,8} = 0$, $a_{23,9} = 1$, $a_{23,10} = 0$, $a_{23,11} = 0$, $a_{23,18} = 0$, $a_{23,27} = 1$
24	$a_{24,2} = 0$, $a_{24,3} = 0$, $a_{24,4} = 0$, $a_{24,7} = 0$, $a_{24,8} = 1$, $a_{24,9} = 1$, $a_{24,10} = 1$, $a_{24,18} = 1$, $a_{24,20} = 0$, $a_{24,27} = 0$
25	$a_{25,2} = 0$, $a_{25,4} = 1$, $a_{25,7} = 0$, $a_{25,11} = 0$, $a_{25,20} = 0$, $a_{25,27} = 1$
26	$a_{26,3} = 0$, $a_{26,4} = 0$, $a_{26,11} = 1$, $a_{26,23} = 1$, $a_{26,27} = 0$
27	$a_{27,2} = 0$, $a_{27,3} = 1$, $a_{27,7} = 0$, $a_{27,20} = 0$, $a_{27,32} = 0$
28	$a_{28,2} = 1$, $a_{28,4} = 0$, $a_{28,7} = 1$, $a_{28,20} = 0$, $a_{28,23} = 1$, $a_{28,24} = 0$, $a_{28,27} = 0$, $a_{28,32} = 0$
29	$a_{29,4} = 1$, $a_{29,7} = 0$, $a_{29,20} = 0$, $a_{29,23} = 1$, $a_{29,24} = 1$, $a_{29,27} = 1$, $a_{29,32} = 0$
30	$a_{30,7} = 0$, $a_{30,16} = 0$, $a_{30,20} = 0$, $a_{30,23} = 1$, $a_{30,24} = 0$, $a_{30,32} = 0$
31	$a_{31,7} = 0$, $a_{31,16} = 1$, $a_{31,20} = 0$, $a_{31,23} = 0$, $a_{31,23} = 0$, $a_{31,24} = 0$, $a_{31,32} = 0$
32	$a_{32,16} = 0$, $a_{32,23} = 0$, $a_{32,24} = 1$, $a_{32,28} = 0$, $a_{32,32} = 1$
33	$a_{33,16} = 0$, $a_{33,23} = 1$, $a_{33,24} = 1$, $a_{33,28} = 1$, $a_{33,32} = 0$
34	$a_{34,13} = 0$, $a_{34,16} = 1$, $a_{34,24} = 0$, $a_{34,28} = 0$
35	$a_{35,16} = 1$, $a_{35,23} = 0$, $a_{35,24} = 1$, $a_{35,28} = 0$

Table 13. (Continued from Table 12)

Step	Conditions of the chaining variable in each step
36	$a_{36,12} = 0$, $a_{36,13} = 0$, $a_{36,16} = 1$, $a_{36,23} = 1$, $a_{36,28} = 1$, $a_{36,32} = 1$
37	$a_{37,6} = 1$, $a_{37,12} = 1$, $a_{37,13} = 1$, $a_{37,16} = 0$, $a_{37,28} = 1$, $a_{37,31} = 0$
38	$a_{38,6} = 0$, $a_{38,12} = 0$, $a_{38,13} = 0$, $a_{38,16} = 1$, $a_{38,24} = 0$, $a_{38,28} = 1$
39	$a_{39,6} = 0$, $a_{39,12} = 0$, $a_{39,13} = 0$, $a_{39,28} = 0$, $a_{39,31} = 0$
40	$a_{40,6} = 0$, $a_{40,12} = 0$, $a_{40,13} = 1$, $a_{40,24} = 0$, $a_{40,28} = 1$, $a_{40,31} = 1$
41	$a_{41,6} = 1$, $a_{41,12} = 1$, $a_{41,13} = 1$, $a_{41,24} = 1$, $a_{41,31} = 0$
42	$a_{42,2} = 1$, $a_{42,6} = 1$, $a_{42,12} = 1$, $a_{42,13} = 0$, $a_{42,24} = 0$, $a_{42,31} = 0$
43	$a_{43,6} = 0$, $a_{43,12} = 1$, $a_{43,13} = 1$, $a_{43,24} = 0$, $a_{43,31} = 1$
44	$a_{44,2} = 0$, $a_{44,6} = 1$, $a_{44,12} = 1$, $a_{44,24} = 1$, $a_{44,31} = 1$
45	$a_{45,2} = 1$, $a_{45,20} = 1$, $a_{45,24} = 1$, $a_{45,31} = 0$
46	$a_{46,2} = 0$, $a_{46,24} = 0$, $a_{46,31} = 1$
47	$a_{47,2} = 0$, $a_{47,20} = 0$, $a_{47,24} = 1$
48	$a_{48,2} = 1$, $a_{48,20} = 0$
49	$a_{49,2} = 1$, $a_{49,20} = 0$
50	$a_{50,2} = 0$, $a_{50,20} = 0$
51	$a_{51,2} = 1$, $a_{51,20} = 1$, $a_{51,23} = 0$
52	$a_{52,16} = 1$, $a_{52,20} = 1$, $a_{52,23} = 0$
53	$a_{53,20} = 0$, $a_{53,23} = 0$
54	$a_{54,20} = 1$, $a_{54,16} = 0$, $a_{54,23} = 0$
55	$a_{55,16} = 1$, $a_{55,23} = 0$
56	$a_{56,16} = 0$, $a_{56,23} = 1$
57	$a_{57,16} = 0$, $a_{57,23} = 1$
58	$a_{58,16} = 1$, $a_{58,23} = 0$
59	$a_{59,12} = 1$, $a_{59,16} = 1$, $a_{59,23} = 1$
60	$a_{60,12} = 0$, $a_{60,16} = 0$
61	$a_{61,12} = 0$, $a_{61,16} = 1$
62	$a_{62,12} = 0$
63	$a_{63,12} = 0$
64	$a_{64,12} = 1$
65	$a_{65,12} = 0$

Collisions and Near-Collisions for Reduced-Round Tiger

John Kelsey[1] and Stefan Lucks[2]

[1] NIST, USA
john.kelsey@nist.gov
[2] University of Mannheim, Germany
http://th.informatik.uni-mannheim.de/people/lucks/

Abstract. We describe a collision-finding attack on 16 rounds of the Tiger hash function requiring the time for about 2^{44} compression function invocations. This extends to a collision-finding attack on 17 rounds of the Tiger hash function in time of about 2^{49} compression function invocations. Another attack generates circular near-collisions, for 20 rounds of Tiger with work less than that of 2^{49} compression function invocations. Since Tiger has only 24 rounds, these attacks may raise some questions about the security of Tiger. In developing these attacks, we adapt the ideas of message modification attacks and neutral bits, developed in the analysis of MD4 family hashes, to a completely different hash function design.

Keywords: Tiger, hash function, collisions, attack.

1 Introduction

In the past two years, a *flood* of cryptanalytic results [5,6,7,8,9,2,3] has washed away most of the practical hash functions used so far. Design-wise, all these hash functions (including MD5, RIPEMD, SHA0, and SHA1) descend from MD4. This has led to a growing interest into alternative hash function designs, which had been mostly overlooked by cryptanalysts so far. One such alternative construction is Tiger, designed by Anderson and Biham in 1996 [1]. Like the MD4-descendants, Tiger iterates an internal *compression function* for hashing arbitrarily long[1] messages. Tiger's compression function, however, is very different from the compression functions of the MD4 family.

Because the compression functions are so different internally, the attacks against the MD4 family would appear unlikely to be directly useful in attacking Tiger. Our analysis bears this out to some extent–the message modification techniques we use differ in important ways from those in [5,6,7,8]. However, we use message modification against Tiger for the same broad purpose as it is used in [7,8]–to control the differences in the first few rounds by the choice of message

---

[1] Tiger appears to restrict messages to 2^{64} bits maximum, based on the size of the message length field.

M.J.B. Robshaw (Ed.): FSE 2006, LNCS 4047, pp. 111–125, 2006.

values, despite having the message differences forced on us by our analysis of the message schedule. Further, the use of neutral parts of the message in [2] is directly applicable to our approach in attacking Tiger. In some sense, this is a hopeful sign; it implies that we may hope to take the attack techniques developed against the MD4 family, and apply them, in suitably altered form, to hash functions built on entirely different lines.

Below, we describe a collision-finding attack on Tiger reduced to 16 rounds. As the full Tiger operates on 24 rounds, this attack gets through two thirds of Tiger, with work equivalent to 2^{44} compression function invocations. Tiger produces a 192-bit hash, so a collision should ideally take 2^{96} such invocations.

We describe how to extend this attack to 17 rounds of Tiger, increasing the work to no more than 2^{49} compression function invocations.

Also, we describe an attack to choose two input chaining values with a small (namely, six bit) Hamming distance, which generates a near-colliding compression function outputs with the same Hamming distance – following, in fact, the same differential pattern as the input. In this sence, we describe our near-collisions as *circular*. (One could also describe them as pseudo-near-collisions.) This third attack gets through more than 80 % of Tiger (20 / 24 rounds), with work equivalent to less than 2^{49} compression function invocations. An ideal 192-bit hash should need approximately

$$\sqrt{2^{192} \Big/ \binom{192}{6}} \approx 2^{80} \quad \text{compression function invocations}$$

for near-collisions with six bits of Hamming distance, instead of $< 2^{49}$.

The remainder of this paper is organized as follows. Section 2 provides a description of Tiger, in sufficient detail to follow our attacks. Section 3 provides an overview over the collision attack and describes some of the details. Sections 4 and 5 deal with the core of the attack: the message modification technique. Section 6 introduces techniques to extend our attack to more than 16 rounds. These are demonstrated by a collision attack against 17 rounds of Tiger, and by a circular near-collision attack against 20 rounds. Section 7 briefly discusses the security of Tiger, and outlines some lessons learned from the attack.

2   High-Level Description of Tiger

Tiger's compression function is based on applying an internal "block cipher like" function, which takes a 192-bit "plaintext" and a 512-bit key to compute a 192-bit "ciphertext". The "block cipher like" function is applied according to the Davies-Meyer construction: a 512-bit message block is used as a key to encrypt the 192-bit chaining value, and then the input chaining value is fed forward to make the whole function non-invertible. In the remainder of this section, we will describe Tiger in sufficient detail to follow the course of our attack. Note that if, for any given input chaining value, we can generate two different messages yielding the same output chaining value, then we have found a collision for Tiger.

Tiger was designed with 64-bit architectures in mind. Accordingly, we will denote a 64-bit unsigned integer as a "word". We will represent a word as a hexadecimal number. Tiger uses arithmetic operations (addition, subtraction and multiplication by small constants), bit-wise XOR, NOT, logical shift operations and S-Box applications. The arithmetic operations over words are modulo 2^{64}. The chaining value is represented internally as three 64-bit words, the message block as eight 64-bit words.

Thus, three words A, B, C describing the input chaining value and eight message words $X_0, \ldots, X_7$ are fed into the compression function, which generates three words A', B', C' describing the output chaining value. The compression function's final output A'', B'', C'' is generated by the feedforward function

$$A'' := A \oplus A',$$
$$B'' := B - B', \text{ and}$$
$$C'' := C + C'.$$

2.1   The Tiger Round Function

In the terminology of [4], Tiger's block cipher like function is a "target-heavy unbalanced Feistel cipher". The block is broken into three words, labeled A, B, and C. Each round, a message word X is XORed into C:

$$C := C \oplus X.$$

Then A and B are modified:

$$A := A - \mathbf{even}(C),$$
$$B := B + \mathbf{odd}(C),$$
$$B := B \times (\text{const}),$$

with a round-dependent constant $(\text{const}) \in \{5, 7, 9\}$. The results are then shifted around, so that A,B,C becomes B,C,A. See Figure 1.

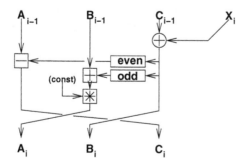

Fig. 1. The round function of Tiger

For the definition of **even** and **odd**, consider the word C being split into eight bytes $C[0], \ldots, C[7]$, with the most significant byte $C[0]$. The functions **even** and **odd** employ four S-Boxes $T_1, \ldots, T_4 : \{0,1\}^8 \rightarrow \{0,1\}^{64}$ as follows:

$$\mathbf{even}(C) := T_1(C[0]) \oplus T_2(C[2]) \oplus T_3(C[4]) \oplus T_4(C[6]) \quad \text{and}$$
$$\mathbf{odd}(C) := T_1(C[7]) \oplus T_2(C[5]) \oplus T_3(C[3]) \oplus T_4(C[1]).$$

The "even bytes" and the "odd bytes" of a word W are defined as

$$W[\mathbf{even}] = (W[0], W[2], W[4], W[6]) \in (\{0,1\}^8)^4 \quad \text{and}$$
$$W[\mathbf{odd}] = (W[7], W[5], W[3], W[1]) \in (\{0,1\}^8)^4.$$

The round function spreads changes around very quickly – a one-bit difference introduced into C in the first round will change about half the bits of the block by the end of the third round. Tiger seems to be much better at this than the members of the MD4 family.

It is easy to produce local collisions for the Tiger round function, using some pattern $(\alpha, \beta, 0, \alpha')$. Here, α is an input difference to the even bytes of the S-boxes, β is an XOR difference which is expected to cancel out the result of that difference on the even function, and α' is α multiplied by (const), being expected to cancel out the original introduced change of α. However, local collisions of this form are surprisingly hard to use in attacks on more than eight rounds of Tiger – the key schedule seems to be quite effective at destroying such patterns.

2.2   The Key Schedule

Tiger consists of 24 rounds. Each round uses one message word X_i as its round key. The first eight round keys $X_0, \ldots, X_7$ are identical to the 512-bit cipher key (or rather, to the 512-bit message block). The remaining 16 round keys are generated by applying the key schedule function:

$$(X_8, \ldots, X_{15}) := \text{KeySchedule}(X_0, \ldots, X_7)$$
$$(X_{16}, \ldots, X_{23}) := \text{KeySchedule}(X_8, \ldots, X_{15})$$

The key schedule uses logical shifts on words, denoted by $\ll$ and $\gg$, e.g.,

- $1111\,5555\,9999\,\text{FFFF} \ll 5 = 222\text{A}\,\text{AAB3}\,333\text{F}\,\text{FFE0}$, and
- $222\text{A}\,\text{AAB3}\,333\text{F}\,\text{FFE0} \gg 9 = 0011\,1555\,5999\,9\text{FFF}$.

Further, it uses the bit-wise NOT function, e.g. for $X = \text{EEEE}\,\text{AAAA}\,6666\,0000$, the negation of X is $\overline{X} = 1111\,5555\,9999\,\text{FFFF}$. The key schedule modifies its input $(x_0, \ldots, x_7)$ in two passes:

first pass

1. $x_0 := x_0 - (x_7 \oplus \text{Const}_1)$
2. $x_1 := x_1 \oplus x_0$
3. $x_2 := x_2 + x_1$
4. $x_3 := x_3 - (x_2 \oplus (\overline{x_1} \ll 19))$
5. $x_4 := x_4 \oplus x_3$
6. $x_5 := x_5 + x_4$
7. $x_6 := x_6 - (x_5 \oplus (\overline{x_4} \gg 23))$
8. $x_7 := x_7 \oplus x_6$

second pass

9. $x_0 := x_0 + x_7$
10. $x_1 := x_1 - (x_0 \oplus (\overline{x_7} \ll 19))$
11. $x_2 := x_2 \oplus x_1$
12. $x_3 := x_3 + x_2$
13. $x_4 := x_4 - (x_3 \oplus \overline{x_2} \gg 23))$
14. $x_5 := x_5 \oplus x_4$
15. $x_6 := x_6 + x_5$
16. $x_7 := x_7 - (x_6 \oplus \text{Const}_2)$

The final values $(x_0, \ldots, x_7)$ are used as the key schedule output. The constants are $\text{Const}_1 = \text{A5A5} \ldots \text{A5A5}$ and $\text{Const}_2 = 0123 \ldots \text{CDEF}$.

3   The Attack

We propose a differential attack on Tiger in three parts. Throughout the attack, we are switching between XOR-differences and additive differences. In general, switching between differences holds with some nonzero probability; for example, an additive difference of 1 can be represented as an XOR difference of 1, with probability $1/2$ of being correct.

3.1   Conventions

Transforming one type of difference into another is typically probabilistic, but for some values, it has probability one.

- If $X - Y = 2^i$, then $\Pr[X \oplus Y = 2^i] = 1/2$. The exception is $i = 63$, where $\Pr[X \oplus Y = 2^i] = 1$.
- Let $I := 2^{63}$. Switching between the additive difference I and the XOR-difference I succeeds with probability 1. In other words, when dealing with a difference I, we need not care what type of difference this actually is. Our attack will make extensive use of this simple fact.
- Note that a difference I in a word W remains the same, even if W is multiplied by some odd constant (const), as done in the Tiger compression function.

We start counting rounds by 0, and we write X_i for the message word input of the i-th round, and A_i, B_i, C_i for the output of round i – which just happens to be the input chaining values for round $i + 1$. Accordingly, the chaining value input for the round 0 (the first round) is A_{-1}, B_{-1}, C_{-1}.

The differences in message words are most usefully seen as XOR-differences, since the message word (or the "round key") X_i is XORed into the state. Additive differences are what we need to know when dealing with the two target words in the round (the two words that get altered), because the arithmetic differences are all mod 2^{64}. For the S-box inputs in the message modification step, XOR differences are most useful.

We will use the following notation for the differences which occur in some word W:

- $\Delta^+(W) = W - W^* \bmod 2^{64}$ for additive differences and
- $\Delta^\oplus(W) = W \oplus W'$ for word-wise differences.

3.2   Outline of the Attack

The attack can be broken into three pieces:

1. Differential characteristic $(I, I, I, I, 0, 0, 0, 0) \to (I, I, 0, 0, 0, 0, 0, 0)$ in the key schedule.
2. Differential characteristic $(I, I, 0) \to (0, 0, 0)$ in rounds 6-9 of the round function. (Because the message words in rounds 10-15 are unchanged, this leads to a collision after 16 rounds.)
3. Message modification to force the difference in the round function after round 6 to $(I, I, 0)$.

3.3   Key Schedule Differences

Consider a difference of the form $(I, I, I, I, 0, 0, 0, 0)$ in the message words. The first pass of the key schedule turns this into an intermediate difference pattern $(I, 0, I, 0, 0, 0, 0, 0)$. The second pass turns this into $(I, I, 0, 0, 0, 0, 0, 0)$. This is the differential pattern we will use for our attack; it holds with probability one, and covers the expanded message words used for rounds 0-15.

The colliding messages will differ only in the high order bits of their first four words. The expanded message words will differ for rounds 8-9, only in their high order bits. Expanded message words 10-15 will have no differences. This means that if the states of the compression functions processing the two messages are equal after round 9, they will remain equal until the end of round 15, yielding a 16-round collision.

3.4   Round Function Differences

Given the key schedule differential characteristic above, we can specify a differential characteristic for the round function from the end of round 6 to the end of round 9, going from $(I, I, 0) \to (0, 0, 0)$ by canceling with the differences in rounds 8-9. The expanded message words from rounds 10-15 have no differences, and thus a collision after round 9 becomes a collision for 16 rounds of Tiger. Figure 2 shows this characteristic.

3.5   Message Modification

The main difficulty of the attack is in the message modification step. Recall that our target difference at the end of round 6 is

$$\Delta^+(A_6) = I, \ \Delta^+(B_6) = I, \Delta^+(C_6) = 0.$$

Independently from the choice of message words, we know $\Delta^+(C_5)$ and $\Delta^+(C_4)$. Also, since $\Delta^+(X_6) = 0$, we need $\Delta^+(C_5) = \Delta^+(B_6) = I$. Similarly, we know the relationship $\Delta^+(C_4) = I + \Delta^+(\mathbf{odd}(B_6))$.

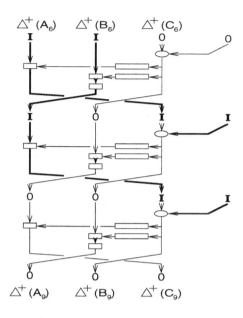

Fig. 2. Probability one characteristic from round 6-9

4   Local Message Modification by Meeting in the Middle

Assume we know inputs $(A_{i-1}, B_{i-1}, C_{i-1})$ and $(A_{i-1}^*, B_{i-1}^*, C_{i-1}^*)$, and some XOR differences in the message words X_i and X_{i+1}. We want to force some additive difference $\Delta^+(C_{i+1})$ to $\delta^* = C_{i+1} - C_{i+1}^*$. As depicted in Figure 3, the difference $\Delta^+(C_{i+1})$ depends on $\Delta^+(B_{i-1})$, the additive output difference of the **odd** function from round i, and the additive output difference of the **even** function from round $i + 1$.

4.1   Plain Message Modifications

First consider the **even** function, which, after computing $B_{i+1} := C_i \oplus X_{i+1}$, evaluates as

$$\mathbf{even}(B_{i+1}) := T_1(B_{i+1}[0]) \oplus T_2(B_{i+1}[2]) \oplus T_3(B_{i+1}[4]) \oplus T_4(B_{i+1}[6]).$$

For any nonzero XOR difference between words B_{i+1} and B_{i+1}^*, we expect about 2^{32} *different* additive output differences of the form $\delta_{\mathbf{even}} = \mathbf{even}(B_{i+1}) - \mathbf{even}(B_{i+1}^*)$. Similarly, when we consider the **odd** function

$$\mathbf{odd}(B_i) := T_1(B_i[7]) \oplus T_2(B_i[5]) \oplus T_3(B_i[3]) \oplus T_4(B_i[1]),$$

we expect close to 2^{32} *different* additive output differences of the form $\delta_{\mathbf{odd}} = \mathbf{odd}(B_i) - \mathbf{odd}(B_i^*)$.

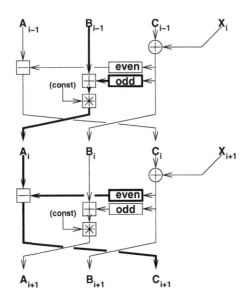

Fig. 3. The information flow from B_{i-1} to C_{i+1}

Thus, if the differences in $B_{i+1}[\textbf{even}]$ and in $B_i[\textbf{odd}]$ both are nonzero, we can apply a meet-in-the-middle (MITM) approach to force

$$(\Delta^+(B_{i-1}) + \delta_{\textbf{odd}}) \times \text{const} - \delta_{\textbf{even}} = \delta^*$$

- Store the 2^{32} candidates for $\delta_{\textbf{odd}}$ in a table.
- For all 2^{32} candidates for $\delta_{\textbf{even}}$, test if $\delta_{\textbf{odd}}$ exists with

$$\delta_{\textbf{even}} = (\Delta^+(B_{i-1}) + \delta_{\textbf{odd}}) \times (\text{const}) - \delta^*,$$

or rather

$$\delta_{\textbf{odd}} = (\delta_{\textbf{even}} + \delta^*)/(\text{const}) - \Delta^+(B_{i-1}) \qquad (1)$$

(note that since (const) is odd, division by (const) mod 2^{64} is well-defined).

This technique takes some 2^{32} evaluations of each of the functions **even** and **odd**, which is equivalent to about 2^{28} evaluations of the compression function – and, of course, some 2^{32} units of storage space.

We estimate that for given $\Delta^+(B_{i-1})$ and δ^*, the meet-in-the-middle approach succeeds with a probability close to $1/2$. In the attack scenario, we will repeat the approach with another $\Delta^+(B_{i-1})$ or another target difference δ^*, if necessary.

Assume $X_i[\textbf{even}]$ has been fixed and the MITM delivered $\delta_{\textbf{even}}$ and $\delta_{\textbf{odd}}$ satisfying Equation 1. We can now determine value for $B_{i+1}[\textbf{even}]$ and $B_i[\textbf{odd}]$ which will produce the desired differences, and thus will map the input difference to the output difference as required. During the MITM step, each candidate additive difference for $\delta_{\textbf{even}}$ is produced by one or more specific values of

$B_{i+1}[\textbf{even}]$, and likewise, each value of δ_{odd} is produced by one or more specific values of $B_i[\textbf{odd}]$.

Finally, we are able to compute 64 local message bits:

$$X_i[\textbf{odd}] := C_{i-1}[\textbf{odd}] \oplus B_i[\textbf{odd}] \quad \text{and}$$
$$X_{i+1}[\textbf{even}] := C_i[\textbf{even}] \oplus B_{i+1}[\textbf{even}].$$

Note that C_i has been defined by fixing $X_i[\textbf{even}]$.

In the attacks below, we use two variations on these ideas.

4.2   Message Modification to Get an XOR Difference

In step 3 of the attack below, we need a specific XOR difference in C_3. However, the meet-in-the-middle technique above takes an additive difference, not an XOR difference, as input. Our solution to this is to throw brute force computation at the problem: For a desired XOR difference of Hamming weight k, we simply go through the meet-in-the-middle search for each additive difference which could be produced by the XOR difference, until we run out of choices or find an additive difference which both matches and yields the desired XOR difference when we compute it forward.

An additive difference which can lead to a given k-bit XOR difference has about a 2^{-k} probability of doing so[2] This means we expect to need to try about 2^k additive differences which are consistent with the k-bit XOR difference before we succeed in finding a match. Since each MITM step succeeds in finding a matching additive difference about half the time, we will need to do a total of 2^{k+1} MITM steps. However, we can optimize this in a simple way, by only redoing one side of the MITM search for each new targeted additive difference. The expected work is thus bounded by 2^{28+k}.

4.3   Message Modification with Constraints

Two of the MITM steps (steps 4 and 5) in the attack below must live with constraints on the selection of message bits. The constraints come from the transition between an XOR difference in C_3 and an additive difference in B_4. Since the XOR difference has k bits active, and the additive difference is consistent with only one set of values for those bits, k bits of message word X_4 are constrained[3].

Constrained message modification is relatively simple: Instead of searching over 2^{32} possible additive differences from each side, we search over a smaller

---

[2] A k-bit XOR difference has either 2^k or 2^{k-1} additive differences consistent with it. For a flipped bit in position j, this represents the choice of whether to add or subtract 2^j. A flipped high order bit always matches both $+2^{63}$ and -2^{63} in mod 2^{64} arithmetic.

[3] For example, an XOR difference of 1 is consistent with an additive difference of either -1 or +1. If the low bit in C_3 is 0, the low bit in C_3^* will be 1, and reaching an additive difference of -1 will require fixing the low bit of X_4 to 1.

number, with the constrained bits of the message fixed to their required values. For the sake of simplicity, we assume that $k/2$ bits are constrained in the even bytes, and $k/2$ in the odd bytes. However, the probability of success is decreased in a corresponding way; with only 2^{28} choices from one side, and 2^{32} from the other, we expect about a 2^{-4} probability of a match. Thus, we expect to have to repeat an MITM search with 4 constrained bits about 16 times.

5   The Global Message Modification Scenario

0. Do a one-time precomputation to find a additive difference L with a low Hamming weight corresponding XOR difference which we can cancel out by our choice of the even bytes of X_6. (Note that the specific value of X_6 is not determined yet; we are simply ensuring that this additive difference will permit a choice of X_6 that will cancel the resulting difference out.) This costs 2^{27} Tiger-16 hash function equivalents, and we expect it to yield an additive difference which is consistent with an 8-bit XOR difference[4].

1. Choose X_0 and $X_1[\textbf{even}]$ to ensure that C_0 and C_1 have useful (that is, nonzero in both the even and odd bytes) differences. Note that at the end of this step, we know $\Delta^\oplus(C_1)$ and $\Delta^\oplus(C_2)$. We use these in the next step. The work here is negligible.

2. Choose $X_1[\textbf{odd}]$ and $X_2[\textbf{even}]$ to ensure that C_2 has a useful difference. Note that at the end of this step, we know $\Delta^\oplus(C_2)$. We use this XOR difference in the next step. The work here is negligible.

3. Do a message modification step to get XOR difference $\Delta^\oplus$ in C_3 which is consistent with the additive difference L. This is described above. The expected work here is about 2^{36} Tiger-16 hash equivalents, and we determine $X_2[\textbf{odd}], X_3[\textbf{even}]$.

4. Do a constrained meet in the middle step, choosing $X_3[\textbf{odd}], X_4[\textbf{even}]$ to get $\Delta C_4 = I$. We expect there to be four constrained bits, meaning that we expect to have to try this 16 times before we get a match. Each failed attempt requires that we go back to step 2. We thus expect to spend about $2^4 2^{36} = 2^{40}$ Tiger-16 equivalents completing this step of the attack.

5. Do a constrained meet in the middle step, choosing $X_4[\textbf{odd}], X_5[\textbf{even}]$ to force $\Delta C_5 = I$. As before, we expect this to be constrained by four bits, and thus to need to be repeated 16 times. Each failure requires that we go back to step 2. This step thus is expected to be completed after doing about $2^4 2^{40} = 2^{44}$ Tiger-16 hash equivalents of work.

6. Given the value of C_5, we use the results of step 0's search to determine the value for the even bytes of X_6. This is negligible work, and never fails.

---

[4] We expect this because in a set of 2^{32} random 64-bit integers, we expect about one with a Hamming weight of 8, since 64-bit integers with Hamming weight 8 make up about 2^{-32} of all 64-bit integers. In this case, a 9-bit XOR difference where one of the active bits is the high-order bit gives identical results in the remainder of the attack.

The result of this is an additive difference in the output of of $I, 0, I$ in the output of round 7. With probability one, this cancels out with the key schedule characteristic, leading to a 16-round collision.

5.1   Neutral Bits

The above attack has specified message words $X_{0,1,2,3,4}$ and the even bytes of message words $X_{5,6}$. This leaves an enormous number of bits of the message which can be varied without interfering with the 16-round collision. After having found the collision, we may freely determine the values for the odd bytes of $X_{5,6}$ and all of X_7. The above attack thus finds 2^{128} 16-round collisions for Tiger.

For example, consider varying the bytes of $X_5[\text{odd}]$. This alters the output of the **odd** function in round 5, and thus the value of A_6. However, since there is no difference active in the odd bytes of B_5, changing the input to the odd function in round 5 adds the same change to A_6 and A_6^*. This leaves the additive difference in A_6 unchanged, which means that the same difference in the **even** function in round 6 will cancel it out. Similarly, a change to the odd bytes of X_5 changes the value of B_5, but doesn't change the additive difference $B_5^* - B_5$, as it adds the same amount to both. The same kind of analysis applies to all the neutral bits.

5.2   Free Bits

The attack also imposes almost no constraints on $X_{0,1}$ or the even bytes of X_2. We need control of about 12 of those bits during the attack. A natural thing to do is to choose $X_{0,1}$ freely at the beginning of the attack in any way that is convenient, and then use the even bytes of X_2 to provide multiple trials for the message modification steps.

6   Going Beyond 16 Rounds

In this section, we will apply the 16-round collision finding technique from above as some subroutine, to attack more rounds of Tiger. Make the following two assumptions

1. The round keys $X_8, \ldots, X_{15}$ observe the characteristic $(I, I, I, I, 0, 0, 0, 0)$.
2. The input difference $(\Delta^+(A_7), \Delta^+(B_7), \Delta^+(C_7))$ to round eight is $(0, 0, 0)$.

If both assumptions hold, we can apply the 16-round technique from above to compute an "intermediate-message" $(X_8, \ldots, X_{15})$, such that we get a collision after round 23. As the key schedule is invertible, the "real message" $(X_0, \ldots, X_7)$ can easily be computed by by running the key schedule backwards.

6.1   A Round Key Differential

Set $I^* := I \gg 23 = 2^{40}$ and $I^{**} := I'^* \gg 23 = 2^{17}$. Assume that the eight message words observe the differential characteristic $(I, I, 0, 0, I + I^*, I + I^*, I^* +$

$I^{**}, 0$). With the probability $\approx 1/16$, we expect the first pass through the message schedule turns this characteristic into $(I, 0, 0, 0, I + I^*, 0, 0, 0)$. If this happens, then we expect the second pass to produce our target characteristic $(I, I, I, I, 0, 0, 0, 0)$ with probability $1/2$.

To summarise, we expect the differential characteristic

$$(I, I, 0, 0, I + I^*, I + I^*, I^* + I^{**}, 0) \longrightarrow (I, I, I, I, 0, 0, 0, 0) \qquad (2)$$

to hold with a probability of about $1/32$. We have verified this experimentically. (Actually, our results seem to indicate a slightly better probability of approximately $1/28$. But for simplicity, we use $1/32 = 2^{-5}$ for our analysis.)

6.2   Attacking 17 Rounds of Tiger

Now we describe an attack on 17 rounds of Tiger, namely rounds 7 to 23:

1. Given the initial value (A_6, B_6, C_6), choose the message word X_7 and apply one round of Tiger to get (A_7, B_7, C_7).
2. Apply the 16-round attack to get a message $(X_8, \ldots, X_{14}, X_{15})$ colliding with $(X_8, \ldots, X_{14}, X_{15}) + (I, I, I, I, 0, 0, 0, 0)$.
   Recall that $(X_8, \ldots, X_{14}, X_{15})$ contains 128 neutral bits, including all the 64 bits of X_{15}. I.e., any choice of X_{15} will produce a collision.
3. Now run the key schedule backwards to get $(X_0, \ldots, X_7)$. As X_7 has already been chosen, we have to observe a little twist here: Given $X_7, \ldots, X_{14}$, but ignoring X_{15}, we compute the remaining seven message words $X_0, \ldots, X_6$ as explained below.
4. Now we check our differential characteristic (2). If it holds, we have found a 17-round collision for rounds 7 to 23 and are done.
   Else, we go back to the first step.

On the average, the characteristic (2) holds at least one in $32 = 2^5$ times. Thus, the attack on 17 rounds of Tiger takes the time of about

$$2^5 * 2^{44} = 2^{49}$$

compression function invocations.

What about step 3 of the attack, i.e., running the key schedule backwards? Given X_7 and $X_8, \ldots, X_{14}$, we have to compute $X_0, \ldots, X_6$. We write $Y_0, \ldots, Y_7$ for the output of the first key schedule pass when computing

$$(X_8, \ldots, X_{14}, X_{15}^*) = \text{KeySchedule}(X_0, \ldots, X_7).$$

(Note that we actually know $X_8, \ldots, X_{14}$, while X_{15}^* is unknown.)

Inverting steps 11–15 of the second pass is straightforward:

$$Y_6 := X_{14} - X_{13}$$
$$Y_5 := X_{13} \oplus X_{12}$$
$$Y_4 := X_{12} + (X_{11} \oplus (\overline{X_{10}} \gg 23))$$
$$Y_3 := X_{11} - X_{10}$$
$$Y_2 := X_{10} \oplus X_9$$

We get the value Y_7 by inverting step 8 of the first pass:

$$Y_7 := X_7 \oplus Y_6$$

Finally, we can invert those steps of second pass which depend on Y_7:

$$Y_6 := X_9 + (X_8 \oplus (\overline{Y_7} \lll 19))$$
$$Y_0 := X_8 - Y_7$$

Inverting steps 1 to 7 of the first pass is quite similar.

6.3   Circular Near-Collisions for 20 Rounds of Tiger

Now we go even further, to 20 rounds of Tiger, at the cost of dealing with a weaker attack model. Instead of a collision-attack, we provide circular near-collisions with small Hamming weight (Hamming weight 6). We attack rounds 4 to 23 (i.e., all but the first four rounds). Hence, we denote the input chaining values by A_3, B_3, C_3. The attack works as follows:

1. Arbitrarily choose the chaining values A_7, B_7, C_7 for round 8.
2. Employ the 16-round attack, to find message words $X_8, \ldots, X_{15}$ such that the output after round 23 collides.
3. Run the key schedule backwards, to compute the "real" message words $X_0, \ldots, X_7$.
   If the characteristic (2) does not hold, go back to step 2.
4. Run the rounds 7, 6, 5, and 4 backwards to compute the initial values A_3, B_3, C_3. The differences in the message words induce the same differences in the initial values, namely

$$\Delta^{\oplus}(A_3) = I + I^* = \Delta^{\oplus}(B_3) \text{ and } \Delta^{\oplus}(C_3) = I^* + I^{**}.$$

5. The feedforward destroys the collision, of course. But with very high probability, it leaves us with a low Hamming weight near-collision. With probability 2^{-3} the feedforward output follows the same differential pattern than the input chaining values:

$$\Delta^{\oplus}(A_{23}) = I + I^* = \Delta^{\oplus}(B_{23}) \text{ and } \Delta^{\oplus}(C_{23}) = I^* + I^{**}.$$

If it doesn't follow this pattern, then randomly vary the neutral bits in $X_{13,14,15}$ until it does hold. We expect to need to try about $2^3 = 8$ sets of neutral bits for this.

Similarly to Section 6.2, we expect to iterate the 16-round attack no more than 2^5 times, on the average. In total, we expect a running time of about 2^{49} Tiger-20 equivalents. Varying the neutral bits in the last step adds negligible cost. Thus, the Tiger-20 near-collision attack costs less work than iterating Tiger-20 2^{40} times.

7   Conclusions and Open Questions

In this paper, we have developed collision attacks:

- an attack against 16 rounds of Tiger, requiring work equivalent to about 2^{44} compression function computations and
- an attack gainst 17 rounds of Tiger, being no more than 32 times slower – work equivalent to about 2^{44} compression function computations.

These attacks heavily use message modification techniques.

We have further exploited this technique for a near-collision attack (with adversarially chosen input chaining values) against the last 20 rounds of Tiger, also for about 2^{49} compression function computations worth of work. These near-collisions are circular, i.e., the input and the output chaining values have identical differences (with a Hamming weight of 6).

7.1   The Security of Tiger

All of our results are based on message modification techniques, which mean that we choose both the XOR differences in the message, and also specific values for most or all of the message bits. This constrains the attack in many ways. For example, we can see no way to adapt our current techniques to collisions against an application using 16-round Tiger in the HMAC construction–our lack of knowledge of the chaining values would make our approach impossible.

Second-preimage attacks on a single compression function computation also appear to be very difficult using our techniques. Both the difference between the colliding message blocks and the specific values of the messages are constrained by our attack; it appears to be very difficult to "work backward" from a specified message block with some hash output to a colliding message block. Second preimages are trivial to find for up to 8 rounds, and appear possible to find for up to 11 rounds using local collisions, but we have not investigated this line of attack in much detail yet.

We are more concerned with the possibility of extending the collision attack to more rounds. As Tiger has only 24 rounds, attaking 16–20 rounds is threatening. A relatively small improvement might make the attack techniques applicable to the full hash function. We definitely do not believe that the attack techniques presented here have been fully exploited in the current attack.

We point out that pseudo-collisions and near-collisions can be more than just certificational weaknesses. Some of the attacks against ciphers from the MD4 family employ pseudo- and near-collisions in attack scenarios with more than one message block, to find plain collisions for the hash function itself (see, e.g., [6]).

7.2   What We've Learned About Tiger

We draw two broad lessons from the analysis so far. First, we believe that Tiger has too few rounds. Message modification techniques allow us to almost completely control what happens in the first third of the hash function at present, allowing us to place differences in the remaining rounds almost without constraint. Second, the use of large S-boxes and mixing between addition and XOR operations is an excellent strategy for building a block cipher, but it works very differently inside a hash function. Large S-boxes tend to have a large set of equally good differentials, but which differential will pass the next round depends on the value of the internal state of the hash function; the attacker facing a block cipher with such large S-boxes must guess which differential to try; the attacker facing a hash function can often choose those values to make his differential work, or at least look inside the state of the hash function to determine the best differential path to try.

7.3 Applicability of the Tools of the MD4 Family Attacks

We have also seen some overlap in the tools used to attack the MD4 family, and our results on reduced-round Tiger. Broadly, we analyze the message expansion for the hash function, and form a differential characteristic which, if entered after round 7, will lead to a collision in the full hash function. We then use message modification to force the hash states processing a pair of messages with our desired difference onto this differential characteristic after round 7. This is quite similar to the techniques used in [7] and [8], though without (yet) the use of advanced message modification techniques. Similarly, a variation on the neutral bit techniques of [2] are used to make our 20-round pseudo-near-collision attack more efficient. While the details of using these attack tools are different for Tiger, the high level similarities in approach suggest that we may be learning generally useful attack techniques against hash functions from the recent results on the MD4 family of hash functions.

Acknowledgements

The authors wish to thank Eli Biham, Lily Chen, Orr Dunkelman, Morris Dworkin, Matt Fanto, and the anonymous referees for useful comments and discussions.

References

1. Anderson, R., Biham, E., Tiger: A Fast New Hash Function, Fast Software Encryption, FSE'96, LNCS 1039, 1996.
2. E. Biham, R. Chen. Near-Collisions of SHA-0. Crypto 04, LNCS 3152.
3. E. Biham, R. Chen, A. Joux, P. Carribault, C. Lemuet, W. Jalby. Collisions of SHA-0 and reduced SHA-1. Eurocrypt 2005, LNCS 3494, 36–57.
4. B. Schneier, J. Kelsey. Unbalanced Feistel Networks and Block Cipher Design. FSE 1996, LNCS, 121–144.
5. X. Wang, X. Lai, D. Feng, H. Cheng, X. Yu. Cryptanalyisis of the hash functions MD4 and RIPEMD. Eurocrypt 2005, LNCS 3494, 1–18.
6. X. Wang, H. Yu. How to break MD5 and other hash functions. Eurocrypt 2005, LNCS 3494, 19–35.
7. X. Wang, H. Yu, Y. L. Yin. Efficient collision search attacks on SHA0. Crypto 2005.
8. X. Wang, Y. L. Yin, H. Yu. Finding collisions in the full SHA1. Crypto 2005.
9. X. Wang, A. Yao, F. Yao. New Collision Search for SHA-1. Presentation at rump session of Crypto 2005 (communicated by A. Shamir).

Analysis of Step-Reduced SHA-256*

Florian Mendel**, Norbert Pramstaller,
Christian Rechberger, and Vincent Rijmen

Institute for Applied Information Processing and Communications (IAIK)
Graz University of Technology, Austria
Christian.Rechberger@iaik.tugraz.at
www.iaik.tugraz.at/research/krypto

Abstract. This is the first article analyzing the security of SHA-256 against fast collision search which considers the recent attacks by Wang *et al.* We show the limits of applying techniques known so far to SHA-256. Next we introduce a new type of perturbation vector which circumvents the identified limits. This new technique is then applied to the unmodified SHA-256. Exploiting the combination of Boolean functions and modular addition together with the newly developed technique allows us to derive collision-producing characteristics for step-reduced SHA-256, which was not possible before. Although our results do not threaten the security of SHA-256, we show that the low probability of a single local collision may give rise to a false sense of security.

1   Introduction

After recent cryptanalytic results on MD5 [19], SHA-1 [2,14,18] and similar hash functions, the resistance of members of the SHA-2 family (*i.e.* SHA-224, SHA-256, SHA-384 and SHA-512) [12] against recent attacks is an important issue.

While SHA-1 and MD5 are currently the most commonly used hash functions worldwide, the direct successor of SHA-1, SHA-256 is in many cases considered to be an upgrade option. However, SHA-256 did not receive as much cryptanalytic scrutiny from the cryptographic community as other hash functions. Even though the underlying design principle did not change since MD4, SHA-256 needs to be considered separately. It is expected to be much stronger than SHA-1, but several questions concerning its collision resistance need to be answered:

– Are the currently known techniques applicable to SHA-256? Which ones and to what extent?
– What about new techniques which are specifically designed to be applied to SHA-256?

In this article, we give preliminary answers to these questions. To put our contribution into perspective, we first survey existing approaches and previous results.

---

* The work in this paper has been supported by CRYPTREC.
** This author is supported by the Austrian Science Fund (FWF) project P18138.

M.J.B. Robshaw (Ed.): FSE 2006, LNCS 4047, pp. 126–143, 2006.

1.1 Outline of Existing Approaches

The basic approach for efficient collision search of the predecessors of SHA-256, SHA-0 and SHA-1, can be described as follows:

1. Identify local collisions in the state-update transformation.
2. Search for low-weight perturbation vectors by searching for low-weight expanded messages. In the approach by Chabaud and Joux [4] the perturbation vectors need to satisfy some additional properties which were dropped later on by Wang *et al.* [18] by using more complicated techniques.
3. Build the difference vector by interleaving the local collisions as described by the perturbation vector. Note that in [14] and [13] an approach is described which combines the three steps above.
4. The complexity of the collision search attack is related to the probability with which the characteristic described by these interleaved local collisions is followed.
5. By adjusting message bits for the chosen characteristic and allowing small variations in the characteristic, the computational effort for the collision search is decreased.

1.2 Survey of Existing Results on SHA-256

Being standardized by NIST in 2000 [12], the first published independent analysis of members of the SHA-2 family was done by Gilbert and Handschuh [6]. They show that there exists a 9-step local collision with probability 2^{-66}. Later on, the result was improved by Hawkes *et al.* [7]. By considering modular differences, they give a new maximal probability of 2^{-39}.

In [9] SHA-256 is analyzed in encryption mode. Attacks based on related-key assumptions for up to 37 steps are presented there. In [21], all modular additions are replaced by XOR. For this variant, a search for pseudo-collisions which is faster than brute force search for up to 34 steps faster is described.

In [11] a variant of SHA-256 is analyzed where all Σ and σ are removed. The conclusion is that collisions can be found much faster than by brute force search for this variant. Additionally, some low-weight expanded message differences for a GF(2)-linearized message expansion are given. The work shows that the approach used by Chabaud and Joux [4] in their analysis of SHA-0 is extensible to that particular variant of SHA-256.

1.3 Our Contribution

So far, nobody described ways to do collision search attacks for the unmodified SHA-256 or step-reduced variants thereof. After a short description of SHA-256 in Sect. 2, we address this issue in several ways.

First we analyze the message expansion of SHA-256 and show that its properties prevent the efficient extension of the techniques used by Chabaud and Joux [4] on SHA-0 or by Wang *et al.* in the analysis of SHA-1 [18] (see Sect. 3.1).

Table 1. Notation

notation	description
$A_i \ldots H_i$	state variables at step i of the compression function
$A \oplus B$	bit-wise XOR of state variable A and B
$A + B$	addition of state variable A and B modulo 2^{32}
A'	XOR difference in state variable A
M_t	input message word t (32 bits), $t \geq 1$
W_t	expanded input message word t (32 bits), $t \geq 1$
$ROTR^n(A)$	bit-rotation of A by n positions to the right
$SHR^n(A)$	bit-shift of A by n positions to the right
N	number of steps of the compression function

To illustrate our point, we define a variant SHA-256-3R, where the shift operation in the message expansion is replaced by a rotation. We show how to use low-weight perturbation vectors in a collision search for this variant in Sect. 3.2.

Next we focus on the unmodified SHA-256. To overcome the limitations we identified before, we introduce the idea to drop the requirement of having a perturbation vector which is a valid expanded message. In Sect. 3.3, we develop a way to derive this new type of perturbation vector for SHA-256, which can be used to find collision-producing characteristics with high probability. The price we have to pay is an increase of the search space for this new type of perturbation vector.

Some heuristics, which were developed in the case of SHA-1 to reduce the search space, do not apply to SHA-256. In Sect. 3.4 we describe a new way to reduce the search space.

As an example, a 19-step collision for unmodified SHA-224 is presented in Sect. 3.5, including a detailed description of the characteristic being used. There are two lessons to be learned from this example. Firstly, compared to independently multiplying probabilities for local collisions, interleaving them dramatically increases their overall probability. This contrasts with observations on older members of the SHA family. Secondly, techniques that exploit the combination of Boolean functions and modular addition in the state update are shown to be applicable in this example of a SHA-224 collision.

Additionally, we briefly survey methods to speed up the collision search for SHA-256 in Sect. 4. These methods and refinements thereof are subsequently used in some examples. These examples include the fast generation of 18-step collisions and 22-step pseudo-collisions with non-zero message difference on a standard PC.

2   Description of SHA-256

In the remainder of this article we use the notation given in Table 1. A complete description of SHA-256 can be found in [12]. We briefly review parts of the specification needed subsequently.

SHA-256 is an iterated cryptographic hash function based on a compression function that updates the eight 32-bit state variables $A, \ldots, H$ according to the values of 16 32-bit words $M_0, \ldots, M_{15}$ of the message. The compression function consists of 64 identical steps as presented in Fig. 1. The step transformation employs the bitwise Boolean functions f_{MAJ} and f_{IF}, and two GF(2)-linear functions

$$\Sigma_0(x) = ROTR^2(x) \oplus ROTR^{13}(x) \oplus ROTR^{22}(x),$$
$$\Sigma_1(x) = ROTR^6(x) \oplus ROTR^{11}(x) \oplus ROTR^{25}(x).$$

The i-th step uses a fixed constant K_i and the i-th word W_i of the expanded message. The message expansion works as follows. An input message is padded and split into 512-bit message blocks. Let ME denote the message expansion function. ME takes as input a vector M with 16 coordinates and outputs a vector W with N coordinates. The coordinates W_i of the expanded vector are generated from the initial message M according to the following formula:

$$W_i = \begin{cases} M_i & \text{for } 0 \leq i < 16 \\ \sigma_1(W_{i-2}) + W_{i-7} + \sigma_0(W_{i-15}) + W_{i-16} & \text{for } 16 \leq i < N \end{cases} \quad (1)$$

Taking a value for N different to 64 results in a step-reduced (or extended) variant of the hash function. The functions $\sigma_0(x)$ and $\sigma_1(x)$ are defined as follows: $\sigma_0(x) = ROTR^7(x) \oplus ROTR^{18}(x) \oplus SHR^3(x)$ and $\sigma_1(x) = ROTR^{17}(x) \oplus ROTR^{19}(x) \oplus SHR^{10}(x)$.

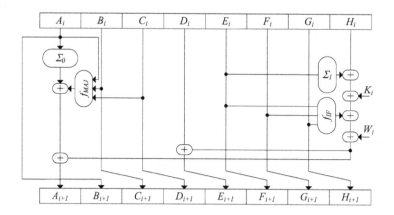

Fig. 1. One step of the state update transformation of SHA-256

3 Finding Collision Producing Characteristics for Step-Reduced SHA-256

Finding collision-producing characteristics for SHA-256 with a high probability is difficult. While searching for high probability characteristics, $GF(2)$-linear approximations that hold with high probability are useful. In the case of the addition mod 2^{32}, bit-wise XOR of the inputs has probability 1 for the LSB and probability 0.5 for all other bits. In the case of the Boolean functions f_{IF} and f_{MAJ}, several approximations are possible which all hold with probability 0.5. Superficially comparing the results in [7] and [6][1] would lead to the preliminary conclusion that the notion of modular differences instead of XOR differences offers a significant advantage. However, we argue that this is not the case. Using the XOR differences as presented in [6] and looking at possible characteristics for a local collision, we estimate that the probability of a single local collision (depending on the bit position of the perturbation) can be higher than 2^{-39}. This is by assuming unknown state variables at the beginning of the local collision. Since this compares favorably with the best results known so far, we will stick to XOR differences..

Compared to SHA-1, where the corresponding probabilities are between 2^{-2} and 2^{-5}, the probability for a local collision in SHA-256 is still very low. However, we will show by means of an example, that by interleaving several local collisions to build a collision-producing characteristic, the combined probability is much higher than the product of the single probabilities. This effect also occurs in the case of SHA-1 [18], but with much less impact on the overall complexity of the attack. However, before we arrive there, we need to discuss how to find suitable ways for interleaving these local collisions.

3.1   Why Existing Approaches Do Not Work

In this section we discuss to which extent the methods that are used in the analysis of SHA-0 and SHA-1 are applicable to SHA-256 as well. A very high-level description of the so-called perturbation-correction method to find collisions for SHA-0 and SHA-1 used in [4,18] could be the following:

1. Find a vector d such that the perturbation vector $d' = \mathrm{ME}(d)$ has a low Hamming weight.
2. Determine the correction vectors c'_u which ensure that the expanded message difference $e' = d' + \sum_u c'_u$ results in a collision for the linearized hash function. The mapping from d' to the c'_u depends on the properties of the state update transformation alone.
3. Determine the vectors c_u such that $c'_u = \mathrm{ME}(c_u)$. Construct the message difference as $e = d + \sum_u c_u$.
4. Determine M and M^* such that the differences in the real hash function follow the characteristic built for the linearized hash functions. We will refer to this characteristic as *L-characteristic*.

For all the hash functions of the SHA family, the vectors c'_u can be computed as $c'_u = \mathrm{R}_{r_u} \circ \mathrm{T}_u(d')$. The map $\mathrm{R}_{r_u}(x')$ *rotates* every coordinate of the vector x' over the constant amount r_u. The map $\mathrm{T}_u(x')$ *translates* the coordinates of the vector x' over u positions to the right, dropping the leftmost u coordinates and filling in u zeroes on the left. The values (u, r_u) depend on the

---

[1] Probability of 2^{-39} vs. 2^{-66} for a single local collision in SHA-256.

state update transformation. For instance, for the case of SHA-1, the values are $(1,5),(2,0),(3,-2),(4,-2)$ and $(5,-2)$.

The message expansion of a hash function is not surjective. We call x' a *valid expanded message* if there exists a value x such that $x' = \text{ME}(x)$. Additional conditions can be imposed on d in order to ensure that the vectors c'_u are valid expanded messages. In particular, we need the two following conditions.

Condition 1: $R_{r_u}(\text{ME}(d))$ needs to be a valid expanded message, for all values r_u that occur.

Condition 2: $T_u(\text{ME}(d))$ needs to be a valid expanded message, for all values u that occur.

It can easily be verified that for the message expansion of SHA-1, Condition 1 is satisfied for all d and for all r_u. Condition 2 can be satisfied by ensuring that "the backwards expanded difference equals zero in the first 5 steps" [4].

For the case of SHA-256 with linearized message expansion (all modular additions are replaced by XOR), Condition 2 can easily be satisfied by requiring that the backwards expanded difference equals zero in the first 8 steps. Contrary to SHA-1, satisfying Condition 1 imposes severe restrictions on d.

It has been observed before [13,14] that the perturbation-correction method imposes overly strict requirements. Indeed, instead of requiring that d' and each of the c'_u are valid expanded messages, it suffices to demand that the sum $e' = d' + \sum_u c'_u$ is a valid expanded message. For SHA-1, this observation doesn't lead to improved results. However, for SHA-256, it does as we will show in Sect. 3.3.

We show that for SHA-256, Condition 1 cannot always be met by proving the following Theorem.

Theorem 1. *For SHA-256, not all perturbation vectors d satisfying Condition 1 lead to a perturbation-correction vector e' which is a valid expanded message.*

The proof is given in Appendix A and shows first that this holds for a variant of SHA-256 with linearized message expansion and then extends this result to unmodified SHA-256.

The implication of this result is as follows: when we try to extend the standard perturbation-correction method, which is at the core of every analysis of SHA-0 and SHA-1 including those of Wang *et al.*, to analyze SHA-256, we cannot prevent the fact that there will be *unwanted differences* due to the message expansion. For later reference, we term them "ghost differences of type 2".

Theorem 1 also shows that the additional degrees of freedom we have due to the $GF(2)$ non-linearity of the message expansion are not sufficient to always correct this undesired behavior. In other words, by applying the standard perturbation-correction method, we are facing impossible differentials in the message expansion.

Implications of Theorem 1 on the collision search complexity. The major improvement of Wang *et al.*, which eventually lead to the break of SHA-1, was the ability to deal with a different kind of ghost-difference. By dropping

Condition 2, unwanted differences appear in the first 5 steps of SHA-1. We term them "ghost differences of type 1". In the following, we expand on that.

In the case of SHA-1 [18], the (near-)collision-producing characteristic is actually a concatenation of a low-probability general characteristic with a high probability L-characteristic. By means of the general characteristic in the first steps, these "ghost differences of type 1" are incorporated. This general characteristic has a very low probability, but this fact is compensated by message modification, which "bypasses" the probability of the chosen characteristic for more than 20 steps.

What would happen if we drop Condition 2 in the case of SHA-256? The "ghost differences of type 1" as described above will now appear up to step 8. However, starting from step 17 until step N, there will also be "ghost differences of type 2". Even if it would be possible to incorporate them in an even more complex general characteristic covering all N steps, the impact of this approach on the attack complexity would be severe.

The attack complexity is determined by the probability with which the part of the characteristic is followed that is not covered by message modification techniques. Since the low-probability general characteristic needs to be followed for all steps now, message modification cannot prevent its influence on the attack complexity anymore. It is by no means clear that such a general characteristic for all N steps of SHA-256 is even possible. Even if it is, the probability to follow this general characteristic up to step N is likely to be prohibitively low. Therefore, an other approach will be needed.

3.2   A Short Detour: SHA-256-3R

We show that by making a small change in SHA-256, the basic perturbation-correction approach can be applied again. We name this variant SHA-256-3R and change the message expansion of SHA-256 in the following way: The functions $\sigma_0(x)$ and $\sigma_1(x)$ are replaced by the following:

$$\sigma_0(x) = ROTR^7(x) \oplus ROTR^{18}(x) \oplus ROTR^3(x)$$
$$\sigma_1(x) = ROTR^{17}(x) \oplus ROTR^{19}(x) \oplus ROTR^{10}(x)$$

SHR is replaced by $ROTR$ which has the effect that Condition 1 imposes no restrictions anymore. Table 2 gives us the starting point for our analysis. It shows a remarkably low-weight perturbation vector which satisfies the following requirements. Firstly, the last 8 perturbation words are all-zero, which means that we can finish all the needed corrections. Secondly, the backwards expansion is all-zero for the first 8 steps which prevents "ghost differences of type 1" in our perturbation-correction vector. These requirements are enough to build a collision-producing characteristic which is constructed by interleaved local collisions as described by the perturbation vector. It is given in Appendix B. Note that this characteristic is an L-characteristic.

Most of the local collisions will be completed within the first 16 steps. The last local collision will be completed at step 27. Due to the small change in the message expansion, we do not have any "ghost differences of type 2". Showing

Table 2. Low-weight expanded message for the XOR-linearized 31-step message expansion of SHA-256 which can be used as a perturbation vector for SHA-256-3R

d_i'			
$i = 1$	80000000	$i = 11$	0
$i = 2$	11002000	$i = 12$	0
$i = 3$	80000000	$i = 13$	0
$i = 4$	14044aa8	$i = 14$	0
$i = 5$	00205000	$i = 15$	0
$i = 6$	0	$i = 16$	0
$i = 7$	0	$i = 17$	0a020000
$i = 9$	0	$i = 18$	0
$i = 9$	11002000	$i = 19$	80000000
$i = 10$	80000000	$i = 20 \ldots 31$	0

this fulfills the purpose of this detour, hence we stop the analysis of this L-characteristic of SHA-256-3R here.

3.3 Extending the Rijmen-Oswald Approach to Unmodified SHA-256

In the next two subsections, we show that despite the findings in the previous section it is still possible to find perturbation vectors of low-weight which lead to collision-producing characteristics. It will turn out that these perturbation vectors are no longer valid expanded messages by themselves. Thus we will have a *new type of perturbation vector* for SHA-256.

The approach to find such a new type of perturbation vector for SHA-256 is outlined below. The underlying idea is originally proposed in [14] and extended in [13]. Basically, it works as follows. First, we build a linearized version of the message expansion and the state update transformation. Then, we construct a generator matrix G which describes all possible state variables that result in a collision for this linearized version. By searching for low-weight codewords(see [3,10,15]) in the linear code described by G, we are actually searching for L-characteristics with high probability.

In the case of SHA-256, the dimension of G is $512 \times (9 * 32 * N)$ where N denotes the number of steps. Note that we can cut the parts representing the state variables B, C, D, F, G, H and thus reducing the *length* of the code without loosing information. However, without a way to reduce the *size* of the code and not excluding low-weight codewords, a search for L-characteristics with high probability is not feasible.

3.4 Reducing the Search Space to Find Useful L-Characteristics

In the analysis of SHA-1 [2,8,14,18], it was possible to reduce the search space for perturbation vectors or general collision producing characteristics by applying the following observation. Low-weight expanded messages for SHA-1 have the property that non-zero bits occur in bands, *i.e.* the non-zero values are concentrated on a few bit positions in every word. This can be explained by the weak avalanche effect of the SHA-1 message expansion. This heuristic does not apply to SHA-256. The functions $\Sigma_0, \Sigma_1, \sigma_0$ and σ_1 effectively prevent such a structure in L-characteristics. Therefore, the search space and hence the size of the code needs to be reduced by other means.

Another way of looking at the search for low-weight codewords in the code describing L-characteristics is as follows. Searching for low-weight codewords maps to searching for low-weight solutions in a homogeneous system of equations in GF(2). Actually, the corresponding check matrix H of the code described by G is a representation of the coefficients of this system. The variables refer to all message bits and state variable bits in the linearized variant. The system of equations described by H is under-defined, *i.e.* there are more variables than equations.

Forcing bits to zero or one can also be seen as adding new equations, where we simply set this bit to that particular value. The generator matrix G as described in 3.3 gives us 512 degrees of freedom, which means we can add up to 511 equations to H. By forcing those bits to zero which we expect to be zero in an L-characteristic, we eventually arrive at a system of equations where it is feasible to search for low-weight solutions. Note that this is a rather rough way to reduce the search space which does not work for larger number of steps N.

3.5   Example of a Collision-Producing L-Characteristic for 19-Step SHA-224

In Table 3, we give an L-characteristic of a 1-near-collision for 19 steps of SHA-256 which is a 19-step collision for SHA-224 at the same time. Note that the only difference between SHA-224 and SHA-256 is that at the output, the rightmost 32 bits are discarded. By applying the techniques described in Sect.3.4, we reduced the size of the code to 64, which led to our results.

The perturbation vector which is used as a building block for this characteristic is the vector A' in Table 3. The perturbation vector is not a valid expanded message. Note that this perturbation vector can be word-wise rotated without loosing its property of leading to a perturbation-correction vector which is al-

Table 3. Example of a 19-step SHA-224 collision. All-zero differences are denoted by a single 0 to improve readability.

Step	W'	A'	B'	C'	D'	E'	F'	G'	H'
1-4	0	0	0	0	0	0	0	0	0
05	85009008	85009008	0	0	0	85009008	0	0	0
06	a14cae12	a1442610	85009008	0	0	02000802	85009008	0	0
07	0	0	a1442610	85009008	0	084c4120	02000802	85009008	0
08	8200a8a8	00000020	0	a1442610	85009008	00000020	084c4120	02000802	85009008
09	85009008	85009008	00000020	0	a1442610	01008008	00000020	084c4120	02000802
10	0	0	85009008	00000020	0	02000802	01008008	00000020	084c4120
11	0	0	0	85009008	00000020	0	02000802	01008008	00000020
12	0	00000020	0	0	85009008	0	0	02000802	01008008
13	0	0	00000020	0	0	84001000	0	0	02000802
14	00088802	0	0	00000020	0	0	84001000	0	0
15	0	0	0	0	00000020	0	0	84001000	0
16	0	0	0	0	0	00000020	0	0	84001000
17	0	0	0	0	0	0	00000020	0	0
18	0	0	0	0	0	0	0	00000020	0
19	0	0	0	0	0	0	0	0	00000020

ways a valid difference between expanded message. Thus, we can rotate our L-characteristic to maximize the number of MSBs involved such that the probability of this L-characteristic is maximized. The first perturbations start at step 5, and there are 23 of them in total. The local collision for SHA-256 as originally described in [6] needs 24 single-bit differences in the message. Using this as an upper bound, we would expect up to 552 single-bit differences in the expanded messages. Note that the actual value 37, the weight of the message difference, is far below this upper bound.

In this particular example, more than 200 conditions on the state variables need to be met to follow the given L-characteristic. Assuming random independent trials, each perturbation would on average contribute a factor of 2^{-10} instead of about 2^{-40} to the overall probability. Hence, the fact that a single local collision in SHA-256 has a comparatively low probability may give a false feeling of security. Note that the actual collision search complexity can be further reduced by techniques mentioned in Sect. 4.

3.6   Adjustments to Circumvent Impossible Characteristics

In this section we take a closer look at the presented L-characteristic. The best probabilities for local collisions are achieved by approximating the differential behavior of the functions f_{MAJ} and f_{IF} by 0. On average, both approximations hold with probability 0.5. However, in certain cases, the probability for this approximation is 0.

Translating these properties into the sequence of states of the SHA-256 compression function gives rise to the following observations.

Observation 1. *Whenever we have 3 non-zero differences in consecutive variables of the state $(A'_r, A'_{r+1}, A'_{r+2})$ at the same bit position, the chosen linear approximation fails to predict any subsequent difference.*

Observation 2. *Whenever we have 2 non-zero differences followed by one zero differences in consecutive variables of the state $(E'_r, E'_{r+1}, E'_{r+2})$ at the same bit position, the chosen linear approximation fails to predict any subsequent difference.*

In order to prevent these cases, we would need to exclude all of them from our search space. However by doing this, low-weight solutions might be excluded. By using the degrees of freedom we have in our characteristic, *i.e.* various ways in which differences can propagate through the Boolean functions and the modular addition, we observe the following. It turns out to be possible to circumvent these impossible characteristics by choosing a slightly different characteristic for the same differential. Note that a similar strategy was used in the analysis of SHA-1 [2,18].

This suggests that the additional complexity of the SHA-256 state update transformation does not prevent us from using a similar approach. To illustrate this property, we take the 19-step L-characteristic presented in the previous subsection. Indeed, we have a single case of two consecutive words which have a difference at the same bit position. This happens in E_7 and E_8 at bit position 5. Thus Observation 2 applies. The result is that the function f_{IF} accepts $(0, 1, 1)$ as input difference at bit position 5 in step 9. Hence the output of f_{IF} will flip with probability 1.

The easiest way to cancel out this additional difference is by using other differences in the same step. At the output of Σ_1, we have a difference in bit 4. By a simple carry extension we can produce a change in the carry caused by this difference. The result will be that in contrast to the prediction of our L-characteristic, the difference in bit 4 will cause bit 5 to flip as well. However, this additional difference due to the carry extension will now cancel the additional difference at the output of f_{IF} in this step. Eventually, the path described by the L-characteristic can be followed without the need to circumvent additional impossible characteristics.

4   Increasing the Performance of Collision Search for SHA-256

In this section we briefly cover ways to speed up the collision search for members of the SHA-2 family once a suitable characteristic is found. For their predecessors SHA-1 and MD5, two competing approaches can be found in the literature. One approach has been termed message modification. It was first introduced in [16,19]. A variant of the technique was also used in the most recent analysis of SHA-0 [20] and SHA-1 [17,18].

The second approach was introduced in [1] and later on applied in [2]. It extends the idea of [5] to the hash function SHA-0. So-called neutral bits in the input message are used to circumvent the probabilistic behavior of the first steps of SHA-0. Within certain limits, both approaches can be extended to the case of SHA-256. Subsequently, we briefly discuss to which extent this is possible.

In the first 16 steps of SHA-256, the conditions on the state variables can be directly rewritten to conditions on the message words. The procedure can be described as follows:

$$
\begin{aligned}
A_{N+1} &= f_1(A_N, \ldots, H_N) + K_N + W_N \\
E_{N+1} &= f_2(A_N, \ldots, H_N) + K_N + W_N
\end{aligned}
\tag{2}
$$

Next, adjust A_{N+1} and E_{N+1} accordingly to meet the conditions derived for the characteristic. Then calculate

$$
\begin{aligned}
W_N &= A_{N+1} - f_1(A_N, \ldots, H_N) - K_N \\
W_N &= E_{N+1} - f_2(A_N, \ldots, H_N) - K_N
\end{aligned}
\tag{3}
$$

Note that by applying these formulas, each new message word is calculated twice. Hence it is possible that changes in the message bits contradict each other. In these cases, adjusting message words which are input in the steps before the contradiction occurs is necessary. A high-level algorithm to deal with this issue is given below.

Algorithm 1. Way to fulfill contradicting conditions in SHA-256

Require: Contradicting Conditions in A_i and E_i
Ensure: Condition in A_i and E_i are fulfilled
  Fulfill Condition in E_i by adjusting W_i as described in Equation (3).
  while Condition in A_i is not fulfilled or any previously fulfilled conditions are affected
  do
    Go back to step $x \in \{0 \ldots i-1\}$ and check if W_x can be adjusted such that the
    condition in A_i is fulfilled
  end while

Note that such methods were not needed in any of the predecessors of the SHA-2 family, because there can be no contradictions in fulfilling conditions in the first 16 steps.

In order to illustrate the technique, let's assume that by applying the simple message modification rules described by (2) we are getting a contradiction in bit 4 of W_6. We set this bit such that the condition on state variable E_6 is met (Step 1). In order to fulfill the condition on state variable A_6 (*i.e.* bit 4 should have opposite value) we simply go back one step and flip bit 4 in W_5 (Step 3). That way, bit 4 in A_5 is flipped. However A_5 is not directly influencing A_6, but via Σ_0 and the f_{MAJ}-function. The effect is twofold.

- Firstly, depending on the other inputs of the f_{MAJ}-function, the output might not change. In this case, B_5 or C_5 need to be updated by going back and adjusting the input word at the respective step. In general, every message word before the step where the contradiction occurred might be a candidate for message modification. However, the risk that other conditions are affected by these adjustments increases with the number of backward-steps.
- Secondly, due to Σ_0, three other bit positions are also affected with every message word adjustment. These might in turn affect other conditions and might even cancel out the desired effect of the flipped bit at the input via carry propagation.

If it turns out that it cannot be prevented that other conditions are affected with these adjustments, another choice in step 2 needs to be made.

To sum up, compared to SHA-1 or MD4/MD5, message modification is more complex due to the fact that two state variables are updated at the same time. After step 16, chances that existing conditions are affected by message modification increase. In the Appendices C and D we show simple examples of the application of these techniques. Table 4 summarizes them.

Table 4. Summary of examples

function	steps	type	local collisions	probability
SHA-256-3R	31	collision	25	–
SHA-256	18	collision	1	~ 1 using neutral bits
SHA-224	19	collision	23	$< 2^{-200}$ before message modification
SHA-256	19	1-near-collision	23	$< 2^{-200}$ before message modification
SHA-256	22	pseudo-collision	–	~ 1 using neutral bits

5   Conclusions and Future Work

When the attack techniques that were used successfully against SHA-1, are applied to SHA-256, several problems arise. Firstly, the shift operations in the message expansion of SHA-256 severely limit the usefulness of the perturbation-correction approach. To circumvent this obstacle, we introduced a new type of perturbation vector. We showed that it is still possible to find low-weight difference vectors that may result in a collision, but the search space increases dramatically. In order to find collisions for versions with more than 20 steps, we need new heuristics to reduce the search space.

Secondly, the increased Hamming weight of the difference and the presence of two nonlinear Boolean functions in each step make it very difficult to avoid consecutive ones in the inputs of the Boolean functions. Hence we have to deal with the fact that the linear approximations for these functions often won't hold. We have presented some ideas on how to deal with this problem.

Thirdly, the very low probability for one local collision in SHA-256 may give rise to a false feeling of security. We have shown with examples that the interleaving of local collisions results in many canceled differences. Hence the probability of n interleaved local collisions is typically significantly larger than the probability of one local collision, raised to the power n. We need to develop better ways of estimating this probability.

Acknowledgements

We would like to thank Christophe De Cannière and Krystian Matusiewicz for helpful comments.

References

1. Eli Biham and Rafi Chen. Near-Collisions of SHA-0. In Matthew K. Franklin, editor, *Advances in Cryptology - CRYPTO 2004, 24th Annual International Cryptology Conference, Santa Barbara, California, USA, August 15-19, 2004, Proceedings*, volume 3152 of *LNCS*, pages 290–305. Springer, 2004.
2. Eli Biham, Rafi Chen, Antoine Joux, Patrick Carribault, Christophe Lemuet, and William Jalby. Collisions of SHA-0 and Reduced SHA-1. In Ronald Cramer, editor, *Advances in Cryptology - EUROCRYPT 2005: 24th Annual International Conference on the Theory and Applications of Cryptographic Techniques, Aarhus, Denmark, May 22-26, 2005. Proceedings*, volume 3494 of *LNCS*, pages 36–57. Springer, 2005.
3. Anne Canteaut and Florent Chabaud. A New Algorithm for Finding Minimum-Weight Words in a Linear Code: Application to McEliece's Cryptosystem and to Narrow-Sense BCH Codes of Length 511. IEEE Transactions on Information Theory, 44(1):367–378, 1998.
4. Florent Chabaud and Antoine Joux. Differential Collisions in SHA-0. In Hugo Krawczyk, editor, *Advances in Cryptology - CRYPTO '98, 18th Annual International Cryptology Conference, Santa Barbara, California, USA, August 23-27, 1998, Proceedings*, volume 1462, pages 56–71. Springer, 1998.

5. David Chaum and Jan-Hendrik Evertse. Crytanalysis of DES with a Reduced Number of Rounds: Sequences of Linear Factors in Block Ciphers. In Hugh C. Williams, editor, *Advances in Cryptology - CRYPTO '85, Santa Barbara, California, USA, August 18-22, 1985, Proceedings*, volume 218 of *LNCS*, pages 1–16. Springer, 1986.

6. Henri Gilbert and Helena Handschuh. Security analysis of SHA-256 and sisters. In Mitsuru Matsui and Robert Zuccherato, editors, *Selected Areas in Cryptography, 10th Annual International Workshop, SAC 2003, Ottawa, Canada, August 14-15, 2003, Revised Papers*, volume 3006 of *Lecture Notes in Computer Science*, pages 175–193. Springer, 2003.

7. Philip Hawkes, Michael Paddon, and Gregory G. Rose. On corrective patterns for the SHA-2 family. Cryptology ePrint Archive, Report 2004/207, August 2004. http://eprint.iacr.org/.

8. Charanjit S. Jutla and Anindya C. Patthak. A Matching Lower Bound on the Minimum Weight of SHA-1 Expansion Code. Cryptology ePrint Archive, Report 2005/266, 2005. http://eprint.iacr.org/.

9. Jongsung Kim, Guil Kim, Sangjin Lee, Jongin Lim, and Jung Hwan Song. Related-Key Attacks on Reduced Rounds of SHACAL-2. In Anne Canteaut and Kapalee Viswanathan, editors, *Progress in Cryptology - INDOCRYPT 2004, 5th International Conference on Cryptology in India, Chennai, India, December 20-22, 2004, Proceedings*, volume 3348 of *LNCS*, pages 175–190. Springer, 2004.

10. Jeffrey S. Leon. A probabilistic algorithm for computing minimum weights of large error-correcting codes. *IEEE Transactions on Information Theory*, 34(5):1354–1359, 1988.

11. Krystian Matusiewicz, Josef Pieprzyk, Norbert Pramstaller, Christian Rechberger, and Vincent Rijmen. Analysis of simplified variants of SHA-256. In *Proceedings of WEWoRC 2005*, LNI P-74, pages 123–134, 2005.

12. National Institute of Standards and Technology (NIST). FIPS-180-2: Secure Hash Standard, August 2002. Available online at http://www.itl.nist.gov/fipspubs/.

13. Norbert Pramstaller, Christian Rechberger, and Vincent Rijmen. Exploiting Coding Theory for Collision Attacks on SHA-1. In Nigel P. Smart, editor, *Cryptography and Coding, 10th IMA International Conference, Cirencester, UK, December 19-21, 2005, Proceedings*, volume 3796 of *LNCS*, pages 78–95. Springer, 2005.

14. Vincent Rijmen and Elisabeth Oswald. Update on SHA-1. In Alfred Menezes, editor, *Topics in Cryptology - CT-RSA 2005, The Cryptographers' Track at the RSA Conference 2005, San Francisco, CA, USA, February 14-18, 2005, Proceedings*, volume 3376 of *LNCS*, pages 58–71. Springer, 2005.

15. Jacques Stern. A method for finding codewords of small weight. In G. Cohen and J. Wolfmann, editors, *Coding Theory and Applications, 3rd International Colloquium, Toulon, France, November, 1988, Proceedings*, volume 388 of *LNCS*, pages 106–113. Springer, 1989.

16. Xiaoyun Wang, Xuejia Lai, Dengguo Feng, Hui Chen, and Xiuyuan Yu. Cryptanalysis of the Hash Functions MD4 and RIPEMD. In Ronald Cramer, editor, *Advances in Cryptology - EUROCRYPT 2005: 24th Annual International Conference on the Theory and Applications of Cryptographic Techniques, Aarhus, Denmark, May 22-26, 2005. Proceedings*, volume 3494 of *LNCS*, pages 1–18. Springer, 2005.

17. Xiaoyun Wang, Andrew Yao, and Frances Yao. New Collision Search for SHA-1, August 2005. Presented at rump session of CRYPTO 2005.

18. Xiaoyun Wang, Yiqun Lisa Yin, and Hongbo Yu. Finding Collisions in the Full SHA-1. In Victor Shoup, editor, *Advances in Cryptology - CRYPTO 2005, 25th Annual International Cryptology Conference, Santa Barbara, California, USA, August 14-18, 2005, Proceedings*, volume 3621 of *LNCS*, pages 17–36. Springer, 2005.
19. Xiaoyun Wang and Hongbo Yu. How to Break MD5 and Other Hash Functions. In Ronald Cramer, editor, *Advances in Cryptology - EUROCRYPT 2005: 24th Annual International Conference on the Theory and Applications of Cryptographic Techniques, Aarhus, Denmark, May 22-26, 2005. Proceedings*, volume 3494 of *LNCS*, pages 19–35. Springer, 2005.
20. Xiaoyun Wang, Hongbo Yu, and Yiqun Lisa Yin. Efficient Collision Search Attacks on SHA-0. In Victor Shoup, editor, *Advances in Cryptology - CRYPTO 2005, 25th Annual International Cryptology Conference, Santa Barbara, California, USA, August 14-18, 2005, Proceedings*, volume 3621 of *LNCS*, pages 1–16. Springer, 2005.
21. Hirotaka Yoshida and Alex Biryukov. Analysis of a SHA-256 variant. In Bart Preneel and Stafford Tavares, editors, *Selected Areas in Cryptography (SAC 2005), Kingston, Ontario, Canada, August 11-12, 2005, Proceedings to appear*, LNCS. Springer, 2005.

A   Proof of Theorem 1

Before proofing Theorem 1, we need the following Observation.

Observation 3. *Let (V',X',Y',Z') be the XOR-difference of 4 inputs of an addition mod 2^n. Let R' denote the XOR-difference of the result of this addition. $\forall j: 32 > j > i$ where the following situation occurs:*

$$V_i' = X_i' = Y_i' = Z_i' = R_i'$$

the following relation must hold:

$$V_{i+1}' \oplus X_{i+1}' \oplus Y_{i+1}' \oplus Z_{i+1}' \oplus R_{i+1}' = R_i'$$

Now we can give a constructive proof for Theorem 1.

Proof. Let's first consider the case of a GF(2)-linearized variant of ME, ME_{lin}. In order to proof Theorem 1 for this variant, it suffices to show that for a particular d, the vector e' is not a valid expanded message.

We choose d to be the perturbation vector shown in Table 2. Note that A' contains the perturbation vector in this table. Vector 17 of e' (the sum of the perturbations and corrections for W_{17}) is 1b022000. However, applying the recurrence relation for the SHA-256 message expansion, W_{17} turns out to be 8b022000. For the GF(2)-linearized variant of the SHA-256 message expansion in Equation 1, the proof would already be finished. Let's now consider unmodified SHA-256.

We can build up on the previously proved part on the linearized variant, but need to consider the additional degrees of freedom we have due to carries. Here we need to show that no two expansions of messages m and m^* can exist such that $W_{17} \oplus W_{17}^* = $ 1b022000.

W_{17}' can be rewritten as $(V + X + Y + Z) \oplus (V^* + X^* + Y^* + Z^*)$ Considering the recurrence relation given in Equation 1 and inserting the value from Table 2 we get $V' = \sigma_1(W_{15}) = 81609048$, $X' = W_{10} = 04f61081$, $Y' = \sigma_0(W_2) = 8e94a0c9$, $Z' = W_1 = 80000000$.

Now we apply Observation 3 and see that we get the following contradiction at bit-position 32(MSB). We have $V'_{31} = X'_{31} = Y'_{31} = Z'_{31} = R'_{31} = 0$, thus we require $V'_{32} \oplus X'_{32} \oplus Y'_{32} \oplus Z'_{32} \oplus R'_{32} = 0'$ which is not the case. Thus we have shown that even by using the additional degrees of freedom in the message expansion(*i.e.* the carry effect), we can never arrive at the desired difference 1b022000 in W_{17}.   □

B   L-Characteristic for a 31-Step Collision of SHA-256-3R

The L-characteristic for 31-step SHA-256-3R including the message difference used in Sect. 3.2 is given in Table 5.

Table 5. L-characteristic for a 31-step collision in SHA-256-3R

Step	W'	A'	B'	C'	D'	E'	F'	G'	H'
01	00000001	00000001	0	0	0	00000001	0	0	0
02	66284480	22004000	00000001	0	0	62084400	00000001	0	0
03	8c2760a2	00000001	22004000	00000001	0	0981008b	62084400	00000001	0
04	95c7e0f6	28089550	00000001	22004000	00000001	68009150	0981008b	62084400	00000001
05	e9732fd2	0040a000	28089550	00000001	22004000	829685b1	68009150	0981008b	62084400
06	5be0be03	0	0040a000	28089550	00000001	20906a04	829685b1	68009150	0981008b
07	11b2513e	0	0	0040a000	28089550	00000001	20906a04	829685b1	68009150
08	6c2091d0	0	0	0	0040a000	28089550	00000001	20906a04	829685b1
09	4e794ee2	22004000	0	0	0	2240e000	28089550	00000001	20906a04
10	09ec2102	00000001	22004000	0	0	0981008b	2240e000	28089550	00000001
11	bdcf75a7	0	00000001	22004000	0	40080400	0981008b	2240e000	28089550
12	ad02b460	0	0	00000001	22004000	0	40080400	0981008b	2240e000
13	2240e000	0	0	0	00000001	22004000	0	40080400	0981008b
14	092d4192	0	0	0	0	00000001	22004000	0	40080400
15	44280480	0	0	0	0	0	00000001	22004000	0
16	0	0	0	0	0	0	0	00000001	22004000
17	36044000	14040000	0	0	0	14040000	0	0	00000001
18	175330fb	0	14040000	0	0	1501a070	14040000	0	0
19	4e869ebe	00000001	0	14040000	0	00000001	1501a070	14040000	0
20	44280480	0	00000001	0	14040000	40080400	00000001	1501a070	14040000
21	910e2130	0	0	00000001	0	14040000	40080400	00000001	1501a070
22	175330fa	0	0	0	00000001	0	14040000	40080400	00000001
23	00000001	0	0	0	0	00000001	0	14040000	40080400
24	44280480	0	0	0	0	0	00000001	0	14040000
25	14040000	0	0	0	0	0	0	00000001	0
26	0	0	0	0	0	0	0	0	00000001
27	00000001	0	0	0	0	0	0	0	0
28–31	0	0	0	0	0	0	0	0	0

C   Example of an 18-Step Collision for SHA-256

In Table 7 we give an example of an 18-step collision for SHA-256. We used a combination of the message modification technique described in Sect. 4 and the search for neutral bits to

 - find the first 18-step collision in much less than a minute
 - generate millions of them by using a large set of 2-neutral bits

The L-characteristic for this 18-step collision in SHA-256 including the message difference is given in Table 6. Since there are no conditions on the IVs, every IV including the standard-IV can be used. By adding more steps to this characteristic, near-collisions for more than 18-step can be derived in a straightforward manner. Note however that the weight of the difference at the output will be higher than one, thus a 1-near-collision as presented in Sect. 3.5 cannot be derived that way.

Table 6. L-characteristic of an 18-step collision in SHA-256

Step	W'	A'	B'	C'	D'	E'	F'	G'	H'
01–03	0	0	0	0	0	0	0	0	0
04	80000000	80000000	0	0	0	80000000	0	0	0
05	22140240	0	80000000	0	0	20040200	80000000	0	0
06	42851098	0	0	80000000	0	0	20040200	80000000	0
07	0	0	0	0	80000000	0	0	20040200	80000000
08	80000000	0	0	0	0	80000000	0	0	20040200
09	22140240	0	0	0	0	0	80000000	0	0
10	0	0	0	0	0	0	0	80000000	0
11	0	0	0	0	0	0	0	0	80000000
12	80000000	0	0	0	0	0	0	0	0
13–18	0	0	0	0	0	0	0	0	0

Table 7. Example of an 18-step collision using the standard IV

i	M_i							
1–8	02679857	0183b9a1	005de4f5	0266ee0c	0d1442f0	06373a71	c445dec2	12542ec1
9–16	0982b61a	205a614c	2495a094	166ae4ac	15917909	1178f05a	0aae5a46	178058c6

D   Pseudo-collisions for the Compression Function of SHA-256

We give an example of a pseudo-collision for the compression function of step-reduced SHA-256. The attacker has more freedom in such a setting: In addition to choose different messages M and M^*, he is also allowed to choose different IVs for the compression function. The goal is to find (M, M^*, IV, IV^*) such that $compress(M, IV) = compress(M^*, IV^*)$. Note that this type of pseudo-collision is different from the one described in [21].

The difference to Sect. 3.3 is that we have more degrees of freedom since we do not require the starting difference to be all-zero. To derive the actual collision, we used the same techniques as in Appendix C.

Note that this serves as an example. More steps can be achieved by extending the given characteristic in the backwards direction. In the example of a pseudo-collision given in Table 9, we need a different IV. $IV_{new} = IV_{standard} \oplus IV_{Corr}$. IV_{Corr} is given in Table 10. The corresponding L-characteristic is given in Table 8. The required difference in the IV for this pseudo-collision is given in this table as well. Note that this L-characteristic is similar to the 23-step related-key characteristic used in [9].

Table 8. L-characteristic for a 22-step pseudo-collision in SHA-256

Step	W'	A'	B'	C'	D'	E'	F'	G'	H'
IV'	0	00000200	0	0	50090088	0	0	20880000	10080080
01	0	0	00000200	0	0	40010008	0	0	20880000
02	0	0	0	00000200	0	0	40010008	0	0
03	0	0	0	0	00000200	0	0	40010008	0
04	0	0	0	0	0	00000200	0	0	40010008
05	0	0	0	0	0	0	00000200	0	0
06	0	0	0	0	0	0	0	00000200	0
07	0	0	0	0	0	0	0	0	00000200
08	00000200	0	0	0	0	0	0	0	0
09-22	0	0	0	0	0	0	0	0	0

Table 9. 22-step Pseudo-Collision with $M' \neq 0$

i	M_i							
1-8	39b1309b	048a8b67	02e0fc89	1dd4b937	02784cbd	1527473f	0134eb90	023f18aa
9-16	008a6849	063fbdbc	2e06da49	0f2e9e2a	085d407e	1686fa83	03ad81fe	091da09b

Table 10. IV Correction for 22-step Pseudo-Collision with $M' \neq 0$

i	IV_{Corr}				
1-5	DCBD1A68	00000000	00000000	00000080	60810000

Improved Linear Distinguishers for SNOW 2.0

Kaisa Nyberg[1,2] and Johan Wallén[1]

[1] Helsinki University of Technology
[2] Nokia ResearchCenter, Finland
kaisa.nyberg@nokia.com, johan.wallen@tkk.fi

Abstract. In this paper we present new and more accurate estimates of the biases of the linear approximation of the FSM of the stream cipher SNOW 2.0. Based on improved bias estimates we also find a new linear distinguisher with bias $2^{-86.9}$ that is significantly stronger than the previously found ones by Watanabe et al. (2003) and makes it possible to distinguish the output keystream of SNOW 2.0 of length 2^{174} words from a truly random sequence with workload 2^{174}. This attack is also stronger than the recent distinguishing attack by Maximov and Johansson (2005). We also investigate the diffusion properties of the MixColumn transformation used in the FSM of SNOW 2.0 and present some evidence why much more efficient distinguishers may not exist.

Keywords: Stream cipher, SNOW 2.0, linear masking method, modular addition.

1 Introduction

Key stream generators are widely used in practise for random number generation and data encryption as stream ciphers. The history of this type of cryptographic primitive has not always been glorious. Most recently, algebraic cryptanalysis method has been successfully applied to a number of stream ciphers. On the other hand, there is no scientific evidence that stream ciphers are inherently less secure than block ciphers. To strengthen the scientific foundations of the security of stream ciphers the ECRYPT NoE launched in November 2004 a new multi-year project eSTREAM, the ECRYPT Stream Cipher Project, to identify new stream ciphers that might become suitable for widespread adoption [8].

In this paper new results of the strength of the stream cipher SNOW 2.0 against linear approximation are presented. SNOW 2.0 was proposed by Ekdahl and Johansson in [3] as a strengthened version of SNOW 1.0, which was a NESSIE candidate. Currently SNOW 2.0 is considered as one of the most efficient stream ciphers. It is used for benchmarking the performance of stream ciphers by the eSTREAM project. SNOW 2.0 has also been taken as a starting point for the ETSI project on a design of a new UMTS encryption algorithm [4].

Linear methods have been widely used to analyse stream ciphers. In addition to the traditional methods such as linear complexity and correlation analysis, attacks based on linear cryptanalysis method have been succesfully launched against stream ciphers. One of the reasons why SNOW 1.0 was rejected by the NESSIE project was its vulnerability against a distinguishing attack using linear cryptanalysis [2,3].

M.J.B. Robshaw (Ed.): FSE 2006, LNCS 4047, pp. 144–162, 2006.

Distinguishing attacks using linear cryptanalysis (linear masking) were previously applied to SNOW 2.0 by Watanabe et al., to see if the designers of the algorithm learnt their lesson [11]. An efficient distinguisher can be used to detect statistical bias in the key stream, and in some cases, also derive the key or initial state of the key stream generator. In this paper we show that the estimates of the strength of the linear approximations given in [11] were not accurate. Their best masking was estimated to have bias $2^{-112.25}$, while the true value is closer to $2^{-107.26}$. Further we find a linear masking of the FSM of SNOW 2.0 with bias $2^{-86.89}$. Using this masking a distinguishing attack on SNOW 2.0 can be given which requires 2^{179} bits of the key stream and 2^{174} operations. This attack also superceeds the attack by Maximov an Johansson in [7].

The paper is structured as follows. In Section 2 we present the details of SNOW 2.0 as needed in the investigations of this paper. Section 3 explains the linear masking method on SNOW 2.0 and summarises our results. In Section 4 we analyse assumptions under which the bias values in [11] were computed, and show that the assumptions do not hold. We also give examples of large deviations from correct values and investigate the behaviour of linear approximation of modular addition with three inputs. The main too is an algorithm for computing the correlations for modular addition with an arbitrary number of inputs, which we present in Annex A. In Section 5 we present our observations about the structure of SNOW 2.0 and other results from mask searches. Finally, in Section 6 we give some results about resistance against linear distinguishing attacks for SNOW 3G, which is a modification of SNOW 2.0 by ETSI SAGE intended to become a second encryption algorithm for the UMTS system. A draft version of SNOW 3G can be found in [4]. The description of the final version of SNOW 3G and rationale of its design can be found in the design and development report [5].

2   The Stream Cipher SNOW 2.0

The structure of SNOW 2.0 is depicted in Figure 1. The running engine is a linear feedback shift register (LFSR) consisting of 16 words of length 32 bits each. The LFSR is defined over $GF(2^{32})$ with feedback polynomial

$$\alpha x^{16} + x^{14} + \alpha^{-1} x^5 + 1 \in GF(2^{32})[x]$$

where $\alpha \in GF(2^{32})$ is a root of the polynomial

$$x^4 + \beta^{23} x^3 + \beta^{245} x^2 + \beta^{48} x + \beta^{239} \in GF(2^8)[x]$$

and β is a root of the polynomial

$$x^8 + x^7 + x^5 + x^3 + 1 \in GF(2)[x].$$

The bitwise xor of two 32-bit blocks is denoted by $\oplus$ and addition modulo 2^{32} is denoted by $\boxplus$. The LFSR feeds into a finite state machine (FSM). The FSM has two 32-bit registers $R1$ and $R2$. The state of the LFSR at time t is denoted by $(s_{t+15}, \ldots, s_t)$. The input to the FSM is s_{t+15} and s_{t+5} and the output F_t of the FSM is calculated as

$$F_t = (s_{t+15} \boxplus R1_t) \oplus R2_t,$$

for all $t \geq 0$, where we have denoted by $R1_t$ and $R2_t$ the contents of the registers $R1$ and $R2$, respectively, at time t. Then the output z_t of the keystream generator is given as

$$z_t = F_t \oplus s_t.$$

The contents of $R1$ is updated as $s_{t+5} \boxplus R2_t$ and the contents of $R2$ is updated as $S(R1_t)$ where the transformation S is composed of four parallel AES S-boxes followed by the AES MixColumn transformation. For the purposes of this paper only the details of the LFSR and the FSM are needed. For a complete description of SNOW 2.0 we refer to the paper [3] by Ekdahl and Johansson.

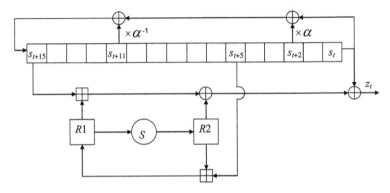

Fig. 1. SNOW 2.0

3   The Linear Masking Method on SNOW 2.0

3.1   Linear Masking of the FSM

We denote $\mathbf{F}_2 = GF(2)$. Let n be a non-negative integer. Given two vectors $x = (a_{n-1}, \ldots, a_0)$ and $y = (b_{n-1}, \ldots, b_0) \in \mathbf{F}_2^n$, let $x \cdot y$ denote the standard inner product $x \cdot y = a_{n-1}b_{n-1} \oplus \ldots \oplus a_0 b_0$. A constant vector which is used to compute inner product with inputs (outputs) of a function is called a linear input (output) mask of the function. Given a linear mask $\Gamma \in \mathbf{F}_2^n$ and an element $\alpha \in \mathbf{F}_2^n$, we denote by $\Gamma \alpha$ the linear mask, which satisfies the following equality

$$\Gamma \alpha \cdot x = \Gamma \cdot \alpha x, \text{ for all } x \in \mathbf{F}_2^n,$$

where the product αx is taken in $GF(2^{32})$. Let m and n be positive integers. Given a functional dependency $F : \mathbf{F}_2^n \rightarrow \mathbf{F}_2^m$, a linear input mask $\Lambda \in \mathbf{F}_2^n$ and a linear output mask $\Gamma \in \mathbf{F}_2^m$, the strength of the linear approximate relation $\Gamma \cdot F(x) = \Lambda \cdot x$, for $x \in \mathbf{F}_2^n$, is measured using its correlation

$$\begin{aligned} \mathrm{cor}_F(\Lambda, \Gamma) &= \mathrm{cor}(\Gamma \cdot F(x) \oplus \Lambda \cdot x) \\ &= 2^{-n}(\#\{x \in \mathbf{F}_2^n : \Gamma \cdot F(x) \oplus \Lambda \cdot x = 0\} - \#\{x \in \mathbf{F}_2^n : \Gamma \cdot F(x) \oplus \Lambda \cdot x = 1\}). \end{aligned}$$

For the purposes of this paper we use a derived value $\epsilon_F(\Lambda, \Gamma) = |\mathrm{cor}_F(\Lambda, \Gamma)/2|$ and call it the bias of the linear approximate relation $\Gamma \cdot F(x) = \Lambda \cdot x$.

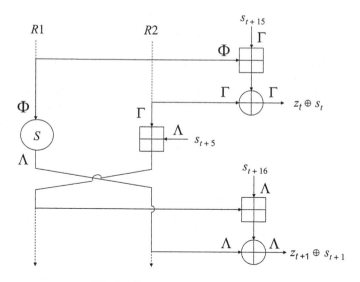

Fig. 2. Linear masking of SNOW 2.0

The linear masking method was applied to SNOW 2.0 in [11]. The linear approximation of the FSM of SNOW 2.0 used in [11] is depicted in Figure 2 in a slightly generalised form. In [11], it was always assumed that the output masks Γ at time t and Λ time $t + 1$ are equal, for all $t \geq 0$. In case when they are allowed to be different, it is straightforward to verify that the main distinguishing equation, [11], Equation (12), takes the following form

$$\Gamma \cdot (z_{t+16} \oplus z_{t+2}) \oplus \Gamma \alpha \cdot z_t \oplus \Gamma \alpha^{-1} \cdot z_{t+11} \oplus$$
$$\Lambda \cdot (z_{t+17} \oplus z_{t+3}) \oplus \Lambda \alpha \cdot z_{t+1} \oplus \Lambda \alpha^{-1} \cdot z_{t+12} = 0, \tag{1}$$

for all $t \geq 0$. This relation is obtained by using the approximation depicted in Figure 2 four times: firstly, two times with the mask pair Γ, Λ at time $t + 2$ and $t + 16$, then once with the mask pair $\Gamma \alpha$, $\Lambda \alpha$ at time t, and finally once with the mask pair $\Gamma \alpha^{-1}$, $\Lambda \alpha^{-1}$ at time $t+11$. Given the biases, these four approximations can be combined and the total bias value computed using the Piling Up Lemma [6]. Similarly as in [11] we denote by $\epsilon_{FSM}(\Lambda, \Gamma)$ the bias of the linear approximate relation of Figure 2. Hence the total bias $\epsilon(\Lambda, \Gamma)$ of the linear distinguisher (1) is calculated as

$$\epsilon(\Lambda, \Gamma) = 8\epsilon_{FSM}(\Lambda, \Gamma)^2 \epsilon_{FSM}(\Lambda \alpha, \Gamma \alpha)\epsilon_{FSM}(\Lambda \alpha^{-1}, \Gamma \alpha^{-1}).$$

We also introduce a new mask Φ, see Figure 2, whose role will be explained in subsection 4.2.

3.2 Our Results

We implemented a new wider mask search over the FSM SNOW 2.0 to achieve more accurate and improved estimates of the total bias of the linear distinguisher (1). In particular,

- we allow output masks Γ and Λ be different; and
- we improve the accuracy of the estimates of the bias values.

The effect of the first change turns out not to be significant. Suitable candidates for Γ were searched by first identifying Γ such that it performs reasonably well with Λ in the linear approximation. Here algorithms from [10] were used. Still, for a given Λ, the total bias of the distinguisher of Equation (1) is usually higher with $\Gamma = \Lambda$ than with any $\Gamma \neq \Lambda$. In some cases higher biases where obtained with $\Gamma \neq \Lambda$ but the achieved bias values were far from the best. Two show that such cases exist we give an example in Table 1.

Table 1. Two masks Λ with higher bias with $\Gamma \neq \Lambda$

Λ	Γ	$\epsilon(\Lambda, \Gamma)$
0x04400240	0x04400240	0
0x04400240	0x04400360	$2^{-122.29}$
0x08400280	0x08400280	$2^{-140.67}$
0x08400280	0x084003a0	$2^{-124.41}$

Table 2. The mask $\Lambda = \Gamma$ with the highest bias $2^{-86.89}$ for the linear distinguisher (1)

	mask value	ϵ_{FSM}
Λ	0x00018001	$2^{-15.496}$
$\Lambda\alpha$	0xc7000180	$2^{-27.676}$
$\Lambda\alpha^{-1}$	0x0180015c	$2^{-31.221}$

Table 3. Improved estimates of the biases for the best mask in [11]

	mask value	ϵ_{FSM} estimate in [11]	ϵ_{FSM} our estimate
Λ	0x0303600c	$2^{-27.61}$	$2^{-24.48}$
$\Lambda\alpha$	0x0c030360	$2^{-27.61}$	$2^{-24.49}$
$\Lambda\alpha^{-1}$	0x03600c63	$2^{-32.42}$	$2^{-36.82}$

The strongest linear approximation of the FSM of SNOW 2.0 found in our search is using the distinguisher (1) with $\Lambda = \Gamma = $ 0x00018001. The values of the biases of the linear approximation of the FSM are given in Table 2. They result in the total bias value of $2^{-86.89}$. This value is significantly higher than the bias value $2^{-112.25}$ achieved using the best linear mask 0x0303600 reported in [11]. The difference of results is due to the fact that Watanabe et al., used different and less accurate estimates of bias values as will be explained in more detail in Section 4. In Table 3 we give the new and more accurate bias values for the best mask in [11] which show that its strength was originally underestimated. Our new estimate of the total bias is $2^{-107.26}$.

We also looked at other linear approximations of SNOW 2.0 FSM than the one depicted in Figure 1. These will be discussed in Section 5.

4   Improved Approximation Over Dependent Functions

In [11] the biases of the linear approximations of the FSM were calculated under the assumption that all linear approximations over various nonlinear functions involved in the approximation depicted in Figure 2, the S-box ensemble S and the three additions modulo 2^{32}, are independent. Such an assumption is often reasonable in practise since there is no evidence of dependency although it is not possible to prove the opposite either. However, in the linear approximation of the FSM of SNOW 2.0, see Figure 2, there are two cases where combinations of two subsequent approximations are strongly dependent. The first such case is when one approximates over two subsequent additions modulo 2^{32}, that is, the output from the first addition is an input to the second addition. We will show in the next subsection that this must be handled as addition modulo 2^{32} with three inputs and not as two independent additions with two inputs. The second case is due to the fact that the value in register $R1$ is both an input to the modular addition $\boxplus$ and an input to the S-box ensemble S.

4.1   Linear Approximation Over Two Subsequent Modular Additions

In this section we investigate the behaviour of two consecutive modular additions with two inputs each, where the output from the first addition is input to the second one. Clearly such a composition is equivalent to one modular addition with three inputs. Previously, results and algorithms for computing biases of linear approximations have been presented only for the case with two inputs, see [10]. The basic algorithm for computing the bias for given input and output masks can be straightforwardly generalised to the case with an arbitrary finite number of inputs, and is given in Annex A. Our results show that the behaviour of modular addition under linear approximation depends to a large extent on the number of inputs. As a first result we demonstrate in Table 4 the reasons why the best linear mask found by us was not found by Watanabe et al. We denote by ϵ_+ the bias of linear approximation of modulo 2^{32} addition with two inputs and by ϵ_{++} the same value with three inputs using the same given mask value for all input and output masks. The value $2\epsilon_+^2$ in the middle is the one used in [11] in place of ϵ_{++}.

Table 4. Biases of linear approximation of addition with 2 inputs and 3 inputs for the best mask

	mask value	ϵ_+	$2\epsilon_+^2$	ϵ_{++}
Λ	0x00018001	2^{-2}	2^{-3}	$2^{-2.58}$
$\Lambda\alpha$	0xc7000180	2^{-26}	2^{-51}	$2^{-6.75}$
$\Lambda\alpha^{-1}$	0x0180015c	2^{-7}	2^{-13}	$2^{-7.71}$

It is also interesting to observe how differently linear approximation with one-bit masks behave over modular addition. In the two input case, the strength of the linear approximation degrades when moving towards the most significant bits. For addition with three inputs the bias values are almost the same in all positions as shown in Table 5. Moreover, we observed that linear approximation over modular addition with three inputs is more flexible and gives better bias values also when not all input masks

Table 5. Biases of linear approximation of addition with 2 inputs and 3 inputs for some 1-bit masks

mask value	ϵ_+	ϵ_{++}
0x00000010	2^{-5}	$2^{-2.61}$
0x00000100	2^{-9}	$2^{-2.59}$
0x00001000	2^{-13}	$2^{-2.58}$
0x00010000	2^{-17}	$2^{-2.58}$
0x00100000	2^{-21}	$2^{-2.58}$
0x01000000	2^{-25}	$2^{-2.58}$

are the same. The same holds in general for very sparse masks. Therefore we made an exhaustive search over all masks Λ with at most five non-zero bits. The results are given in Section 5.

4.2   Linear Approximation Over Composition of Different Functions

The modular addition $\boxplus$ and an input to the S-box ensemble S have the contents of register $R1$ as common inputs. Since S is invertible, we can compose these functions as follows:

$$f : y, z \mapsto S^{-1}(y) \boxplus z,$$

for all y that are output from the S-box ensemble and for all $z = s_{t+15}$. The task is to compute the correlation between the following linear combination of inputs and linear combination of outputs

$$\mathrm{cor}(\Gamma \cdot f(y, z) \oplus \Gamma \cdot z \oplus \Lambda \cdot y).$$

By applying a well-known theorem about correlations over composed functions, see e.g. [9], Theorem 3, we get that the correlation can be computed as a sum of partial correlations over all intermediate linear masks Φ as follows:

$$
\begin{aligned}
&\mathrm{cor}(\Gamma \cdot f(y, z) \oplus \Gamma \cdot z \oplus \Lambda \cdot y) \\
&= \sum_{\Phi} \mathrm{cor}(\Gamma \cdot (w \boxplus z) \oplus \Phi \cdot w \oplus \Gamma \cdot z)\mathrm{cor}(\Phi \cdot S^{-1}(y) \oplus \Lambda \cdot y) \quad (2)\\
&= \sum_{\Phi} \mathrm{cor}(\Gamma \cdot (w \boxplus z) \oplus \Phi \cdot w \oplus \Gamma \cdot z)\mathrm{cor}(\Phi \cdot x \oplus \Lambda \cdot S(x)).
\end{aligned}
$$

Considering the addition modulo 2^{32} and the S-box ensemble S as independent functions is equivalent of taking just one term in the sum (2). Moreover, in [11] this one term was selected with $\Phi = \Lambda = \Gamma$. We observed that this may cause large deviations from the true value. On the other hand, including all terms of the sum would mean unnecessarily large amount of work. It turns out that including all terms with $\mathrm{cor}(\Gamma \cdot (w \boxplus z) \oplus \Phi \cdot w \oplus \Gamma \cdot z) \geq 2^{-24}$ yields sufficiently accurate estimates of the total correlation over the composed function. To search for all such linear masks Φ we used the algorithms by Wallén [10] (see Annex A). This explains the role of the linear mask Φ in Figure 2.

5   More Searches

5.1   Reducing the Number of Active S-Boxes

One strategy to increase the total bias of the linear approximation would be to limit the number of active S-boxes in the S-box ensemble S. Given an output mask Λ of S let us denote by Ω the mask such that $\Omega \cdot x = \Lambda \cdot Mx$, for all 32-bit values x, where M denotes the MixColumn transformation of the AES. Our best mask $\Lambda = \text{0x00018001}$ corresponds to the mask $\Omega = \text{0x0041c01}$. It means that only three S-boxes are active in the approximation of the FSM. However, in linear approximation with $\Lambda\alpha$ and $\Lambda\alpha^{-1}$ all four S-boxes are active. This means that in our best approximation of relation (1) the total number of active S-boxes is 14.

The MixColumn transformation is known to have good diffusion properties, more precicely, the total number of nonzero octets in (Ω, Λ) mask pairs is at least five. For linear approximation of SNOW 2.0 the diffusion properties of MixColumn must be investigated in combination with multiplication by α and α^{-1}. More precisely, it would be interesting to know exactly how many S-boxes at least are involved in the linear approximation (1). For this purpose, we studied all masks Λ such that the corresponding Ω has at most two non-zero octets. For each such Λ we computed the masks $\Lambda\alpha$ and $\Lambda\alpha^{-1}$ and their related Ω-masks, for which the number of non-zero octets was determined. Finally the total number of non-zero octets involved in the four FSM approximations in (1) was computed for each Λ. The same search was performed also for all Λ such that the input mask to M corresponding to $\Lambda\alpha$ ($\Lambda\alpha^{-1}$, respectively) has at most two non-zero octets. The minimum number of active S-boxes was found to be 7, and there are four masks Λ having this property. They are $\text{0x64ad5846}, \text{0xad584664}, \text{0x55bcc50d}$ and 0x0d55bcc5, and their respective one-octet input masks to MixColumn M are: $\text{0xd7000000}, \text{0x000000d7}, \text{0x00210000}$ and 0x00002100. However, none of these four masks Λ has a second mask Γ with a non-zero total bias over approximation (1). This follows from the results of a wider search we explain next.

We made a comprehensive search over all masks Λ such that the input mask Ω to M has at most two non-zero octets. This means limiting the search for such masks Λ that in approximation over the S-box-ensemble S at most two S-boxes are active. For any such Λ there was no Γ such that the linear approximate relation (1) would have a non-zero bias. This is obviously a strength in the structure of SNOW 2.0. We can only give a heuristic explanation of the reasons why this happens. Assume Λ is such that only two S-boxes are active. Then one can find an approximation over the FSM with a pretty good bias. Then Γ typically has two or three nonzero octets. Four non-zero octets may be possible in theory (we could not find any examples) but then in two of the octets only the least or the most significant bit is non-zero. In other words, the mask is sparse. Also when modified by α (or α^{-1}) the sparse structure is preserved. On the other hand, when Λ is modified by α (or α^{-1}) then almost always all four S-boxes will be active, and consequenty, the mask Φ has four nonzero octets. Therefore in about all cases, if not all, the mask $\Gamma\alpha$ and the mask Φ that fits over the S-box ensemble with $\Lambda\alpha$ have different structure. The same holds for the approximation of the FSM with $\Gamma\alpha^{-1}$

and $\Lambda \alpha^{-1}$. Since both of them should work to make the entire approximation work, the chances are negligible.

5.2 Sparse Masks

As explained above we were not able to significantly reduce the number of S-boxes that are active in the linear approximate relation (1). On the other hand, we observed that the modular addition with three inputs can be efficiently approximated using sparse masks. This is also well exemplified by the best linear distinguisher we found, which is based on a three-bit linear mask. Motivated by this observation we made a complete search over all masks Λ with at most five non-zero bits, allowing as in all our searches the mask Γ to be different from Λ. For one and two-bit masks there were no results. For the three-bit masks it turned out that we already found the best one. The best masks with four or five nonzero bits and their respective bias values are given in Table 6. The total bias of the linear distinguisher (1) with the four-bit mask $\Lambda = \Gamma = 0x40100060$ is $2^{-89.95}$, and with the five-bit mask $\Lambda = \Gamma = 0x00040701$ it is $2^{-89.25}$.

Table 6. The best four-bit and five-bit linear masks Λ

	mask value	ϵ_{FSM}
Λ	0x40100060	$2^{-18.49}$
$\Lambda\alpha$	0x02401000	$2^{-26.94}$
$\Lambda\alpha^{-1}$	0x10006029	$2^{-29.02}$
Λ	0x00040701	$2^{-18.72}$
$\Lambda\alpha$	0x75000407	$2^{-27.47}$
$\Lambda\alpha^{-1}$	0x04070100	$2^{-27.35}$

5.3 Three-Round Distinguisher

Also other more complex distinguishers were investigated. In particular, we looked at the distiguisher which involves output at time $t - 1$, t and $t + 1$, two instances of the S-box ensemble S and five (or four, if $\Pi = 0$) modular additions $\boxplus$ out of which two collapse into one addition with three inputs, see Figure 4 in Annex B. The resulting linear approximative relation is given in equation (3) in Annex B. Such a three-round distinguisher could compete with the two-round one only if the number of active S-boxes could be significantly reduced. However, this does not seem to be possible, for the same reason why the two-round distinguisher does not have non-zero bias with small number of active S-boxes as explained above in Section 5.1. Moreover, in approximations for (3), approximation over the latter S-box ensemble involves at least seven active S-boxes. The absolute minimum for the first S-box ensemble is four active S-boxes. Since the largest achievable bias of linear approximation over the AES S-box is 2^{-4} we get a theoretical upperbound of $2^{10}(2^{-4})^{11} = 2^{-34}$ to the bias of (3) for SNOW 2.0. The largest bias for the approximation (3) we have seen in practise is $2^{-202.17}$. In this case, the masks Γ, Π and Λ had 2,3 and 4 non-zero octets, respecively. In the first S-box ensemble, totally 6 S-boxes were active, and in the second S-box ensemble 10 S-boxes were active.

6   Linear Distinguishers for SN0W 3G

During its still relatively short lifetime SNOW 2.0 has gained confidence as demonstrated by the fact that it has been selected as a starting point for a few new designs,
see [8]. Most prominently, a draft for a new encryption algorithm for the UMTS system
was recently made public in [5]. It is called SNOW 3G in and is depicted in Figure 3.
This design preserves all features of SNOW 2.0, e.g. $S1 = S$, but adds a third register $R3$ to the FSM and a transform denoted by $S2$. The function $S2$ has been selected
to strengthen the FSM against algebraic cryptanalysis as response to the concerns expressed in [1,5].

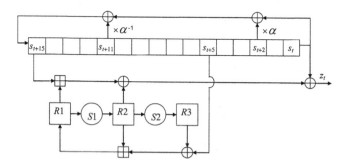

Fig. 3. SNOW 3G

The first, later rejected, choice for $S2$ was a non-bijective 32-to-32-bit S-box [4]. It
was constructed from a single eight-to-one bit Boolean function $V8$ by selecting for
each output bit a set of eight input bits. Then the output bit value is computed from the
selected input bits using $V8$. The input sets are selected in such a way that any two sets
has at most three bits in common. The Boolean function $V8$ is not balanced, hence $S2$
is not a bijection.

The simplest distinguisher for this version of SNOW 3G is obtained by using linear
masking over two and half rounds of the FSM and it is depicted in Figure 5. The linear
approximate relation is the same as in (1). In addition to the linear distinguisher depicted
in Figure 2 it involves approximations over $S2$, where the input mask is all-zero. The
mask search is very similar to the two round distinguisher for SNOW 2.0. We just need
to add the bias of approximation over $S2$ to it. Hence it is no surprise that the best Λ,
Γ pair we found for this distinguisher is the same as for SNOW 2.0, that is, $\Lambda = \Gamma =$
0x00018001. The total bias of the linear approximation of Figure 5 is $2^{-137.01}$.

In the final version of SNOW 3G the transformation $S2$ is otherwise identical to $S1$
but the AES S-box is replaced by a bijective mapping derived from a Dickson polynomial. This S-box has maximum linear bias of 2^{-3}. A three-round linear distinguisher
for SNOW 3G is depicted in Figure 6. We showed in Section 5.3 that the minimum
number of S-boxes in the approximations over $S2$ is at least seven. The same holds for
the second instance of $S1$. However, the input and output masks of the first instance of
$S1$ are not modified by α or α^{-1}. Nevertheless, at least four active S-boxes are needed.
Hence there are always at least eleven active AES S-boxes and seven active $S2$ S-boxes,

giving an upper bound of $2^{17}(2^{-4})^{11}(2^{-3})^7 = 2^{-48}$ to the bias of any three-round linear approximation of SNOW 3G. This bound is not tight. The true bias values will most likely be significantly reduced due to the biases of linear approximations over the modular additions.

7   Conclusions

It is well known that the Piling Up Lemma cannot be applied to combine linear approximations over consecutive functions in cipher constructions unless there is some evidence that the output from the first function is practically independent of the input to the second function. We showed that in [11] the Piling Up Lemma was used in case where the output from the first function is identical to the input to the second function. We showed not only how to compute correctly the estimates of the bias values but also implemented wide mask searches to find new and significantly stronger distinguishers that escaped the searches by Watanabe et al.

Some mask searches that were limited to certain types of linear masks failed to produce any results with non-zero bias. For example, we could demonstrate that it is impossible to significantly reduce the number of active S-boxes when approximating over the S-box ensemble S of SNOW 2.0. The same holds to other more complex distinguishers of SNOW 2.0 as well as to the recently presented new SNOW variant SNOW 3G, and is preserved as long as the feedback polynomial does not have a low degree multiple, which is a trinomial or a four-term polynomial with only two different coefficients. This gives some evidence about the strength of the SNOW design against cryptanalysis using the linear masking method.

Acknowledgements

We wish to thank Jukka Valkonen for implementing all mask searches needed to complete this work and the Krypto project of the Finnish Defence Forces for making it possible. We also wish to thank Emilia Käsper for providing optimised implementations of some specific parts. Finally, we owe many thanks to the members of the SAGE group, and to Matt Robshaw, in particular, for invaluable discussions.

References

1. Olivier Billet and Henri Gilbert. Resistance of SNOW 2.0 against algebraic attacks. In *CT-RSA 2005*, pages 19–28, 2004.
2. Don Coppersmith, Shai Halevi, and Charanjit Jutla. Cryptanalysis of stream ciphers with linear masking. In *Advances in Cryptology - Crypto 2002, LNCS 2442, Springer-Verlag* , pages 515–532, 2002.
3. Patrick Ekdahl and Thomas Johansson. A new version of the stream cipher SNOW. In *Selected Areas in Cryptography, SAC 2002, LNCS 1233, Springer-Verlag* , pages 37–46, 2002.
4. ETSI/SAGE. Specification of the 3GPP confidentiality and integrity algorithms UEA2 & UIA2. Document 2: SNOW 3G specification, draft version 0.5, August 2005. http://www.3gpp.org/ftp/tsg_sa/WG3_Security/TSGS3_40_Slovenia/ Docs/S3050579%20.zip.

5. ETSI/SAGE. Specification of the 3GPP confidentiality and integrity algorithms UEA2 & UIA2. Document 5: Design and evaluation report, version: 1.0 http://www.3gpp.org/ftp/tsg_sa/WG3_Security/TSGS3_42_Bangalore/Docs/S3060180.zip.

6. Mitsuru Matsui. Linear Cryptanalysis Method for DES Cipher. In *Advances in Cryptology, Eurocrypt 1993, LNCS 765, Springer-Verlag* , pages 386–397, 1994.

7. Alexander Maximov and Thomas Johansson. Fast Computation of Large Distributions and Its Cryptographic Applications. In *Asiacrypt 2005, LNCS 3788, Springer-Verlag* , pages 313–332, 2005.

8. ECRYPT NoE. eSTREAM, the ECRYPT stream cipher project, 2005. http://www.ecrypt.eu.org/stream/.

9. Kaisa Nyberg. Correlation theorems in cryptanalysis. *Discrete Applied Mathematics*, pages 177–188, 111 2001.

10. Johan Wallén. Linear Approximations of Addition Modulo 2^n. In *Fast Software Encryption, FSE 2003, LNCS 2887, Springer-Verlag* , pages 261–273, 2003.

11. Dai Watanabe, Alex Biryukov, and Christophe De Cannière. A Distinguishing Attack of SNOW 2.0 with Linear Masking Method. In *Selected Areas in Cryptography, SAC 2003, LNCS 3006, Springer-Verlag* , pages 222–233, 2004.

A   Linear Approximation of Addition Modulo 2^n

A.1   Notation

We identify the integers in $\{0, \ldots, 2^n - 1\}$ with the vectors in $\mathbf{F}_2^n$ by the natural correspondence that identifies the integer whose binary expansion is $\sum_{i=0}^{n-1} a_i 2^i$ with the vector $(a_{n-1}, \ldots, a_1, a_0)$. Given k n-bit integers $x^{(h)}$, $h = 1, \ldots, k$, the sum $(x^{(1)} + \cdots + x^{(k)}) \bmod 2^n$ carries over to a function from $(\mathbf{F}_2^n)^k$ to $\mathbf{F}_2^n$. Addition in $\mathbf{F}_2$ and $\mathbf{F}_2^n$ is always denoted by $\oplus$.

For vectors $x = (a_{n-1}, \ldots, a_0)$ and $y = (b_{n-1}, \ldots, b_0) \in \mathbf{F}_2^n$, let $x \cdot y$ denote the standard inner product $x \cdot y = a_{n-1}b_{n-1} \oplus \cdots \oplus a_0 b_0$. For k tuples of vectors in $\mathbf{F}_2^n$, $x^{(1)}, \ldots, x^{(k)}$ and $y^{(1)}, \ldots, y^{(k)}$, we set $(x^{(1)}, \ldots, x^{(k)}) \cdot (y^{(1)}, \ldots, y^{(k)}) = x^{(1)} \cdot y^{(1)} \oplus \cdots \oplus x^{(k)} \cdot y^{(k)}$. A linear approximation of the sum modulo 2^n with k inputs is an approximate relation of the form

$$ u \cdot (x^{(1)} \boxplus \cdots \boxplus x^{(k)}) = (x^{(1)}, \ldots, x^{(k)}) \cdot (w^{(1)}, \ldots, w^{(k)}) $$

where $u \in \mathbf{F}_2^n$ and $w^{(h)} \in \mathbf{F}_2^n$, $h = 1, \ldots, k$ are the mask vectors. The strength of the approximation is measured by the correlation

$$ \begin{aligned} &\text{cor}(u; w^{(1)}, \ldots, w^{(k)}) \\ &= 2\mathbf{Pr}[u \cdot (x^{(1)} \boxplus \ldots \boxplus x^{(k)}) = (x^{(1)}, \ldots, x^{(k)}) \cdot (w^{(1)}, \ldots, w^{(k)})] - 1, \end{aligned} $$

where the probability is taken over uniformly distributed $x^{(1)}, \ldots, x^{(k)}$.

A.2   Linear Representation

We will derive a linear representation for the correlation of linear approximations of addition modulo 2^n. Towards this end, we write the linear approximation with mask vectors $u = (u_{n-1}, \ldots, u_0)$ and $w^{(1)}, \ldots, w^{(k)}$, where $w^{(h)} = (w_{n-1}^{(h)}, \ldots, w_0^{(h)})$, as a word $z_{n-1} \ldots z_1, z_0$ over the alphabet $\{0, \ldots, 2^{k+1} - 1\}$, where $z_i = u_i 2^k + \sum_{h=1}^{k} w_i^{(h)} 2^{h-1}$. We will then show that there are 2^{k+1} $k \times k$ matrices over rationals, a row vector L and a column vector C such that

$$\text{cor}(u; w^{(1)}, \ldots, w^{(k)}) = L A_{z_{n-1}} \ldots A_{z_1} A_{z_0} C,$$

for all n and all linear approximations $(u; w^{(1)}, \ldots, w^{(k)})$ of addition modulo 2^n of k n-bit integers. We say that the matrices L, $A_r, r = 0, \ldots, 2^{k+1} - 1$, and C form a linear representation of the correlation with dimension k.

For a vector $x \in \mathbf{F}_2^n$ (or integer $x \in \{0, \ldots, 2^n - 1\}$), we let $w_H(x)$ denote the Hamming weight of x, that is, $w_H(x)$ is a non-negative integer less than or equal to n, which is the number of non-zero components of x.

Theorem 1. *Let $k > 1$ be a fixed integer. Let L be the row vector of dimension k with all entries equal to 1, and let C be the column vector of dimension k with a single 1 in row 0 and zero otherwise. Let $A_0, \ldots, A_{2^{k+1}-1}$ be the $k \times k$ matrices*

$$(A_r)_{d,c} = 2^{-k}(|\{x \in \mathbf{F}_2^k : u \cdot g(x,c) = x \cdot v, f(x,c) = d\}| -$$
$$|\{x \in \mathbf{F}_2^k : u \cdot g(x,c) \neq x \cdot v, f(x,c) = d\}|),$$

where

$$r = u2^k + \sum_{h=1}^{k} v_h 2^{h-1}, \quad v = (v_1, \ldots, v_k), \quad x = (x_1, \ldots, x_k),$$

$c, d \in \{0, \ldots, k-1\}$,
$f: \{0, \ldots, k-1\}^2 \to \{0, \ldots, k-1\}, f(x,c) = \lfloor (w_H(x) + c)/2 \rfloor$,
$g: \{0, \ldots, k-1\}^2 \to \{0,1\}, g(x,c) = (w_H(x) + c) \bmod 2$.

Let $n \geq 1$ be an integer and let $(u; w^{(1)}, \ldots, w^{(k)})$ be a linear approximation of addition modulo 2^n with k inputs. Let $z = z_{n-1} \ldots z_1, z_0$ be the word associated with the approximation. We then have

$$\text{cor}(u; w^{(1)}, \ldots, w^{(k)}) = L A_{z_{n-1}} \cdots A_{z_1} A_{z_0} C.$$

Note that the functions f and g are the carry and sum functions for the basic school-book method for adding k integers in binary.

Proof. We denote by $(x^{(1)}, \ldots, x^{(k)})$ the n-bit integers that are added modulo 2^n. We use the simple school-book method. We set the first carry bit $c_0 = 0$. Then the carries c_i and the sum bits s_i at each step $i = 0, \ldots, n-1$ are computed as follows

$$s_i = g((x_i^{(1)}, \ldots, x_i^{(k)}), c_i),$$
$$c_{i+1} = f((x_i^{(1)}, \ldots, x_i^{(k)}), c_i)$$

We set $b_0 = 0$ and, for all $j = 1, \ldots, n$, let

$$b_j = \bigoplus_{i=0}^{j-1} (u_i s_i \oplus w_i^{(1)} x_i^{(1)} \oplus \cdots \oplus w_i^{(k)} x_i^{(k)}).$$

Let $P(z, j)$ be the column vector

$$P(z, j)_c = \mathbf{Pr}[b_j = 0, c_j = c] - \mathbf{Pr}[b_j = 1, c_j = c]$$

for $j = 0, \ldots, n$ and $c = 0, \ldots, k - 1$. Let $M(z, i)$ be the $k \times k$ matrix

$$M(z, i)_{d,c} = \mathbf{Pr}[(u_i s_i \oplus w_i^{(1)} x_i^{(1)} \oplus \cdots \oplus w_i^{(k)} x_i^{(k)}) = 0 \text{ and } c_{i+1} = d \,|\, c_i = c] -$$
$$\mathbf{Pr}[(u_i s_i \oplus w_i^{(1)} x_i^{(1)} \oplus \cdots \oplus w_i^{(k)} x_i^{(k)}) = 1 \text{ and } c_{i+1} = d \,|\, c_i = c],$$

for $i = 0, \ldots, n - 1$. Then we have

$$\sum_{c=0}^{k-1} M(z, i)_{d,c} P(z, i)_c = P(z, i + 1)_d,$$

and thus

$$P(z, i + 1) = M(z, i) P(z, i).$$

Note that

$$P(z, 0)_0 = \mathbf{Pr}[b_0 = 0, c_0 = 0] - \mathbf{Pr}[b_0 = 1, c_0 = 0] = 1, \text{ and}$$
$$P(z, 0)_c = \mathbf{Pr}[b_0 = 0, c_0 = c] - \mathbf{Pr}[b_0 = 1, c_0 = c] = 0, \text{ for } c \neq 0.$$

At the other end we have

$$LP(z, n) = \sum_{c=0}^{k-1} (\mathbf{Pr}[b_n = 0, c_n = c] - \mathbf{Pr}[b_n = 1, c_n = c])$$
$$= \mathbf{Pr}[b_n = 0] - \mathbf{Pr}[b_n = 1] = \mathrm{cor}(u; w^{(1)}, \ldots, w^{(k)})$$

as desired. Since $A_{z_i} = M(z, i)$, it follows that

$$\mathrm{cor}(u; w^{(1)}, \ldots, w^{(k)}) = LA_{z_{n-1}} \cdots A_{z_1} A_{z_0} C. \qquad \square$$

The correlation of a linear approximation of addition modulo 2^n with k inputs can thus be computed by doing n multiplications of a $k \times k$ matrix and a column vector, and n additional additions. For a fixed k, this is a linear -time algorithm, and for small k efficient in practice. Note that the number of matrices to be stored in memory is 2^{k+1}. We remark that an analogous method can be used to compute the differential probability of addition modulo 2^n with k inputs.

Using Theorem 1 we get the following matrices for $k = 3$.

$$A_0 = \frac{1}{8}\begin{pmatrix} 4 & 1 & 0 \\ 4 & 6 & 4 \\ 0 & 1 & 4 \end{pmatrix},$$

$$A_1 = A_2 = A_4 = -A_8 = \frac{1}{8}\begin{pmatrix} 2 & 1 & 0 \\ -2 & 0 & 2 \\ 0 & -1 & -2 \end{pmatrix},$$

$$A_3 = A_5 = A_6 = -A_9 = -A_{10} = -A_{12} = \frac{1}{8}\begin{pmatrix} 0 & 1 & 0 \\ 0 & -2 & 0 \\ 0 & 1 & 0 \end{pmatrix},$$

$$A_7 = -A_{11} = -A_{13} = -A_{14} = \frac{1}{8}\begin{pmatrix} -2 & 1 & 0 \\ 2 & 0 & -2 \\ 0 & -1 & 2 \end{pmatrix}, \text{ and}$$

$$A_{15} = \frac{1}{8}\begin{pmatrix} 4 & -1 & 0 \\ 4 & -6 & 4 \\ 0 & -1 & 4 \end{pmatrix}.$$

A.3   Searching for Masks for a Given Correlation

In this section, we briefly describe the method used to search for all relevant masks for addition modulo 2^n with two inputs. Using Theorem 1, we get a linear representation $L', A_0', \ldots, A_7', C$ of dimension 2 for the correlation of linear approximations of addition modulo 2^n with two inputs. The matrix A_0' has the Jordan form $\mathrm{diag}(1, 1/2) = H_2 A_0' H_2^{-1}$, where $H_2 = \left(\begin{smallmatrix} 1 & 1 \\ 1 & -1 \end{smallmatrix}\right)$ is the 2×2 Hadamard matrix. We get a new linear representation by making the change of basis $L = R' H_2^{-1}$, $A_i = H_2 A_i H_2^{-1}$ and $C = H_2 C'$. This gives the matrices $L = \begin{pmatrix} 1 & 0 \end{pmatrix}$, $C = \begin{pmatrix} 1 & 1 \end{pmatrix}^t$,

$$A_0 = \frac{1}{2}\begin{pmatrix} 2 & 0 \\ 0 & 1 \end{pmatrix}, \quad A_1 = A_2 = -A_4 = \frac{1}{2}\begin{pmatrix} 0 & 0 \\ 1 & 0 \end{pmatrix},$$

$$A_7 = \frac{1}{2}\begin{pmatrix} 0 & 2 \\ 1 & 0 \end{pmatrix}, \quad -A_3 = A_5 = A_6 = \frac{1}{2}\begin{pmatrix} 0 & 0 \\ 0 & 1 \end{pmatrix}.$$

Let $e_0 = \begin{pmatrix} 1 & 0 \end{pmatrix}$ and $e_1 = \begin{pmatrix} 0 & 1 \end{pmatrix}$. Then $e_0 A_0 = e_0$, $e_0 A_7 = e_1$, $e_0 A_i = 0$ for $i \neq 0, 7$, $e_1 A_0 = e_1 A_5 = e_1 A_6 = \frac{1}{2}e_1$, $e_1 A_1 = e_1 A_2 = e_1 A_7 = \frac{1}{2}e_0$, $e_1 A_3 = -\frac{1}{2}e_1$ and $e_1 A_4 = -\frac{1}{2}e_0$. It follows that the computation of $L A_{w_{n-1}} \cdots A_{w_0} C$ by multiplication from left to right can be described by the following automaton.

When reading w from left to right, if the automaton ends up in state 0, $LA_{w_{n-1}} \cdots A_{w_0} C = 0$. If the automaton ends up in state e_0 or e_1, $LA_{w_{n-1}} \cdots A_{w_0} C = \pm 2^{-k}$, where k is the number of transitions marked by a solid arrow, and the sign is determined by the number of occurrences of $\{3, 4\}$: $LA_{w_{n-1}} \cdots A_{w_0} C > 0$ if and only if the number of occurrences is even. For example, when $w = 73620_8$, we have the state transitions

$$\left(e_0\right) \xrightarrow{7} \left(e_1\right) \xrightarrow{3} \left(e_1\right) \xrightarrow{6} \left(e_1\right) \xrightarrow{2} \left(e_0\right) \xrightarrow{0} \left(e_0\right)$$

and thus $LA_7 A_3 A_6 A_2 A_0 = -2^{-3}$. Clearly, $LA_{w_{n-1}} \cdots A_{w_0} C = 0$ if and only if w matches the regular expression

$$\left(0 + 7(0 + 3 + 5 + 6)^*(1 + 2 + 4 + 7)\right)^* (1 + 2 + 3 + 4 + 5 + 6)\Sigma^* \ ,$$

where $\Sigma = 0 + 1 + \cdots + 7$.

Let $S^0(n, k)$ and $S^1(n, k)$ denote the formal languages

$$S^0(n, k) = \{w \mid |w| = n, e_0 A_{w_{n-1}} \cdots A_{w_0} = \pm 2^{-k} e_0\} \quad \text{and}$$
$$S^1(n, k) = \{w \mid |w| = n, e_0 A_{w_{n-1}} \cdots A_{w_0} = \pm 2^{-k} e_1\}$$

for $n > 0$. Then $S^0(n, k) + S^1(n, k)$ is the set of words of length $n > 0$ corresponding to linear approximations of addition with two inputs that have correlation $\pm 2^{-k}$. The languages are clearly given recursively by (juxtaposition denotes concatenation, and $+$ denotes union)

$$S^0(n, k) = S^0(n - 1, k)0 + S^1(n - 1, k - 1)(1 + 2 + 4 + 7) \quad \text{and}$$
$$S^1(n, k) = S^0(n - 1, k)7 + S^1(n - 1, k - 1)(0 + 3 + 5 + 6)$$

for all $0 \leq k < n$. The base cases are $S^0(1, 0) = 0$ and $S^1(1, 0) = 7$. If $k < 0$ or $k \geq n$, $S^0(n, k) = S^1(n, k) = \emptyset$.

These recursive descriptions immediately give an efficient algorithm for finding all input and output masks for addition with a given correlation. Moreover, one or two of the three masks can optionally be fixed. Using generating functions, it is also straightforward to determine the distribution of the correlation coefficients—that is, count the number of input/output masks with a given correlation. These results where proved using different methods in [10].

Unfortunately, there does not seem to be any simple way to obtain the same results for addition with three inputs, since it seems impossible to obtain an equally simple linear representation with a change of basis.

B   Other Linear Distinguishers—Figures and Equations

B.1   A Three-Round Linear Distinguisher for SNOW 2.0

A three-round linear distinguisher for SNOW 2.0 is depicted in Figure 4 and the corresponding linear approximate relation is given by equation (3).

$$\Gamma \cdot (z_{t+15} \oplus z_{t+1}) \oplus \Gamma \alpha \cdot z_{t-1} \oplus \Gamma \alpha^{-1} \cdot z_{t+10} \oplus$$

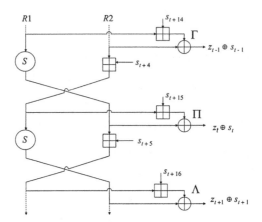

Fig. 4. Linear masking of SNOW 2.0 over three rounds

$$\Pi \cdot (z_{t+16} \oplus z_{t+2}) \oplus \Pi \alpha \cdot z_t \oplus \Pi \alpha^{-1} \cdot z_{t+11} \oplus$$
$$\Lambda \cdot (z_{t+17} \oplus z_{t+3}) \oplus \Lambda \alpha \cdot z_{t+1} \oplus \Lambda \alpha^{-1} \cdot z_{t+12} = 0. \tag{3}$$

B.2   A Two-and-Half-Round Linear Distinguisher for SNOW 3G with Non-bijective $S2$

In this distinguisher it is assumed that the linear masking by Λ of the output from $S2$ is approximated by zero, see Figure 5. Then the linear approximate relation of this distinguisher is identical to (1).

B.3   A Three-Round Linear Distinguisher for SNOW 3G

The resulting linear approximate relation involving keystream terms z_i only, is the same as (3). This can be seen as follows. Let x denote the input to the first (in time) instance

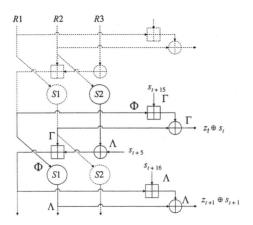

Fig. 5. Linear masking of SNOW 3G with non-bijective $S2$

of $S1$, y denote the input to $S2$, and w the input to the second instance of $S1$. In addition to the mask values given in Figure 6 we denote by Δ and Ψ the input and output masks for the first $S1$, Θ the input mask to the second $S1$, and finally, by Σ_1, Σ_2 and Σ_3 the three masks used to mask s_{t+14}, s_{t+15} and s_{t+16}, respectively. Then we have the following approximate relations:

$$\Psi \cdot S1(x) = \Delta \cdot x \tag{4}$$
$$\Phi \cdot S2(y) = \Gamma \cdot y$$
$$\Lambda \cdot S1(w) = \Theta \cdot w$$
$$\Gamma \cdot (s_{t+14} \boxplus x) = \Sigma_1 \cdot s_{t+14} \oplus \Delta \cdot x$$
$$\Pi \cdot (s_{t+15} \boxplus w) = \Sigma_2 \cdot s_{t+15} \oplus \Theta \cdot w$$
$$\Lambda \cdot (s_{t+16} \boxplus S1(x) \boxplus (s_{t+5} \oplus S2(y)))$$
$$= \Sigma_3 \cdot s_{t+16} \oplus (\Phi \oplus \Pi) \cdot S1(x) \oplus \Phi \cdot (s_{t+15} \oplus S2(y)).$$

The three auxiliary variables x, y and w cancel due to the following three equalities:

$$x \boxplus s_{t+14} = y \oplus z_{t-1} \oplus s_{t-1}$$
$$z \boxplus s_{t+15} = S1(x) \oplus z_t \oplus s_{t-1}$$
$$S1(x) \boxplus (S2(y) \oplus s_{t+5})) \boxplus s_{t+16} = S1(w) \oplus z_{t+1} \oplus s_{t+1}$$

Then we have:

$$\Gamma \cdot (z_{t-1} \oplus s_{t-1}) \oplus \Pi \cdot (z_t \oplus s_t) \oplus \Lambda \cdot (z_{t+1} \oplus s_{t+1})$$
$$\oplus \Sigma_1 \cdot s_{t+14} \oplus \Sigma_2 \oplus s_{t+15} \oplus \Sigma_3 \cdot s_{t+16} \oplus \Phi \cdot s_{t+5} = 0,$$

or what is the same:

$$\Gamma \cdot z_{t-1} \oplus \oplus \Pi \cdot z_t \oplus \Lambda \cdot z_{t+1}$$
$$= \Gamma \cdot s_{t-1}) \oplus \Pi \cdot s_t \oplus \Lambda \cdot s_{t+1} \oplus \Sigma_1 \oplus \Phi \cdot s_{t+5} \cdot s_{t+14} \oplus \Sigma_2 \oplus s_{t+15} \oplus \Sigma_3 \cdot s_{t+16}.$$

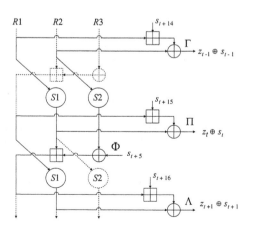

Fig. 6. Linear masking of SNOW 3G

This equation is used four times. First two times for $t = t + 2$ and $t = t + 16$. Then once with $t = t$ and with all z_i and s_i variables multiplied by α. Finally, the equation is used for $t = t + 11$ with all z_i and s_i variables multiplied by α^{-1}. Since the $\alpha s_t \oplus s_{t+2} \oplus \alpha^{-1} s_{t+11} \oplus s_{t+16} = 0$, for all t, the s_i variables cancel and we get:

$$\Gamma \cdot z_{t+1} \oplus \Pi \cdot z_{t+2} \oplus \Lambda \cdot z_{t+3} \oplus \Gamma \cdot z_{t+15} \oplus \Pi \cdot z_{t+16} \oplus \Lambda \cdot z_{t+17} \oplus \Gamma \cdot \alpha z_{t-1}$$
$$\oplus \Pi \cdot \alpha z_t \oplus \Lambda \cdot \alpha z_{t+1} \oplus \Gamma \cdot \alpha^{-1} z_{t+10} \oplus \Pi \cdot \alpha^{-1} z_{t+11} \oplus \Lambda \cdot \alpha^{-1} z_{t+12} = 0$$

which is the same as (3). When all s_i variables are multiplied by α the approximations over invidual functions take the following forms:

$$\Psi \cdot S1(x) = \Delta \cdot x$$
$$\Phi \alpha \cdot S2(y) = \Gamma \alpha \cdot y$$
$$\Lambda \alpha \cdot S1(z) = \Theta \cdot z$$
$$\Gamma \alpha \cdot (s_{t+14} \boxplus x) = \Sigma_1 \alpha \cdot s_{t+14} \oplus \Delta \cdot x$$
$$\Pi \alpha \cdot (s_{t+15} \boxplus z) = \Sigma_2 \alpha \cdot s_{t+15} \oplus \Theta \cdot z$$
$$\Lambda \alpha \cdot (s_{t+16} \boxplus S1(x) \boxplus (s_{t+5} \oplus S2(y)))$$
$$= \Sigma_3 \alpha \cdot s_{t+16} \oplus (\Phi \alpha \oplus \Pi \alpha) \cdot S1(x) \oplus \Phi \alpha \cdot (s_{t+15} \oplus S2(y)).$$

To get approximations with multiplication by α^{-1} just replace α by α^{-1}. In both cases the masks Ψ, Δ and Θ can be chosen independently of the masks denoted using the same symbol for approximations (4).

Reducing the Space Complexity of BDD-Based Attacks on Keystream Generators

Matthias Krause and Dirk Stegemann

Theoretical Computer Science
University of Mannheim, Germany
{kraus, stegemann}@th.informatik.uni-mannheim.de

Abstract. The main application of stream ciphers is online-encryption of arbitrarily long data, for example when transmitting speech data between a Bluetooth headset and a mobile GSM phone or between the phone and a GSM base station. Many practically used and intensively discussed stream ciphers such as the E_0 generator used in Bluetooth and the GSM cipher A5/1 consist of a small number of linear feedback shift registers (LFSRs) that transform a secret key $x \in \{0,1\}^n$ into an output keystream of arbitrary length. In 2002, Krause proposed a Binary Decision Diagram (BDD) based attack on this type of ciphers, which in the case of E_0 is the best short-keystream attack known so far. However, BDD-attacks generally require a large amount of memory. In this paper, we show how to substantially reduce the memory consumption by divide-and-conquer strategies and present the first comprehensive experimental results for the BDD-attack on reduced versions of E_0, A5/1 and the self-shrinking generator.

Keywords: Stream cipher, cryptanalysis, BDD, Bluetooth E_0, GSM A5/1, self-shrinking generator.

1 Introduction

The main purpose of LFSR-based keystream generators is online encryption of bitstreams $p \in \{0,1\}^*$ that have to be sent over an insecure channel, e.g., for encrypting speech data to be transmitted from and to a mobile phone over the air interface. The output keystream $y \in \{0,1\}^*$ of the generator is bitwise XORed to the plaintext stream p in order to obtain the ciphertext stream $c \in \{0,1\}^*$, i.e., $c_i = p_i \oplus y_i$ for all i. Based on a secret initial state $x \in \{0,1\}^n$, which has to be exchanged between sender and legal receiver in advance, the receiver can compute the keystream y from x in the same way as the sender computed it and decrypt the message using the above rule.

We consider the special type of LFSR-based keystream generators that consist of a linear bitstream generator with a small number of Linear Feedback Shift Registers (LFSRs) and a non-linear compression function $C : \{0,1\}^* \rightarrow \{0,1\}^*$. From the secret key x, the LFSRs produce an internal bitstream $z \in \{0,1\}^*$, which is then transformed into the output keystream y via $y = C(z)$. Practical examples for this design include the E_0 generator, which is used as a building

M.J.B. Robshaw (Ed.): FSE 2006, LNCS 4047, pp. 163–178, 2006.

block for the cipher used in the Bluetooth standard for short-range wireless communication [4], the A5/1 generator from the widely used GSM standard for mobile telephones [5], and the self-shrinking generator [14].

Currently, the best attacks on E_0 in terms of time and memory requirements are algebraic attacks [1,6] and correlation attacks [12,11], but both types rely on the rather unrealistic assumption that a large amount of output keystream is available. The correlation attacks presented in [12,11] additionally depend on the linearity of the key-schedule and other specific properties of the Bluetooth encryption system that could easily be altered in future versions of the cipher. Particularly, in [2] it was shown that small changes of the cipher design could completely avert the correlation attack in [12] and significantly worsen the efficiency of algebraic attacks on E_0.

The attack by Krause [10], which we consider in this paper, is a generic attack in the sense that it does not depend on specific design properties of the respective cipher. It only relies on the assumptions that the generator's output behaves pseudorandomly and that the test whether a given internal bitstream z produces a sample keystream can be represented in a Free Binary Decision Diagram (FBDD) of size polynomial in the length of z. In the case of E_0, the BDD-attack can easily be extended to an attack on the whole Bluetooth cipher. Another major advantage of the attack is that it reconstructs the secret key from the shortest information-theoretically possible prefix of the keystream; in the case of E_0 and A5/1, the first keystream frame already suffices to obtain all the information that is needed to compute the initial state, whereas both algebraic attacks and correlation attacks depend on the unrealistic number of at least 2^{23} available keystream frames. In fact, the BDD-attack is the best short-keystream attack on E_0 that is known so far.

Unlike both algebraic and correlation attacks, BDD-attacks can also be applied to irregularly clocked keystream generators like the A5/1 generator, for which the BDD-attack is one of the best generic attacks that do not depend on special properties of the GSM encryption system.

However, one drawback of the BDD-attack is its high memory consumption. We will approach this problem by presenting various efficiently parallelizable divide-and-conquer strategies (DCS) for E_0 and A5/1 that substantially reduce the memory requirements and allow us to tackle much larger keylengths with fixed computational resources. In the case of E_0, our DCS lowers the attack's memory requirements by a factor of 2^{25} and additionally yields a slight theoretical improvement of the theoretical runtime. Hence, we obtain the best attack on E_0 under realistic assumptions.

In [10], the application of the basic BDD-based attack to E_0, A5/1 and the self-shrinking generator were already theoretically described, but with rather pessimistic assumptions on the time and space requirements. We present the first comprehensive experimental results for the BDD-attack on reduced versions of these ciphers, showing that the performance in practice does not substantially deviate from the theoretical figures.

This paper is organized as follows. In Sect. 2, we introduce some notations, give an overview of Binary Decision Diagrams and their algorithmic properties, and review the original BDD-based attack presented in [10]. The impact of the BDD-attack on the keystream generators E_0, A5/1 and the self-shrinking generator are described together with their basic definitions in Sect. 3. Section 4 introduces our divide-and-conquer strategies for the attacks on E_0 and A5/1, and Sect. 5 presents our experimental results. Finally, Sect. 6 concludes the paper.

2 Preliminaries

2.1 LFSR-Based Keystream Generators

In order to establish a consistent notation, we restate the definitions of linear feedback shift registers, linear bitstream generators and LFSR-based keystream generators and their basic properties.

Definition 1. *A Linear Feedback Shift Register (LFSR) of length n with a coefficient vector $c = (c_1, \ldots, c_n) \in \{0,1\}^n$ takes an initial state $x = (x_0, \ldots, x_{n-1}) \in \{0,1\}^n$ as input and produces a bitstream $l(x) = l_0(x), l_1(x), \ldots, l_i(x), \ldots$ according to*

$$l_i(x) := \begin{cases} x_i & \text{for } 0 \le i \le n-1 \\ \bigoplus_{k=0}^{n-1} c_{k+1} \cdot l_{i-n+k}(x) & \text{for } i > n-1 \end{cases}.$$

Note that each output bit of an LFSR is a linear combination of the initial state bits and that for each position i, there exists a subset $D(i) \subseteq \{0, \ldots, n-1\}$ such that $l_i(x) = \bigoplus_{j \in D(i)} x_j$. We call $D(i)$ the *domain* of i.

In practice, an LFSR is implemented in hardware with n binary register cells that are connected by a feedback channel.

Definition 2. *A Linear Bitstream Generator L consists of $k \ge 1$ parallel LFSRs L^r of length n_r, $r \in \{0, \ldots, k-1\}$, and $n_0 + \ldots + n_{k-1} = n$. L produces a bitstream $L(x) = L_0(x), L_1(x), \ldots, L_i(x), \ldots$ where*

$$L_i(x) := l_{s(i)}^{r(i)}\left(x^{r(i)}\right) \quad \text{where} \quad \begin{array}{l} r(i) = i \bmod k \\ s(i) = i \text{ div } k \end{array},$$

i.e., the i-th output bit of L corresponds to the $s(i)$-th output bit of LFSR $L^{r(i)}$. The initial states x^p of the LFSRs L^p, $p \in \{0, \ldots, k-1\}$, form the initial state $x \in \{0,1\}^n$ of L. For $i \ge 1$, we denote by $L_{\le i}(x)$ the i-extension of x, i.e., the first i output bits $L_0(x), \ldots, L_{i-1}(x)$ that L produces from x.

Definition 3. *An LFSR-based (k, l)-keystream generator (or (k, l)-combiner) $K = (L, C)$ consists of a linear bitstream generator L with k LFSRs and a non-linear compression function $C : \{0,1\}^* \to \{0,1\}^*$ with l memory bits. From the secret key $x \in \{0,1\}^n$ that L is initialized with, K computes an internal bitstream $z = L(x)$ and transforms z into the output keystream via $y = C(z) = y_0, y_1, \ldots, y_i, \ldots$. The compression function C computes the keystream in an* *online manner, i.e., there exists a function $\delta : \mathbb{N} \to \mathbb{N}$ with $\delta(i) < \delta(j)$ for $i < j$, such that $y_i = C(z_0, \ldots, z_{\delta(i)-1})$, i.e., y_i only depends on the first $\delta(i)$ bits of z. Moreover, C reads the internal bits in the order in which they are produced by the LFSRs, i.e., for $s > 0$ and all $r \in \{0, \ldots, k-1\}$, $L_{k \cdot s + r}(x)$ is not read before $L_{k \cdot (s-1)+r}(x)$.*

We call a number $i \geq 1$ a key position in $L(x)$ if $L_i(x)$ corresponds to one of the x-bits and a non-key position otherwise. Correspondingly, we denote by $KP(i)$ the set of key positions in $\{0, \ldots, i-1\}$ and by $KB(z) \in \{0,1\}^{|KP(i)|}$ the bits at the key positions in $L(x)$. Let n_{min} denote the maximum i for which all $i' \leq i$ are key positions and $n_{\max}$ the minimum i for which all $i' > i$ are non-key positions.

In the context of the previous definitions, we can characterize the well-known regularly clocked combiners with memory (or shortly *regular (k,l)-combiners*), which consist of k LFSRs and an l-bit memory unit, in the following way.

Definition 4. *We call an LFSR-based (k,l)-keystream generator* regular, *if y_i only depends on the internal bits $(z_{ki}, \ldots, z_{(k+1)i-1})$, i.e., the $(i+1)$-st output bits of the LFSRs, and the state of the memory bits in iteration i.*

Definition 5. *Let γ denote the* best-case compression ratio *$\gamma \in (0,1]$, i.e., γm is the maximum number of keybits that C produces from internal bitstreams of length m. For a randomly chosen and uniformly distributed internal bitstream $Z^{(m)} \in \{0,1\}^m$ and a random keystream Y, we define the* average information *that Y reveals about $Z^{(m)}$ as $\alpha := \frac{1}{m} I\left(Z^{(m)}, Y\right) \in (0,1]$.[1]*

For a randomly chosen and uniformly distributed internal bitstream $z \in \{0,1\}^m$, the probability of the keybits $C(z)$ being a prefix of a given keystream $y \in \{0,1\}^*$ can be expressed as

$$Prob_z[C(z) \text{ is prefix of } y] =$$

$$\sum_{i=0}^{\lceil \gamma m \rceil} Prob_{z \in \{0,1\}^m}[|C(z)| = i] \cdot Prob_{z \in \{0,1\}^m, |C(z)|=i}[C(z) = (y_0, \ldots, y_{i-1})] \ .$$

Concerning this probability, we will make the following assumption.

Assumption 1 (Independence Assumption). *For all $m \geq 1$, a randomly chosen, uniformly distributed internal bitstream $z \in \{0,1\}^m$, and all keystreams $y \in \{0,1\}^*$, we have $Prob_z[C(z) \text{ is prefix of } y] = p_C(m)$, i.e., the probability of $C(z)$ being a prefix of y is the same for all y.*

As shown in [10], Assumption 1 yields $\alpha = -\frac{1}{m} \log_2 p_C(m)$.

From a straightforward calculation (c.f. [10] for details), we obtain

Observation 1. *For a regular (k,l)-combiner, it is $\alpha = \gamma = \frac{1}{k}$.*

Finally, we assume the keystream y to behave pseudorandomly.

---

[1] Recall that for two random variables A and B, the value $I(A, B) = H(A) - H(A|B)$ defines the information that B reveals about A.

Assumption 2 (Pseudorandomness Assumption). *For all keystreams y and all $m \leq \lceil \alpha^{-1} n \rceil$ it holds that $Prob_z[C(z)$ is prefix of $y] \approx Prob_x[C(L_{\leq m}(x))$ is prefix of $y]$, where z and x denote randomly chosen, uniformly distributed elements of $\{0,1\}^m$ and $\{0,1\}^{|KP(m)|}$, respectively.*

Note that a severe violation of Assumption 2 would imply a vulnerability of K via a correlation attack.

2.2 Binary Decision Diagrams (BDDs)

We briefly review the definitions of Binary Decision Diagrams and those algorithmic properties that are used in the BDD-based attack.

Definition 6. *A Binary Decision Diagram (BDD) over a set of variables $X_n = \{x_1, \ldots, x_n\}$ is a directed, acyclic graph $G = (V, E)$ with $E \subseteq V \times V \times \{0,1\}$. Each inner node v has exactly two outgoing edges, a 0-edge $(v, v_0, 0)$ and a 1-edge $(v, v_1, 1)$ leading to the 0-successor v_0 and the 1-successor v_1, respectively. A BDD contains exactly two nodes with outdegree 0, the sinks s_0 and s_1. Each inner node v is assigned a label $v.label \in X_n$, whereas the two sinks are labeled $s_0.label = 0$ and $s_1.label = 1$. There is exacly one node with indegree 0, the root of the BDD. We define the size of a BDD to be the number of nodes in G, i.e., $|G| := |V|$. Each node $v \in V$ represents a Boolean Function $f_v \in B_n = \{f | f : \{0,1\}^n \to \{0,1\}\}$ in the following manner: For an input $a = (a_1, \ldots, a_n) \in \{0,1\}^n$, the computation of $f_v(a)$ starts in v. In a node with label x_i, the outgoing edge with label a_i is chosen, until one of the sinks is reached. The value $f_v(a)$ is then given by the label of this sink.*

Definition 7. *For a BDD G over X_n, let $G^{-1}(1) \subseteq \{0,1\}^n$ denote the set of inputs accepted by G, i.e., all inputs $a \in \{0,1\}^n$ such that $f_{root}(v) = 1$.*

Definition 8. *An oracle graph $G_0 = (V, E)$ over a set of variables $X_n = \{x_1, \ldots, x_n\}$ is a modified BDD that contains only one sink s, labeled $*$, and for all $x_i \in X_n$ and all paths P from the root in G to the sink, there exists at most one node in P that is labeled x_i.*

Definition 9. *A Free Binary Decision Diagram with respect to an oracle graph G_0 (a G_0-FBDD for short) over a set of variables $X_n = \{x_1, \ldots, x_n\}$ is a BDD in which the following property holds for all inputs $a \in \{0,1\}^n$. Let the list $G_0(a)$ contain the variables from X_n in the order in which they are tested on the path defined by a in G_0. Similarly, let the list $G(a)$ contain the variables from X_n in the order of testing in G. If x_i and x_j are both contained in $G(a)$, then they occur in $G(a)$ in the same order as in $G_0(a)$. We call a BDD G an FBDD, if there exists an oracle graph G_0 such that G is a G_0-FBDD.*

Figure 1 shows examples for an oracle graph G_0 and a G_0-FBDD.

Definition 10. *An FBDD G is called Ordered Binary Decision Diagram (OBDD) if there exists an oracle graph G_0 such that G is a G_0-FBDD and G_0 is degenerated into a linear list.*

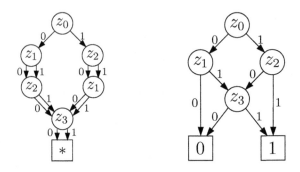

Fig. 1. An oracle graph G_0 over $\{z_0, \ldots, z_3\}$ and a G_0-FBDD

FBDDs possess several algorithmic properties that will prove useful in our context. Let G_0 denote an oracle graph over $X_n = \{x_1, \ldots, x_n\}$ and let the G_0-FBDDs G_f, G_g and G_h represent Boolean functions $f, g, h : \{0,1\}^n \rightarrow \{0,1\}$. Then, there exists an algorithm MIN that computes in time $O(|G_f|)$ the (uniquely determined) minimal G_0-FBDD G of size $|G| \leq n \cdot |G_f^{-1}(1)|$ that represents f. In time $O(|G_0| \cdot |G_f| \cdot |G_g| \cdot |G_h|)$, we can compute a minimal G_0-FBDD G with $|G| \leq |G_0| \cdot |G_f| \cdot |G_g| \cdot |G_h|$ that represents the function $f \wedge g \wedge h$. Additionally, it is possible to enumerate all elementes in $G_f^{-1}(1)$ in time $O\left(n \cdot |G_f^{-1}(1)|\right)$. We refer the reader to [18] for details on the corresponding algorithms.

2.3   BDD-Based Attack

The original BDD-based attack in [10], which we are going to describe in this section, assumes a known-plaintext scenario, i.e., the attacker manages to obtain a few plaintext-ciphertext pairs $(p_1, c_1), \ldots, (p_t, c_t) \in \{0,1\}^2$. Since the c_i were computed as $c_i = p_i \oplus y_i$ based on the output $y_0, \ldots, y_t \in \{0,1\}$ of an LFSR-based keystream generator $K = (L, C)$, he can compute the first t keybits as $y_i = p_i \oplus c_i$. From this prefix of the keystream, he wants to reconstruct the secret initial state x of L.

We observe that for any internal bitstream $z \in \{0,1\}^m$ that yields a prefix of the observed keystream-piece y, the following conditions must hold.

Condition 1. z is an m-extension of the key bits in z, i.e., $L_{\leq m}(KB(z)) = z$.

Condition 2. $C(z)$ is a prefix of y.

We call any z of length m that satisfies these conditions an m-candidate. The idea is now to start with $m = n_{min}$ and to dynamically compute the m-candidates for $m > n_{min}$, until only one m-candidate is left. The smallest m for which this will be most likely the case follows directly from the following Lemma.

Lemma 1. *Under Assumption 2, it holds for all keystreams y and all $m \leq \lceil \alpha^{-1} n \rceil$ that $|\{x \in \{0,1\}^n : C(L_{\leq m}(x)) \text{ is prefix of } y\}| \approx 2^{n^* - \alpha m} \leq 2^{n - \alpha m}$, where $n^* = |KP(m)|$. Hence, there exist approximately $2^{n - \alpha m}$ m-candidates.*

Lemma 1 implies that there will be only one m-candidate for $m \geq \lceil \alpha^{-1}n \rceil$. The key bits in this m-candidate form the secret initial state that the attacker is looking for.

In order to compute and represent the intermediate m-candidates efficiently, we use the following BDD-based approach. For $m \geq 1$, let G_m^C denote the oracle graph over $\{z_0, \ldots, z_{m-1}\}$ that determines for each internal bitstream $z = (z_0', \ldots, z_{m-1}') \in \{0,1\}^m$ the order in which the bits of z are read by the compression function C. Bitstreams z fulfilling conditions 1 and 2 will be represented in the minimal G_m^C-FBDDs R_m and Q_m^y, respectively. Then, the G_m^C-FBDD $P_m^y = \mathrm{MIN}(Q_m^y \wedge R_m)$ accepts exactly the m-candidates.

The cost of this strategy essentially depends on the sizes of the intermediate results P_m^y, which can be determined as follows.

Assumption 3 (FBDD Assumption). *For all $m \geq n_{min}$, it holds that $|G_m^C| \in m^{O(1)}$, $|Q_m| \in m^{O(1)}$, and $|R_m| \leq |G_m^C| 2^{m-n^*}$.*

Lemma 2. *If K fulfills Assumption 3, then*

$$|P_m^y| \leq \max_{1 \leq m \leq \lceil \alpha^{-1}n \rceil} \left\{ \min \left\{ p(m) \cdot 2^{m-n^*}, m \cdot 2^{n^*-\alpha m} \right\} \right\} \leq p(m) \cdot 2^{r^*(m)} \quad ,$$

where $p(m) = |G_m^C|^2 \cdot |Q_m^y|$ and $r^*(m) = \frac{1-\alpha}{1+\alpha}n^*$.

From this bound on $|P_m^y|$, one can straightforwardly derive the time, space and data requirements of the BDD-based attack.

Theorem 1. *Let $K = (L, C)$ be an LFSR-based keystream generator with initial state $x \in \{0,1\}^n$, information ratio α and best-case compression ratio γ. If K fulfills the Independence Assumption, the Pseudorandomness Assumption and the FBDD Assumption, an initial state $\tilde{x}$ with $C(L(\tilde{x})) = y$ for a given keystream $y = C(L(x))$ can be computed in time and with space $n^{O(1)}2^{\frac{1-\alpha}{1+\alpha}n}$ from the first $\lceil \gamma \alpha^{-1} n \rceil$ consecutive bits of y.*

3   Applications

We now survey the impact of the basic BDD-attack on the self-shrinking generator, the E_0 generator, and the A5/1 generator and compare it to other attacks on these ciphers.

The self-shrinking generator was introduced by Meier and Staffelbach [14]. It consists of only one LFSR and no memory. [10] showed that for the self-shrinking Generator, we have $\alpha \approx 0.2075$ and $\gamma = 0.5$ as well as $|Q_m| \leq m^2$ for $m \geq 1$.

Corollary 1. *From a prefix of length $\lceil 2.41n \rceil$ of a keystream $y = C(L(x))$ produced by a self-shrinking generator of keylength n, an initial state $\tilde{x}$ with $C(L(\tilde{x})) = y$ can be computed in time and with space $n^{O(1)}2^{0.6563n}$.*

This is the best short-keystream attack on the self-shrinking generator known so far. It slightly improves the bounds of $2^{0.75n}$ and $2^{0.694n}$ that were obtained in [14] and [19], respectively. The long-keystream attack in [15] needs at least $2^{0.3n}$ keystream bits in order to compute the initial state in less than $2^{0.6563n}$ polynomial-time operations.

The E_0 keystream generator from the short-range wireless communication standard Bluetooth [4] is a regular $(4,4)$-combiner with key length 128; its LFSRs have lengths $39, 33, 31, 25$. Therefore, we have $\alpha = \gamma = \frac{1}{4}$, and [10] showed that $|Q_m| \leq 32m$. Hence, we obtain from Theorem 1:

Corollary 2. *From a prefix of length n of a keystream $y = C(L(x))$ produced by an E_0 keystream generator of keylength n, an initial state $\tilde{x}$ with $C(L(\tilde{x})) = y$ can be computed in time and with space $n^{O(1)} 2^{0.6n} = n^{O(1)} 2^{76.8}$ for $n = 128$.*

The attack on E_0 by Fluhrer and Lucks [8] trades off time and necessary keystream bits. For the minimum number of 132 available keystream bits the attack needs 2^{84} polynomial time operations. The best currently known long-keystream attacks against E_0 are algebraic attacks [1] and correlation attacks [12,11]. These attacks all need a large amount of keystream (2^{28} to 2^{39} in the case of correlation attacks), and even in terms of time and memory requirements, [11] is the only feasible attack among them.

The A5/1 generator is used in the GSM standard for mobile telephones. According to [5], who obtained its design by reverse engineering, the generator consists of 3 LFSRs R_0, R_1, R_2 of lengths n_0, n_1, n_2, respectively, and a clock control ensuring that the keybits do not linearly depend on the initial states of the LFSRs. For each $r \in \{0,1,2\}$, a register cell q_{N^r}, $N^r \in \{\lceil \frac{n_r}{2} \rceil - 1, \lceil \frac{n_r}{2} \rceil\}$, is selected in LFSR R_r as input for the clock control. The GSM standard uses the parameters $(n_0, n_1, n_2) = (19, 22, 23)$ and $(N^0, N^1, N^2) = (11, 12, 13)$.

In order to write the generator in a $K = (L, C)$ fashion, we simulate its linear bitstream generator by six LFSRs $L^0, \ldots, L^5$. L^0, L^1, and L^2 are used exclusively for producing the keybits and correspond to R_0, R_1 and R_2 in the original generator, and the control values are computed from the outputs of L^3, L^4 and L^5, which correspond to L^0, L^1 and L^2 shifted by N^0, N^1 and N^2.

In [10], it was shown that in the case of A5/1, $\alpha = 0.2193$ and $\gamma = \frac{1}{4}$ as well as $|G_m^C| \in O(m^3)$ and $|Q_m| \in O(m^4)$. Plugging these values into the statement of Theorem 1 yields

Corollary 3. *From a prefix of length $\lceil 1.14n \rceil$ of a keystream $y = C(L(x))$ produced by an A5/1 keystream generator of keylength n, an initial state $\tilde{x}$ with $C(L(\tilde{x})) = y$ can be computed in time and with space $n^{O(1)} 2^{0.6403n} = n^{O(1)} 2^{41}$ for $n = 64$.*

We note that since $\lceil 1.14n \rceil = 73$ and the framelength in GSM is 114 Bits for each direction, we only need the first frame, i.e., the first around 4.6 milliseconds of a conversation in order to reconstruct the initial state.

The first short-keystream attack on A5/1 was given by Golić in [9] and needs 2^{42} polynomial time operations. Afterwards, several long-keystream attacks on A5/1 were proposed. [3] presents an attack that breaks A5/1 from 2^{15} known keystream bits within minutes after a preprocessing step of 2^{48} operations. Due to exploits of the linearity of the initialization procedure, the attack described in [7] and its refinement in [13] manage to break the cipher within minutes, requiring only few seconds of conversation and little computational resources.

4   Divide-and-Conquer Strategies

One obvious disadvantage of BDD-based attacks is the high memory consumption that is essentially determined by the size of the intermediate results P_m^y. For an LFSR-based keystream generator with keylength n, one possible approach to this problem is to divide the search space, more precisely the set $B_n = \{0,1\}^n$ of possible initial states of L, into segments and to apply BDD-based attacks to the segments individually. We denote a segmentation of B_n by the pair (f, T), where T is the finite set of segment labels and $f : \{0,1\}^* \to T$ a partial function that assigns a segment to each possible internal bitstream. For a given keystream y and each $t \in T$, we perform a BDD-based search on the set $B_n^t = \{x \in B_n | f(L(x)) = t\}$ in order to find an initial state $\tilde{x} \in B_n^t$ such that $C(L(\tilde{x})) = y$.

Similarly to the general attack described in the previous section, we denote by $Q_m^{y,t}$ the minimal G_0^C-FBDD that decides whether $C(z) = y$ and $f(z) = t$, and by R_m^t the minimal G_m^C-FBDD that accepts for $f(z) = t$ exactly those internal bitstreams z that are m-extensions of $KB(z)$. Moreover, let S_m^t be the minimal G_m^C-FBDD that decides for $f(z) = t$ whether $z_{m-1} = L_{m-1}(KB(z))$ and define $P_m^{y,t} := MIN(Q_m^{y,t} \wedge R_m^t)$. We can then apply the same algorithm for dynamically computing $P_{n_{min}}^{y,t}, P_{n_{min}+1}^{y,t}, \ldots$ as in the original case. Consequently, we obtain $n^{O(1)}2^{w^*}$ as time and space requirements for the BDD-based search on B_n^t, with w^* computed analogously to r^* in Lemma 2. For the overall attack, i.e., performing the BDD-based search on B_n^t for all $t \in T$, we get a memory consumption in the order of $n^{O(1)}2^{w^*}$ and a runtime of $n^{O(1)} \cdot |T| \cdot 2^{w^*}$ if the attacks are executed sequentially. Since the B_n^t are disjoint, the overall attack is efficiently parallelizable, and the $|T|$ factor can be further reduced.

We note that in general, we will only gain from a divide-and-conquer strategy (DCS) if $|T|$ is not too large and $w^* \leq r^*$. For the latter to be the case, the $|Q_m^{y,t}|$ have to be negligibly small and $|R_m^t|$ must be significantly smaller than $|R_m|$.

We consider now DCS that define a subset $V \subseteq KP(n_{max})$ of the initial state bits of L to be constant. We call a position $m \geq 1$ a V-*determined* position if $m \in V$ or if its domain $D(m)$ is a subset of V. For an internal bitstream z, let $t \in \{0,1\}^{|V|}$ denote the values of z at the positions in V. Then, the segmentation of B_n is given by $(f_V, T(V))$, where $f_V(z) = t$ and $T(V) = \{0,1\}^{|V|}$.

The FBDDs $Q_m^{y,t}$ can be obtained from Q_m^y by setting constant the variables that correspond to the V-determined positions. Hence, $|Q_m^{y,t}| \leq |Q_m^y|$. Moreover, since the test whether $z_{m-1} = L_{m-1}(KB(z))$ can be omitted for the V-determined positions, we have $|R_m^t| \leq |G_m^C|2^{r(m,V)}$, where $r(m, V)$ denotes the number of non-V-determined positions in $\{n_{min} + 1, \ldots, m\}$. Note that the original attack corresponds to the case $V = \emptyset$ and therefore $r(m, V) \leq m - n^*$, hence $|R_m^t| \leq |R_m|$.

4.1 DCS for Regular (k, l)-Combiners

We consider two examples that are applicable to regular (k, l)-combiners like the E_0 keystream generator.

First, we define V to contain exactly the positions of the first s output bits of each LFSR. In the worst case, there are no V-determined positions besides the positions in V. We only need to consider the assignments to the positions in V that are consistent with y. By Lemma 1, we have $|T(V)| \approx |\{0,1\}^{(1-\alpha)ks}| = 2^{(k-1)s}$. For $t \in T(V)$, the effort of a BDD-based search of the corresponding segment is equivalent to the effort for the original BDD-attack on a (k, l)-combiner of keylength $(n - ks)$, i.e., $w^* = \frac{k-1}{k+1}(n - ks)$. For the overall runtime, we obtain

$$n^{O(1)} \cdot 2^{(k-1)s + \frac{k-1}{k+1}(n-ks)} \in n^{O(1)} 2^{\frac{k-1}{k+1}n + \frac{k-1}{k+1}s} ,$$

which is by a factor of $2^{\frac{k-1}{k+1}s}$ worse than the original attack . On the other hand, the required memory is reduced by a factor of $2^{\frac{k-1}{k+1}ks}$.

As a second example, we choose as V the set of all key positions that belong to the shortest LFSR in L, which we assume w.l.o.g. to be the LFSR L^0. Let $n_0 \leq \frac{n}{k}$ be the length of L^0. Then, $T(V) = \{0,1\}^{n_0}$ is the set of all possible initial states of L^0. Since every k-th position of an internal bitstream z is V-determined, w^* corresponds to the performance of the original BDD-attack on a $(k-1, l)$ combiner of keylength $n - n_0$, i.e., $w^* = \frac{k-2}{k}(n - n_0)$. For the overall runtime, we obtain $2^{n_0 + \frac{k-2}{k}(n-n_0)}$. It is easy to see that for $n_0 \leq \frac{n}{k+1}$, we have

$$n_0 + \frac{k-2}{k}(n - n_0) \leq \frac{k-1}{k+1}n ,$$

i.e., for sufficiently small n_0, we even obtain a runtime improvement in addition to the significantly reduced space requirements. In the case of the original E_0, we have $n_0 = 25 \leq 25.6 = \frac{128}{4+1}$. Hence, we obtain

Lemma 3. For the E_0 keystream generator with keylength $n = 128$, choosing V to be the set of all key positions that belong to the shortest LFSR yields a runtime of the BDD-based attack of $n^{O(1)}2^{25 + \frac{1}{2}103} = 2^{76.5}$ and a memory consumption of $2^{51.5}$.

Compared to the original BDD-attack, we have improved the memory consumption by a factor of about 2^{25} and the runtime by a factor of $2^{0.3}$.

4.2 DCS for the A5/1 Generator

In the following, we compute the information rate of the A5/1 generator with respect to a family of choices for the set V, particularly those defined by setting one or several LFSRs or half-LFSRs to be constant. As stated in Sect. 3, in the unmodified definition of the A5/1 generator, each of the three LFSRs is divided into two, approximately equally long halfs, a value-half consisting of the output cell and the cells between output and clock-control cell and a control half consisting of the clock-control cell and the rest of the register. Since the value-LFSRs and the control-LFSRs in the modified setting correspond to the value-halfs and the control-halfs in the unmodified case, setting constant LFSRs or half-LFSRs in the original definition is equivalent to fixing the corresponding LFSRs in the modified case.

For all natural $i \geq 1$, let us denote by Y_i and Z_i the random variables corresponding to the i-th output bit and the number of internal bits processed for the production of the i-th output bit, respectively, taken over the probability space of all random internal bitstreams. In all cases, Y_i and Z_i will fulfill the following conditions.

- For all $i > 1$, Z_i is independent of $Z_1, \ldots, Z_{i-1}$, and Y_i is independent of $Y_1, \ldots, Y_{i-1}$.
- It holds that $Pr[Y_i = 0] = Pr[Y_i = 1] = \frac{1}{2}$.
- There are natural numbers $a > b > c$ and probabilities p, q and $r = 1 - p - q$ such that $Pr[Z_i = a] = p$, $Pr[Z_i = b] = q$, and $Pr[Z_i = c] = r$.

Let us denote the situation that Y_i and Z_i fulfill the above conditions as case $[(p, a), (q, b), (r, c)]$. It can be easily checked that the unrestricted A5/1 generator corresponds to case $[(1/4, 6), (3/4, 4), (0, 0)]$. We will see below that all generators derived from the A5/1 generator by setting constant one or more of the six LFSRs correspond to $[(p, a), (q, b), (r, c)]$ for some p, q, r, a, b, c. In these cases, we can compute the information rate α with the help of the following Theorem.

Theorem 2. *In the case* $[(p, a), (q, b), (r, c)]$, *the information rate equals* α, *where* $t = 2^\alpha$ *is the unique positive real solution of* $pt^a + qt^b + rt^c - 2 = 0$.

A proof for Theorem 2 can be found in Appendix A. Note that for the special case $[(1, k), 0, 0]$ the information rate is $1/k$.

In the following, we compute the information rates for restrictions of type $(v_1 v_2 v_3 | c_1 c_2 c_3) \in \{0, 1\}^6$, which means that those value-substreams i for which $v_i = 1$ and control-substreams j for which $c_j = 1$ are set constant. Note that the unrestricted case corresponds to $(000|000)$. We do not consider the case of 5 constant internal substreams as computing the remaining unknown half-LFSR from a given keystream can be done in linear time.

For symmerty reasons, the number of remaining cases resulting from setting constant 1,2,3,4 substreams can be reduced. Firstly, it is easy to see that for all permutations π of $\{1, 2, 3\}$ it holds that restriction $(v_1 v_2 v_3 | c_1 c_2 c_3)$ is equivalent to restiction $(v_{\pi(1)}, v_{\pi(2)} v_{\pi(3)} | c_{\pi(1)} c_{\pi(2)} c_{\pi(3)})$. Furthermore, observe that with respect to restriction $(v|c)$, $v, c \in \{0, 1\}^3$, the number of internal bits $Z(u, V, C)$ processed for the production of the next output bit assuming the current values in the control-substreams are $u \in \{0, 1\}^3$ equals

$$Z(u, v, c) = \sum_{i, c_i = 0} f_i(u) + \sum_{i, v_i = 0} f_i(u) \,, \tag{1}$$

where for $i \in \{1, 2, 3\}$ the Boolean function $f_i : \{0, 1\}^3 \to \{0, 1\}$ is defined to output 1 on u iff the i-th LFSR will be clocked w.r.t. u, i.e.,

$$f_i(u) = (u_i \oplus u_{i+1} \bmod 3 \oplus 1) \vee (u_i \oplus u_{i+2} \bmod 3 \oplus 1) \,.$$

Relation (1) implies that for all $v, c, u \in \{0, 1\}^3$ and $i \in \{1, 2, 3\}$, it holds that $Z(u, v, c) = Z(u, v', c')$, where v', c' are obtained from v, c by exchanging the i-th component. Hence, restriction $(v|c)$ is equivalent to restriction $(v'|c')$. It follows that the relevant cases are the restrictions $(000|100)$, $(100|100)$, $(100|010)$, $(100|110)$, $(000|111)$, $(100|111)$ and $(110|110)$.

The information rates for these cases are summarized in Table 1. The computation of the values can be found in Appendix B.

Table 1. Information rates α

$\log|T|$	restriction	α	w^*
$\frac{2}{3}n$	$(100|111)$	0.6430	$0.2173n$
	$(110|110)$	0.6113	$0.2412n$
$\frac{1}{2}n$	$(000|111)$	0.4386	$0.3902n$
	$(100|110)$	0.4261	$0.4024n$
$\frac{1}{3}n$	$(000|110)$	0.3271	$0.507n$
	$(100|100)$	0.3215	$0.5134n$
$\frac{1}{6}n$	$(000|100)$	0.2622	$0.584n$
0	$(000|000)$	0.2193	$0.6403n$

5   Experimental Results

In order to provide a fast implementation of the FBDD algorithms, an FBDD-library was developed based on the publicly available OBDD package CUDD (see [17]). The experiments were conducted on a standard Linux PC with a 2.7 GHz Intel Xeon processor and 4 GB of RAM. All implementation was done in C using the gcc-compiler version 3.3.5.

Since the runtime of the cryptanalysis fundamentally depends on the maximum size of the intermediate FBDDs P_m^y, we investigate how much experimentally obtained values of $|P_m^y|$ deviate from the theoretical figures.

We first consider the basic BDD-based attack. For the self-shrinking generator, the E_0 generator and the A5/1 generator, we analyzed several thousands of reduced instances with random primitive feedback polynomials and random initial states for various keylengths. For each considered random generator, we computed the actual maximum BDD-size of the intermediate results

$$P_{\max}(n) = \max_{1 \leq m \leq \lceil \alpha^{-1} n \rceil} \{|P_m^y|\} \ ,$$

the theoretical upper bound

$$P_{\max}^t(n) = \max_{1 \leq m \leq \lceil \alpha^{-1} n \rceil} \left\{ \min \left\{ p(m) \cdot 2^{m-n^*}, m \cdot 2^{n^* - \alpha m} \right\} \right\}$$

that was obtained in Lemma 2, as well as the quotient $q(n) = \frac{\log(P_{\max}(n))}{\log(P_{\max}^t(n))}$.

Similarly, we tested for E_0 and A5/1 the divide-and-conquer strategy of setting constant the shortest LFSR (s1), and we considered fixing the first $s = \frac{n_0}{2} \leq \frac{n}{8}$ bits of each of the four LFSRs in E_0 (s2), where n_0 denotes the length of the shortest LFSR. Since the q-values did not noticeably decrease with increasing n in all our simulations, we estimate the attack's performance in dependence of n by multiplying the theoretical figures by $2^{q(n)}$. Particularly, we can obtain conjectures about the attack's performance on real-life instances of E_0 and A5/1 by replacing n with the actual keylengths. Table 2 shows the results of these computations along with details about the conducted experiments.

On average, the attack based on DCS s1 took 87 minutes for E_0 with $n = 37$ and 54 minutes for A5/1 with $n = 30$. The longest keylengths that we were able to tackle with the resources described at the beginning of this section were $n = 46$ for E_0 and $n = 37$ for A5/1. These attacks used up almost all of the available memory and took 60.5 and 25.1 hours to complete on average.

Table 2. Performance of the BDD-based attack in practice

generator	DCS	keylength interval	avg $q(n)$	no. of samples	estimated practical performance			
					Time		Space	
E_0	—	$[19, 37]$	0.85	2000	$2^{0.51n}$	$2^{65.28}$	$2^{0.51n}$	$2^{65.28}$
E_0	s1	$[19, 37]$	0.95	2700	$2^{0.475(n+n_0)}$	$2^{72.68}$	$2^{0.475(n-n_0)}$	$2^{48.93}$
E_0	s2	$[19, 37]$	0.9	2700	$2^{0.54n+0.27n_0}$	$2^{75.87}$	$2^{0.54n-1.08n_0}$	$2^{42.12}$
A5/1	—	$[15, 30]$	0.9	3000	$2^{0.5763n}$	$2^{36.88}$	$2^{0.5763n}$	$2^{36.88}$
A5/1	s1	$[19, 37]$	0.77	2400	$2^{0.3953n+0.77n_0}$	$2^{39.93}$	$2^{0.3953n}$	$2^{25.30}$
SSG	—	$[10, 35]$	0.8	3300	$2^{0.525n}$		$2^{0.525n}$	

6   Conclusion

In this paper, we have presented the first comprehensive experimental results for the BDD-based attack on the self-shrinking generator, the E_0 and the A5/1. Our analysis shows that the performance of the BDD-attack on these generators in practice will not substantially drop below the theoretical upper bounds. We introduced divide-and-conquer strategies based on setting constant several initial state bits of the LFSRs and confirmed experimentally that in this way, the memory consumption of the attack may be reduced at the expense of slightly increasing the runtime. We have only applied a few examples of DCS to the E_0 and the A5/1 generator. In [16], an additional DCS for E_0 is reported which lowers the memory requirements to about 2^{23} while increasing the runtime to $O(2^{83})$. It is an interesting open question if there exist more efficient strategies that are able to simultaneously reduce the runtime by a significant amount.

Acknowledgement

We would like to thank Frederik Armknecht for valuable comments and discussions.

References

1. F. Armknecht and M. Krause. Algebraic attacks on combiners with memory. In *Proc. of CYPTO 2003*, volume 2729 of *LNCS*, pages 162–176. Springer, 2003.
2. F. Armknecht, M. Krause, and D. Stegemann. Design principles for combiners with memory. In *Proc. of INDOCRYPT 2005*, volume 3797 of *LNCS*, pages 104–117. Springer, 2005.
3. A. Biryukov, A. Shamir, and D. Wagner. Real time cryptanalysis of A5/1 on a PC. In *Proc. of Fast Software Encryption 2000*, volume 1978 of *LNCS*, pages 1–13. Springer, 2000.
4. The Bluetooth SIG. *Specification of the Bluetooth System*, February 2001.
5. M. Briceno, I. Goldberg, and D. Wagner. *A pedagogical implementation of A5/1*, May 1999. http://jya.com/a51-pi.htm.
6. N. Courtois. Fast algebraic attacks on stream ciphers with linear feedback. In *Proc. of CRYPTO 2003*, LNCS, pages 177–194. Springer, 2003.
7. P. Ekdahl and T. Johansson. Another attack on A5/1. In *Proc. of International Symposium on Information Theory*, page 160. IEEE, 2001.
8. S. R. Fluhrer and S. Lucks. Analysis of the E_0 encryption system. In *Proc. of SAC 2001*, volume 2259 of *LNCS*, pages 38–48. Springer, 2001.
9. J. Golić. Cryptanalysis of alleged A5 stream cipher. In *Proc. of EUROCRYPT 1997*, volume 1233 of *LNCS*, pages 239–255. Springer, 1997.
10. M. Krause. BDD-based cryptanalysis of keystream generators. In *Proc. of EURO-CRYPT 2002*, volume 2332 of *LNCS*, pages 222–237. Springer, 2002.
11. Y. Lu, W. Meier, and S. Vaudenay. The conditional correlation attack: A practical attack on bluetooth encryption. In *Proc. of CRYPTO 2005*, volume 3621 of *LNCS*, pages 97–117. Springer, 2005.
12. Y. Lu and S. Vaudenay. Cryptanalysis of the bluetooth keystream generator two-level E0. In *Proc. of ASIACRYPT 2004*, volume 3329 of *LNCS*, pages 483–499. Springer, 2004.
13. A. Maximov, T. Johansson, and S. Babbage. An improved correlation attack on A5/1. In *Proc. of SAC 2004*, volume 3357 of *LNCS*, pages 1–18. Springer, 2004.
14. W. Meier and O. Staffelbach. The self-shrinking generator. In *Proc. of EURO-CRYPT 1994*, volume 950 of *LNCS*, pages 205–214. Springer, 1994.
15. M. J. Mihaljević. A faster cryptanalysis of the self-shrinking generator. In *Proc. of ACISP 1996*, volume 1172 of *LNCS*, pages 192–189. Springer, 1996.
16. Y. Shaked and A. Wool. Cryptanalysis of the bluetooth E_0 cipher using OBDDs. Technical report, Cryptology ePrint Archive, Report 2006/072, 2006.
17. F. Somenzi. *CUDD: CU decision diagram package*. University of Colorado, Boulder, CO, USA, March 2001. http://vlsi.colorado.edu/~fabio/.
18. I. Wegener. *Branching Programs and Binary Decision Diagrams: Theory and Applications*. SIAM Monographs on Discrete Mathematics and Applications, 2000.
19. E. Zenner, M. Krause, and S. Lucks. Improved cryptanalysis of the self-shrinking generator. In *Proc. of ACISP 2001*, volume 2119 of *LNCS*, pages 21–35. Springer, 2001.

A   Proof of Theorem 2

In order to prove Theorem 2, we need the following technical result that was proved in [10].

Lemma 4. *For all natural $N \geq 1$, probabilities $p \in (0,1)$ and real $\beta > 0$ it holds that $\sum_{i=0}^{N} \binom{N}{i} p^i (1-p)^{N-i} 2^{\beta i} = (1 - p + p2^{\beta})^{N}$.*

Since we can obtain the information rate α from $\alpha = -\frac{1}{m} \log_2 p_C(m)$, we now compute the probability $p_C(m) = Prob_z[C(z)$ is prefix of $y]$ for the cases that parts of the LFSRs are set constant.

Case $[(p,a),(q,b),(r,c)]$ implies that on all random internal bitstreams of length m, m divisible by a, at least m/a output bits are produced. The number of internal bits remaining from m internal bits after the production of m/a output bits can be computed as

$$m - aU - bV - c\left(\frac{m}{a} - U - V\right) = \frac{a-c}{a}m - (a-c)U - (b-c)V ,$$

where U and V denote the number of output bits among the first m/a output bits for which a, resp. b internal bits are processed. Note that U is $(p, m/a)$-binomially distributed and that V, under the condition that $U = i$, is $(q/(q+r), m/a - i)$-binomially distributed. We obtain the following relation for $p_C(m)$.

$$p_C(m) = 2^{-\frac{m}{a}} \sum_{i=0}^{\frac{m}{a}} \sum_{j=0}^{\frac{m}{a}-i} Pr[U=i, V=j] p\left(\frac{a-c}{a}m - (a-c)i - (b-c)j\right) \text{ ,i.e.,}$$

$$2^{-\alpha m} = 2^{-\frac{m}{a}} \sum_{i=0}^{\frac{m}{a}} \binom{\frac{m}{a}}{i} p^i (1-p)^{\frac{m}{a}-i}$$

$$\sum_{j=0}^{\frac{m}{a}-i} \binom{\frac{m}{a}-i}{j} \left(\frac{q}{q+r}\right)^j \left(\frac{r}{q+r}\right)^{\frac{m}{a}-i-j} \cdot 2^{-\alpha(\frac{a-c}{a}m - (a-c)i - (b-c)j)} \text{ ,i.e.,}$$

$$2^{(1-a\alpha+(a-c)\alpha)\frac{m}{a}} = \sum_{i=0}^{\frac{m}{a}} \binom{\frac{m}{a}}{i} p^i (1-p)^{\frac{m}{a}-i} \cdot 2^{(a-c)\alpha i}$$

$$\sum_{j=0}^{\frac{m}{a}-i} \binom{\frac{m}{a}-i}{j} \left(\frac{q}{1-p}\right)^j \left(\frac{r}{1-p}\right)^{\frac{m}{a}-i-j} \cdot 2^{(b-c)\alpha j} .$$

Now, we apply Lemma 4 to the inner sum and obtain

$$2^{(1-n\alpha)\frac{m}{a}} = \sum_{i=0}^{\frac{m}{a}} \binom{\frac{m}{a}}{i} p^i (1-p)^{\frac{m}{a}-i} \cdot 2^{(a-c)\alpha i} \cdot \left(\frac{r}{1-p} + \frac{q}{1-p} 2^{(b-c)\alpha}\right)^{\frac{m}{a}-i} .$$

Setting $s = \frac{r}{1-p} + \frac{q}{1-p} 2^{(b-c)\alpha}$, we get

$$\left(\frac{2}{s2^{c\alpha}}\right)^{\frac{m}{a}} = \sum_{i=0}^{\frac{m}{a}} \binom{\frac{m}{a}}{i} p^i (1-p)^{\frac{m}{a}-i} \cdot 2^{((a-c)\alpha - \log(s))i} = \left(1 - p + p2^{(a-c)\alpha - \log(s)}\right)^{\frac{m}{a}} .$$

Consequently, setting $t = 2^\alpha$, we obtain

$$\frac{2}{st^c} = 1 - p + p\frac{t^{a-c}}{s} \quad \Leftrightarrow \quad 2 = (1-p)st^c + pt^a.$$

$s = \frac{r}{1-p} + \frac{q}{1-p}t^{b-c}$ implies $2 = rt^c + qt^b + pt^a$, which yields the Theorem.

B   Computation of α for the Considered DCS for A5/1

In order to compute the remaining α values, we only need to compute the corresponding cases of the form $[(p, a), (q, b), (r, c)]$ for the given restrictions on the LFSRs.

We first consider the restriction $(100|100)$. If the actual content of the output cells of the two non-constant control substreams is 00 or 11, then 4 internal bits will be processed, otherwise 2 internal bits will be processed. Hence, the corresponding case is $[(1/2, 4), (1/2, 2), 0]$ and therefore $\alpha \approx 0.3215$.

Under restriction $(100|010)$, 4 internal bits will be processed if the actual content of the output cell of the constant control substream is $b \in \{0, 1\}$ and the actual content of the two non-constant control substreams is bb. If it is $\bar{b}\bar{b}$ then 2, and in all remaining cases 3 internal bits will be processed. Therefore, the corresponding case is $[(1/4, 4), (1/2, 3), (1/4, 2)]$ and $\alpha \approx 0.3271$.

If we assume restriction $(110|110)$, 2 internal bits will be processed if the assignments to the output cells of the constant control substreams is 01 or 10 or if all 3 output cells of the control-substreams coincide. If the assignment to the output cells of the constant control substreams is bb for some $b \in \{0, 1\}$ and the random assignment to the remaining control is output cell is $\neg b$, then the next output bit depends only on the constant assignments, and no internal bit will be processed. This implies that, in contrast to the above cases, $p_C(m)$ and α are not independent of the constant substreams and the given keystream. Therefore, we compute only the average information rate over all possible assignments to the constant control and output substreams. According to the above observation, the probability that 2 internal bits are processed for the next output bit is 3/4, and the probability that 0 internal bits are processed for the next ouput bit is 1/4. In total, we obtain $[(3/4, 4), (1/4, 0), 0]$ and therefore $\alpha \approx 0.6113$.

We can handle the remaining cases with similar arguments.

Breaking the ICE - Finding Multicollisions in Iterated Concatenated and Expanded (ICE) Hash Functions

Jonathan J. Hoch and Adi Shamir

Department of Computer Science and Applied Mathematics
The Weizmann Institute of Science, Israel

Abstract. The security of hash functions has recently become one of
the hottest topics in the design and analysis of cryptographic primitives.
Since almost all the hash functions used today (including the MD and
SHA families) have an iterated design, it is important to study the general security properties of such functions. At Crypto 2004 Joux showed
that in any iterated hash function it is relatively easy to find exponential
sized multicollisions, and thus the concatenation of several hash functions
does not increase their security. However, in his proof it was essential
that each message block is used at most once. In 2005 Nandi and Stinson extended the technique to handle iterated hash functions in which
each message block is used at most twice. In this paper we consider the
general case and prove that even if we allow each iterated hash function
to scan the input multiple times in an arbitrary expanded order, their
concatenation is not stronger than a single function. Finally, we extend
the result to tree-based hash functions with arbitrary tree structures.

Keywords: Hash functions, iterated hash functions, tree based hash
functions, multicollisions, cryptanalysis.

1 Introduction

The recent discovery of major flaws in almost all the hash functions proposed so
far ([18], [5], [1]) made the analysis of the security properties of these functions
extremely important. Some researchers (e.g., Jutla and Patthak [6]) proposed
clever ways to strengthen the internal components of standard hash functions in
order to make them provably resistant against some types of attacks. A different line of research (which was extensively studied and formalized in Preneel's
pioneering work [11]) considered the structural properties of various types of
hash functions, assuming that the primitive operations (such as compression
functions on fixed length inputs) are perfectly secure. This is similar to the
structural study of various modes of operation of encryption schemes, ignoring
their internal details.

One of the most surprising results in this area was the recent discovery by
Joux [5] of an efficient attack on Iterated Concatenated (IC) hash functions. An
iterated hash function has a constant size state, which is mixed with a constant

M.J.B. Robshaw (Ed.): FSE 2006, LNCS 4047, pp. 179–194, 2006.

size input by a compression function f to generate the next state. A message of unbounded size is hashed by dividing it into a sequence of message blocks, and providing them one by one to the compression function. The initial state is a fixed IV, and the last state is the output of the hash function. A concatenated hash function starts from several IV's, applies a different compression function to the original message in each chain, and concatenates the final states of all the chains to get a longer output. To prove that multiple chains of compression functions are not much stronger than a single chain, Joux showed how to generate a 2^k−multicollision (i.e., 2^k different messages which are all mapped to the same output value by the hash function) with complexity $k2^{\frac{n}{2}}$. This is only slightly larger than the $2^{\frac{n}{2}}$ complexity of finding one pairwise collision in the underlying compression function via the birthday paradox, and much smaller than the 2^k2^n complexity of finding such a multicollision in a random non-iterated hash function. He then showed how to use multicollisions in F_1 in order to find collisions in the concatenated hash function $F_1(M)\|F_2(M)$ with complexity $O(n2^{\frac{n}{2}})$, which is much smaller than the 2^n complexity of the birthday paradox applied to the $2n$−bit concatenated state. Other possible applications of multicollisions are in the MicroMint micropayment scheme [14] and in distinguishing iterated hash functions from random functions.

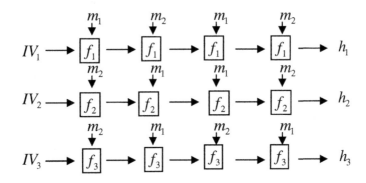

Fig. 1. An example of an ICE hash function, where the output is $h_1\|h_2\|h_3$

One of the simplest ways to overcome Joux's multicollision attack is to use message expansion which forces the iterated hash function to process each message block more than once. For example, the hash function can scan the original message blocks forwards, then backwards, then the even numbered blocks, and finally the odd numbered blocks, before producing the output. In addition, a concatenated hash function can use a different expanded order with each compression function, before concatenating their outputs (see Fig 1). We can assume that the expansion phase increases the total number of message blocks by at most a constant factor s, since higher expansion rates (e.g., quadratic) will make it too expensive to hash long messages, and thus lead to impractical constructions. We call such a generalized scheme an Iterated Concatenated and Expanded (ICE) hash function. Joux's original technique could not handle such functions, since a pair of message blocks which create a collision in a compression function at one point is very unlikely to create another collision later when they are mixed with a different state.

This difficulty was partially resolved in 2005 by Nandi & Stinson [10]. They considered the special case of ICE hash functions in which each message block is used at most twice in the expanded message, and extended Joux's original technique in a highly specialized way to handle this slightly larger class of hash functions.

In this paper we consider the general case of an arbitrary expansion rate s, and show how to find in any ICE hash function whose individual compression functions have $n-$bit states an $O(2^n)$ sized multicollision, using messages whose length is polynomial in n for any constant s. This shows that the Joux multicollision technique is much more powerful and the ICE hash construction is considerably less secure than originally believed.

1.1   Outline of This Paper

The new proof technique is based on careful analysis of the structural properties of sets of words of the form $M' = m_{\alpha_1} m_{\alpha_2} ... m_{\alpha_e}$ which can be derived from the original message $M = m_1 m_2 ... m_l$ by replicating and reordering the message blocks m_i during the expansion phase, when $e \le sl$. The proof is quite involved, and uses a series of combinatorial lemmas. To make it easier to follow, we first give an overview of the various steps.

The first step is to show that the case of expansion by a total factor s can be reduced to the case of an expansion in which each message block appears at most $q = 2s$ times. The next step of the proof is to reduce such expanded words to the form $\pi_1(M) \| \pi_2(M) ... \| \pi_k(M)$ where $k \le q$ and each π_i is a permutation which contains each message block exactly once. We then show how to construct arbitrarily large multicollisions when the expanded sequence consists of k successive permutations of the message blocks. Finally we show how to use such multicollisions in order to find collisions in the concatenation of several hash functions defined by different sequences.

In section 2 we deal with expansion schemes which can be represented as a sequence of permutations. Section 3 generalizes the proof to any ICE hash function with a constant expansion rate. Section 4 shows how to construct multicollisions when the iterative compression structure is replaced by a tree-like compression scheme. Section 5 summarizes our results and presents some open problems.

2   The Successive Permutations Case

Throughout the paper we denote the set of the first l integers by $L = \{1, 2, ..., l\}$ where $l = |M|$ is the length of the original (unexpanded) message. Where no message is clear from the context, l can be an arbitrary integer. We start by proving a useful lemma:

Lemma 1. *Let B and C be two permuted sequences of the elements of L. Divide B into k consecutive groups of the same size $(\frac{l}{k})$ and name the groups $B_1, ..., B_k$, and divide C into k consecutive groups of the same size $(\frac{l}{k})$ and name the groups $C_1, ..., C_k$. Then for $x > 0$ and $l \geq k^3 x$ there exists a perfect matching of B_i's and C_j's such that $B_i \bigcap C_j \geq x$.*

Proof. We will use the fact that B and C are partitioned into a small number of large disjoint sets, which are likely to have large intersections. We construct the following bipartite graph: $V = \{B_1, ..., B_k, C_1, ..., C_k\}$ and $(B_i, C_j) \in E$ iff $B_i \bigcap C_j \geq x$. According to Hall's matching theorem it is enough to show that any subset of B_i's of size t has at least t neighbors in C, in order to prove that there exists a perfect matching between B and C. Without loss of generality, let $A = B_1 \bigcup ... \bigcup B_t$ be all the elements from a subset of B_i's. Assume for the sake of contradiction that this subset has at most $t - 1$ neighbors in C. This means that at most $t - 1$ C_j's intersect these B_i's with an intersection of x or more. The maximal number of elements from A which are 'covered' by these elements is $(t - 1)\frac{l}{k}$. In addition there are $k - t + 1$ C_i's which intersect each of the B_i's in A by less than x. Since there are t B_i's in A, the maximal number of elements in A covered by the remaining C_i's is less than $(k - t + 1)tx$. So the total number of elements in A covered by any element from C is less than $(t-1)\frac{l}{k} + (k-t+1)tx$. However, the total number of elements in A is $t\frac{l}{k}$. Taking $l \geq k^3 x$ we have $t\frac{l}{k} \geq txk^2 \geq (t-1)xk^2 + (k-t+1)tx$ for any t. Thus we have a contradiction (not all the elements of A are 'covered') and we conclude that any subset of t B_i's must have at least t neighbors among the C_j's. Hence the conditions from Hall's theorem are fulfilled and there exists a perfect matching between the B_i's and the C_j's.   □

Definition 1. *An interval $I = [i_1, i_2]$ is a continuous set of indices $1 \leq i_1 \leq i_2 \leq l$. Then for any sequence α of elements from L, $\alpha[I]$ denotes the subsequence of α defined by $(\alpha_{i_1}, \alpha_{i_1+1}, ..., \alpha_{i_2})$.*

Definition 2. *Let α be some sequence over L and let $X \subseteq L$ $\alpha|_X$ is constructed as follows: First we take β to be the subsequence of α containing only elements from X. Then we set all consecutive appearances of the same value to a single appearance. For example, if $\alpha = 1, 2, 3, 3, 2, 4, 2, 3$ and $X = \{2, 3\}$ then we first set $\beta = 2, 3, 3, 2, 2, 3$ and then set $\alpha|_X = 2, 3, 2, 3$.*

We now state another useful lemma.

Lemma 2. *Let α be a sequence over L and let X be a subset of elements of L. If we can construct a 2^k Joux multicollision against the hash function based on $\alpha|_X$ then we can construct a 2^k Joux multicollision against the hash function based on α.*

Proof. Let h_0 be the initial hash value. In a Joux multicollision, starting from the initial hash we have a series of intermediate hash values $(h_1, h_2, ..., h_k)$ such that h_i is reachable from h_{i-1} by two different choices for the relevant message

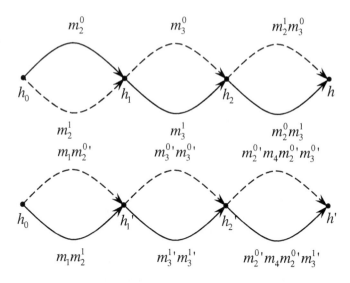

Fig. 2. The Joux multicollision in $\alpha|_X = 2,3,2,3$ (top part) and in $\alpha = 1,2,3,3,2,4,$ $2,3$ (bottom part, where $X = \{2,3\}$). Notice how the message blocks are different in $\alpha|_X$ and in α and that all message blocks not in X are set to a constant value. The dotted and solid lines describe the two collision paths in the final 2-collision when $n = 2$.

blocks. Now let $J_1, J_2, .., J_k$ be the indices of the intervals of message blocks used for the Joux multicollision such that $F(h_{i-1}, M(J_i)) = h_i$ where $M(J_i)$ is the sequence of message blocks corresponding to the indices in the interval J_i. The interval J_i in $\alpha|_X$ corresponds to an interval I_i in the original sequence α such that $\alpha[I_i]|_X = \alpha|_X[J_i]$. Now starting from J_1, we have that there are at least $2^{\frac{n}{2}}$ different messages that can be constructed by changing the message blocks indexed by the indices in J_1, since I_1 includes all of those indices, we can set all other message blocks to a fixed constant and varying only the message blocks indexed by J_1, construct a collision in $F(h_0, I_1^i) = h_1'$ with I_1^0 and I_1^1. The same goes for J_2 and I_2 and so on until J_k and I_k. The important thing to notice is that even when the possible combinations that are used in J_i are not all the combinations, i.e. there are some restrictions stemming from previous use of the message blocks, we still have at least $2^{\frac{n}{2}}$ possible combinations in J_i (which is sufficient for finding a collision with high probability among the different intermediate hash values) and therefore also in I_i. At the culmination of this process we have constructed a 2^k Joux multicollision in the hash function based on α.   □

To ease the understanding of the general case of successive permutations, we first give a proof for the special case of 3 successive permutations $\alpha = \pi_1(L)\|\pi_2(L)\|\pi_3$ (L) which is the simplest case which is not treated in [10]. We start by taking a message M of length $\frac{k^3 n^2}{4}$. We now look at the message blocks $\pi_2(L)$ and group them into consecutive groups of size $k\frac{n^2}{4}$. We call the first group B_1 and the last group B_k where 2^k is the size of the multi-collision we are constructing. Similarly we group the message blocks $\pi_3(L)$ into consecutive groups of the same size and name the groups $C_1, ..., C_k$. We use lemma 1 in order to pair each B_i with a unique C_j such that $B_i \cap C_j \geq \frac{n^2}{4}$. We now choose from each pair $\frac{n^2}{4}$ message block indices from the intersection and call the union of all the intersections *active indices*, the rest of the message block indices will be called *inactive indices*. Note that since π_2 and π_3 are permutations, each active index occurs in a single pair of B_i and C_j. Let X be the set of all the active indices. According to lemma 2 it suffices to show that we can construct a 2^k Joux multicollision in $\beta = \alpha|_X$. We construct a Joux multicollision on the message blocks indexed by the first part of β (which is taken from $\pi_1(L)$), starting from the initial IV. We then construct a multicollision on the message blocks indexed by the section of β which is taken from $\pi_2(L)$ using intervals containing $\frac{n}{2}$ message blocks each. Finally we construct a multicollision in the message blocks indexed by the section of β which is taken from $\pi_3(L)$ by using intervals containing $\frac{n^2}{4}$ message blocks (which correspond to the C_i's). Notice that the final stage of the construction works because the elements in a specific C_i are all contained in the same interval B_j (and in no other B_t) and thus do not affect the intermediate hash values outside this interval. While the basic idea of using larger and larger blocks in not new (for example, it was used by Joux [5] to compute preimages in generic hash functions), our results generalize the technique and show its real power.

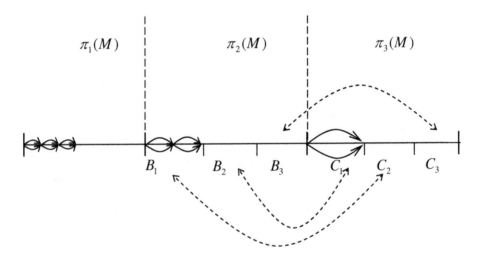

Fig. 3. Multicollision in 3 successive permutations. The dotted lines represent the matching between the B_i's and the C_j's. The solid lines show the collisions built along the way. The collisions in the leftmost section are collisions over single message blocks. The collisions in the middle section are over intervals containing $\frac{n}{2}$ message blocks. The collisions in the rightmost section are over intervals containing $\frac{n^2}{4}$ message blocks.

We now prove the general case of successive permutations, by using messages whose length is polynomial in n for any constant expansion rate s.

Theorem 1. *Let α be a sequence of the form $\pi_1(L)\|\pi_2(L)...\|\pi_q(L)$. We can construct a 2^k Joux multicollision against the hash function based on α whenever $l = |M| \geq k^3 n^{3(q-3)+2}$.*

Proof. We start by dividing the last two permutation copies, $\pi_{q-1}(L)$ and $\pi_q(L)$, into k equal length intervals each. We then find a perfect matching between the two sets of intervals as in the 3 permutations case. However, this time we seek an intersection of size $n^{3(q-3)+2}$. After we have our new set of active indices X (which is the disjoint union of the indices from all the intersections), we turn to look at $\alpha|_X$. In this new sequence we examine the permutations $\pi_{q-2}(L)$ and $\pi_{q-1}(L)$. We divide them into kn intervals of equal length and use our lemma to find a perfect matching with an intersection size of $n^{3(q-4)+2}$. We then divide the permutations $\pi_{q-3}(L)$ and $\pi_{q-2}(L)$ into kn^2 intervals and find a perfect matching with an intersection size of $n^{3(q-5)+2}$. We continue downsizing our list of active indices in the same manner until we have found a perfect matching with an intersection size of n^2 between $\pi_3(L)$ and $\pi_2(L)$. The size of X, the set of active indices, starts with $|X| = k^3 n^{3(q-3)+2}$. After the first step we have $kn^{3(q-3)+2}$ remaining active indices, after the second step we have kn segments, each with $n^{3(q-4)+2}$, and after $q-2$ steps we have an intersection size of n^2 for each of the kn^{q-3} segments.

The next stage is to build a Joux 2^k multicollision in the hash function based on $\beta = \alpha|_X$ where X is the final (smallest) set of indices. As in the three permutations case, we start by constructing a Joux multicollision on the $\pi_1(L)$ part of the sequence. We then use intervals of n message blocks to construct a multicollision in the $\pi_2(L)$ part and in general use intervals of size n^{i-1} in the i-th permutation. Since we have k blocks of size n^{q-1} in the last permutation, the process terminates with a 2^k multicollision in the hash function based on β. Using lemma 2, we get a 2^k multicollision in the hash function based on α as required. □

3   Solving the General Case

We now show how to reduce the general case to the successive permutations case. First we state some definitions and prove a useful lemma.

Definition 3. *Let α be a sequence over L:*

$$freq(x, \alpha) = |\{i : \alpha_i = x\}| \tag{1}$$

$$freq(\alpha) = max\{freq(x, \alpha) : x \in L\} \tag{2}$$

Definition 4. *Let* $T = t_1, ..., t_t$ *be a (not necessary contiguous) sequence of indices in* α. *Then :*

$$\alpha[T] = \alpha_{t_1}, ..., \alpha_{t_t} \tag{3}$$

In particular if $T = [t_1, t_2]$ *is an interval then the definition coincides with definition 1.*

Definition 5. *Given any subsequence* $\alpha[T]$ *of* α, *we define*

$$S(\alpha[T]) = |\{x \in L : freq(x, \alpha[T]) \geq 1\}| \tag{4}$$

Definition 6. *A set of disjoint intervals* $I_1, ..., I_j$ *is called* independent over α *if there exists a set of distinct elements* $x_1, ..., x_j$ *in* α *such that all the appearances of* x_i *in* α *are in* $\alpha[I_i]$.

We will call a set $x_1, ..., x_j$ *of distinct elements in* α independent *if there exist independent intervals* $I_1, ..., I_j$ *such that all appearances of* x_i *are in* $\alpha[I_i]$.

Definition 7. $Ind(\alpha)$ *is the largest* j *such that there exists a set* $I_1, ..., I_j$ *which is independent over* α.

For example $\alpha = 1, 2, 1, 3, 2, 4, 2, 4$ has $Ind(\alpha) = 3$ by taking the independent elements $1, 3, 4$. We can see for example that the smallest interval containing all the appearances of 4 does not contain either 1 or 3. However, we cannot chose $1, 2, 3, 4$ as independent elements since they are interleaved in α.

Definition 8. *A left-end interval is an interval of the form* $I = [1, i]$ *for some integer* i.

In Nandi and Stinson's paper[10] the authors proved and used the following lemma (translated into our notation):

Lemma 3. *Let* α *be a sequence of elements from* L *with* $freq(\alpha) \leq 2$ *and* $S(\alpha) = l$. *Suppose that* $l \geq MN$. *Then at least one of the following holds:*

1. $Ind(\alpha) \geq M$, *or*
2. *there exists a left-end interval* I *such that* $Ind(\alpha[I]) \geq N$.

The generalization we wish to prove in order to handle arbitrary ICE hash functions is as follows:

Lemma 4. *Let* α *be a sequence of elements from* L *with* $freq(\alpha) \leq q$ *and* $S(\alpha) = l$. *Suppose that* $l \geq MN$. *Then at least one of the following holds:*

1. $Ind(\alpha) \geq M$, *or*
2. *there exists a left-end interval* I *and a subset* $X \subseteq L$ *s.t.* $freq(\beta) \leq q - 1$ *and* $S(\beta) \geq \frac{N}{q-1}$ *where* $\beta = \alpha[I]|_X$.

Proof. The proof follows the same general lines as in [10], and uses induction on l. For the left-end interval $I = [1, N]$ either $freq(\alpha[I]) \leq q - 1$ or there exists an element x_1, which appears q times in the sequence $\alpha[I]$. If the former holds then we have N elements in $\alpha[I]$ and each one of them can occur at most $q-1$ times, and thus the number of distinct elements $S(\alpha[I])$ is at least $\frac{N}{q-1}$. We set $X = L$ and $\beta = \alpha[I]|_X = \alpha[I]$ and we are done. So we assume that there exists an element x_1, which appears q times in the sequence $\alpha[I]$. We remove all elements from α which appear in $\alpha[I]$ and call the new sequence $\alpha_1 = \alpha[I_1]$ for some set of indices I_1.

Note that $S(\alpha_1) \geq MN - N = (M-1)N$ since we have removed at most N distinct elements from α. By the induction hypothesis, either $Ind(\alpha_1) \geq M-1$ or there exists a left-end interval J and a subset X of L such that $freq(\beta[J]) \leq q-1$ and $S(\beta[J]) \geq \frac{N}{q-1}$ where $\beta = \alpha[J]|_X$. In the latter case we simply take X and β as provided from the lemma and set the interval I to be the shortest left-end interval containing J. In the former case let $I_2, ..., I_M$ be an independent set of intervals over α_1 containing the independent indices $x_2, .., x_M$. These intervals can be mapped to independent intervals $J_2, ..., J_M$ over α where J_i is the minimal interval containing all the occurrences of x_i for $i = 2, .., M$. Notice that $x_1 \notin J_i$ for $i = 2..M$ since all appearances of x_1 are before the first index of α_1 so we can add an interval $J_1 = 1..N$ to the list of independent intervals and now we have that $Ind(\alpha) \geq M$ as required.   □

Now we prove one final lemma before turning to prove our main theorem. We want to prove by induction on q the following claim:

Lemma 5. *For any integer x, given a sequence α with $freq(\alpha) \leq q$ and $S(\alpha)$ large enough, we can find a subset of indices X, $|X| \geq x$ such that $\alpha|_X$ is in the form of up to q successive permutations over the same set of indices X.*

Proof. Let $f_q(x)$ be the minimal alphabet size of a sequence α with $freq(\alpha) \leq q$ that ensures that there is a subset of indices X, $|X| \geq x$ such that $\alpha|_X$ is in the form of successive permutations. We will prove that $f_q(x) \leq C_q x^{D_q}$, for some constants C_q, D_q which increase with q.

We start by claiming that $f_1(x) = x$ (i.e., $C_1 = D_1 = 1$), since any sequence α with $S(\alpha) = x$ and $freq(\alpha) = 1$ is a single permutation of all the indices that occur in α. For notational purposes we will define $f_0(x) = 0$ for all x. Now assume that we have proven the inequality $f_k(x) \leq C_k x^{D_k}$ for all $k < q$. Given a sequence α such that $S(\alpha) \geq x(q-1)f_{q-1}(f_1(x) + f_2(x) + ... + f_{q-1}(x))$, we apply lemma 4 with $M = x$ and $N = (q-1)f_{q-1}(f_1(x) + f_2(x) + ... + f_{q-1}(x))$. There are now two cases. In the first we have $Ind(\alpha) \geq x$, and let X be the set of all independent indices. By definition we have $|X| = Ind(\alpha) \geq x$ and $\alpha|_X$ is a single permutation of the indices in X (since $freq(\alpha|_X) = 1$). In the second case we have a left-end interval I and a subset X' such that $freq(\alpha[I]|_{X'}) \leq q-1$ and $S(\alpha[I]|_{X'}) \geq \frac{N}{q-1} = f_{q-1}(f_1(x) + ... + f_{q-1}(x))$. Now using the inductive hypothesis on $\alpha[I]|_{X'}$ we get a subset X'' such that $|X''| \geq f_1(x) + ... + f_{q-1}(x)$ and $\alpha[I]|_{X''}$ is in successive permutations form with at most $q-1$ permutations. Using the pigeonhole principle we see that there must exist an $0 \leq i \leq q-1$ such that at least $f_i(x)$ indices appear exactly i times in the remainder of $\alpha|_{X''}$. We set X''' to be that subset of indices and apply our induction hypothesis on the remainder of $\alpha|_{X'''}$ (after the interval I). We remain with a subset X, $|X| \geq x$

such that $\alpha|_X$ is in successive permutations form with at most i permutations. Now notice that each index appeared at most q times in α so the number of permutations is at most q. We have shown that

$$f_q(x) \leq x(q-1)f_{q-1}(f_1(x) + f_2(x) + \dots + f_{q-1}(x)) \tag{5}$$
$$\leq x(q-1)f_{q-1}((q-1)f_{q-1}(x)) \tag{6}$$
$$\leq x(q-1)f_{q-1}((q-1)C_{q-1}x^{D_{q-1}}) \tag{7}$$
$$\leq x(q-1)C_{q-1}(q-1)^{D_{q-1}}C_{q-1}^{D_{q-1}}x^{D_{q-1}^2} = C_q x^{D_q} \tag{8}$$

for $C_q = (q-1)^{D_{q-1}+1}C_{q-1}^{D_{q-1}+1}$ and $D_q = D_{q-1}^2 + 1$. This proves the induction hypothesis for q. □

Finally we put all the building blocks together to prove the theorem:

Theorem 2. *Let α be any sequence over L with $|\alpha| \leq sl$ (where $l = |L|$ and s is the constant expansion factor). Then we can compute a 2^k multicollision in the hash function based on α with time complexity $O(poly(n, k)2^{\frac{n}{2}})$.*

Proof. We start with a sequence α over L of length at most sl. There must be a subset of $\frac{l}{2}$ indices, each appearing at most $q = 2s$ times in α. Since otherwise we would have more than $\frac{l}{2}$ indices each appearing at least $2s$ times, giving more than $\frac{l}{2}2s = sl$ elements in the sequence. Let X be the set of these indices. According to lemma 2 it is enough to show that we can construct a Joux multicollision against the hash function based on $\alpha|_X$. Notice that $freq(\alpha|_X) \leq q$. We now apply lemma 5 and we get a subset X', $|X'| \geq k^3 n^{3(q-3)+2}$ such that $\alpha|_{X'}$ is in successive permutations form and $freq(\alpha|_{X'}) \leq q$. According to theorem 1, we can now construct a 2^k multicollision in the hash function based on $\alpha|_{X'}$ and according to lemma 2, we can construct a multicollision in the hash function based on α. □

3.1 Constructing a Collision in an ICE Hash Function

Constructing a collision in a concatenation of two iterated and expanded functions is done by following the recipe presented by Joux. We first construct a $2^{\frac{n}{2}}$ multicollision in the first function and then rely on the birthday paradox to find a collision among the $2^{\frac{n}{2}}$ values of the second hash function on the messages used in the multicollision. However, generalizing the result for 3 or more functions is not as easy.

As you recall the intermediate hash values of an iterated and expanded hash function based on a sequence α are calculated by $h_i = f(h_{i-1}, m_{\alpha_i})$. However, we have not used in our proof the fact that the compression function f is the same in each step. In fact, we do not need this fact and can generalize the calculation of the intermediate hash values to $h_i = f(i, h_{i-1}, m_{\alpha_i})$. We will now show how to construct a collision in an ICE hash function based on three sequences $\alpha_1, \alpha_2, \alpha_3$ and corresponding hash functions F_1, F_2, F_3. The construction we show is easily generalized to an arbitrary number of hash functions.

We will look at the sequence $\alpha = \alpha_1 \| \alpha_2$. The first step is to find a set X such that $\alpha_2|_X$ is in successive permutations form. We then find a subset $X' \subseteq X$ such that $\alpha_1|_{X'}$ is in successive permutations form. Notice that $\alpha_2|_{X'}$ will still be in successive permutations form. We now construct a $2^{\frac{n}{2}}$ Joux multicollision in the sequence $\alpha|_{X'}$ which is also in successive permutations form (as the concatenation of two such sequences). The important point is that the sequence of intervals $I_1, ..., I_k$ which form the multicollision, does not have any interval which spans the border between α_1 and α_2. Taking this sequence of intervals we can now construct a $2^{\frac{n}{2}}$ simultaneous multicollision in the hash functions F_1 and F_2. With such a large multicollision we can find with high probability a pair of messages which hash to the same value also under F_3. Thus we have found a collision in the ICE hash function $F_1(M) \| F_2(M) \| F_3(M)$ with complexity $O(poly(n)2^{\frac{n}{2}})$ instead of the expected $2^{\frac{3n}{2}}$ from the birthday paradox. A simple extension of the idea can handle the concatenation of any constant number of hash functions.

4   Tree Based Hash Functions

We now turn our attention to a more general model for constructing hash functions which we call TCE (Tree based, Concatenated, and Expanded). As in the iterated case we will base our analysis on the model presented in [10]. A tree based hash function uses a binary tree $G = (V, E)$ where the leaves are at the top and the root at the bottom. The leaves are labeled by message block indices or constant values. Given a message M, $F_G(M)$ is computed as follows: the label for each non-leaf x is computed by applying the compression function f to the two nodes directly above x. The label of the root is the output of the hash function. Note that tree based hash functions include iterated hash functions as a special case, by using trees with a single IV to root path, and hanging all the messages blocks off this path. In [10] the authors treated the special case in which every index appears at most twice in the leaves of the tree. We generalize this result to any constant number of appearances.

Definition 9. *Let $v \in V$ be a vertex in G, $W(v)$ is the set of all leaves in the subtree rooted at v.*

Definition 10. *If v is a leaf then $\rho(v)$ is its label (the index of the corresponding message block), and $\rho(v_1, ..., v_k)$ is the sequence $\rho(v_1)...\rho(v_k)$.*

In the following definitions we redefine some of the notations used in the iterated case to apply to trees. When using the definitions we will sometimes abuse notation and use interchangeably a tree G and its root r. For example we write $Ind(v)$ when meaning $Ind(G')$ where G' is the subtree rooted at v.

Definition 11. *Let r be the root of G. An independent vertex sequence is an **ordered** sequence of vertices $v_1, ..., v_k$ such that there exists a sequence of leaves $w_1, ..., w_k$ satisfying the following conditions:*

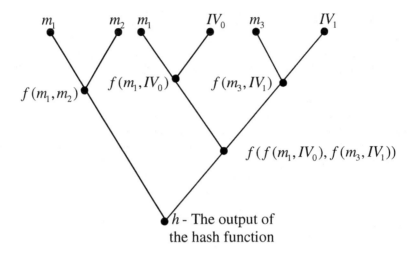

Fig. 4. A example of a TCE

1. All appearances of $\rho(w_i)$ are in $\rho(W(v_i))$
2. $j < i \Longrightarrow \rho(w_i) \notin \rho(W(v_j))$
3. $v_k = r$

The maximal length of an independent vertex sequence in G is denoted $Ind(G)$.

Definition 12. Let r be the root of G.

1. $S(G)$ is the number of distinct labels in $\rho(W(r))$
2. $freq(G) = freq(\rho(W(r)))$ where $\rho(W(r))$ is treated as a sequence.

Definition 13. Let G be a tree with a whose leaves are labeled by elements from L and let $X \subseteq L$. $G|_X$ is the pruned tree resulting from the following process:

1. Delete from G all the original leaves which have labels not in X.
2. Repeatedly delete from G any newly created leaf which is unlabeled.

Before we start the technical proof, we will give an overview of what is coming and show the correspondence between the proof of the tree-based case and the proof of the iterated case. As in the previous proof, we first want to reduce the general case to a case equivalent to the successive permutations case.

Definition 14. A tree G is in 'successive permutations' form (with r 'permutations') if we have a set of vertices $v_1, ..., v_r$ s.t. $S(v_1) = ... = S(v_r) = S(G)$ and $Ind(W(v_i) \setminus \bigcup_{j<i} W(v_j)) = S(G)$.

Each vertex v_i corresponds to a permutation in the iterated case, and contains in its leaves all the variables. Furthermore, if we look at the subtree rooted at v_i and remove all smaller subtrees rooted at v_j, then we can construct an independent sequence using all the indices with the root at v_i. The definition is best understood if we think about the iterated case as a special case of a tree with a single path from IV to the root.

The next step is to show that we can construct a multicollision in this special tree structure. The proof will be very similar to the one in the iterated case but with one additional component. In a tree we have to ensure that when taking a set of indices and trying to get a collision by changing their values, they actually have a common root which is compatible with the other groups of indices. We will later prove a lemma to this effect, and use it to prove the tree version of lemma 4 and then the tree version of lemma 5.

We start by proving the reduction from the general case to the 'successive permutations' case.

Lemma 6. *Given a tree G with $S(G) \leq 2MN$ and $freq(G) \leq q$, at least one of the following claims is true:*

1. *$Ind(G) \leq M$ or*
2. *there exists a node v and a subset X such that $freq(G|_X(v)) \leq q - 1$ and $S(G|_X(v)) \leq N$*

Proof. We extend the proof of a similar lemma from [10]. We first note that in any binary tree G such that $S(G) \geq 2N$, there exists a vertex v such that $N \leq S(v) \leq 2N$. We now prove the result by induction on $l = S(G)$. For the basis of the induction we take $M = 1$ and we have that $Ind(G) \geq 1$ is always true. Now assume that we have proved the lemma for all values less than l. Since $S(G) \geq 2MN \geq 2N$ we know from the observation above that there exists a vertex v such that $N \leq S(v) \leq 2N$. Now if $freq(v) \leq q - 1$ then we are done, since we set $X = L$ and we have that $S(G|_X(v)) \geq N$ and $freq(G|_X(v)) \leq q-1$. Otherwise we have an element x_1 such that x_1 appears q times in $\rho(W(v))$. We define $G' = G \setminus \{v\}$ (removing the subtree rooted at v) and set X to be $\rho(G')$. $S(G') \geq 2MN - 2N = 2(M-1)N$, so by the induction hypothesis we have that either $Ind(G') \geq M - 1$ or there exists v' and X' such that $S(G'|_{X'}(v')) \geq N$ and $freq(G'|_{X'}(v')) \leq q-1$. If the later happens then we simply set $X = X'$ and $v = v'$ and we are done. Otherwise we have an independent sequence $x_2, ..., x_M$ in G' but since x_1 appears only in $W(v)$, we can add x_1 to our list of independent indices and have $Ind(G) \geq M$. □

One of the differences between the iterated case and the tree case is that in a tree it is not sufficient to find a group of message blocks which can be varied independently in order to find a collision. In a tree these blocks must have a common root in which the collision will be formed. In the following lemma we prove that we can always find suitable groups of message blocks in the tree.

Lemma 7. *Given a tree G s.t. $freq(G) = 1$ and $S(G) \geq (2k - 1)x$ we can find k distinct nodes, $v_1, ..., v_k$, such that $W(v_i) \not\subseteq W(v_j)$ whenever $i > j$, and $S(W(v_i) \setminus \bigcup_{j<i} W(v_j)) \geq x$.*

Proof. We prove the claim by induction on k. For $k = 1$ we have a tree G with $freq(G) = 1$ and $S(G) \geq x$, and setting v_1 to the root of G satisfies the lemma. Now we assume that we have proved the lemma for all positive integers less than k. Given a tree G with $S(G) \geq (2k-1)x \geq 2x$ we find a node v' with $x \leq S(v') \leq 2x$. Let G' be the subtree rooted at v' and let the tree $G'' = G \setminus G'$ be the result of removing all nodes of G' from G. Notice that $S(G') \geq 2(k-1)x - 2x = (2(k-1)-1)x$. Using the induction hypothesis on G'' we have $k-1$ vertices $v_2, ..., v_k$ such that $W(v_i) \not\subseteq W(v_j)$ whenever $i > j$ and $S(W(v_i) \setminus \bigcup_{j<i} W(v_j)) \geq x$. Now setting $v_1 = v'$ we get a full set $v_1, ..., v_k$ as required since $S(G') \geq x$.   □

Theorem 3. *Given a tree based hash function F_G based on the tree G, we can find a 2^k multicollision whenever G is in 'successive permutations' form and $S(G) \geq 2^{\frac{q(q-1)}{2}} k^3 n^{3(q-3)+2}$.*

Proof. The idea of the proof is the same as in the iterated case, the only difference is that we have to make sure that when choosing a group of message blocks, they indeed have a common root (high enough in the tree) where they can form a collision. The main step in the iterated case was finding a perfect matching between two permutations. In the tree case we also need to make sure that each segment of indices has a common root which doesn't interfere with the other segments of indices. After finding the first matching, we have to find k distinct nodes as in lemma 7 in each 'permutation' copy. We start by finding such a sequence in the first 'permutation' copy, and this reduces the number of active indices by a multiplicative factor of 2. From the remaining indices we need to find a vertex sequence in each of the other 'permutation' copies such that all the variables will be the same. This way we lose a total factor of 2^q for the q 'permutations'. In the second step we have kn segments in the matching. This time however we don't care what happens in the last permutation since we only need the larger structure of k segments. So this time we lose a factor of 2^{q-1} to make sure that all the segments of indices have the required common roots. Continuing for the $q-1$ steps we see that we lose a factor of $2^{\frac{q(q-1)}{2}}$. So the required size of $S(G)$ is the same as in the iterated case except for a factor of $2^{\frac{q(q-1)}{2}}$. Once we have $S(G) \geq 2^{\frac{q(q-1)}{2}} k^3 n^{3(q-3)+2}$ we can carry out the same construction as in the iterated case, where between the steps we use lemma 7 to ensure that the remaining indices have the required structure.

As in the iterated case we need a lemma saying that it is ok to set unselected message blocks to constants.

Lemma 8. *Let G be a tree over L, and let X be a subset of indices. If we can construct a 2^k Joux multicollision against the hash function based on $G|_X$ then we can construct a 2^k Joux multicollision against the hash function based on G.*

The proof of this lemma follows the same lines as in the iterated case. We have one more lemma to prove in order to create all the building blocks needed for the general case.

Lemma 9. *Given a tree G with $freq(G) \leq q$ we can find a subset of indices X such that $G|_X$ is in the form of 'successive permutations'.*

The proof is practically the same as in the sequential case and is omitted here due to space limitations. The only difference is that lemma 6 is used instead of lemma 4. We can now sketch the proof for the general case.

Theorem 4. *Given a tree based hash function F_G based on the tree G. We can find a 2^k multicollision whenever there exists a constant q such that $freq(G) \leq q$ and $S(G) \geq poly_q(n, k)$ in time complexity $O(poly(n, k) 2^{\frac{n}{2}})$.*

Proof. We start with a tree G over L with $freq(G) \leq q$. We now apply lemma 9 and we get a subset X, $|X| \geq 2^{\frac{q(q-1)}{2}} k^3 n^{3(q-3)}$ such that $G|_X$ is in 'successive permutations' form and $freq(G|_{X'}) \leq q$. We can now construct a 2^k Joux multicollision in the hash function based on $G|_{X'}$ and according to lemma 8, we can construct a Joux multicollision in the hash function based on G.   □

Due to space limitations we omit the full description of finding a collision in a TCE hash function, which is a concatenation of the outputs of several trees. However we can use a procedure analogous to the one used in the iterated case to show that we can find a collision in a general TCE hash function in time $O(poly(n) 2^{\frac{n}{2}})$.

5   Summary

We have shown that a large class of natural hash functions (ICE and its generalization TCE) is vulnerable to a multicollision attack, and we hope that the techniques developed here will help in creating multicollision attacks against even more complicated types of hash functions. For example, a different type of message expansion which would be interesting to examine can use linear mixing of the message blocks, instead of pure repetition of the message blocks. Other research directions are to find other countermeasures against the Joux multicollision attack such as the scheme suggested by Lucks [9], or finding additional uses of multicollisions as building blocks in more general attacks as in [5], [7] and [8].

Acknowledgments

The authors would like to thank Mridul Nandi and Douglas Stinson whose paper[10] motivated our research and contributed to its development. In addition, we would like to thank the anonymous referees for helping clarify the presentation and pointing out some minor errors.

References

1. E. Biham, R. Chen, A. Joux, P. Carribault, C. Lemuet & W. Jalby, *Collisions of SHA-0 and Reduced SHA-1*, Eurocrypt 2005
2. J. Daemen, R. Govaerts & J. Vandewalle, *A Framework for the Design of One-Way Hash Functions Including Cryptanalysis of Damgrd's One-Way Function Based on a Cellular Automaton*, Asiacrypt 1991

3. A. De Santis & M. Yung, *On the Design of Provably Secure Cryptographic Hash*, Eurocrypt 1990

4. H. Gilbert & H. Handschuh, *Security Analysis of SHA-256 and Sisters*, Selected Areas in Cryptography 2003 NIST Cryptographic Hash Workshop 2005

5. A. Joux, *Multicollisions in Iterated Hash Functions*, Crypto 2004.

6. C. Jutla & A. Patthak, *A Simple and Provably Good Code for SHA Message Expansion*, IACR preprint archive

7. J. Kelsey & B. Schneier, *Second Preimages on n-bit Hash Functions for Much Less than 2^n Work*, Eurocrypt 2005.

8. J. Kelsey & T. Kohno, *Herding Hash Functions and the Nostradamus Attack*, NIST Cryptographic Hash Workshop 2005

9. S. Lucks, *Design Principles for Iterated Hash Functions* IACR preprint archive

10. M. Nandi & D. R. Stinson, *Multicollision Attacks on a Class of Hash Functions* , IACR preprint archive

11. B. Preneel, *Analysis and design of cryptographic hash functions*, PhD thesis, Katholieke Universiteit Leuven (Belgium), 1993.

12. B. Preneel, R. Govaerts & J. Vandewalle, *Hash Functions Based on Block Ciphers: A Synthetic Approach*, Crypto 1993

13. B. Preneel, *Design Principles for Dedicated Hash Functions*, Fast Software Encryption 1993

14. R. Rivest & A. Shamir, *PayWord and MicroMint: Two simple micropayment schemes*, CryptoBytes, volume 2, number 1

15. P. Rogaway & T. Shrimpton, *Cryptographic Hash-Function Basics: Definitions, Implications, and Separations for Preimage Resistance, Second-Preimage Resistance, and Collision Resistance*, Fast Software Encryption 2004

16. X. Wang, X. Lai, D. Feng, H. Chen & X. Yu, *Cryptanalysis of the Hash Functions MD4 and RIPEMD*, EUROCRYPT 2005

17. X. Wang, H. Yu & Y. Yin, *Efficient Collision Search Attacks on SHA-0* ,Crypto 2005

18. X.Wang, Y. Yin & H. Yu, *Finding Collisions in the Full SHA-1 Collision Search Attacks on SHA1*, Crypto 2005

A New Dedicated 256-Bit Hash Function: FORK-256

Deukjo Hong[1], Donghoon Chang[1], Jaechul Sung[2], Sangjin Lee[1],
Seokhie Hong[1], Jaesang Lee[1], Dukjae Moon[3], and Sungtaek Chee[3]

[1] Center for Information Security Technologies(CIST),
Korea University, Seoul, Korea
{hongdj, pointchang, sangjin, hsh, jslee}@cist.korea.ac.kr
[2] Department of Mathematics, University of Seoul, Seoul, Korea
jcsung@uos.ac.kr
[3] National Security Research Institute
{djmoon, chee}@etri.re.kr

Abstract. This paper describes a new software-efficient 256-bit hash function, FORK-256. Recently proposed attacks on MD5 and SHA-1 motivate a new hash function design. It is designed not only to have higher security but also to be faster than SHA-256. The performance of the new hash function is at least 30% better than that of SHA-256 in software. And it is secure against any known cryptographic attacks on hash functions.

Keywords: 256-bit Hash Function, FORK-256.

1 Introduction

For cryptographic hash function, the following properties are required:

- **preimage resistance:** it is computationally infeasible to find any input which hashes to any pre-specified output.
- **second preimage resistance:** it is computationally infeasible to find any second input which has the same output as any specified input.
- **collision resistance:** it is computationally infeasible to find a collision, i.e. two distinct inputs that hash to the same result.

For an ideal hash function with an m-bit output, finding a preimage or a second preimage requires about 2^m operations and the fastest way to find a collision is a birthday attack which needs approximately $2^{m/2}$ operations.

Most dedicated hash functions which have iterative process use the Merkle-Damgård construction [6,10] in order to hash inputs of arbitrary length. They work as follows. Let HASH be a hash function. The message M is padded to a multiple of the block length and subsequently divided into n blocks $M_0, \cdots, M_{n-1}$. Then HASH can be described as follows:

$$CV_0 = IV; \quad CV_{i+1} = \mathsf{COMP}(CV_i, M_i), 0 \leq i \leq n-1; \quad \mathsf{HASH}(M) = CV_n,$$

M.J.B. Robshaw (Ed.): FSE 2006, LNCS 4047, pp. 195–209, 2006.

where COMP is the compression function of HASH, CV_i's are chaining variables, and IV is a fixed initial value.

The most popular method of designing compression functions of dedicated hash functions is a serial successive iteration of a small step function, as like round functions of block ciphers. Many hash functions such as MD4 [12], MD5 [13], HAVAL [19], SHA-family [11], etc., follow that idea. Attacks on hash functions have been focused on vanishing the difference of intermediate values caused by the difference of messages. On the other hand, a hash function has been considered secure if it is computationally hard to vanish such difference in its compression function. Usually, the lower the probability of the differential characteristic is, the harder the attack is. Therefore a step function is regarded as a good candidate if it causes a good avalanche effect in the serial structure. A function which has a good diffusion property can not be so light in general. However, most step functions have been developed to be light for efficiency. This may be why MD4-type hash functions including SHA-1 are vulnerable to Wang et al.'s collision-finding attack [15,16,17,18].

RIPEMD-family [9] has somewhat different approach for designing a secure hash function. The attacker who tries to break members of RIPEMD-family should aim simultaneously at two ways where the message difference passes. This design strategy is still successful because so far there is not any effective attack on RIPEMD-family except the first proposal of RIPEMD. However, RIPEMD-family have heavier compression functions than hash functions with serial structure. For example, the first proposal of RIPEMD consists of two lines of MD4. Total number of steps is twice as many as that of MD4. Also, the number of steps of RIPEMD-160 is almost twice as many as that of SHA-0.

In this paper, we propose a new dedicated hash function FORK-256. According to the above observation, we determined the design goals (of compression function) as follows.

- It should have a 256-bit output because the security of 2^{128} operations is recommended for symmetric key cryptography as the computing power increases.
- Its structure should be resistant against known attacks including Wang et al.'s attack [1,2,3,4,5,7,8,14,15,16,17,18].
- The performance should be as competitive as that of SHA-256.

2   Description of FORK-256

In this section, we will describe FORK-256. These are basic notations used in FORK-256.

$$\boxplus : \text{addition mod } 2^{32}$$
$$\oplus : \text{XOR (eXclusive OR)}$$
$$A^{\lll s} : s\text{-bit left rotation for a 32-bit string } A$$
$$|A|_{512} : \text{the number of 512-bit blocks in a string } A$$

2.1   Construction of FORK-256

FORK-256 employs Merkle-Damgård construction with the compression function $\mathsf{FORK256COMP}(\cdot,\cdot)$ and the padding method $\mathsf{PAD}(\cdot)$ as follows: For the initial value $CV_0 = IV$ and the message M,

```
FORK256HASH(CV₀, M)
    n ← |PAD(M)|₅₁₂;
    Partition |PAD(M)|₅₁₂ into n 512-bit blocks M₀, ···, Mₙ₋₁;
    For i = 0 to n − 1 {
        CVᵢ₊₁ ← FORK256COMP(CVᵢ, M);
    }
    Return CVₙ;
```

2.2   Message Block Length and Padding

The message block length of the compression function $\mathsf{FORK256COMP}$ is 512-bit. PAD pads a message by appending a single bit 1 next to the least significant bit of the message, followed by zero or more bit 0's until the length of the message is 448 modulo 512, and then appends to the message the 64-bit original message length modulo 2^{64}.

2.3   Structure of FORK-256 Compression Function

Fig. 1 depicts the outline of the compression function $\mathsf{FORK256COMP}$. The name 'FORK' was originated from the figure. $\mathsf{FORK256COMP}$ hashes a 512-bit string to a 256-bit string. It consists of four parallel branch functions, BRANCH_1, BRANCH_2, BRANCH_3, and BRANCH_4. Let $CV_i = (CV_i[0], CV_i[1], \cdots, CV_i[7])$ where $CV_i[j]$ is a 32-bit word. The initial value CV_0 is set as follows:

$CV_0[0] = \texttt{0x6a09e667}$	$CV_0[1] = \texttt{0xbb67ae85}$
$CV_0[2] = \texttt{0x3c6ef372}$	$CV_0[3] = \texttt{0xa54ff53a}$
$CV_0[4] = \texttt{0x510e527f}$	$CV_0[5] = \texttt{0x9b05688c}$
$CV_0[6] = \texttt{0x1f83d9ab}$	$CV_0[7] = \texttt{0x5be0cd19}$

Let us see the computing procedure of the i-th iteration of $\mathsf{FORK256COMP}$. The message block M_i is partitioned to 16 32-bit Words $(M_i[0], \cdots M_i[15])$. Let $R_j^{(s)} = (R_j^{(s)}[0], \cdots, R_j^{(s)}[7])$ for $1 \leq j \leq 4$ and $0 \leq s \leq 8$ where each $R_j^{(s)}[t]$ is a 32-bit word for $0 \leq t \leq 7$. $R_j^{(8)}$ is the output of BRANCH_j on the inputs CV_i and M_i, for $1 \leq j \leq 4$ and computed as follows:

$$R_j^{(8)} = \mathsf{BRANCH}_j(CV_i, M_i), \quad \text{for } 1 \leq j \leq 4$$

where $R_j^{(s)}$'s are used in computation of BRANCH_j for $1 \leq j \leq 4$ and $0 \leq s \leq 7$. Consequently, $CV_{i+1} = (CV_{i+1}[0], \cdots, CV_{i+1}[7])$ is the output of the i-th iteration of $\mathsf{FORK256COMP}$ and computed as follows:

$$CV_{i+1}[t] = CV_i[t] \boxplus ((R_1^{(8)}[t] \boxplus R_2^{(8)}[t]) \oplus (R_3^{(8)}[t] \boxplus R_4^{(8)}[t])), \quad \text{for } 0 \leq t \leq 7.$$

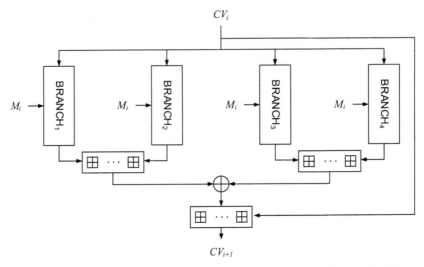

Fig. 1. Outline of the FORK-256 compression function, FORK256COMP

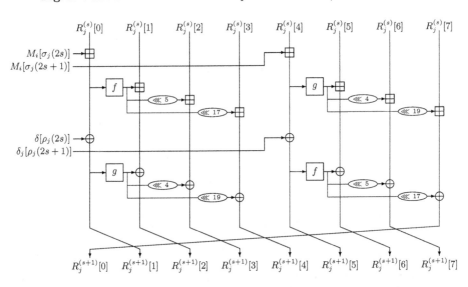

Fig. 2. Step function of FORK-256, STEP $(0 \leq s \leq 7, 1 \leq j \leq 4)$

2.4 Branch Function

Each BRANCH$_j$ for $1 \leq j \leq 4$ is computed on the inputs CV_i and M_i as follows:

$$
\begin{array}{l}
\text{BRANCH}_j(CV_i, M_i) \\
\quad R_j^{(0)} \leftarrow CV_i; \\
\quad \text{For } s = 0 \text{ to } 7 \{ \\
\qquad R_j^{(s+1)} \leftarrow \text{STEP}(R_j^{(s)}, M_i[\sigma_j(2s)], M_i[\sigma_j(2s+1)], \delta[\rho_j(2s)], \delta[\rho_j(2s+1)]); \\
\quad \text{Return } R_j^{(8)};
\end{array}
$$

Message Word Ordering. Each BRANCH_j for $1 \leq j \leq 4$ uses the message words $M_i[0], \cdots, M_i[15]$ with different order σ_j.

Table 1. Message word ordering

s	0	1	2	3	4	5	6	7	8	9	10	11	12	13	14	15
$\sigma_1(s)$	0	1	2	3	4	5	6	7	8	9	10	11	12	13	14	15
$\sigma_2(s)$	14	15	11	9	8	10	3	4	2	13	0	5	6	7	12	1
$\sigma_3(s)$	7	6	10	14	13	2	9	12	11	4	15	8	5	0	1	3
$\sigma_4(s)$	5	12	1	8	15	0	13	11	3	10	9	2	7	14	4	6

Constants. FORK256COMP totally uses sixteen constants:

$$
\begin{array}{ll}
\delta[0] \ = \text{0x428a2f98} & \delta[1] \ = \text{0x71374491} \\
\delta[2] \ = \text{0xb5c0fbcf} & \delta[3] \ = \text{0xe9b5dba5} \\
\delta[4] \ = \text{0x3956c25b} & \delta[5] \ = \text{0x59f111f1} \\
\delta[6] \ = \text{0x923f82a4} & \delta[7] \ = \text{0xab1c5ed5} \\
\delta[8] \ = \text{0xd807aa98} & \delta[9] \ = \text{0x12835b01} \\
\delta[10] = \text{0x243185be} & \delta[11] = \text{0x550c7dc3} \\
\delta[12] = \text{0x72be5d74} & \delta[13] = \text{0x80deb1fe} \\
\delta[14] = \text{0x9bdc06a7} & \delta[15] = \text{0xc19bf174}
\end{array}
$$

These constants are used in each BRANCH_j with different order ρ_j for $1 \leq j \leq 4$.

Table 2. Constant ordering

s	0	1	2	3	4	5	6	7	8	9	10	11	12	13	14	15
$\rho_1(s)$	0	1	2	3	4	5	6	7	8	9	10	11	12	13	14	15
$\rho_2(s)$	15	14	13	12	11	10	9	8	7	6	5	4	3	2	1	0
$\rho_3(s)$	1	0	3	2	5	4	7	6	9	8	11	10	13	12	15	14
$\rho_4(s)$	14	15	12	13	10	11	8	9	6	7	4	5	2	3	0	1

Step Function. In the s-th step of BRANCH_j for $1 \leq j \leq 4$ and $0 \leq s \leq 7$, STEP outputs $R_j^{(s+1)}$ on the inputs $R_j^{(s)}, M_i[\sigma_j(2s)], M_i[\sigma_j(2s+1)], \delta[\rho_j(2s)]$, and $\delta[\rho_j(2s+1)]$, and $R_j^{(s+1)}$ is computed as follows (See Fig. 2):

$$R_j^{(s+1)}[0] = R_j^{(s)}[7] \boxplus g(R_j^{(s)}[4] \boxplus M_i[\sigma_j(2s+1)])^{\lll 21}$$
$$\oplus f(R_j^{(s)}[4] \boxplus M_i[\sigma_j(2s+1)] \boxplus \delta[\rho_j(2s+1)])^{\lll 17},$$
$$R_j^{(s+1)}[1] = R_j^{(s)}[0] \boxplus M_i[\sigma_j(2s)] \boxplus \delta[\rho_j(2s)],$$
$$R_j^{(s+1)}[2] = R_j^{(s)}[1] \boxplus f(R_j^{(s)}[0] \boxplus M_i[\sigma_j(2s)])$$
$$\oplus g(R_j^{(s)}[0] \boxplus M_i[\sigma_j(2s)] \boxplus \delta[\rho_j(2s)]),$$

$$R_j^{(s+1)}[3] = R_j^{(s)}[2] \boxplus f(R_j^{(s)}[0] \boxplus M_i[\sigma_j(2s)])^{\lll 5}$$
$$\oplus g(R_j^{(s)}[0] \boxplus M_i[\sigma_j(2s)] \boxplus \delta[\rho_j(2s)])^{\lll 9},$$

$$R_j^{(s+1)}[4] = R_j^{(s)}[3] \boxplus f(R_j^{(s)}[0] \boxplus M_i[\sigma_j(2s)])^{\lll 17}$$
$$\oplus g(R_j^{(s)}[0] \boxplus M_i[\sigma_j(2s)] \boxplus \delta[\rho_j(2s)])^{\lll 21},$$

$$R_j^{(s+1)}[5] = R_j^{(s)}[4] \boxplus M_i[\sigma_j(2s+1)] \boxplus \delta[\rho_j(2s+1)],$$

$$R_j^{(s+1)}[6] = R_j^{(s)}[5] \boxplus g(R_j^{(s)}[4] \boxplus M_i[\sigma_j(2s+1)])$$
$$\oplus f(R_j^{(s)}[4] \boxplus M_i[\sigma_j(2s+1)] \boxplus \delta[\rho_j(2s+1)]),$$

$$R_j^{(s+1)}[7] = R_j^{(s)}[6] \boxplus g(R_j^{(s)}[4] \boxplus M_i[\sigma_j(2s+1)])^{\lll 9}$$
$$\oplus f(R_j^{(s)}[4] \boxplus M_i[\sigma_j(2s+1)] \boxplus \delta[\rho_j(2s+1)])^{\lll 5},$$

where f and g are nonlinear functions as follows:

$$f(x) = x \boxplus (x^{\lll 7} \oplus x^{\lll 22}),$$
$$g(x) = x \oplus (x^{\lll 13} \boxplus x^{\lll 27}).$$

3   Design Strategy

3.1   Motivation for Our Proposal

In 2004, Wang et al.'s attacks on MD4, MD5, HAVAL, and RIPEMD [15,16] and SHA-0/1 [17,18] brought a big impact on the field of symmetric key cryptography including hash function. However, RIPEMD-128/160 are the algorithms which are still secure against their attacks. No attacks on them are found so far.

They were designed to have two parallel lines, which is different from MD4, MD5 and SHA-family. This makes an attacker take into account two lines simultaneously. However, since each line needs almost same operation of MD5 and SHA algorithms, its efficiency was degenerated almost half of them. This motivates our design. We use four lines instead of two.

In order to overcome disadvantage of RIPEMD algorithms, we manage to reduce operations for step functions of each line. The message reordering of each branch is deliberately designed to be resistant against Wang et al.'s attack and differential attacks. The function f and g in each step are chosen to have good avalanche effects.

3.2   Design Principle

Structure. The compression function FORK256COMP consists of 4 Branches. In the security aspect, we can give the security against known attacks with the different message ordering in branches. For example, RIPEMD, which consists of 2 branches, was fully attacked by Wang et al. [15] because RIPEMD has the same message ordering in 2 branches. On the other hand, in case of RIPEMD-128/160, there is no attack result because RIPEMD-128/160 have different message ordering in branches. In the implementation aspect, FORK-256 can be implemented efficiently because the message ordering is simpler than the message expansion such as that of SHA-256.

Constants. Each BRANCH$_j$ uses 16 different constants $\delta[t]$ for $0 \leq t \leq 15$. By using constants we pursue the goal to disturb the attacker who tries to find a good differential characteristic with a relatively high probability. So, we prefer the constants which represent the first 32 bits of the fractional parts of the cube roots of the first sixteen four prime numbers.

Nonlinear Functions. The nonlinear functions f and g of FORK256COMP output one word on the input of one word, and their outputs are XORed or added modulo 2^{32} to the multiple words in the chaining variables. Due to this property, f and g propagate the difference of a message word to the chaining variables.

Shift Rotations in Nonlinear Functions. If the addition is changed into the bitwise xor operation in f and g, nonlinear functions are generalized as $x \oplus (x \lll s_1 \oplus x \lll s_2)$. We consider all $465(=\binom{31}{2})$ cases for s_1 and s_2 and want to define shift rotations satisfying the following 7 conditions. HW(w) denotes the Hamming Weight of a 32-bit word w. Let x be an input of f or g and let y be $f(x)$ or $g(x)$.

- The branch number of f and g is 4.
- If HW(x) = 2, then HW(y) $\geq$ 4.
- If HW(x) = 3, then HW(y) $\geq$ 3.
- If HW(x) = 4, then HW(y) $\geq$ 4.
- If HW(y) = 1, then HW(x) $\geq$ 17.
- If HW(y) = 2, then HW(x) $\geq$ 14.
- The interval of shift rotations are greater than or equal to 4.

We determined the shift rotations such that f and g satisfy the above conditions.

Message Word Ordering. We adopt the message word ordering instead of the message word extension. If an attacker constructs an intended differential characteristics for one branch function, the ordering of message words will cause unintended differential patterns in the other branch functions. This is the core part of the security in the compression function. We considered the following four conditions in defining the message word ordering.

- Balance of upper (step 0~3) and lower (step 4~7) parts : each word should be applied twice to upper and lower parts, respectively.
- Balance of left and right parts : each word should be applied twice to left and right parts, respectively.
- Balance of sums of indices:
  - Each word should be applied four times and indexed by 0~15.
  - Total sum of indices is 480. Therefore, the average of sum of indices applied to each word is 30.

- We searched the ordering so that the sum of indices corresponding to each word is around 25~35.
 - Conditions which do not have same differential patterns in all branches:
   - Specific differential pattern used at a branch may be applied to other branches.
   - Therefore, except the case of giving a same difference to all words, we try to find an ordering such that there is no same differential patterns in all branches.

Shift Rotations and Rank. In the step function, 5 and 17, the values of shift rotation, are fixed. Then we search all the case and find candidate values (corresponding to 9 and 21) so that the rank of the linearly-changed step function is maximized. The maximum of the rank is 252. Finally we select 9 and 21 among candidate values so that differences generated from the outputs of f and g functions do not overlap when a message word inserted at a step function has an one-bit difference.

4   Security Analysis of FORK-256

4.1   Collision-Finding Attack

We can analyze the collision-finding attacker's behavior in the aspect of message difference. Let $\Delta R_j^{(8)}$ be the output difference of BRANCH_j for $1 \leq j \leq 4$. Usually, the attacker expects the following event for finding collisions:

$$(\Delta R_1^{(8)}[t] \boxplus \Delta R_2^{(8)}[t]) \oplus (\Delta R_3^{(8)}[t] \boxplus \Delta R_4^{(8)}[t]) = 0, \quad \text{for } 0 \leq t \leq 7.$$

For this, he can take several strategies:

1. The attacker constructs a differential characteristic with high probability for a branch function, say BRANCH_1, and then expects that $\Delta R_3^{(8)}[t] \boxplus \Delta R_4^{(8)}[t] \boxminus \Delta R_2^{(8)}[t]$ is equal to $\Delta R_1^{(8)}[t]$ for $0 \leq t \leq 7$.
2. The attacker constructs two distinct differential characteristics, and expects that $\Delta R_1^{(8)} = -\Delta R_2^{(8)}$ and $\Delta R_3^{(8)} = -\Delta R_4^{(8)}$ for $0 \leq t \leq 7$.
3. The attacker inserts the message difference which yields same message difference pattern in four branches, and expects that same differential characteristic occurs simultaneously in four branches. Then the output difference of the compression function vanishes if the hamming weight of the output difference of each branch is small. This is because the final output is generated with using $\oplus$ and $\boxplus$ by turns.

Let us see the first strategy. If we assume that the outputs of each branch function is random, the probability of the event is almost close to 2^{-256}. It is also difficult for the attacker to mount any attack following the second strategy because he should find such differential pattern of the message words.

Third strategy is relatively easy for the attacker to perform. For example, if he inserts the same difference to all the message words, then the same message difference pattern occurs in every branches. However, the message word reordering was designed so that the third strategy is satisfied only if the attacker inserts the same difference to all the message words. Under the assumption that every step is independent, we can compute the upper bound of the probability that such kind of differential characteristic occurs, which frustrates the attacker.

4.2   Attacks Using Inner Collision Patterns

When the attacker inserts the differences to the message words, the event that the difference of the intermediate value becomes zero often occurs. It is called *inner collision*. We call a differential characteristic which causes an inner collision with a probability, *inner collision pattern*.

Note that an inner collision is not a real collision, but the notion of inner collision pattern is important in cryptanalysis of hash function because it can be repeatedly used to yield a real collision with a high probability. The main idea of attacks on SHA-0 and SHA-1 is also the repetition of an inner collision pattern. So, in hash functions with a serial structure it is related to the resistance against collision-finding attack how many time an inner collision can be repeated.

Let us focus on only one branch function. We omit the subscript index in the variables. We can construct 5-step inner collision pattern easily. Let $\Delta R^{(s)}[t]$ be the difference of $R^{(s)}[t]$, and let $\Delta M[\sigma(i)]$ be the difference of $M[\sigma(i)]$. Table 3 and 4 show two among 5-step inner collision patterns in one branch function, which we found. They holds with the probability 2^{-40}, respectively.

If we apply these patterns to BRANCH_1, the output difference $\Delta R_1^{(8)}$ will be zero with the probability 2^{-40}. As mentioned in the previous subsection, however, it is hard to use the pattern for the attack on FORK-256 because the following events seldom occurs: either that the computation of the output differences of

Table 3. Case 1. 5-step inner collision pattern in a branch: The numbers in the entries of the table denotes the bits in which the difference is 1

s	0	1	2	3	4
$\Delta M[\sigma(2s)]$	31				
$\Delta M[\sigma(2s+1)]$		1,6,15,16,20,23	3,4,8,11,21,26	6,12,21,26	31
$\Delta R^{(s)}[0]$					
$\Delta R^{(s)}[1]$		31			
$\Delta R^{(s)}[2]$		6,12,21,26	31		
$\Delta R^{(s)}[3]$		3,4,8,11,21,26	6,12,21,26	31	
$\Delta R^{(s)}[4]$		1,6,15,16	3,4,8,11,20,23	6,12,21,26,21,26	31
$\Delta R^{(s)}[5]$					
$\Delta R^{(s)}[6]$					
$\Delta R^{(s)}[7]$					
Prob.	2^{-10}	2^{-16}	2^{-10}	2^{-4}	1

Table 4. Case 2. 5-step inner collision pattern in a branch: The numbers in the entries of the table denotes the bits in which the difference is 1

s	0	1	2	3	4
$\Delta M[\sigma(2s)]$		1,6,15,16,20,23	3,4,8,11,21,26	6,12,21,26	31
$\Delta M[\sigma(2s+1)]$	31				
$\Delta R^{(s)}[0]$		1,6,15,16,20,23	3,4,8,11,21,26	6,12,21,26	31
$\Delta R^{(s)}[1]$					
$\Delta R^{(s)}[2]$					
$\Delta R^{(s)}[3]$					
$\Delta R^{(s)}[4]$					
$\Delta R^{(s)}[5]$		31			
$\Delta R^{(s)}[6]$		6,12,21,26	31		
$\Delta R^{(s)}[7]$		3,4,8,11,21,26	6,12,21,26	31	
Prob.	2^{-10}	2^{-16}	2^{-10}	2^{-4}	1

the other branches is zero or that the other branches have the same differential pattern in the message words as BRANCH$_1$.

5   Efficiency and Performance

In this section we compare the total number of operations and the performance of FORK-256 and SHA-256. The total number of operations is compared in the Table 5.

Table 5. Number of operations used in FORK-256 and SHA-256

operation	FORK-256	SHA-256
addition (+)	472	600
bitwise operation $(\oplus, \wedge, \vee)$	328	1024
shift $(\ll, \gg)$	-	96
shift rotation $(\lll, \ggg)$	512	576

Table 6. Performance of FORK-256 and SHA-256

environment	FORK-256		SHA-256	
	Mbps	Cycle/Byte	Mbps	Cycle/Byte
Pen3/WinXP/VC	192.010	31.413	132.469	44.581
Pen4/WinXP/VC	521.111	28.755	318.721	46.372

Table 6 shows the performance of FORK-256 and SHA-256 in the two environments. The environments are denoted by *CPU/OS/Compiler* and the following notations are used in description of environments for simplicity.

Pen3 : Pentium III, 801 MHz
Pen4 : Pentium IV, 2.0 GHz
WinXP : Microsoft Windows XP Professional ver 2002
VC : Microsoft Visual C++ Ver 6.0

6 Summary

In this paper we have proposed a new dedicated 256-bit hash function FORK-256. The main features are the followings;

- The structure of the compression function consists of 4 parallel branch functions.
- Nonlinear functions f and g are quite different from the boolean functions which have been used in existing hash functions, and updates multiple words.
- The ordering of the message words is simple but well organized such that it is very difficult for any attacker to find good inner collision patterns.

According to our security analysis, FORK-256 looks resistant against existing attacks including Wang et al.'s attack, but we encourage the readers to give any further security analysis. Finally, our performance test shows that the performance of FORK-256 is faster than that of SHA-256, and we expect that the difference between the performance of FORK-256 and SHA-256 would increase after optimization.

Acknowledgement

This research was supported by the MIC(Ministry of Information and Communication), Korea, under the ITRC(Information Technology Research Center) support program supervised by the IITA(Institute of Information Technology Assessment).

References

1. E. Biham and R. Chen, "Near-Collisions of SHA-0," *Advances in Cryptology – CRYPTO 2004*, LNCS 3152, Springer-Verlag, pp. 290–305, 2004.
2. E. Biham, R. Chen, A. Joux, P. Carribault, C. Lemuet and W. Jalby, "Collisions of SHA-0 and Reduced SHA-1," *Advances in Cryptology – EUROCRYPT 2005*, LNCS 3494, Springer-Verlag, pp. 36–57, 2005.
3. B. den Boer and A. Bosselaers, "An Attack on the Last Two Rounds of MD4," *Advances in Cryptology – CRYPTO'91*, LNCS 576, Springer-Verlag, pp. 194–203, 1992.
4. B. den Boer and A. Bosselaers, "Collisions for the Compression Function of MD5," *Advances in Cryptology – CRYPTO'93*, LNCS 765, Springer-Verlag, pp. 293–304, 1994.
5. F. Chabaud and A. Joux, "Differential Collisions in SHA-0," *Advances in Cryptology – CRYPTO'98*, LNCS 1462, Springer-Verlag, pp. 56–71, 1998.

6. I. Damgård, "A Design Priciple for Hash Functions," *Advances in Cryptology – CRYPTO'89*, LNCS 435, Springer-Verlag, pp. 416–427, 1989.

7. H. Dobbertin, "RIPEMD with Two-Round Compress Function is Not Collision-Free," *Journal of Cryptology* 10:1, pp. 51–70, 1997.

8. H. Dobbertin, "Cryptanalysis of MD4," *Journal of Cryptology* 11:4, pp. 253–271, 1998.

9. H. Dobbertin, A. Bosselaers and B. Preneel, "RIPEMD-160, a strengthened version of RIPEMD," *FSE'96*, LNCS 1039, Springer-Verlag, pp. 71–82, 1996.

10. R. C. Merkle, "One way hash functions and DES," *Advances in Cryptology – CRYPTO'89*, LNCS 435, Springer-Verlag, pages 428–446, 1989.

11. NIST/NSA, "FIPS 180-2: Secure Hash Standard (SHS)", August 2002 (change notice: February 2004).

12. R. L. Rivest, "The MD4 Message Digest Algorithm," *Advances in Cryptology – CRYPTO'90*, LNCS 537, Springer-Verlag, pp. 303–311, 1991.

13. R. L. Rivest, "The MD5 Message-Digest Algorithm," IETF Request for Comments, RFC 1321, April 1992.

14. B. Van Rompay, A. Biryukov, B. Preneel and J. Vandewalle, "Cryptanalysis of 3-pass HAVAL," *Advances in Cryptology – ASIACRYPT 2003*, LNCS 2894, Springer-Verlag, pp. 228–245, 2003.

15. X. Wang, X. Lai, D. Feng, H. Chen and X. Yu, "Cryptanalysis of the Hash Functions MD4 and RIPEMD," *Advances in Cryptology – EUROCRYPT 2005*, LNCS 3494, Springer-Verlag, pp. 1–18, 2005.

16. X. Wang and H. Yu, "How to Break MD5 and Other Hash Functions," *Advances in Cryptology – EUROCRYPT 2005*, LNCS 3494, Springer-Verlag, pp. 19–35, 2005.

17. X. Wang, H. Yu and Y. L. Yin, "Efficient Collision Search Attacks on SHA-0," *Advances in Cryptology – CRYPTO 2005*, LNCS 3621, Springer-Verlag, pp. 1–16, 2005.

18. X. Wang, Y. L. Yin and H. Yu, "Finding Collisions in the Full SHA-1," *Advances in Cryptology – CRYPTO 2005*, LNCS 3621, Springer-Verlag, pp. 17-36, 2005.

19. Y. Zheng, J. Pieprzyk and J. Seberry, "HAVAL – A One-Way Hashing Algorithm with Variable Length of Output," *Advances in Cryptology – AUSCRYPT'92*, LNCS 718, Springer-Verlag, pp. 83–104, 1993.

A   Source Code

```
unsigned int delta[16] = {
        0x428a2f98, 0x71374491, 0xb5c0fbcf, 0xe9b5dba5,
        0x3956c25b, 0x59f111f1, 0x923f82a4, 0xab1c5ed5,
        0xd807aa98, 0x12835b01, 0x243185be, 0x550c7dc3,
        0x72be5d74, 0x80deb1fe, 0x9bdc06a7, 0xc19bf174
        };

#define ROL(x, n)    ( ( (x) << n ) | ( (x) >> (32-n) ) )

#define f(x)    (x + (ROL(x,7)^ROL(x,22)))

#define g(x)    (x ^ (ROL(x,13)+ROL(x,27)))


#define step(A,B,C,D,E,F,G,H,M1,M2,D1,D2)    \
    temp1 = E+M2;    \
    temp2 = g(temp1);    temp3 = f(temp1+D2);    \
```

```
    H = (H + ROL(temp2,21)) ^ ROL(temp3,17);    \
    G = (G + ROL(temp2,9)) ^ ROL(temp3,5);  \
    F = (F + temp2) ^ temp3;                    \
    E = temp1 +D2;  \
    temp1 = A+M1;   \
    temp2 = f(temp1);   temp3 = g(temp1+D1);        \
    D = (D + ROL(temp2,17)) ^ ROL(temp3,21);        \
    C = (C + ROL(temp2,5)) ^ ROL(temp3,9);  \
    B = (B + temp2) ^ temp3;                    \
    A = temp1 + D1;


static void FORK256_Compression_Function(unsigned int *CV, unsigned
int *M) {
    unsigned long R1[8],R2[8],R3[8],R4[8];
    unsigned long temp1, temp2, temp3;

    R1[0] = R2[0] = R3[0] = R4[0] = CV[0];
    R1[1] = R2[1] = R3[1] = R4[1] = CV[1];
    R1[2] = R2[2] = R3[2] = R4[2] = CV[2];
    R1[3] = R2[3] = R3[3] = R4[3] = CV[3];
    R1[4] = R2[4] = R3[4] = R4[4] = CV[4];
    R1[5] = R2[5] = R3[5] = R4[5] = CV[5];
    R1[6] = R2[6] = R3[6] = R4[6] = CV[6];
    R1[7] = R2[7] = R3[7] = R4[7] = CV[7];


    // BRANCH1(CV,M)
    step(R1[0],R1[1],R1[2],R1[3],R1[4],R1[5],R1[6],R1[7],M[0],M[1],delta[0],delta[1]);
    step(R1[7],R1[0],R1[1],R1[2],R1[3],R1[4],R1[5],R1[6],M[2],M[3],delta[2],delta[3]);
    step(R1[6],R1[7],R1[0],R1[1],R1[2],R1[3],R1[4],R1[5],M[4],M[5],delta[4],delta[5]);
    step(R1[5],R1[6],R1[7],R1[0],R1[1],R1[2],R1[3],R1[4],M[6],M[7],delta[6],delta[7]);
    step(R1[4],R1[5],R1[6],R1[7],R1[0],R1[1],R1[2],R1[3],M[8],M[9],delta[8],delta[9]);
    step(R1[3],R1[4],R1[5],R1[6],R1[7],R1[0],R1[1],R1[2],M[10],M[11],delta[10],delta[11]);
    step(R1[2],R1[3],R1[4],R1[5],R1[6],R1[7],R1[0],R1[1],M[12],M[13],delta[12],delta[13]);
    step(R1[1],R1[2],R1[3],R1[4],R1[5],R1[6],R1[7],R1[0],M[14],M[15],delta[14],delta[15]);

    // BRANCH2(CV,M)
    step(R2[0],R2[1],R2[2],R2[3],R2[4],R2[5],R2[6],R2[7],M[14],M[15],delta[15],delta[14]);
    step(R2[7],R2[0],R2[1],R2[2],R2[3],R2[4],R2[5],R2[6],M[11],M[9],delta[13],delta[12]);
    step(R2[6],R2[7],R2[0],R2[1],R2[2],R2[3],R2[4],R2[5],M[8],M[10],delta[11],delta[10]);
    step(R2[5],R2[6],R2[7],R2[0],R2[1],R2[2],R2[3],R2[4],M[3],M[4],delta[9],delta[8]);
    step(R2[4],R2[5],R2[6],R2[7],R2[0],R2[1],R2[2],R2[3],M[2],M[13],delta[7],delta[6]);
    step(R2[3],R2[4],R2[5],R2[6],R2[7],R2[0],R2[1],R2[2],M[0],M[5],delta[5],delta[4]);
    step(R2[2],R2[3],R2[4],R2[5],R2[6],R2[7],R2[0],R2[1],M[6],M[7],delta[3],delta[2]);
    step(R2[1],R2[2],R2[3],R2[4],R2[5],R2[6],R2[7],R2[0],M[12],M[1],delta[1],delta[0]);

    // BRANCH3(CV,M)
    step(R3[0],R3[1],R3[2],R3[3],R3[4],R3[5],R3[6],R3[7],M[7],M[6],delta[1],delta[0]);
    step(R3[7],R3[0],R3[1],R3[2],R3[3],R3[4],R3[5],R3[6],M[10],M[14],delta[3],delta[2]);
    step(R3[6],R3[7],R3[0],R3[1],R3[2],R3[3],R3[4],R3[5],M[13],M[2],delta[5],delta[4]);
    step(R3[5],R3[6],R3[7],R3[0],R3[1],R3[2],R3[3],R3[4],M[9],M[12],delta[7],delta[6]);
    step(R3[4],R3[5],R3[6],R3[7],R3[0],R3[1],R3[2],R3[3],M[11],M[4],delta[9],delta[8]);
    step(R3[3],R3[4],R3[5],R3[6],R3[7],R3[0],R3[1],R3[2],M[15],M[8],delta[11],delta[10]);
    step(R3[2],R3[3],R3[4],R3[5],R3[6],R3[7],R3[0],R3[1],M[5],M[0],delta[13],delta[12]);
    step(R3[1],R3[2],R3[3],R3[4],R3[5],R3[6],R3[7],R3[0],M[1],M[3],delta[15],delta[14]);

    // BRANCH4(CV,M)
    step(R4[0],R4[1],R4[2],R4[3],R4[4],R4[5],R4[6],R4[7],M[5],M[12],delta[14],delta[15]);
    step(R4[7],R4[0],R4[1],R4[2],R4[3],R4[4],R4[5],R4[6],M[1],M[8],delta[12],delta[13]);
    step(R4[6],R4[7],R4[0],R4[1],R4[2],R4[3],R4[4],R4[5],M[15],M[0],delta[10],delta[11]);
    step(R4[5],R4[6],R4[7],R4[0],R4[1],R4[2],R4[3],R4[4],M[13],M[11],delta[8],delta[9]);
    step(R4[4],R4[5],R4[6],R4[7],R4[0],R4[1],R4[2],R4[3],M[3],M[10],delta[6],delta[7]);
    step(R4[3],R4[4],R4[5],R4[6],R4[7],R4[0],R4[1],R4[2],M[9],M[2],delta[4],delta[5]);
    step(R4[2],R4[3],R4[4],R4[5],R4[6],R4[7],R4[0],R4[1],M[7],M[14],delta[2],delta[3]);
    step(R4[1],R4[2],R4[3],R4[4],R4[5],R4[6],R4[7],R4[0],M[4],M[6],delta[0],delta[1]);
```

```
// output
CV[0] = CV[0] + ((R1[0] + R2[0]) ^ (R3[0] + R4[0]));
CV[1] = CV[1] + ((R1[1] + R2[1]) ^ (R3[1] + R4[1]));
CV[2] = CV[2] + ((R1[2] + R2[2]) ^ (R3[2] + R4[2]));
CV[3] = CV[3] + ((R1[3] + R2[3]) ^ (R3[3] + R4[3]));
CV[4] = CV[4] + ((R1[4] + R2[4]) ^ (R3[4] + R4[4]));
CV[5] = CV[5] + ((R1[5] + R2[5]) ^ (R3[5] + R4[5]));
CV[6] = CV[6] + ((R1[6] + R2[6]) ^ (R3[6] + R4[6]));
CV[7] = CV[7] + ((R1[7] + R2[7]) ^ (R3[7] + R4[7]));
}
```

B   Test Vector

Message M (1 block)

4105ba8c d8423ce8 ac484680 07ee1d40 bc18d07a 89fc027c 5ee37091 cd1824f0
878de230 dbbaf0fc da7e4408 c6c05bc0 33065020 7367cfc5 f4aa5c78 e1cbc780

Output of Compression Function CV_1

ebcc5b3d d3715534 a6a7a68a e6022b02 49c676ed 639a34b0 b8d978c2 cfdf1a2b

Intermediate Values

BRANCH$_1$

$R_1^{(0)}$ = 6a09e667 bb67ae85 3c6ef372 a54ff53a 510e527f 9b05688c 1f83d9ab 5be0cd19
$R_1^{(1)}$ = 574faabb ed99d08b 55559509 ca832197 cc3e5d3d 9a87d3f8 a53a7eff e5b76844
$R_1^{(2)}$ = 15b6cd3d b958ed0a bc5ec9da 0685ff8e eecd75a9 bde25622 730387f0 8cd537f4
$R_1^{(3)}$ = b37a2f3c 0b266012 421e26a6 c78f6e0b 1cd85800 d2ba8a16 7449f6c0 0f8c7a01
$R_1^{(4)}$ = 31be4596 a49d2271 6ee14e1a e33ff108 11f5f01a 950cdbc5 5dcd1a2a 32aa199f
$R_1^{(5)}$ = 62fd9d8b 9153d25e 4a23586e 9b599483 cf29e3af 00343c17 f33f23cb 9c903e62
$R_1^{(6)}$ = d36228e4 61ad6751 fe55bb69 94720b3c 8a810aa7 eaf6bd32 737155e2 b96a93e9
$R_1^{(7)}$ = 7a779e32 7926d678 3aec6bdd 0e208057 c349f555 7ec78c6a 91ebeb68 1fc96600
$R_1^{(8)}$ = 85c3c25b 0afe0151 60d37e53 93df1ad6 390f9cea 66b1ae49 71de5de6 17ae42cd

BRANCH$_2$

$R_2^{(0)}$ = 6a09e667 bb67ae85 3c6ef372 a54ff53a 510e527f 9b05688c 1f83d9ab 5be0cd19
$R_2^{(1)}$ = 09a80c1a 20503453 b7ce65dc 686c5844 8f7b750a ceb620a6 e84808f4 13a2716f
$R_2^{(2)}$ = e21fd29c 514719d8 47c2c8b0 116c12a7 42ddee6f ddf4c37a 3b2884ee 1b6552ca
$R_2^{(3)}$ = 608f85bc beba328f da492019 ce8cc5ac e939ee3d 418db835 0d4088c0 a4515753
$R_2^{(4)}$ = 9d819935 7b00fdfd d9947c55 0dfccfd7 817088d7 7d5a694f 8da6b62e 3b63944f
$R_2^{(5)}$ = f22fa55e f4e63e8a 2516289f 77d9b888 dc500533 8717db40 6158e3e7 0e922286
$R_2^{(6)}$ = 13ca89c4 8d2671db afbc022b 9580fdfe 356e2f63 9fa2ca0a d2199dee 455937e5
$R_2^{(7)}$ = b8d0fc67 5c63d5fa d2b45236 fad40792 759b52ab b8475022 1cf6c001 6a0cf5f2
$R_2^{(8)}$ = 08283ecb 5d0e9118 da92c996 9316c47c 26167358 9067bf2b 33a76294 a2c36255

BRANCH$_3$

$R_3^{(0)}$ = 6a09e667 bb67ae85 3c6ef372 a54ff53a 510e527f 9b05688c 1f83d9ab 5be0cd19
$R_3^{(1)}$ = 46f81ba6 a8594fe8 f0348c97 749c040f 8e6801dc f27bf2a8 275472bf 0866407e
$R_3^{(2)}$ = 56a9eac1 0b2c3b53 0e98c271 ec010b6c 448475b5 38d35a23 455b10c5 4c819e3b

$R_3^{(3)}$ = 38cd29dc 2402cc77 48018a70 26a5dcf2 3da527e9 2a237e90 2f4dc6a8 33bd5b6f
$R_3^{(4)}$ = a28f637c bfa479ad 68059737 374a7e75 b5e5b8c6 02eafaad 15799680 ae2d5da0
$R_3^{(5)}$ = 64607852 7bd31a3d a54f54b2 4013d658 1fbcbc0a 4a0633d8 972027f7 40a519ed
$R_3^{(6)}$ = b27cf46d 9b38bd95 fb3978fd d52a18c8 1cdbd155 cb7c23f8 d3ce2cdd 5e6705b2
$R_3^{(7)}$ = 317ce148 bd57a8e7 d3b60337 f0dd8789 1a925421 d09fe955 c626a195 8d38ed5d
$R_3^{(8)}$ = 72ec7187 cb5b0fa4 59b04096 55b45924 d54c20ad be5c7808 ec104b46 08d57f3d

BRANCH$_4$

$R_4^{(0)}$ = 6a09e667 bb67ae85 3c6ef372 a54ff53a 510e527f 9b05688c 1f83d9ab 5be0cd19
$R_4^{(1)}$ = ce371d88 8fe1ef8a f4e6891a dd47fbec 8655e369 45b09413 8d2e660f 968ed897
$R_4^{(2)}$ = 015a57e3 1937b7e4 d82e18fe 374895df 3e1357d6 8ec27797 81e87c75 627d168a
$R_4^{(3)}$ = f2619dce 0757a521 b3dc348f a91771d4 00a58535 d4259025 37fc2a18 c5a9d37a
$R_4^{(4)}$ = dc4ebcd3 3dd1182b acb226cd 3ed1c4a9 f6191a1b d9e93bf6 62752a33 d29d946e
$R_4^{(5)}$ = ad2c36d3 767c5cb7 8d977401 ebd447de a0e6e49b 7bb3bcf8 d7b3eadc 71c2d2a4
$R_4^{(6)}$ = b871dbb2 c23dea2a aebfcf21 6de34a20 41d677c5 a7203d0c 14c00db6 d5b6d5ce
$R_4^{(7)}$ = a6072510 3b4afc71 e74b9db3 5120200b b1167426 2036afe2 ddcd1ac5 096735bb
$R_4^{(8)}$ = 99420469 a4aa2522 f7aeb45b 10939176 d252137f 81312948 50c01427 c0ba68f3

Some Plausible Constructions of Double-Block-Length Hash Functions

Shoichi Hirose

Faculty of Engineering, The University of Fukui, Fukui 910-8507 Japan
`hirose@fuee.fukui-u.ac.jp`

Abstract. In this article, it is discussed how to construct a compression function with $2n$-bit output using a component function with n-bit output. The component function is either a smaller compression function or a block cipher. Some constructions are presented which compose collision-resistant hash functions: Any collision-finding attack on them is at most as efficient as the birthday attack in the random oracle model or in the ideal cipher model. A new security notion is also introduced, which we call indistinguishability in the iteration, with a construction satisfying the notion.

1 Introduction

A cryptographic hash function is a function which maps an input of arbitrary length to an output of fixed length. It satisfies preimage resistance, second-preimage resistance and collision resistance. It is one of the most important primitives in cryptography [19]. For simplicity, a cryptographic hash function is called a hash function in this article.

A hash function usually consists of iteration of a compression function with fixed input/output length. This type of hash function is called an iterated hash function. There has been an interest in constructing a compression function from component functions with smaller output length. Many schemes have been presented following the approach [4,10,11,13,14,15,17,20]. It is typical for constructions using block ciphers. For example, suppose that AES is used for construction. The block length of AES is 128 bits, and a hash function with 128-bit output is no longer secure against the birthday attack. Thus, it is desired to construct a compression function with larger output length.

In this article, we study how to construct a compression function with $2n$-bit output using a component function with n-bit output. A hash function with such a compression function is called a double-block-length (DBL) hash function (as opposed to a single-block-length (SBL) hash function, where the compression function has n-bit output). The component function may be either a block cipher or a smaller compression function.

We first discuss constructions using a smaller compression function. We focus on the constructions formalized by Nandi [22]. In his formalization, the compression function is of the form $F(x) = (f(x), f(p(x)))$, where f is a component

M.J.B. Robshaw (Ed.): FSE 2006, LNCS 4047, pp. 210–225, 2006.

compression function and p is a permutation such that both p and p^{-1} are easy to compute. We show that any collision-finding attack on a hash function with the compression function F is at most as efficient as the birthday attack if f is a random oracle and p satisfies some properties. Our properties for p are easy to be satisfied; for example, they are satisfied by the permutation $p(x) = x \oplus c$, where $\oplus$ is bit-wise addition and c is a non-zero constant.

Similar results are in fact already obtained by Nandi [21], whose analysis actually applies to a broader range of hash functions than our analysis. However, our results are sharper. We give a significantly better upper bound on the probability of finding a collision as a function of the number of queries made by the adversary.

A new security notion for a compression function is also introduced, which we call indistinguishability in the iteration. It is really weaker than the notion proposed in [5]. However, it may be valuable in practice. Loosely speaking, a compression function $F(x) = (f(x), f(p(x)))$ where f is a random oracle is called indistinguishable in the iteration if F cannot be distinguished from a random oracle in the iterated hash function. We give sufficient conditions on p for F to be indistinguishable in the iteration.

Second, we discuss constructions using a block cipher. A compression function composed of a block cipher is presented and its collision resistance is analyzed. We show that any collision-finding attack on a hash function composed of the compression function is at most as efficient as the birthday attack if the block cipher used is ideal. A block cipher is ideal if it is assumed to be a keyed invertible random permutation. The compression function presented in this article is quite simple but has not been explicitly discussed previously.

In [10], it is shown that a collision-resistant hash function can be easily composed of a compression function using two distinct block ciphers. It is well-known that two distinct block ciphers can be obtained from a block cipher by fixing, for example, one key bit by 0 and 1. However, it is preferable in practice that fixing key bits is avoided. Moreover, fixing one bit may not be sufficient and more bits may be required to be fixed. Our new construction does not involve any fixing of key bits by constants.

The technique in [3] is used in the security proofs in this article. However, the analysis is more complicated than the one in [3] since the relation of two component-compression-function/block-cipher calls in a compression function need to be taken into account.

The rest of this article is organized as follows. Section 2 includes notations, definitions and a brief review of the related works. Section 3 discusses compression functions composed of a smaller compression function, including the results on collision resistance and our new notion of indistinguishability in the iteration. Section 4 exhibits a block-cipher-based compression function whose associated hash function has optimal collision resistance; the proof of collision resistance is given in the appendix. Section 5 gives a concluding remark which mentions a recent collision attack on the scheme in Sect. 4.

2 Preliminaries

2.1 Iterated Hash Function

A hash function $H : \{0,1\}^* \to \{0,1\}^\ell$ usually consists of a compression function $F : \{0,1\}^\ell \times \{0,1\}^{\ell'} \to \{0,1\}^\ell$ and a fixed initial value $h_0 \in \{0,1\}^\ell$. An input m is divided into the ℓ'-bit blocks $m_1, m_2, \ldots, m_l$. Then,

$$h_i = F(h_{i-1}, m_i)$$

is computed successively for $1 \leq i \leq l$ and $h_l = H(m)$. H is called an iterated hash function.

Before being divided into the blocks, unambiguous padding is applied to the input. The length of the padded input is a multiple of ℓ'. In this article, Merkle-Damgård strengthening [6,20] is assumed for padding. Thus, the last block contains the length of the input.

2.2 Random Oracle Model and Ideal Cipher Model

Random Oracle Model. Let $\boldsymbol{F}_{n',n} = \{f \mid f : \{0,1\}^{n'} \to \{0,1\}^n\}$. In the random oracle model, the function f is assumed to be randomly selected from $\boldsymbol{F}_{n',n}$. The computation of f is simulated by the following oracle.

The oracle f first receives an input x_i as a query. Then, it returns a randomly selected output y_i if the query has never been asked before. It keeps a table of pairs of queries and replies, and it returns the same reply to the same query.

Ideal Cipher Model. A block cipher with the block length n and the key length κ is called an (n, κ) block cipher. Let $e : \{0,1\}^\kappa \times \{0,1\}^n \to \{0,1\}^n$ be an (n, κ) block cipher. Then, $e(k, \cdot)$ is a permutation for every $k \in \{0,1\}^\kappa$, and it is easy to compute both $e(k, \cdot)$ and $e(k, \cdot)^{-1}$.

Let $\boldsymbol{B}_{n,\kappa}$ be the set of all (n, κ) block ciphers. In the ideal cipher model, e is assumed to be randomly selected from $\boldsymbol{B}_{n,\kappa}$. The encryption e and the decryption e^{-1} are simulated by the following two oracles.

The encryption oracle e first receives a pair of a key and a plaintext as a query. Then, it returns a randomly selected ciphertext. On the other hand, the decryption oracle e^{-1} first receives a pair of a key and a ciphertext as a query. Then, it returns a randomly selected plaintext. The oracles e and e^{-1} share a table of triplets of keys, plaintexts and ciphertexts, (k_i, x_i, y_i)'s, which are produced by the queries and the corresponding replies. Referring to the table, they select a reply to a new query under the restriction that $e(k, \cdot)$ is a permutation for every k. They also add the triplet produced by the query and the reply to the table.

2.3 DBL Hash Function

An iterated hash function whose compression function is composed of a block cipher is called a single-block-length (SBL) hash function if its output length is equal to the block length of the block cipher. It is called a double-block-length (DBL) hash function if its output length is twice larger than the block length.

Let F be a compression function composed of a block cipher. For an iterated hash function composed of F, the rate r defined below is often used as a measure of efficiency:

$$r = \frac{|m_i|}{(\text{the number of block-cipher calls in } F) \times n} .$$

In this article, we also call an iterated hash function a DBL hash function if its compression function F is composed of a smaller compression function f and its output length is twice larger than the output length of f.

2.4   Related Work

Knudsen and Preneel studied the schemes to construct secure compression functions with longer outputs from secure ones based on error-correcting codes [13,14,15]. It is an open question whether optimally collision-resistant compression functions are constructed by their schemes. A hash/compression function is optimally collision-resistant if any attack to find its collision is at most as efficient as the birthday attack.

Our work is largely motivated by the recent works by Lucks [18] and Nandi [22]. Nandi generalized the results by Lucks and by Hirose [10]. He discussed how to construct DBL hash functions and presented optimally collision-resistant ones. However, their security analysis is not so sharp as ours, which is mentioned in Sect. 1.

Coron, Dodis, Malinaud and Puniya [5] discussed how to construct a random oracle with arbitrary input length given a random oracle with fixed input length.

As is reviewed in the following, there are many papers on hash functions composed of block ciphers.

Preneel, Govaerts and Vandewalle [25] discussed the security of SBL hash functions against several generic attacks. They considered SBL hash functions with compression functions represented by $h_i = e(k, x) \oplus z$, where e is an (n, n) block cipher, $k, x, z \in \{h_{i-1}, m_i, h_{i-1} \oplus m_i, c\}$ and c is a constant. They concluded that 12 out of $64 (= 4^3)$ hash functions are secure against the attacks. However, they did not provide any formal proofs.

Black, Rogaway and Shrimpton [3] presented a detailed investigation of provable security of SBL hash functions given in [25] in the ideal cipher model. The most important result in their paper is that 20 hash functions including the 12 mentioned above is optimally collision-resistant.

Knudsen, Lai and Preneel [16] discussed the insecurity of DBL hash functions with the rate 1 composed of (n, n) block ciphers. Hohl, Lai, Meier and Waldvogel [11] discussed the security of compression functions of DBL hash functions with the rate $1/2$. On the other hand, the security of DBL hash functions with the rate 1 composed of $(n, 2n)$ block ciphers was discussed by Satoh, Haga and Kurosawa [26] and by Hattori, Hirose and Yoshida [8]. These works presented no construction for DBL hash functions with optimal collision resistance,

Many schemes with the rates less than 1 were also presented. Merkle [20] presented three DBL hash functions composed of DES with the rates at most 0.276. They are optimally collision-resistant in the ideal cipher model. MDC-2 and MDC-4 [4] are also DBL hash functions composed of DES with the rates 1/2 and 1/4, respectively. Lai and Massey proposed the tandem/abreast Davies-Meyer [17]. They consist of an $(n, 2n)$ block cipher and their rates are 1/2. It is an open question whether the four schemes are optimally collision-resistant or not.

Hirose [10] presented a large class of DBL hash functions with the rate 1/2, which are composed of $(n, 2n)$ block ciphers. They were shown to be optimally collision-resistant in the ideal cipher model. However, his construction requires two independent block ciphers, which makes the results less attractive.

Nandi, Lee, Sakurai and Lee [23] also proposed an interesting construction with the rate 2/3. However, they are not optimally collision-resistant. Knudsen and Muller [12] presented some attacks against it and illustrated its weaknesses, none of which contradicts the security proof in [23].

Black, Cochran and Shrimpton [2] showed that it is impossible to construct a highly efficient block-cipher-based hash function provably secure in the ideal cipher model. A block-cipher-based hash function is highly efficient if it makes exactly one block-cipher call for each message block and all block-cipher calls use a single key.

Gauravaram, Millan and May proposed a new approach based on iterated halving to design a hash function with a block cipher [7].

3   DBL Hash Function in the Random Oracle Model

3.1   Compression Function

In this section, we consider the DBL hash functions with compression functions given in the following definition.

Definition 1. *Let* $F : \{0,1\}^{2n} \times \{0,1\}^b \to \{0,1\}^{2n}$ *be a compression function such that* $(g_i, h_i) = F(g_{i-1}, h_{i-1}, m_i)$*, where* $g_i, h_i \in \{0,1\}^n$ *and* $m_i \in \{0,1\}^b$*.* F *consists of* $f : \{0,1\}^{2n} \times \{0,1\}^b \to \{0,1\}^n$ *and a permutation* $p : \{0,1\}^{2n+b} \to \{0,1\}^{2n+b}$ *as follows:*

$$\begin{cases} g_i = F_U(g_{i-1}, h_{i-1}, m_i) = f(g_{i-1}, h_{i-1}, m_i) \\ h_i = F_L(g_{i-1}, h_{i-1}, m_i) = f(p(g_{i-1}, h_{i-1}, m_i)) \end{cases}$$

p satisfies the following properties:

- *It is easy to compute both p and p^{-1},*
- *$p(p(\cdot))$ is an identity permutation,*
- *p has no fixed points, that is, $p(g_{i-1}, h_{i-1}, m_i) \neq (g_{i-1}, h_{i-1}, m_i)$ for any (g_{i-1}, h_{i-1}, m_i).*

3.2   Collision Resistance

We will analyze the collision resistance of DBL hash functions composed of F under the assumption that f is a random oracle.

Two queries to the oracle f are required to compute the output of F for an input. For this compression function, a query to f for F_U or F_L uniquely determines the query to f for the other since p is a permutation. Moreover, for every $w \in \{0,1\}^{2n+b}$, $f(w)$ and $f(p(w))$ are only used to compute $F(w)$ and $F(p(w))$, and $w \neq p(w)$ from the properties for p in Definition 1. Thus, it is reasonable to assume that a pair of queries w and $p(w)$ to f are asked at a time.

Definition 2. *A pair of distinct inputs w, w' to F are called a matching pair if $w' = p(w)$. Otherwise, they are called a non-matching pair.*

Notice that $w' = p(w)$ iff $w = p(w')$ since $p(p(\cdot))$ is an identity permutation.

Definition. Insecurity is quantified by success probability of an optimal resource-bounded adversary. The resource is the number of the queries to f in the random oracle model.

For a set S, let $z \leftarrow_R S$ represent random sampling from S under the uniform distribution. For a probabilistic algorithm $\mathcal{M}$, let $z \leftarrow_R \mathcal{M}$ mean that z is an output of $\mathcal{M}$ and its distribution is based on the random choices of $\mathcal{M}$.

Let H be a DBL hash function composed of a compression function F in Definition 1. The following experiment $\texttt{FindColHF}(\mathcal{A}, H)$ is introduced to quantify the collision resistance of H. The adversary $\mathcal{A}$ with the oracle f is a collision-finding algorithm of H.

> $\texttt{FindColHF}(\mathcal{A}, H)$
> $\quad f \leftarrow_R \mathbf{F}_{2n+b,n}$;
> $\quad (m, m') \leftarrow_R \mathcal{A}^f$;
> $\quad$ if $m \neq m' \wedge H(m) = H(m')$ return 1; else return 0;

$\texttt{FindColHF}(\mathcal{A}, H)$ returns 1 iff $\mathcal{A}$ finds a collision. Let $\mathbf{Adv}_H^{\text{coll}}(\mathcal{A})$ be the probability that $\texttt{FindColHF}(\mathcal{A}, H)$ returns 1. The probability is taken over the uniform distribution on $\mathbf{F}_{2n+b,n}$ and random choices of $\mathcal{A}$.

Definition 3. *For $q \geq 1$, let*

$$\mathbf{Adv}_H^{\text{coll}}(q) = \max_{\mathcal{A}} \left\{ \mathbf{Adv}_H^{\text{coll}}(\mathcal{A}) \right\} ,$$

where $\mathcal{A}$ makes at most q pairs of queries to f in total.

Without loss of generality, it is assumed that $\mathcal{A}$ does not ask the same query twice. $\mathcal{A}$ can keep pairs of queries and their corresponding answers by himself.

Analysis. The following theorem shows the collision resistance of a hash function composed of F in Definition 1.

Theorem 1. *Let H be a hash function composed of a compression function F specified in Definition 1. Then, for every $1 \leq q \leq 2^n$,*

$$\mathbf{Adv}_H^{\mathrm{coll}}(q) \leq \left(\frac{q}{2^n}\right)^2 + \frac{q}{2^n} \; .$$

Proof. Let $\mathcal{A}$ be a collision-finding algorithm of H with the oracle f. $\mathcal{A}$ asks q pairs of queries to f in total. Suppose that $\mathcal{A}$ finds a colliding pair m, m' of H. Then, it is easy to find a colliding pair of inputs for F without any additional queries to the oracle. Moreover, a pair of inputs to F are either matching or non-matching, so are the colliding pair of inputs for F.

For $2 \leq j \leq q$, let C_j be the event that a colliding pair of non-matching inputs are found for F with the j-th pair of queries. Namely, it is the event that

$$(f(w_j), f(p(w_j))) \in \{(f(w_{j'}), f(p(w_{j'}))), (f(p(w_{j'})), f(w_{j'}))\}$$

for some $j' < j$, where w_j and $p(w_j)$ are the j-th pair of queries. Since both $f(w_j)$ and $f(p(w_j))$ are randomly selected by the oracle,

$$\Pr[\mathsf{C}_j] \leq \frac{2(j-1)}{2^{2n}} \; .$$

Let C be the event that a colliding pair of non-matching inputs are found for F with q pairs of queries. Then,

$$\Pr[\mathsf{C}] = \Pr[\mathsf{C}_2 \vee \mathsf{C}_3 \vee \cdots \vee \mathsf{C}_q] \leq \sum_{j=2}^{q} \Pr[\mathsf{C}_j] \leq \left(\frac{q}{2^n}\right)^2 \; .$$

For $1 \leq j \leq q$, let $\mathsf{C}_j^{\mathrm{m}}$ be the event that a colliding pair of matching inputs are found for F with the j-th pair of queries, that is, $f(w_j) = f(p(w_j))$. Thus,

$$\Pr[\mathsf{C}_j^{\mathrm{m}}] = \frac{1}{2^n} \; .$$

Let C^{m} be the event that a colliding pair of matching inputs are found for F with q pairs of queries. Then,

$$\Pr[\mathsf{C}^{\mathrm{m}}] = \Pr[\mathsf{C}_1^{\mathrm{m}} \vee \mathsf{C}_2^{\mathrm{m}} \vee \cdots \vee \mathsf{C}_q^{\mathrm{m}}] \leq \sum_{j=1}^{q} \Pr[\mathsf{C}_j^{\mathrm{m}}] = \frac{q}{2^n} \; .$$

Thus, if $q \leq 2^n$, then

$$\mathbf{Adv}_H^{\mathrm{coll}}(\mathcal{A}) \leq \Pr[\mathsf{C} \vee \mathsf{C}^{\mathrm{m}}] \leq \Pr[\mathsf{C}] + \Pr[\mathsf{C}^{\mathrm{m}}] \leq \left(\frac{q}{2^n}\right)^2 + \frac{q}{2^n} \; ,$$

which holds for any $\mathcal{A}$.   □

From Theorem 1, any constant probability of success in finding a collision of H requires $\Omega(2^n)$ queries.

A better bound can be obtained with more restricted permutations.

Theorem 2. *Let H be a hash function composed of a compression function F specified in Definition 1. Suppose that the permutation p is represented by $p(g, h, m) = (p_{cv}(g, h), p_m(m))$, where $p_{cv} : \{0, 1\}^{2n} \to \{0, 1\}^{2n}$ and $p_m : \{0, 1\}^b \to \{0, 1\}^b$. Suppose that p_{cv} has no fixed points and that $p_{cv}(g, h) \neq (h, g)$ for any (g, h). Then, for every $1 \leq q \leq 2^n$,*

$$\mathbf{Adv}_H^{\mathrm{coll}}(q) \leq 3 \left(\frac{q}{2^n} \right)^2 .$$

Proof. Let $\mathcal{A}$ be a collision-finding algorithm of H with the oracle f. $\mathcal{A}$ asks q pairs of queries to f in total. Suppose that $\mathcal{A}$ finds a colliding pair m, m' of H. Then, it is easy to find a colliding pair of inputs for F without any additional queries. Moreover, a pair of inputs to F are either matching or non-matching, so are the colliding pair of inputs for F.

Let C be the event that a colliding pair of non-matching inputs are found for F with q pairs of queries. Then, as in the proof of Theorem 1,

$$\Pr[\mathsf{C}] \leq \left(\frac{q}{2^n} \right)^2 .$$

Suppose that a colliding pair of matching inputs are obtained for F from the collision of H found by $\mathcal{A}$. Let (g, h, m) and (g', h', m') be the colliding pair. Then, $(g, h) = p_{cv}(g', h')$ (and $(g', h') = p_{cv}(g, h)$). (g, h) and (g', h') are both outputs of F, or at most one of them is the initial value (g_0, h_0) of H since $(g, h) \neq (g', h')$. Thus, a pair of inputs w and w' are also found for F from the collision of H such that $F(w) = p_{cv}(F(w'))$ or $F(w) = p_{cv}(g_0, h_0)$.

Suppose that $(g, h) = F(w)$ and $(g', h') = F(w')$. Then, a pair of w and w' are non-matching since $(g, h) = p_{cv}(g', h') \neq (h', g')$.

For $1 \leq j \leq q$, let $\hat{\mathsf{C}}_j^m$ be the event that, for the j-th pair of queries w_j and $p(w_j)$,

$$F(w_j) \in \{p_{cv}(g_0, h_0)\} \cup \bigcup_{1 \leq j' < j} \{p_{cv}(F(w_{j'})), p_{cv}(F(p(w_{j'})))\}$$

or

$$F(p(w_j)) \in \{p_{cv}(g_0, h_0)\} \cup \bigcup_{1 \leq j' < j} \{p_{cv}(F(w_{j'})), p_{cv}(F(p(w_{j'})))\} .$$

Thus,

$$\Pr[\hat{\mathsf{C}}_j^m] \leq \frac{2(2j - 1)}{2^{2n}} .$$

Let $\hat{\mathsf{C}}^m = \hat{\mathsf{C}}_1^m \vee \cdots \vee \hat{\mathsf{C}}_q^m$. Then,

$$\Pr[\hat{\mathsf{C}}^m] \leq \sum_{j=1}^q \Pr[\hat{\mathsf{C}}_j^m] \leq \sum_{j=1}^q \frac{2(2j - 1)}{2^{2n}} = 2 \left(\frac{q}{2^n} \right)^2 .$$

Thus,

$$\mathbf{Adv}_H^{\mathrm{coll}}(\mathcal{A}) \le \Pr[\mathsf{C} \vee \hat{\mathsf{C}}^{\mathrm{m}}] \le \Pr[\mathsf{C}] + \Pr[\hat{\mathsf{C}}^{\mathrm{m}}] \le 3 \left(\frac{q}{2^n}\right)^2 ,$$

for $1 \le q \le 2^n$, which holds for any $\mathcal{A}$.   □

For $q < 2^{n-1}$, Theorem 2 gives a smaller upper bound than Theorem 1. The difference between their upper bounds is significant. For example, let $n = 128$ and $q = 2^{80}$. Then, the upper bound of Theorem 1 is about 2^{-48}, while the upper bound of Theorem 2 is less than 2^{-94}.

Example 1. Here is an example of the permutation p satisfying the conditions given in Theorem 2:

$$p(g, h, m) = (g \oplus c_1, h \oplus c_2, m) ,$$

where c_1 and c_2 are distinct constants in $\{0, 1\}^n$.

3.3   Indistinguishability in the Iteration

We introduce a new security notion which is called indistinguishability in the iteration.

Definition. Let F be a compression function specified in Definition 1. The following experiment $\mathtt{DistinguishCF}(\mathcal{A}, F)$ is introduced to quantify the indistinguishability in the iteration of F. The adversary $\mathcal{A}$ is a distinguishing algorithm of F. $\mathcal{A}$ has an oracle $\mathcal{O}$. In this experiment, a randomly chosen bit $d \in \{0, 1\}$ is given to $\mathcal{O}$ first. If $d = 1$, then $\mathcal{O}$ chooses $f \in F_{2n+b,n}$ randomly in advance. Then, $\mathcal{O}$ returns $F(w) = (f(w), f(p(w)))$ to each query w from $\mathcal{A}$. If $d = 0$, then $\mathcal{O}$ chooses $R \in F_{2n+b,2n}$ randomly in advance. Then, $\mathcal{O}$ returns $R(w)$ to each query w from $\mathcal{A}$. $\mathcal{A}$ makes a chosen message attack and tries to tell whether $\mathcal{O}$ uses F or R. However, $\mathcal{A}$ is only allowed to select his j-th query $w_j = (w_j^{(1)}, w_j^{(2)}, w_j^{(3)})$ from

$$\left\{ (w^{(1)}, w^{(2)}, w^{(3)}) \ \middle| \ (w^{(1)}, w^{(2)}) \in \bigcup_{\ell=0}^{j-1} (v_\ell^{(1)}, v_\ell^{(2)}) \wedge w^{(3)} \in \{0, 1\}^b \right\} ,$$

where $(v_\ell^{(1)}, v_\ell^{(2)})$ is $\mathcal{O}$'s answer to the ℓ-th query for $\ell \ge 1$ and $(v_0^{(1)}, v_0^{(2)})$ is some fixed initial value of a hash function H. F is assumed to be used only in the iteration of H.

$\mathtt{DistinguishCF}(\mathcal{A}, F)$
  $d \leftarrow_R \{0, 1\}$;
  $d' \leftarrow_R \mathcal{A}^{\mathcal{O}(d)}$;
  if $d = d'$ return 1; else return 0;

Let $\mathbf{Succ}_F^{\mathrm{ind\text{-}it}}(\mathcal{A})$ be the probability that $\mathtt{DistinguishCF}(\mathcal{A}, F)$ returns 1. Without loss of generality, it can be assumed that $\mathbf{Succ}_F^{\mathrm{ind\text{-}it}}(\mathcal{A}) \ge 1/2$ because the probability that $d = d'$ is $1/2$ even if $\mathcal{A}$ chooses d' randomly. It can also be assumed that $\mathcal{A}$ does not ask the same query twice. Let

$$\mathbf{Adv}_F^{\text{ind-it}}(\mathcal{A}) \overset{\text{def}}{=} \mathbf{Succ}_F^{\text{ind-it}}(\mathcal{A}) - 1/2 \ .$$

Definition 4. *For $q \geq 1$, let*

$$\mathbf{Adv}_F^{\text{ind-it}}(q) = \max_{\mathcal{A}} \left\{ \mathbf{Adv}_F^{\text{ind-it}}(\mathcal{A}) \right\} \ ,$$

where $\mathcal{A}$ makes at most q queries to $\mathcal{O}$.

As long as $\mathbf{Adv}_F^{\text{ind-it}}(q)$ is small enough, the compression function F behaves like a random function in the iterated hash function. The following theorem presents an upper bound on $\mathbf{Adv}_F^{\text{ind-it}}(q)$ with additional restriction on the permutation p.

Theorem 3. *Let F be a compression function specified in Definition 1. Suppose that the permutation p is represented by $p(g, h, m) = (p_{\text{cv}}(g, h), p_{\text{m}}(m))$, where $p_{\text{cv}} : \{0, 1\}^{2n} \to \{0, 1\}^{2n}$ and $p_{\text{m}} : \{0, 1\}^b \to \{0, 1\}^b$. Suppose that p_{cv} has no fixed points. Then, for every $1 \leq q \leq 2^n$,*

$$\mathbf{Adv}_F^{\text{ind-it}}(q) \leq \frac{1}{2} \left(\frac{q}{2^n} \right)^2 \ .$$

Proof. Let $\mathcal{A}$ be the optimal distinguishing algorithm for F which makes q queries. Let w_j be $\mathcal{A}$'s j-th query to $\mathcal{O}$ and $T = \{w_j \mid 1 \leq j \leq q\} \cap \{p(w_j) \mid 1 \leq j \leq q\}$. Suppose that $d = 1$. Then, $\mathcal{O}$ returns $F(w_j) = (f(w_j), f(p(w_j)))$ for w_j. If $T = \phi$, then F is completely indistinguishable from R. It is because each one of $f(w_j)$ and $f(p(w_j))$ for $1 \leq j \leq q$ appears only once and it is chosen randomly by $\mathcal{O}$.

Let Empty be the event that $T = \phi$. Then,

$$\begin{aligned}
\mathbf{Succ}_F^{\text{ind-it}}(\mathcal{A}) = \Pr[d = d'] &= \Pr[d = d' \wedge \text{Empty}] + \Pr[d = d' \wedge \neg\text{Empty}] \\
&= \Pr[d = d' \mid \text{Empty}] \Pr[\text{Empty}] + \Pr[d = d' \mid \neg\text{Empty}] \Pr[\neg\text{Empty}] \\
&\leq \frac{1}{2} + \Pr[\neg\text{Empty}] \ .
\end{aligned}$$

Let v_j be the initial value if $j = 0$ and the answer of $\mathcal{O}$ to the j-th query by $\mathcal{A}$ if $j \geq 1$. For $1 \leq j \leq q$, let C'_j be the event that $v_j \in \{p_{\text{cv}}(v_\ell) \mid 0 \leq \ell \leq j - 1\}$. Then,

$$\Pr[C'_j] \leq \frac{j}{2^{2n}} \ .$$

For $1 \leq q \leq 2^n$,

$$\Pr[\neg\text{Empty}] \leq \Pr[C'_1 \vee \cdots \vee C'_{q-1}] \leq \sum_{j=1}^{q-1} \Pr[C'_j] \leq \frac{1}{2} \left(\frac{q}{2^n} \right)^2$$

which implies that $\mathbf{Adv}_F^{\text{ind-it}}(q) \leq (q/2^n)^2/2$.   □

4   DBL Hash Function in the Ideal Cipher Model

4.1   Compression Function

In this section, the collision resistance of a DBL hash function composed of a compression function using a block cipher is analyzed. The compression function specified in the following definition is considered.

Definition 5. *Let $F : \{0,1\}^{2n} \times \{0,1\}^b \rightarrow \{0,1\}^{2n}$ be a compression function such that $(g_i, h_i) = F(g_{i-1}, h_{i-1}, m_i)$, where $g_i, h_i \in \{0,1\}^n$ and $m_i \in \{0,1\}^b$. F consists of a $(n, n+b)$ block cipher e as follows:*

$$\begin{cases} g_i = F_U(g_{i-1}, h_{i-1}, m_i) = e(h_{i-1}\|m_i, g_{i-1}) \oplus g_{i-1} \\ h_i = F_L(g_{i-1}, h_{i-1}, m_i) = e(h_{i-1}\|m_i, g_{i-1} \oplus c) \oplus g_{i-1} \oplus c , \end{cases}$$

where $\|$ represents concatenation and $c \in \{0,1\}^n - \{0^n\}$ is a constant.

The compression function in Definition 5 is also shown in Fig. 1. It is one of the compression functions specified in Definition 1 and its f and p are specified as follows:

$$f(g_{i-1}, h_{i-1}, m_i) = e(h_{i-1}\|m_i, g_{i-1}) \oplus g_{i-1} ,$$
$$p(g_{i-1}, h_{i-1}, m_i) = (g_{i-1} \oplus c, h_{i-1}, m_i) .$$

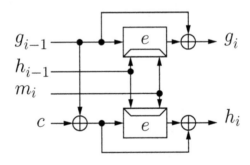

Fig. 1. A compression function considered in Sect. 4.2

F requires two invocations of e to produce an output. However, these two invocations need only one key scheduling of e. If F is implemented using the AES with 192-bit key-length, then $n = 128$, $b = 64$ and the rate is $1/4$. If implemented using the AES with 256-bit key-length, then $n = b = 128$ and the rate is $1/2$.

4.2   Collision Resistance

Let F be a compression function specified in Definition 5. Two queries to the oracles e and e^{-1} in total are required to compute the output of F for an input. It is easy to see from Fig. 1 that a query to e or e^{-1} and the corresponding reply for F_U (F_L) uniquely determine the query to e for F_L (F_U). Moreover, these two queries are only used to compute the outputs of F for a matching pair of inputs. Thus, it is assumed that a pair of queries to e, e^{-1} required to compute an output of F are asked at a time.

Definition. The following experiment $\texttt{FindColHF}(\mathcal{A}, H)$ is similar to the one in Sect. 3 except that the adversary $\mathcal{A}$ is a collision-finding algorithm with the oracles e, e^{-1}.

$\texttt{FindColHF}(\mathcal{A}, H)$
$\quad e \leftarrow_R \boldsymbol{B}_{n,n+b};$
$\quad (m, m') \leftarrow_R \mathcal{A}^{e, e^{-1}};$
$\quad \text{if } m \neq m' \wedge H(m) = H(m') \text{ return 1; else return 0;}$

Let $\mathbf{Adv}_H^{\mathrm{coll}}(\mathcal{A})$ be the probability that $\texttt{FindColHF}(\mathcal{A}, H)$ returns 1. The probability is taken over the uniform distribution on $\boldsymbol{B}_{n,n+b}$ and random choices of $\mathcal{A}$.

Definition 6. *For $q \geq 1$, let*

$$\mathbf{Adv}_H^{\mathrm{coll}}(q) = \max_{\mathcal{A}} \left\{ \mathbf{Adv}_H^{\mathrm{coll}}(\mathcal{A}) \right\} ,$$

where $\mathcal{A}$ makes at most q pairs of queries to e, e^{-1} in total.

Without loss of generality, it is assumed that $\mathcal{A}$ asks at most only once on a triplet of a key, a plaintext and a ciphertext obtained by a query and the corresponding reply.

Analysis. The following theorem shows the collision resistance of a hash function composed of F in Definition 5.

Theorem 4. *Let H be a hash function composed of the compression function F specified in Definition 5. Then, for every $1 \leq q \leq 2^{n-2}$,*

$$\mathbf{Adv}_H^{\mathrm{coll}}(q) \leq 3 \left(\frac{q}{2^{n-1}} \right)^2 .$$

The proof of Theorem 4 is given in the appendix.

5   Concluding Remark

In this article, some plausible constructions have been proposed for DBL hash functions.

Recently, Pramstaller and Rijmen presented a collision attack on the scheme in Sect. 4 with DESX as an underlying block cipher [24]. Their result does not contradict Theorem 4. It is a warning that we should be careful when we choose an underlying block cipher. It also shows a limitation of the random oracle/ideal cipher model. Related topics are discussed in [1,9].

Acknowledgements. The author would like to thank the anonymous reviewers for their helpful comments.

References

1. J. Black. The ideal-cipher model, revisited: An uninstantiable blockcipher-based hash function. Cryptology ePrint Archive, Report 2005/210, 2005. `http://eprint.iacr.org/`. Also appear in this proceedings.
2. J. Black, M. Cochran, and T. Shrimpton. On the impossibility of highly efficient blockcipher-based hash functions. In *EUROCRYPT 2005 Proceedings, Lecture Notes in Computer Science 3494*, pages 526–541, 2005.
3. J. Black, P. Rogaway, and T. Shrimpton. Black-box analysis of the block-cipher-based hash-function constructions from PGV. In *CRYPTO 2002 Proceedings, Lecture Notes in Computer Science 2442*, pages 320–335, 2002.
4. B. O. Brachtl, D. Coppersmith, M. M. Hyden, S. M. Matyas Jr., C. H. W. Meyer, J. Oseas, S. Pilpel, and M. Schilling. Data authentication using modification detection codes based on a public one-way encryption function, mar 1990. U. S. Patent # 4,908,861.
5. J.-S. Coron, Y. Dodis, C. Malinaud, and P. Puniya. Merkle-damgård revisited: How to construct a hash function. In *CRYPTO 2005 Proceedings, Lecture Notes in Computer Science 3621*, pages 430–448, 2005.
6. I. Damgård. Collision free hash functions and public key signature schemes. In *EUROCRYPT '87 Proceedings, Lecture Notes in Computer Science 304*, pages 203–216, 1988.
7. P. Gauravaram, W. Millan, and L. May. CRUSH: A new cryptographic hash function using iterated halving technique. In *Proceedings of Cryptographic Algorithms and their Uses 2004*, pages 28–39, 2004.
8. M. Hattori, S. Hirose, and S. Yoshida. Analysis of double block length hash functions. In *Proceedings of the 9th IMA International Conference on Cryptography and Coding, Lecture Notes in Computer Science 2898*, pages 290–302, 2003.
9. S. Hirose. Secure block ciphers are not sufficient for one-way hash functions in the Preneel-Govaerts-Vandewalle model. In *Proceedings of the 9th Selected Areas in Cryptography (SAC 2002), Lecture Notes in Computer Science 2595*, pages 339–352, 2002.
10. S. Hirose. Provably secure double-block-length hash functions in a black-box model. In *Proceedings of the 7th Internatinal Conference on Information Security and Cryptology (ICISC 2004), Lecture Notes in Computer Science 3506*, pages 330–342, 2005.
11. W. Hohl, X. Lai, T. Meier, and C. Waldvogel. Security of iterated hash functions based on block ciphers. In *CRYPTO '93 Proceedings, Lecture Notes in Computer Science 773*, pages 379–390, 1994.
12. L. Knudsen and F. Muller. Some attacks against a double length hash proposal. In *ASIACRYPT 2005 Proceedings, Lecture Notes in Computer Science 3788*, pages 462–473, 2005.
13. L. Knudsen and B. Preneel. Hash functions based on block ciphers and quaternary codes. In *ASIACRYPT '96 Proceedings, Lecture Notes in Computer Science 1163*, pages 77–90, 1996.
14. L. Knudsen and B. Preneel. Fast and secure hashing based on codes. In *CRYPTO '97 Proceedings, Lecture Notes in Computer Science 1294*, pages 485–498, 1997.
15. L. Knudsen and B. Preneel. Construction of secure and fast hash functions using nonbinary error-correcting codes. *IEEE Transactions on Information Theory*, 48(9):2524–2539, 2002.

16. L. R. Knudsen, X. Lai, and B. Preneel. Attacks on fast double block length hash functions. *Journal of Cryptology*, 11(1):59–72, 1998.
17. X. Lai and J. L. Massey. Hash function based on block ciphers. In *EUROCRYPT '92 Proceedings, Lecture Notes in Computer Science 658*, pages 55–70, 1993.
18. S. Lucks. A failure-friendly design principle for hash functions. In *ASIACRYPT 2005 Proceedings, Lecture Notes in Computer Science 3788*, pages 474–494, 2005.
19. A. J. Menezes, P. C. van Oorschot, and S. A. Vanstone. *Handbook of Applied Cryptography*. CRC Press, 1996.
20. R. C. Merkle. One way hash functions and DES. In *CRYPTO '89 Proceedings, Lecture Notes in Computer Science 435*, pages 428–446, 1990.
21. M. Nandi. *Design of Iteration on Hash Functions and Its Cryptanalysis*. PhD thesis, Indian Statistical Institute, 2005.
22. M. Nandi. Towards optimal double-length hash functions. In *Proceedings of the 6th International Conference on Cryptology in India (INDOCRYPT 2005), Lecture Notes in Computer Science 3797*, pages 77–89, 2005.
23. M. Nandi, W. Lee, K. Sakurai, and S. Lee. Security analysis of a 2/3-rate double length compression function in the black-box model. In *Proceedings of the 12th Fast Software Encryption (FSE 2005), Lecture Notes in Computer Science 35571*, pages 243–254, 2005.
24. N. Pramstaller and V. Rijmen. A collision attack on a double-block-length hash proposal. Cryptology ePrint Archive, Report 2006/116, 2006. `http://eprint.iacr.org/`.
25. B. Preneel, R. Govaerts, and J. Vandewalle. Hash functions based on block ciphers: A synthetic approach. In *CRYPTO '93 Proceedings, Lecture Notes in Computer Science 773*, pages 368–378, 1994.
26. T. Satoh, M. Haga, and K. Kurosawa. Towards secure and fast hash functions. *IEICE Transactions on Fundamentals*, E82-A(1):55–62, 1999.

A   Proof of Theorem 4

Let $\mathcal{A}$ be a collision-finding algorithm of H with oracles e, e^{-1}. $\mathcal{A}$ asks q pairs of queries to e, e^{-1} in total.

Since $g_i = e(h_{i-1} \| m_i, g_{i-1}) \oplus g_{i-1}$, g_i depends both on the plaintext and the ciphertext of e and one of them is fixed by a query and the other is determined randomly by the answer from the oracle. Thus, g_i is randomly determined by the answer. h_i is also randomly determined by the other answer.

Let $(k_{1,j} \| k_{2,j}, x_j, y_j)$ and $(k_{1,j} \| k_{2,j}, x_j \oplus c, z_j)$ be the triplets of e obtained by the j-th pair of queries and the corresponding answers.

For every $2 \leq j \leq q$, let C_j be the event that a colliding pair of non-matching inputs are found for F with the j-th pair of queries. Namely, it is the event that, for some $j' < j$,

$$F(x_j, k_{1,j}, k_{2,j}) = F(x_{j'}, k_{1,j'}, k_{2,j'}) \text{ or } F(x_{j'} \oplus c, k_{1,j'}, k_{2,j'})$$

or

$$F(x_j \oplus c, k_{1,j}, k_{2,j}) = F(x_{j'}, k_{1,j'}, k_{2,j'}) \text{ or } F(x_{j'} \oplus c, k_{1,j'}, k_{2,j'}) \ ,$$

which is equivalent to

$$(y_j \oplus x_j, z_j \oplus x_j \oplus c) = (y_{j'} \oplus x_{j'}, z_{j'} \oplus x_{j'} \oplus c) \text{ or } (z_{j'} \oplus x_{j'} \oplus c, y_{j'} \oplus x_{j'}) \ .$$

Thus,

$$\Pr[C_j] \leq \frac{2(j-1)}{(2^n - (2j-2))(2^n - (2j-1))} \leq \frac{2(j-1)}{(2^n - (2j-1))^2} \ .$$

Let C be the event that a colliding pair of non-matching inputs are found for F with q pairs of queries. Then,

$$\Pr[C] \leq \sum_{j=2}^{q} \Pr[C_j] \leq \sum_{j=2}^{q} \frac{2(j-1)}{(2^n - (2j-1))^2} \ .$$

Suppose that a colliding pair of matching inputs are obtained for F from the collision of H found by $\mathcal{A}$. Let (g, h, m) and (g', h', m') be the colliding pair of F. Then, $(g, h) = (g' \oplus c, h')$. (g, h) and (g', h') are both outputs of F or at most one of them is the initial value (g_0, h_0) of H. Thus, a pair of inputs w and w' are also found for F from the collision of H such that

$$(F_U(w), F_L(w)) = (F_U(w') \oplus c, F_L(w')) \text{ or } (g_0 \oplus c, h_0) \ .$$

Suppose that $(F_U(w), F_L(w)) = (F_U(w') \oplus c, F_L(w'))$. Then, a pair of w and w' are non-matching since

$$(F_U(w), F_L(w)) = (F_U(w') \oplus c, F_L(w')) \neq (F_L(w'), F_U(w')) \ .$$

For $1 \leq j \leq q$, let $(k_{1,j} \| k_{2,j}, x_j, y_j)$ and $(k_{1,j} \| k_{2,j}, x_j \oplus c, z_j)$ be the pair of triplets of e obtained by the j-th pair of queries and the corresponding answers. Let $\hat{C}_j^m$ be the event that $F(x_j, k_{1,j}, k_{2,j}) \in V$ or $F(x_j \oplus c, k_{1,j}, k_{2,j}) \in V$, where

$$V = \{(g_0 \oplus c, h_0)\} \cup \bigcup_{1 \leq j' < j} \{(F_U(x_{j'}, k_{1,j'}, k_{2,j'}) \oplus c, F_L(x_{j'}, k_{1,j'}, k_{2,j'}))\} \cup$$

$$\bigcup_{1 \leq j' < j} \{(F_U(x_{j'} \oplus c, k_{1,j'}, k_{2,j'}) \oplus c, F_L(x_{j'} \oplus c, k_{1,j'}, k_{2,j'}))\} \ .$$

Thus,

$$\Pr[\hat{C}_j^m] \leq \frac{2(2j-1)}{(2^n - (2j-2))(2^n - (2j-1))} \leq \frac{2(2j-1)}{(2^n - (2j-1))^2} \ .$$

Let $\hat{C}^m = \hat{C}_1^m \vee \cdots \vee \hat{C}_q^m$. Then,

$$\Pr[\hat{C}^m] \leq \sum_{j=1}^{q} \Pr[\hat{C}_j^m] \leq \sum_{j=1}^{q} \frac{2(2j-1)}{(2^n - (2j-1))^2} \ .$$

Thus, if $q \leq 2^{n-2}$, then

$$\mathbf{Adv}_H^{\text{coll}}(\mathcal{A}) \leq \Pr[\mathsf{C} \vee \hat{\mathsf{C}}^m] \leq \Pr[\mathsf{C}] + \Pr[\hat{\mathsf{C}}^m]$$

$$\leq \sum_{j=2}^{q} \frac{2(j-1)}{(2^n - (2j-1))^2} + \sum_{j=1}^{q} \frac{2(2j-1)}{(2^n - (2j-1))^2}$$

$$\leq \sum_{j=1}^{q} \frac{6j-4}{(2^n - (2j-1))^2}$$

$$\leq \frac{3q^2 - q}{(2^{n-1})^2} \leq 3 \left(\frac{q}{2^{n-1}} \right)^2$$

which holds for any $\mathcal{A}$.   □

Provably Secure MACs from Differentially-Uniform Permutations and AES-Based Implementations

Kazuhiko Minematsu and Yukiyasu Tsunoo

NEC Corporation, 1753 Shimonumabe, Nakahara-Ku, Kawasaki 211-8666, Japan
`k-minematsu@ah.jp.nec.com`

Abstract. We propose message authentication codes (MACs) that combine a block cipher and an additional (keyed or unkeyed) permutation. Our MACs are provably secure if the block cipher is pseudorandom and the additional permutation has a small differential probability. We also demonstrate that our MACs are easily implemented with AES and its 4-round version to obtain MACs that are provably secure and 1.4 to 2.5 times faster than the previous MAC modes of AES such as the CBC-MAC-AES.

Keywords: MAC, Block cipher, AES, Differentially-uniform permutation.

1 Introduction

Message Authentication Codes (MACs) are symmetric cryptographic functions that ensure the authenticities of messages. The CBC-MAC and its variants (such as EMAC [4], XCBC [8], and OMAC [16]) are well-known modes for block ciphers to provide MACs. They are provably secure and efficient; they operate at almost the same throughput as that of the underlying block cipher. However, what can we do if we want a MAC that is faster than these MAC modes to keep the additional implementation as small as possible? In other words, can we have a block cipher based MAC faster than the CBC-MAC?

In this paper, we give a solution to this problem. We propose MACs that combine a block cipher and its component, typically a reduced-round version of the block cipher. These kinds of MACs can easily be implemented on any platform where the block cipher has already been implemented. The additional program size would be quite small. A similar approach called ALRED [11] was recently proposed by Daemen and Rijmen. It was interesting because of its efficiency (in terms of authentication tag generation and preprocessing). It was also shown that ALRED was secure against some attacks. Compared to ALRED, our schemes are secure in a stronger security model: if one can distinguish our MACs from uniform random function, then the underlying block cipher can be distinguished from uniform random permutation. A MAC with this property is called a provably secure MAC.

M.J.B. Robshaw (Ed.): FSE 2006, LNCS 4047, pp. 226–241, 2006.

Formally, our MACs combine an n-bit block pseudorandom function (PRF) and an n-bit *auxiliary permutation* (AXP), which is an unkeyed or keyed permutation. AXPs are naturally expected to be faster than the block cipher, since they do not need to be cryptographically strong: they are only required to be ϵ-*differentially-uniform*, i.e., their maximum differential probability (MDP) or maximum expected differential probability (MEDP) is at most ϵ. Since we have assumed that the AXP is derived from the block cipher we intend to use, not all block ciphers can be used for our MACs. However, a keyed permutation with small MEDP can be obtained as a reduced-round of well-designed block ciphers, since such permutations are essential components of the block ciphers that are secure against differential cryptanalysis. For example, the MEDP of the 4-round AES with independent round keys is very small [26,18] and thus our proposals can be securely implemented using AES and 4-round AES.

We propose two approaches. They have different characteristics regarding the amount of preprocessing, memory consumption, and the speed for long and short messages. The first approach is based on the modified tree hash (MTH), which was proposed by Boesgaard et al. [9]. It was an improvement on the well-known tree hash [10,27]. Although they used the Length Annotation (LA) [19] to handle variable message lengths, we demonstrate that this is redundant. Removing LA from MTH improves efficiency, particularly for short messages.

The second uses chaining of the block cipher and the AXP. This is similar to the CBC-MAC, but the block cipher is only called for every d message blocks, where d is a parameter that determines the amount of preprocessing and MAC speed. This scheme is provably secure if the AXP is ϵ-differentially-uniform (for small ϵ) and satisfies an additional weak condition.

If our MACs are built using AES and its 4-round, we have MACs 1.4 to 2.5 times faster than the CBC-MAC-AES, depending on the scheme we use. The key length is short (one block cipher key, K, or K and an additional one-block key), and only one block cipher keyscheduling is needed. Their preprocessing times are moderate (log-order of the message length for the first approach, and constant for the second). We also show a software implementation of our MACs and comparisons between other MACs.

2   Preliminaries

Notations. $\{0,1\}$ is denoted by Σ and n-bit space is denoted by Σ^n. $(\Sigma^n)^{\leq m}$ denotes the set of binary sequences whose lengths are multiples of n and at most nm. $(\Sigma^n)^+$ is the set of all binary sequences whose lengths are multiples of n, and Σ^* is the set of all finite-length binary sequences. If X is distributed independently and uniformly over set $\mathcal{X}$, we write $X \in_U \mathcal{X}$. If F is a keyed function with domain $\mathcal{X}$, and range $\mathcal{Y}$, and key $K \in_U \mathcal{K}$, then we write $F : \mathcal{X} \to \mathcal{Y}$ and there is function $f : \mathcal{K} \times \mathcal{X} \to \mathcal{Y}$ such that $\Pr[F(x) = y] = |\{k \in \mathcal{K} : f(k,x) = y\}|/|\mathcal{K}|$. If we want to emphasize that F's key is K, F_K is written and if K is fixed to k, then F_k denotes function $f(k,*)$. Keyed and fixed functions are written by upper and lower case letters, respectively.

Definition 1. *Keyed function* $F \in_U \{f : \Sigma^n \to \Sigma^m\}$ *is called an n-bit to m-bit uniformly random function (URF) and denoted by* $R_{n,m}$. *If* F *is uniform over all n-bit permutations, it is called an n-bit uniformly random permutation (URP) and denoted by* P_n. *Specifically,* $R_{*,n}$ *denotes the Variable-Input-Length (VIL)-URF such that* $R_{*,n} \in_U \{f : \Sigma^* \to \Sigma^n\}$. *Here, VIL means that it accepts inputs of all lengths.*

We express the elements of field $GF(2^n)$ by the n-bit coefficient vectors of the polynomials in the field. We alternatively represent n-bit coefficient vectors by integers $0, 1, \ldots, 2^n - 1$, e.g., 2 corresponds to the coefficient vector $(00 \ldots 010)$ and 1 denotes $(00 \ldots 01)$, i.e., the identity element.

Definition 2. *Let* f *be a permutation on group* $\mathcal{X}$ *and* F_K *be a keyed permutation on* $\mathcal{X}$ *with key* $K \in_U \mathcal{K}$. *The maximum differential probability (MDP) for* f, *denoted by* $\mathrm{MDP}(f)$, *is* $\max_{a \neq 0, b} \Pr(f(X) - f(X + a) = b)$, *where* $X \in_U \mathcal{X}$. *Similarly, the maximum expected differential probability (MEDP) of* F_K *is defined as* $\mathrm{MEDP}(F_K) \overset{\mathrm{def}}{=} \max_{a \neq 0, b} \Pr(F_K(X) - F_K(X + a) = b)$, *which can also be written as* $\max_{a \neq 0, b} \sum_{k \in \mathcal{K}} \Pr(F_k(X) - F_k(X + a) = b)/|\mathcal{K}|$.

If $\mathcal{X}$ is a field with characteristic 2 (say, $GF(2^n)$), then the addition and subtraction in Def. 2 correspond to the bitwise XOR operation, i.e., $\oplus$. In this case, MDP is always no less than $2/2^n$. However, this does not hold true for MEDP.

Definition 3. *Let* H *be a keyed function:* $(\Sigma^n)^{\leq l} \to \Sigma^n$. *The maximum collision probability of* H *for a pair of m-block and m'-block input is defined as*

$$\mathrm{Coll}_H(m, m') \overset{\mathrm{def}}{=} \max_{x \in (\Sigma^n)^m, x' \in (\Sigma^n)^{m'}, x \neq x'} \Pr(H(x) = H(x')), \text{ where } m, m' \leq l.$$

If the collision probability is no more than ϵ for all possible inputs, it is called an ϵ-almost universal (ϵ-AU) hash function.

Security Notions. We used a standard security notion for symmetric cryptography [5,6,13].

Definition 4. *Let* F *and* G *be two keyed functions. Let us assume that the oracle has implemented* H, *which is identical to one of* F *or* G. *An adversary,* $\mathcal{A}$, *guesses if* H *is* F *or* G *using Chosen-plaintext attack (CPA). The maximum CPA-advantage in distinguishing* F *from* G *is defined as*

$$\mathrm{Adv}_{F,G}^{\mathrm{cpa}}(q, t, \sigma) \overset{\mathrm{def}}{=} \max_{\mathcal{A}:(q,t,\sigma)\text{-CPA}} \left| \Pr(\mathcal{A}^F = 1) - \Pr(\mathcal{A}^G = 1) \right|, \tag{1}$$

where $\mathcal{A}^F = 1$ denotes that $\mathcal{A}$'s guess is 1, which indicates one of F or G, and (q, t, σ)-CPA denotes a CPA that uses q queries with time complexity t, and the total length of q queries is at most σn bits. Instead of σ, we can use ρ, which denotes the maximum length (in n-bit block) of each query, to limit the adversary's resources. We omit σ and ρ if F and G have fixed input length, since they have been determined from q in this case. Also, we omit t if we consider *attacks without computational restrictions. Especially, if $F : \Sigma^m \to \Sigma^n$ we have* $\mathrm{Adv}_F^{\mathrm{prf}}(q,t) \overset{\mathrm{def}}{=} \mathrm{Adv}_{F,R_{m,n}}^{\mathrm{cpa}}(q,t)$. *Similarly, if F is an n-bit keyed permutation, then* $\mathrm{Adv}_F^{\mathrm{prp}}(q,t) \overset{\mathrm{def}}{=} \mathrm{Adv}_{F,P_n}^{\mathrm{cpa}}(q,t)$. *Finally, if F is a VIL keyed function:$\Sigma^* \to \Sigma^n$, then* $\mathrm{Adv}_F^{\mathrm{vilprf}}(q,t,\kappa) \overset{\mathrm{def}}{=} \mathrm{Adv}_{F,R_{*,n}}^{\mathrm{cpa}}(q,t,\kappa)$, *where κ is σ or ρ.*

If $\mathrm{Adv}_{F,R_{m,n}}^{\mathrm{cpa}}(q,t)$ is small for any sufficiently large q and t, F is called the pseudorandom function (PRF) [13]. The pseudorandom permutation (PRP) and VIL-PRF are defined similarly. As a VIL-PRF is also a secure VIL-MAC (e.g., see Proposition 2.7 of [5]), we focus on building VIL-PRFs.

3   Basic Idea

Let g be a (possibly keyed) function with n-bit domain and X be an n-bit random variable. Then, $g^{\oplus X}$ denotes a function such that $g^{\oplus X}(a) = g(a \oplus X)$. All our MACs are based on the following function.

Lemma 1. *We define the Add-Permute-Add (APA) function: $(\Sigma^n)^{\leq 2} \to \Sigma^n$ as follows.*

$$\mathrm{APA}_{K,F}(x) = \begin{cases} x & \text{if } x \in \Sigma^n, \\ F^{\oplus K}(x_1) \oplus x_2 & \text{if } x = (x_1, x_2) \in (\Sigma^n)^2, \end{cases} \quad (2)$$

where $K \in_U \Sigma^n$ and F is an n-bit (keyed or fixed) permutation. Then, $\mathrm{APA}_{K,F}$ is ϵ-AU if F's MEDP (for the case of keyed permutation) or MDP (for the case of fixed permutation) is at most ϵ.

Proof. Let $x = (x_1, x_2)$ and $x' = (x_1', x_2')$ be two different inputs to $\mathrm{APA}_{K,F}$. If $x_1 \neq x_1'$, then the output collision means $F^{\oplus K}(x_1) \oplus F^{\oplus K}(x_1') = x_2 \oplus x_2'$, which cannot occur with a probability larger than ϵ. If $x_1 = x_1'$, which implies $x_2 \neq x_2'$, then clearly the collision probability is zero. Moreover, the probability of $x_1 = F^{\oplus K}(x_1') \oplus x_2'$ is $1/2^n$, since F is invertible. Thus, the maximum collision probability is at most ϵ.

An example of a permutation with small MDP is the following.

Example 1. Let inv be an n-bit permutation such that $\mathrm{inv}(x) = x^{-1}$ for $x \neq 0$ and $\mathrm{inv}(0) = 0$, where x^{-1} satisfies $x \cdot x^{-1} = 1$. Then, $\mathrm{MDP}(\mathrm{inv}) = 4/2^n$ [25].

As this example shows, a fixed permutation with small MDP does exist. However, this cannot be efficiently implemented if input is large (say, more than 32-bit). In contrast, a keyed permutation with small MEDP is more practical. For example, the 4-round AES with independent round keys has an MEDP of less than 2^{-113} [18]. In later sections, this fact enables us to implement our MACs using AES.

As stated in the Introduction, all our MACs combine an n-bit block cipher, E_K, and an n-bit additional keyed or fixed permutation, which is called the auxiliary permutation (AXP). If the AXP has a key, we denote it by G_U, where key $U \in_U \mathcal{U}$. The sequence of m AXPs are denoted by $\mathbf{G} = (G_{U_1}, \ldots, G_{U_m})$. We will abbreviate G_{U_i} to G_i unless it is confusing. Hereafter, we will usually assume that the AXP is a keyed permutation. Since a fixed permutation can be seen as a keyed permutation with a single-point key space, this provides general descriptions of our schemes.

4   Building Variable Input Length Universal Hash

4.1   Modified Tree Hash

As Lemma 1 shows, a double input length AU hash function can be built using one invocation of a differentially-uniform permutation and an n-bit random key. The simplest way to expand the input length of this AU hash is using the well-known tree hash. The original tree hash proposed by Wegman and Carter [27] required some redundant calls of AU hash when the length of an input was not $2^l n$ for some positive integer l. An improvement to remove these redundant calls was proposed by Boesgaard et al., which is as follows.

Definition 5. *(Modified Tree Hash (MTH) [9], the binary case)*
Let $\mathbf{H} = (H_1, H_2, \ldots,)$ *be an infinite sequence of keyed functions:* $(\Sigma^n)^2 \to \Sigma^n$.
Let $x = (x_1, \ldots, x_m) \in (\Sigma^n)^m$. *For all* $i \geq 1$, *let* L_{H_i} *be a function defined as:*

$$L_{H_i}(x) = \begin{cases} H_i(x_1, x_2) \| H_i(x_3, x_4) \| \ldots \| H_i(x_{m-1}, x_m) & \text{if } m \bmod 2 = 0, \\ H_i(x_1, x_2) \| H_i(x_3, x_4) \| \ldots \| H_i(x_{m-2}, x_{m-1}) \| x_m & \text{if } m \bmod 2 = 1. \end{cases}$$

The output of the modified tree hash using $\mathbf{H}$ *for input* x *is*

$$\text{MTH}_{\mathbf{H}}(x) = L_{H_b} \circ L_{H_{b-1}} \circ \cdots \circ L_{H_1}(x), \text{ where } b = \lceil \log_2 m \rceil.$$

Here, $\circ$ denotes the serial composition (i.e., $F_2 \circ F_1(x) = F_2(F_1(x))$).

Collision Probability of MTH. The collision probability of MTH for equal length inputs was proved [9]. To handle inputs with unequal lengths, Boesgaard et al. suggested using a technique called the Length Annotation (LA), i.e., appending the length information of x to $\text{MTH}_{\mathbf{H}}(x)$. However, we here prove that LA is not needed, if some additional conditions are satisfied.

Lemma 2. *In Def. 5, if each* H_i *is independent* ϵ-*AU and satisfies* $\Pr(H_i(x) = y) = 1/2^n$ *for any* $x \in (\Sigma^n)^2$ *and* $y \in \Sigma^n$, *then*

$$\text{Coll}_{\text{MTH}_{\mathbf{H}}}(m, m') \leq \max\{\lceil \log_2 m \rceil, \lceil \log_2 m' \rceil\} \cdot \epsilon, \text{ for any } (m, m'). \quad (3)$$

Moreover, if $H_i = \text{APA}_{K_i, G_{U_i}}$ *and* K_i *and* U_i *are independent and random, then Eq.* (3) *holds, where* ϵ *is the MEDP of* G_{U_i} *and* $1/2^n \leq \epsilon$.

Proof. Let us prove the first claim. We start with the case for inputs with equal lengths. Let us abbreviate $\max_{2^{c-1} < i \leq 2^c} \text{Coll}_{\text{MTH}_{\mathbf{H}}}(i, i)$ to $p_c^=$. Clearly $p_0^= = 0$ and $p_1^= \leq \epsilon$ hold. Assume the claim holds for $c = i - 1$ for some $i \geq 1$. Let

$x = (x_1, \ldots, x_m)$ and $x' = (x'_1, \ldots, x'_m)$ be two m-block inputs where $2^{i-1} < m \leq 2^i$. Let S, T, and V denote $\text{MTH}_\mathbf{H}(x_1, \ldots, x_{2^{i-1}})$, $\text{MTH}_\mathbf{H}(x_{2^{i-1}+1}, \ldots, x_m)$, and $\text{MTH}_\mathbf{H}(x)$, respectively. For x', S', T', and V' are similarly defined. If the first 2^{i-1}-block prefixes of x and x' are identical, then $P(S = S') = 1$ and we have

$$P(V = V') \leq P(T = T', S = S') + P(V = V'|T \neq T', S = S') \leq (i-1)\epsilon + \epsilon = i\epsilon,$$

where the last inequality follows from the assumption and the fact that each H_i is independent. If the first 2^{i-1}-block prefixes are different, we have

$$P(V = V') \leq P(S = S') + P(V = V'|S \neq S') \leq (i-1)\epsilon + \epsilon = i\epsilon.$$

Thus, we have $p_i^= \leq i\epsilon$. Let $p_c^{\neq}$ be the maximum collision probability for two inputs that have unequal lengths and their lengths are at most 2^c blocks. Then, $p_1^{\neq} \leq \epsilon$ follows from the condition of H_i (note that $1/2^n \leq \epsilon$). Let us assume $p_{i-1}^{\neq} \leq (i-1)\epsilon$ holds. Let $x = (x_1, \ldots, x_m)$ and $x' = (x'_1, \ldots, x'_{m'})$ where $m < m'$ and $2^{i-1} < m' \leq 2^i$. If $m < 2^{i-1}$, the computation of V from (S', T') involves the key (for some H_i), $\tilde{K}$, that never appears in the computation of V. If we fix keys other than $\tilde{K}$, the collision of $\text{MTH}_\mathbf{H}(x)$ and $\text{MTH}_\mathbf{H}(x')$ is equivalent to the event that $H_i(s, t) = v$ for some (s, t, v, i), thus occuring with probability $1/2^n$. If $m > 2^{i-1}$, then we prove the collision probability is at most $i\epsilon$ in a similar way to the case of equal length. Therefore we have $p_i^{\neq} \leq i\epsilon$ and the first claim is proved. The second claim follows from the first claim and Lemma 1.

If LA is used, then more AU hash function calls are needed to obtain an n-bit hash value from $(\text{length}(x)\|\text{MTH}_\mathbf{H}(x))$. Therefore, removing LA contributes to faster speed (particularly for short messages) and shorter key length.

4.2 Periodic CBC Hash

The MTH is ideally fast, as its theoretical throughput is almost the same as that of the AXP. However, the amounts of preprocessing and working memories required are proportional to b, where 2^b is the maximum message block length. This implies that MTH is not well suited to constrained (e.g., low-power and/or memory) environments. This problem is common to all tree-based MAC functions. In this section, we focus on building AU hash functions that accept any long block inputs with a small constant amount of preprocessing and memory. Interestingly, our proposal is an iterative procedure similar to the CBC-MAC. Since this is iterative, only a small constant working memory is needed for any input in $(\Sigma^n)^+$, as in the CBC-MAC.

For $i = 1, 2, \ldots, m$, let F_i be an n-bit block keyed function and Z be an n-bit random variable. Let $x = (x_1, \ldots, x_{m+1})$. We define two keyed functions: $(\Sigma^n)^+ \to \Sigma^n$ such that

$$\text{Ch}[F_1, \ldots, F_m](x) \stackrel{\text{def}}{=} x_{m+1} \oplus F_m(x_m \oplus F_{m-1}(\ldots F_2(x_2 \oplus F_1(x_1))\ldots)), \quad \text{and}$$

$$\text{Ch}[F_1, \ldots, F_m|Z](x) \stackrel{\text{def}}{=} \text{Ch}[F_1, \ldots, F_m](x'), \quad \text{where } x' = (x_1 \oplus Z, x_2, \ldots, x_{m+1}),$$

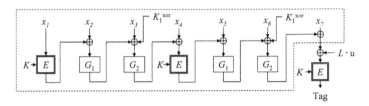

Fig. 1. PC-MAC with $d = 2$. System surrounded by dotted lines denotes $\mathrm{PCH}_2[E_K, \mathbf{G}]$.

i.e., CBC-MAC-like chaining. If the input is longer than $(m + 1)$ blocks, the chaining is iterated using $(F_1, \ldots, F_m)$, and they terminate as soon as the last input block is XORed. Here, the CBC-MAC$[F]$ corresponds to $F \circ \mathrm{Ch}[F]$. For one block input $x = x_1$, the output is itself, i.e., x_1.

Definition 6. *Let E_K be an n-bit block cipher. For $d \geq 0$, let $\mathbf{G} = (G_1, \ldots, G_d)$ be the sequence of d AXPs $\mathbf{G} = (G_1, \ldots, G_d)$ (recall that G_{U_i} has been abbreviated to G_i). We call d the interval. We assume that $(d - 1)$ n-bit keys, denoted by $K_1^{\mathrm{xor}}, \ldots, K_{d-1}^{\mathrm{xor}}$, are available. The Periodic CBC Hash (PCH) with interval d is a keyed function: $(\Sigma^n)^+ \to \Sigma^n$ defined as*

$$\mathrm{PCH}_d[E_K, \mathbf{G}] \stackrel{\mathrm{def}}{=} \mathrm{Ch}[E_K, G_1, G_2^{\oplus K_1^{\mathrm{xor}}}, \ldots, G_d^{\oplus K_{d-1}^{\mathrm{xor}}}].$$

Here, $\mathrm{PCH}_d[E_K, \mathbf{G}]$ terminates as soon as the last input block is XORed.

See Fig. 1 for an example of PCH. If $d = 1$, then no K_i^{xor} is used and E_K and G_1 are called alternately.

Collision Probability of PCH. For any inputs in $(\Sigma^n)^+$, the collision probability of the PCH is small if the AXP has a small differential probability and a small *self-differential* probability, which is defined as follows.

Definition 7. *The maximum self-differential probability (MSDP) of a permutation on group $\mathcal{X}$, f, is defined as $\mathrm{MSDP}(f) \stackrel{\mathrm{def}}{=} \max_{a \in \mathcal{X}} \Pr[X - f(X) = a]$, where $X \in_{\mathrm{U}} \mathcal{X}$. For a keyed permutation, the maximum expected SDP (MESDP) is similarly defined.*

Let $K_{\mathrm{aux}} = (K_1^{\mathrm{xor}}, \ldots, K_{d-1}^{\mathrm{xor}}, U_1, \ldots, U_d)$, where $U_i \in \mathcal{U}$ is the key for G_i. K_{aux} is the key of $\mathbf{G}^{\oplus} \stackrel{\mathrm{def}}{=} (G_1, G_2^{\oplus K_1^{\mathrm{xor}}}, \ldots, G_d^{\oplus K_{d-1}^{\mathrm{xor}}})$ and distributed over $\mathcal{K}_{\mathrm{aux}} \stackrel{\mathrm{def}}{=} (\Sigma^n)^{d-1} \times \mathcal{U}^d$. This determines the operation between two consecutive block cipher calls in PCH. The collision probability of PCH is proved as follows.

Lemma 3. *If $K_{\mathrm{aux}} \in_{\mathrm{U}} \mathcal{K}_{\mathrm{aux}}$, and $\mathrm{MEDP}(G_i) \leq \epsilon_{\mathrm{dp}}$ and $\mathrm{MESDP}(G_i) \leq \epsilon_{\mathrm{sdp}}$ for $i = 1, \ldots, d$, then,*

$$\mathrm{Coll}_{\mathrm{PCH}_d[R, \mathbf{G}]}(m, m') \leq d\epsilon_{\mathrm{dp}} + \epsilon_{\mathrm{sdp}} + \frac{(l + l')^2 + 2}{2^{n+1}}, \tag{4}$$

where $\mathbf{G} = (G_1, \ldots, G_d)$, and R is the n-bit URF, and $l = \lceil \frac{m}{d+1} \rceil$ and $l' = \lceil \frac{m'}{d+1} \rceil$.

Proof. Let $x = (x_1, \ldots, x_m)$ and $x' = (x'_1, \ldots, x'_{m'})$ be two distinct inputs for $\mathrm{PCH}_d[R, \mathbf{G}]$ with $m \leq m'$ and let $V = \mathrm{PCH}_d[R, \mathbf{G}](x)$, $V' = \mathrm{PCH}_d[R, \mathbf{G}](x')$. Let Y_i and Z_i (Y'_i and Z'_i) be the i-th input and output of R for x (for x'). For example, when $d = 1$, then $Y_1 = x_1$, $Z_1 = R(Y_1)$, and $Y_2 = x_3 \oplus G_1(x_2 \oplus Z_1)$. Here, if $m - 1 = c(d + 1)$ for some positive integer c, then Y_{c+1} corresponds to $\mathrm{PCH}_d[R, \mathbf{G}](x)$ and Z_{c+1} does not exist. We also assume that the block length of the longest common prefix (LCP) between x and x' is m_{lcp}. That is, $x_i = x'_i$ for $i = 1, \ldots, m_{\mathrm{lcp}} < m$ and $x_{m_{\mathrm{lcp}}+1} \neq x'_{m_{\mathrm{lcp}}+1}$ or $m_{\mathrm{lcp}} = m$ (if $m < m'$). If $x_1 \neq x'_1$, the LCP is empty and has a length of 0. Let $l_{\mathrm{lcp}} = \lceil \frac{m_{\mathrm{lcp}}}{d+1} \rceil$ (this means that $Y_i = Y'_i$ for $i = 1, \ldots, l_{\mathrm{lcp}}$ with probability 1).

Let D be an event where $Y_\alpha, Y'_\beta, 1 \leq \alpha \leq l, 1 \leq \beta \leq l'$ are distinct, except for the trivial collisions $Y_\gamma = Y'_\gamma$ for $\gamma = 1, \ldots, l_{\mathrm{lcp}}$. In addition, let D_{lcp} denote an event where $Y_i \neq Y_j$ (and $Y'_i \neq Y'_j$) for $1 \leq i < j \leq l_{\mathrm{lcp}}$. If the LCP is empty, we define $\Pr(D_{\mathrm{lcp}}) = 1$. Clearly D_{lcp} is a subevent of D. For any $1 \leq i \leq j \leq d+1$, we have

$$\mathrm{Coll}_{\mathrm{Ch}[\mathbf{G}^\oplus|\mathrm{Rnd}]}(i, j) \leq \begin{cases} (i-1)\epsilon_{\mathrm{dp}} \leq d\epsilon_{\mathrm{dp}} & \text{if } i = j, \\ \epsilon_{\mathrm{sdp}} & \text{if } (i,j) = (1,2), \\ \frac{1}{2^n} & \text{otherwise,} \end{cases} \tag{5}$$

where $\mathrm{Rnd} \in_{\mathsf{U}} \Sigma^n$. In Eq. (5), the first case follows from simple inductive analysis. For the second, note that the collision means $\mathrm{Rnd} \oplus x_1 = x'_2 \oplus G_1(\mathrm{Rnd} \oplus x'_1)$, which occurs with probability (at most) ϵ_{sdp}. The third follows from the fact that the output for the longer input always includes K_j^{xor}, which does not appear in the other output, and that all AXPs are invertible. Here, we show that the probabilities of $\overline{D}$ and $\overline{D_{\mathrm{lcp}}}$ are negligible, where $\overline{D}$ is the negation of D.

Lemma 4. *For any $k_{\mathrm{aux}} \in \mathcal{K}_{\mathrm{aux}}$, we have*

$$\Pr(\overline{D_{\mathrm{lcp}}}) = \Pr(\overline{D_{\mathrm{lcp}}}|K_{\mathrm{aux}} = k_{\mathrm{aux}}) \leq \sum_{i=1}^{l_{\mathrm{lcp}}-1} \frac{i}{2^n} \leq \frac{l_{\mathrm{lcp}}^2}{2^{n+1}}, \quad and \tag{6}$$

$$\Pr(\overline{D}) \leq d\epsilon_{\mathrm{dp}} + \frac{(l+l')^2}{2^{n+1}}. \tag{7}$$

The proof of Lemma 4 is in Appendix A. Next, let us analyze the collision probability of $\mathrm{PCH}_d[R, \mathbf{G}]$. Let x_{last} be the last $m - 1 \bmod (d+1)$ blocks of x. Then, $V = \mathrm{Ch}[\mathbf{G}^\oplus|Z_l](x_{\mathrm{last}})$ if x_{last} is not empty (i.e., $m - 1 \bmod (d+1) \geq 1$), and $V = Y_l$ otherwise. x'_{last} is similarly defined for x'. First, we assume that $l = l' = l_{\mathrm{lcp}}$ does not hold true. Then, the occurrence of D means that Z_l and $Z_{l'}$ are independent and uniformly random even if K_{aux} is fixed. Thus, we have

$$\Pr(V = V') \leq \sum_{k_{\mathrm{aux}} \in \mathcal{K}_{\mathrm{aux}}} \Pr(V = V'|K_{\mathrm{aux}} = k_{\mathrm{aux}}, D) \cdot \Pr(K_{\mathrm{aux}} = k_{\mathrm{aux}}|D) + \Pr(\overline{D}),$$

$$\leq \frac{1}{2^n} \sum_{k_{\mathrm{aux}} \in \mathcal{K}_{\mathrm{aux}}} \Pr(K_{\mathrm{aux}} = k_{\mathrm{aux}}|D) + d\epsilon_{\mathrm{dp}} + \frac{(l+l')^2}{2^{n+1}} \leq d\epsilon_{\mathrm{dp}} + \frac{(l+l')^2 + 2}{2^{n+1}}, \tag{8}$$

unless both x_{last} and x'_{last} are empty. If both are empty, then $\Pr(V = V') \leq P(\overline{D})$ holds. Next, let us assume $l = l' = l_{\text{lcp}}$ holds true. In this case, D is equivalent to D_{lcp} and at least one of x_{last} or x'_{last} is not empty (otherwise we have $x = x'$), and Z_l $(= Z'_{l'})$ is independent and random if D_{lcp} is given. If both x_{last} and x'_{last} are not empty, then $\Pr(V = V'|D_{\text{lcp}})$ equals $\Pr(\text{Ch}[\mathbf{G}^{\oplus}|Z_l](x_{\text{last}}) = \text{Ch}[\mathbf{G}^{\oplus}|Z_l](x'_{\text{last}})|D_{\text{lcp}})$. Here, note that D_{lcp} (or $\overline{D_{\text{lcp}}}$) gives no information on K_{aux}, since Eq. (6) implies $P(K_{\text{aux}} = k_{\text{aux}}|D_{\text{lcp}}) = P(K_{\text{aux}} = k_{\text{aux}})$ for all k_{aux}. From these observations and Eq. (5), and Lemma 4, we have

$$\Pr(V = V') \leq \Pr(V = V'|D_{\text{lcp}}) + \Pr(\overline{D_{\text{lcp}}}) \leq \max\{d\epsilon_{\text{dp}}, \epsilon_{\text{sdp}}, 1/2^n\} + \frac{l_{\text{lcp}}^2}{2^{n+1}}. \quad (9)$$

It is easy to see that Eq. (9) also holds even if one of x_{last} or x'_{last} is empty. Thus, Eq. (9) holds when $l = l' = l_{\text{lcp}}$. We conclude the proof by combining Eqs. (8) and (9).

Relation Between MDP (MEDP) and MSDP (MESDP). It seems that every permutation with a small MDP has a small MSDP, though we have not formally proved this for now. For instance, the inv permutation in Ex. 1 has MSDP $3/2^n$. A more useful fact is that any n-bit keyed permutation that contains an independent key-addition layer has MESDP $1/2^n$, as the output is completely random and independent of the input.

5 Complete Description of Our MACs and Their Securities

The following lemma, proved by Black and Rogaway [8], demonstrates that a VIL-PRF: $\Sigma^* \to \Sigma^n$ can be built with an AU hash: $(\Sigma^n)^+ \to \Sigma^n$ and n-bit PRFs.

Lemma 5. *(Lemma 2 of [8]) Let $H : (\Sigma^n)^+ \to \Sigma^n$ and R, R' be two independent n-bit URFs. We define $\text{CW3}[H, R, R'](x) = R(H(x))$ if the length of x, $|x|$, is a multiple of n, and $R'(H(x \parallel 10^l))$ otherwise, where $|x| \bmod n = n - l - 1$ and 10^i denotes an i-bit sequence $(100\ldots0)$. Then,*

$$\text{Adv}_{\text{CW3}[H,R,R']}^{\text{vilprf}}(q, \sigma) \leq \max_{q, m_1, \ldots, m_q, \Sigma_{s=1}^q m_s = \sigma} \left\{ \sum_{1 \leq i < j \leq q} \text{Coll}_H(m_i, m_j) \right\} \quad (10)$$

holds. In Eq. (10), if σ is substituted with ρ, then the maximum is taken for all $(q, m_1, \ldots, m_q)$ such that $m_s \leq \rho$ for all $s = 1, \ldots, q$.

The Hash-to-MAC (actually Hash-to-PRF) conversion described in Lemma 5 requires two additional n-bit PRFs and thus requires two additional block cipher keyschedulings in practice. However, these keyschedulings can be removed using the idea of tweakable block ciphers [21]. This technique was used to propose the XCBC [8], TMAC [20], and OMAC [16]. In converting our hashing schemes into MACs, we also employed the tweaking technique. Here, we present complete

Preprocessing	Let L be $E_K(0)$.
	Let $\mathbf{U} = (U_1, \ldots, U_b)$ be the first $b\lvert \mathcal{U}\rvert$ bits of $E_K(1) \cdots E_K(a)$.
	Let $\mathbf{H} = (H_1, \ldots, H_b)$, where $H_i = \text{APA}_{E_K(i+a), G_{U_i}}$.
Tag Computation	Input message $x \in \Sigma^*$.
	Let $\text{Tag} = \text{CW3}[\text{MTH}_{\mathbf{H}}, E_K^{\oplus L \cdot \mathbf{u}}, E_K^{\oplus L \cdot \mathbf{u}^2}](x)$.
	Output (x, Tag).

Fig. 2. MT-MAC$_b[E_K\lvert G_U]$. Key of MAC is K, AXP is G_U, and $a = \lceil b\lvert \mathcal{U}\rvert/n\rceil$.

Preprocessing	Let $\mathbf{U} = (U_1, \ldots, U_d)$ be the first $d\lvert \mathcal{U}\rvert$ bits of
	$E_K^{\oplus L}(0) \cdots E_K^{\oplus L}(\hat{a} - 1)$, and let $\mathbf{G} = (G_1, \ldots, G_d)$.
	Let $K_{j-\hat{a}+1}^{\text{xor}}$ be $E_K^{\oplus L}(j)$ for $j = \hat{a}, \ldots, \hat{a} + d - 2$.
Tag Computation	Input message $x \in \Sigma^*$.
	Let $\text{Tag} = \text{CW3}[\text{PCH}_d[E_K, \mathbf{G}], E_K^{\oplus L \cdot \mathbf{u}}, E_K^{\oplus L \cdot \mathbf{u}^2}](x)$.
	Output (x, Tag).

Fig. 3. PC-MAC$_d[E_K, L\lvert G_U]$. Key of MAC is (K, L), AXP is G_U, and $\hat{a} = \lceil d\lvert \mathcal{U}\rvert/n\rceil$.

descriptions of our MACs. The first is based on the MTH and called the MT-MAC. It uses a block cipher E_K and an AXP, G_U, and the maximum message length is $n2^b$ bits. See Fig. 2 for the details of MT-MAC. In Fig. 2, $i + a$ indicates usual integer addition, and $\mathbf{u}$ is an element of $\text{GF}(2^n)$ that is not 0 or 1 and $L \cdot \mathbf{u}$ denotes the multiplication on $\text{GF}(2^n)$. It can be implemented with shift and conditional XOR (e.g., see [16]). The second is based on PCH and called the PC-MAC. The PC-MAC is shown in Fig. 3.

The security of MT-MAC is proved as follows. The proof is in Appendix B.

Theorem 1. *Let* $c = \lceil b\lvert \mathcal{U}\rvert/n\rceil + b + 1$. *Then,*

$$\text{Adv}_{\text{MT-MAC}_b[E_K\lvert G_U]}^{\text{vilprf}}(q, t, \sigma) \leq \text{Adv}_{E_K}^{\text{prp}}(\sigma + c, t') + \frac{(\sigma + c)^2}{2^n} + \epsilon_{\text{dp}}\sigma^2,$$

where $t' = t + O(\sigma)$ *and* $\epsilon_{\text{dp}} = \text{MEDP}(G_U)$.

The security proof of PC-MAC can be similarly obtained, which is as follows.

Theorem 2. *Let* $c = \lceil d\lvert \mathcal{U}\rvert/n\rceil + d$, *where* d *is the interval parameter. Then,*

$$\text{Adv}_{\text{PC-MAC}_d[E_K, L\lvert G_U]}^{\text{vilprf}}(q, t, \rho) \leq \text{Adv}_{E_K}^{\text{prp}}(\rho q + c, t') + \frac{2.5(\rho q + c)^2}{2^n} + (d\epsilon_{\text{dp}} + \epsilon_{\text{sdp}})\frac{q^2}{2},$$

where $t' = t + O(\rho q)$, *and* $\epsilon_{\text{dp}} = \text{MEDP}(G_U)$, *and* $\epsilon_{\text{sdp}} = \text{MESDP}(G_U)$.

The proof of Theorem 2 is in Appendix C.

Security Parameter. In Theorem 2, we used ρ instead of σ, although using σ is generally more preferable than using ρ (see discussion in [15]). If we are forced

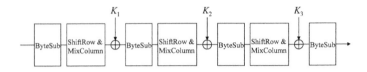

Fig. 4. The simplified 4-round AES. Each K_i is independent and random.

to use σ, the bound of a PC-MAC's CPA-advantage would be $O(\sigma^2 q^2/2^n)$, which seems a bit too loose. It would be nice if we could obtain a smaller bound for the collision probability to obtain a tight security analysis using σ.

Key Length. The PC-MAC uses two keys, the first for the block cipher and the second to make the block cipher tweakable. It is natural to ask whether this can be reduced to one block cipher key without introducing another block cipher keyscheduling. For example, is it secure to let $L = E_K(0)$, just like in the OMAC or MT-MAC? Unfortunately, we do not have a clear answer for now, but at least we found a counterexample if some generalization was applied to the PC-MAC[1]. Nevertheless, we think that a small change can provide a one-key version of the PC-MAC. This is still a problem that needs to be solved.

6   AES-Based Implementation

In this section, we consider the implementation of our MACs using AES. As mentioned earlier, the 4-round AES with independent round keys has MEDP 2^{-113} [18]. Also, MESDP is exactly 2^{-128}, since the 4-round AES contains an independent key-addition layer (see end of Sect. 4.2). Here, the addition of the first round key and the last diffusion layer can be omitted, since they do not affect the differential and self-differential probabilities. Let us denote the simplified 4-round AES in Fig. 4 by 4rAES. If our MACs are implemented with AES and 4rAES, then the securities of the resulting MT-MAC and PC-MAC can be proved by Theorems 1 and 2 with $n = 128$, $|\mathcal{U}| = 384$, $\epsilon_{\mathrm{dp}} = 2^{-113}$, and $\epsilon_{\mathrm{sdp}} = 2^{-128}$.

Some Comparisons. Compared with the previous MAC modes of AES, for example OMAC-AES, our AES-based MACs are faster (MT-MAC is about 2.5 times faster, and PC-MAC is about 1.4 to 2.5 times faster, depending on the interval). Both use the AES encryption, and do not require the AES decryption. Their program sizes are almost the same. Both provide stateless (i.e., no counter

---

[1] For example, even if $L \cdot \mathsf{u}$ is substituted with $\mathrm{inv}(L \oplus \mathsf{u})$ (note that inv is defined in Ex. 1), OMAC will still be secure, as it satisfies the condition for the "OMAC-family" (see [16] for details). However, if $L = E_K(0)$, and the AXP is the inv permutation, and the above substitution is applied to the PC-MAC with interval 1, then the tag for the 3-block input $(0, \mathsf{u}, x_2)$ is $E_K(x_2)$ for any x_2, i.e., direct access to E_K is possible. This means the complete break of the MAC.

Table 1. Summary of AES-based MACs. "Rounds" denotes average AES rounds to process one message block, and "Preproc." denotes AES encryption blocks needed in preprocessing.

MAC	Max.Message Length	Rounds	Preproc.	Key size	Type
MT-MAC$_b$[AES|4rAES]	$n2^b$	4	$4b+1$	128	Tree
PC-MAC$_d$[AES, L|4rAES]	Infinite	$4 + \frac{6}{d+1}$	$4d-1$	256	Iterative
OMAC-AES	Infinite	10	1	128	Iterative

or nonce is used) provably secure VIL-MACs. The computational assumptions we need are the same (i.e., the pseudorandomness of the AES). Drawbacks of our MACs are the amount of preprocessing and slightly-degraded security: many CBC-MAC variants have 64-bit security, i.e., they are secure if q (or σ, ρ) $\ll 2^{64}$, while ours have about 56-bit security. We have summarized the properties of our AES-based MACs below. For comparison, the OMAC-AES is also shown. Table 1 shows only average speed estimates for long messages. However, our MACs are at least as fast as OMAC-AES for any short messages, since the AES rounds needed by our MACs are no more than $10 \cdot m$, when the input is m-block and this holds for all $m \geq 1$.

It may be rather difficult to perform a rigorous comparison between our MACs and the state-of-art CW-MACs, such as UMAC [19], (the MAC part of) GCM [24], and Poly1305 [7], as they use customized functions that can not be derived from AES. For example, CW-MACs are roughly 3 to 5 (or more) times faster than the MAC modes using the optimized AES on software (e.g., see [3]). Therefore our MACs may not be as fast as them on software. Also, some CW-MACs have much shorter keyscheduling time than ours. However, ours can be easily implemented for any platform where an implementation of AES is available. There are many studies on efficient AES implementations for various software and hardware (e.g., see [1]), and we can directly benefit from them. For other comparison items, both provide provably secure VIL-MACs (some CW-MACs are stateful) based on the pseudorandomness of the AES.

The Pelican MAC [12] is an instantiation of the ALRED using AES and its 4-round with all zero round keys. It is similar (but not identical) to the PC-MAC. This is not surprising because the ALRED and the PC-MAC share the same motivation. The Pelican MAC is about 2.5 times faster than the CBC-MAC, thus almost the same speed as that of the MT-MAC or PC-MAC with a long

Table 2. Comparison of AES-based MAC speed on software

MAC	Tag computation (cycle/byte)	Preprocessing (cycles)
MT-MAC ($b = 32$)	12.5	53777 (estimate)
PC-MAC ($d = 1$)	18.5	1651
PC-MAC ($d = 5$)	14.4	8311
PC-MAC ($d = 17$)	13.1	28444
OMAC	25.1	821

interval. Compared to our MACs, the Pelican MAC's preprocessing time is very short (only one block encryption). From the preliminary analysis of the ALRED construction [11], the Pelican MAC's security was proved against attacks that did not invoke internal collisions. In addition, no attack better than the brute force search has not been found for the moment. However, it is still unclear whether the Pelican MAC is a provably secure (i.e., secure against all attacks) VIL-MAC based on the pseudorandomness of the AES.

These comparisons are rough and might be insufficient. As a future work item, we want to do a more comprehensive and quantitative comparison to clarify the effectiveness of our approach.

Implementations. We also implemented our AES-based MACs on software. We used the public-domain C code written by Rijmen et al.[2]. Our implementation was naive and almost no optimization was performed. We did a speed comparison on a Pentium III 1 Ghz, where raw AES encryption ran at about 25 cycle/byte. We can see from Table 2 that our MACs did not achieve the theoretical limit (i.e., 2.5 times faster than OMAC-AES). This is because some overhead was involved in both AES and 4rAES, such as byte/word conversion. The effect overhead has may change according to the platform and AES implementation.

Acknowledgments

We would like to thank Etsuko Tsujihara for implementing MACs and anonymous reviewers for useful and detailed comments.

References

1. http://www.iaik.tu-graz.ac.at/research/krypto/AES/index.php.
2. http://homes.esat.kuleuven.be/~rijmen/rijndael/rijndael-fst-3.0.zip.
3. http://cr.yp.to/streamciphers.html.
4. B. den Boer, J.P. Boly, A. Bosselaers, J. Brandt, D. Chaum, I. Damgård, M. Dichtl, W. Fumy, M. van der Ham, C.J.A. Jansen, P. Landrock, B. Preneel, G. Roelofsen, P. de Rooij, and J. Vandewalle, *RIPE Integrity Primitives*, final report of RACE Integrity Primitives Evaluation. 1995.
5. M. Bellare, J. Kilian, and P. Rogaway. "The Security of the Cipher Block Chaining Message Authentication Code." *Journal of Computer and System Science*, Vol. 61, No. 3, 2000.
6. M. Bellare, A. Desai, E. Jokipii, and P. Rogaway. " A Concrete Security Treatment of Symmetric Encryption." *Proceedings of the 38th Annual Symposium on Foundations of Computer Science*, FOCS '97, pp. 394-403, 1997.
7. D. J. Bernstein. "The Poly1305-AES Message-Authentication Code." *Fast Software Encryption*, FSE'05, LNCS 3557, pp. 32-49, 2005.
8. J. Black and P. Rogaway. "CBC MACs for Arbitrary-Length Messages: The Three-Key Constructions." *Advances in Cryptology- CRYPTO '00*, LNCS 1880, pp. 197-215, 2000.
9. M. Boesgaard, T. Christensen, and E. Zenner. "Badger - A Fast and Provably Secure MAC." *Applied Cryptography and Network Security- ACNS'05*, LNCS 3531, pp. 176-191, 2005.

10. L. Carter and M. Wegman. "Universal Classes of Hash Functions." *Journal of Computer and System Science*, Vol. 18, pp. 143-154, 1979.
11. J Daemen and V. Rijmen. "A New MAC Construction ALRED and a Specific Instance ALPHA-MAC." *Fast Software Encryption*, FSE'05, LNCS 3557, pp. 1-17, 2005.
12. J Daemen and V. Rijmen. "The Pelican MAC Function." *IACR ePrint Archive*, 2005/088.
13. O. Goldreich. "Modern Cryptography, Probabilistic Proofs and Pseudorandomness." Springer-Verlag, Algorithms and Combinatorics, Vol. 17, 1998.
14. S. Halevi and H. Krawczyk. "MMH:Software Message Authentication in the Gbit/second rates." *Fast Software Encryption*, FSE'97, LNCS 1267, pp. 172-189, 1997.
15. T. Iwata and K. Kurosawa. "Stronger Security Bounds for OMAC, TMAC, and XCBC." *Progress in Cryptology- INDOCRYPT'03*, LNCS 2904, pp. 402-415, 2003.
16. T. Iwata and K. Kurosawa. "OMAC: One-Key CBC MAC." *Fast Software Encryption- FSE'03*, LNCS 2887, pp. 129-153, 2003.
17. T. Iwata and K. Kurosawa. "On the Universal Hash Functions in Luby-Rackoff Cipher." *IEICE Transactions*, Volume 87–A, pp. 60-66, 2004.
18. L. Keliher and J. Sui. "Exact Maximum Expected Differential and Linear Probability for 2-Round Advanced Encryption Standard (AES)." *IACR ePrint Archive*, 2005/321.
19. T. Krovetz. "Software-Optimized Universal Hashing and Message Authentication". *PhD dissertation*, available from `http://www.cs.ucdavis.edu/~rogaway/umac`.
20. K. Kurosawa and T. Iwata. "TMAC: Two-Key CBC MAC." *Topics in Cryptology-CT-RSA 2003*, LNCS 2612, pp. 33-49, 2003.
21. M. Liskov, R. L. Rivest, and D. Wagner. "Tweakable Block Ciphers." *Advances in Cryptology- CRYPTO'02*, LNSC 2442, pp. 31-46, 2002.
22. M. Luby and C. Rackoff. "How to Construct Pseudo-random Permutations from Pseudo-random functions." *SIAM J. Computing*, Vol. 17, No. 2, pp. 373-386, 1988.
23. U. Maurer. "Indistinguishability of Random Systems." *Advances in Cryptology-EUROCRYPT'02*, LNCS 2332, pp. 110-132, 2002.
24. D. McGrew and J. Viega. "The Galois/Counter Mode of Operation (GCM)." *Submission to NIST Modes of Operation Process*, 2004.
25. K. Nyberg. "Differentially Uniform Mappings for Cryptography." *Advances in Cryptology- EUROCRYPT'93*, LNCS 765, pp. 55-64, 1994.
26. S. Park, S. H. Sung, S. Lee, and J. Lim. "Improving the Upper Bound on the Maximum Differential and the Maximum Linear Hull Probability for SPN Structures and AES." *Fast Software Encryption*, FSE'03, LNCS 2887, pp. 247-260, 2003.
27. M. Wegman and L. Carter. "New Hash Functions and Their Use in Authentication and Set Equality." *Journal of Computer and System Sciences*, Vol. 22, pp. 265-279, 1981.

A   Proof of Lemma 4

We assumed that LCP was not empty and $l_{\mathrm{lcp}} < l$ (i.e., $Y_{l_{\mathrm{lcp}}+1}$ and $Y'_{l_{\mathrm{lcp}}+1}$ exist). If we fix K_{aux} to k_{aux}, then the operation that accepts Z_i and outputs Y_{i+1} (i.e., $\mathrm{Ch}[\mathbf{G}^{\oplus}]$) is a deterministic n-bit permutation defined by k_{aux} and some $d + 1$ input blocks. It is not hard to see that $\Pr(\overline{D_{\mathrm{lcp}}}|K_{\mathrm{aux}} = k_{\mathrm{aux}})$ is exactly the same as the output collision probability of the OFB mode of R, for all $k_{\mathrm{aux}} \in \mathcal{K}_{\mathrm{aux}}$

and inputs. Then, Eq. (6) follows from this observation and a simple collision analysis. Next, let us prove Eq. (7). Consider the following collision events. (I): $Y_{l_{lcp}+1} = Y'_{l_{lcp}+1}$ and (II): $Y_{l_{lcp}+1} = Y_i, Y'_{l_{lcp}+1} = Y_i$ for $i = 1, \ldots, l_{lcp}$. Here, $Y_{l_{lcp}+1}$ and $Y'_{l_{lcp}+1}$ are two outputs of $\mathrm{Ch}[\mathbf{G}^{\oplus}|Z_{l_{lcp}}]$ with different inputs. Since $Z_{l_{lcp}}$ is independent and uniformly random if D_{lcp} is given, we can use Eq. (5) and obtain $\mathrm{Pr}(Y_{l_{lcp}+1} = Y'_{l_{lcp}+1}|D_{lcp}) \leq d\epsilon_{dp}$. Moreover, $\mathrm{Pr}(Y_{l_{lcp}+1} = Y_i|D_{lcp})$ (or $\mathrm{Pr}(Y'_{l_{lcp}+1} = Y_i|D_{lcp})$) is $1/2^n$ for $i = 1, \ldots, l_{lcp}$, since $Y_{l_{lcp}+1}$ and $Y'_{l_{lcp}+1}$ are permutations of $Z_{l_{lcp}}$. If no collision events of types (I) and (II) have occurred, $Z_{l_{lcp}+1}$ and $Z'_{l_{lcp}+1}$ are independent and completely random, no matter what K_{aux} is. This implies that other collision events consisting of $\overline{D}|D_{lcp}$ occur with probability $1/2^n$. Consequently, all collision events consisting of $\overline{D}|D_{lcp}$ occur with probability $1/2^n$ except for the event $Y_{l_{lcp}+1} = Y'_{l_{lcp}+1}$. By counting these events and using Eq. (6), we have

$$\mathrm{Pr}(\overline{D}) \leq \mathrm{Pr}(\overline{D_{lcp}}) + \mathrm{Pr}(\overline{D}|D_{lcp}),$$
$$\leq \frac{l_{lcp}^2}{2^{n+1}} + d\epsilon_{dp} + \frac{1}{2^n}\left(\binom{l+l'-l_{lcp}}{2} - \binom{l_{lcp}}{2} - 1\right) \leq d\epsilon_{dp} + \frac{(l+l')^2}{2^{n+1}}.$$

For other cases (e.g., the LCP is empty or $l_{lcp} = l$), the above bound also holds true. This proves Eq. (7).

B   Proof of Theorem 1

Let Q be $\mathrm{CW3}[\mathrm{MTH}_{\widetilde{\mathbf{H}}}, R, R']$, where $\widetilde{\mathbf{H}} = (\widetilde{H}_1, \widetilde{H}_2, \ldots, \widetilde{H}_b)$ consists of $\widetilde{H}_i = \mathrm{APA}_{\widetilde{K}_i, G_{\widetilde{U}_i}}$ and $\{\widetilde{U}_i, \widetilde{K}_i\}_{i=1,\ldots,b}$ are independent of each other, and R, R' are independent n-bit URFs. From Lemmas 2 and 5, we have

$$\mathrm{Adv}_Q^{\mathrm{vilprf}}(q, \sigma) \leq \epsilon_{dp} \cdot q\sigma \leq \epsilon_{dp}\sigma^2. \tag{11}$$

Then, we use the following lemma.

Lemma 6. *Let R be an n-bit URF. Let $\mathrm{TE}_1 : \Sigma^n \times \Sigma \to \Sigma^n$, where $\mathrm{TE}_1(x,0) = R^{\oplus L \cdot \mathbf{u}}(x)$ and $\mathrm{TE}_1(x,1) = R^{\oplus L \cdot \mathbf{u}^2}(x)$ for $L = R(0)$. Consider the following two games, Gm1 and Gm2. In Gm1, one can access TE_1 and $R(c_1), R(c_2), \ldots, R(c_a)$ where $c_1, c_2, \ldots, c_a$ are distinct and fixed constants and $c_i \neq 0$ for all i. In Gm2, one can access the URF: $\Sigma^n \times \Sigma \to \Sigma^n$ and a (an)-bit independent and random sequence. Then, $\mathrm{Adv}_{\mathrm{Gm1},\mathrm{Gm2}}^{\mathrm{cpa}}(q)^2$ is at most $\frac{q^2}{2^{n+1}} + \frac{(a+1)q}{2^n}$.*

Proof. (Sketch) Let S_i be the i-th input to R in Gm1, i.e., S_i equals $x_i \oplus L \cdot \mathbf{u}$ if the i-th query is $(x_i, 0)$ and $x_i \oplus L \cdot \mathbf{u}^2$ if the i-th query is $(x_i, 1)$. Let a_i be the event that $S_1, S_2, \ldots, S_i$ are distinct and $S_j \notin \{0, c_1, \ldots, c_a\}$ for $j = 1, \ldots, i$. Then, using the methodology of Maurer [23], $\mathrm{Adv}_{\mathrm{Gm1},\mathrm{Gm2}}^{\mathrm{cpa}}(q)$ is at most the probability

---

[2] This should be interpreted as the maximum CPA-advantage in distinguishing two games using q queries with no computational restriction, where a query is in $\Sigma^n \times \Sigma$.

of $\overline{a_q}$ for all (both adaptive and non-adaptive) adversaries using q queries when Gm1 is considered. All collision events consisting of $\overline{a_q}$ have probability $1/2^n$ or 0. By counting the number of collision events and using the union bound, we conclude the proof.

From Lemma 6, we have $\mathbf{Adv}^{\mathrm{cpa}}_{Q,\mathrm{MT\text{-}MAC}_b[R|G_U]}(q,\sigma) \leq \frac{\sigma^2}{2^{n+1}} + \frac{(c+1)\sigma}{2^n}$. From this observation and Eq. (11), we have

$$\mathbf{Adv}^{\mathrm{vilprf}}_{\mathrm{MT\text{-}MAC}_b[R|G_U]}(q,\sigma) \leq \mathbf{Adv}^{\mathrm{cpa}}_{Q,\mathrm{MT\text{-}MAC}_b[R|G_U]}(q,\sigma) + \epsilon_{\mathrm{dp}}\sigma^2 + \frac{\sigma^2}{2^{n+1}} + \frac{(c+1)\sigma}{2^n}.$$
(12)

Distinguishing $\mathrm{MT\text{-}MAC}_b[R|G_U]$ from $\mathrm{MT\text{-}MAC}_b[E_K|G_U]$ with (q,t,σ) implies distinguishing R from E_K with $\sigma + c$ queries and $t' = t + O(\sigma)$ time. Combining this observation and Eq. (12) and the standard PRF/PRP switching lemma (e.g., see Lemma 1 of [8]) proves the theorem.

C   Proof of Theorem 2

Let Q be $\mathrm{CW3}[\mathrm{PCH}_d[R,\mathbf{G}], R', R'']$ where three n-bit URFs $R, R',$ and R'' are independent and the auxiliary key K_{aux} is generated by the counter mode of another URF, R''', i.e., $K_{\mathrm{aux}} \in_U \mathcal{K}_{\mathrm{aux}}$. Combining Lemmas 5 and 3, we have

$$\mathbf{Adv}^{\mathrm{vilprf}}_Q(q,\rho) \leq \max_{q,m_1,\ldots,m_q,m_s \leq \rho} \sum_{1 \leq i < j \leq q} d\epsilon_{\mathrm{dp}} + \epsilon_{\mathrm{sdp}} + \frac{(\lceil \frac{m_i}{d+1} \rceil + \lceil \frac{m_j}{d+1} \rceil)^2 + 2}{2^{n+1}}$$

$$\leq \left(d\epsilon_{\mathrm{dp}} + \epsilon_{\mathrm{sdp}} + \frac{2\rho^2 + 1}{2^n} \right) \frac{q^2}{2}.$$
(13)

Then, the following lemma is used. It is similar to Lemma 4.1 of [15].

Lemma 7. *Let R be the n-bit URF. Consider the following two games, Gm1 and Gm2. In Gm1, one can access $\mathrm{TE}_2 : \Sigma^n \times \{0,1,2,3\} \to \Sigma^n$ where $\mathrm{TE}_2(x,0) = R(x)$, and $\mathrm{TE}_2(x,i) = R^{\oplus L \cdot \mathbf{u}^{(i-1)}}(x)$ for $i = 1,2,3$, and $L \in_U \Sigma^n$. In Gm2, one can access the URF compatible with TE_2. Then, $\mathbf{Adv}^{\mathrm{cpa}}_{\mathrm{Gm1,Gm2}}(q)$ is at most $\frac{q^2}{2^{n+1}}$.*

As the proof of Lemma 7 is a simple extension of the proof of Lemma 6, we have omitted it here. Note that $\mathrm{PC\text{-}MAC}_d[R, L|G_U]$ invokes R at most $\rho q + c$ times. From this observation and Lemma 7, it is clear that $\mathbf{Adv}^{\mathrm{cpa}}_{\mathrm{PC\text{-}MAC}_d[R,L|G_U],Q}(q,\rho)$ is at most $\frac{(c+\rho q)^2}{2^{n+1}}$. From this and Eq. (13), we have

$$\mathbf{Adv}^{\mathrm{vilprf}}_{\mathrm{PC\text{-}MAC}_d[R,L|G_U]}(q,\rho) \leq \frac{(c+\rho q)^2}{2^{n+1}} + \left(d\epsilon_{\mathrm{dp}} + \epsilon_{\mathrm{sdp}} + \frac{2\rho^2 + 1}{2^n} \right) \frac{q^2}{2}, \quad \text{and}$$

$$\mathbf{Adv}^{\mathrm{vilprf}}_{\mathrm{PC\text{-}MAC}_d[E_K,L|G_U]}(q,t,\rho) \leq \mathbf{Adv}^{\mathrm{prf}}_{E_K}(\rho q + c, t') + \mathbf{Adv}^{\mathrm{vilprf}}_{\mathrm{PC\text{-}MAC}_d[R,L|G_U]}(q,\rho)$$
(14)

where $t' = t + O(\rho q)$. Combining Eq. (14) with the PRF/PRP switching lemma concludes the proof.

Searching for Differential Paths in MD4⋆

Martin Schläffer and Elisabeth Oswald

Institute for Applied Information Processing and Communications (IAIK),
Graz University of Technology, Inffeldgasse 16a, A–8010 Graz, Austria
{martin.schlaeffer, elisabeth.oswald}@iaik.tugraz.at

Abstract. The ground-breaking results of Wang *et al.* have attracted a lot of attention to the collision resistance of hash functions. In their articles, Wang *et al.* give input differences, differential paths and the corresponding conditions that allow to find collisions with a high probability. However, Wang *et al.* do not explain how these paths were found. The common assumption is that they were found by hand with a great deal of intuition.

In this article, we present an algorithm that allows to find paths in an automated way. Our algorithm is successful for MD4. We have found over 1000 differential paths so far. Amongst them, there are paths that have fewer conditions in the second round than the path of Wang *et al.* for MD4. This makes them better suited for the message modification techniques that were also introduced by Wang *et al.*

Keywords: collision search, differential path, MD4.

1   Introduction

The cryptanalysis of hash functions has become a hot topic within the cryptographic community over the last two years. Especially the ground breaking results of Wang *et al.* have drawn significant attention towards the security claims that were made for commonly used hash functions.

During the last two years, most hash functions have succumbed to the attacks of Wang *et al.* At first, the hash functions MD4 (as well as RIPEMD) and MD5 were analyzed by Wang *et al.* in [WLF+05] and [WY05]. Based on the techniques that have been introduced in these two papers, more advanced attacks on SHA-0 and SHA-1 have been published some time later in [WYY05b] and [WYY05a]. In all articles published by Wang *et al.* so far, only little details about the way in which the differences and the conditions were determined, have been published. Except for the article of Hawkes *et al.* [HPR04] that provides some musings on the techniques used for MD5, the PhD thesis of Magnus Daum [Dau05] and an ECRYPT deliverable [ABB+05] that both provide some high level discussions of the techniques of Wang *et al.*, we are not aware of any other article that gives insights into the techniques of Wang *et al.* In particular, there exist, to

---

⋆ The work described in this paper has been supported in part by the European Commission through the IST Programme under Contract IST-2002-507932 ECRYPT.

M.J.B. Robshaw (Ed.): FSE 2006, LNCS 4047, pp. 242–261, 2006.

our knowledge, no insights about the techniques that Wang *et al.* used to find so-called differential paths, *i.e.* to find the specific sequence of differences over a given number of steps that produces a local collision.

It is therefore easy to motivate and to explain the aim of the research that we present in this paper. We have tried to come up with an algorithm that finds differential paths in an automated way. As target for our path-searching-algorithm we picked MD4. The reason for choosing MD4 is also easily motivated; it is the simplest of the well known hashing algorithms and it is the basis for many other algorithms such as MD5, RIPEMD, SHA-0 and SHA-1.

Our algorithm is successful: given a difference for the input message it computes differential paths for MD4 in an automated way. Among the differential paths that we have found so far, there are paths that are even slightly better than the path that Wang *et al.* reported in their original article. Our path has less conditions in the second round.

This article is organized as follows. In Sect. 2 we briefly review the attack by Wang *et al.* on MD4. In Sect. 3, we introduce the notation that we use to describe our algorithm. In Sect. 4, we explain our algorithm and in Sect. 5, we report on the results that we have obtained with it. We conclude this article in Sect. 6. There are several appendices to this article. They give more information about our algorithm and the best path that we found with it.

2 The Wang *et al.* Approach

In this section we outline the approach by Wang *et al.* based on the example of the MD4 hash function. We first review the working principle of MD4 and then we focus on the attack of Wang *et al.*

2.1 The MD4 Hash Function

The MD4 algorithm hashes an input of arbitrary length to a 128-bit value. The algorithm proceeds as follows. The input message M is modified by a specific padding rule to a message with a length that is a multiple of 512. Then, the padded message is subjected to the MD4 compression function. The compression function consists of three rounds having 16 steps each. Each round uses a different Boolean function f_i: in the first round it is the *IF* function, in the second round it is the *MAJ* (majority) function and in the third round it is the *XOR* function.

In every step in MD4, a 32-bit variable r_i is updated according to the rule given in (1). Later in this article, we use the notation that the j-th bit of r_i is denoted by $r_{i,j}$. In (1), the operator $+$ denotes the addition modulo 2^{32} and

Table 1. The order of message words in MD4

i	w_i
0...15	0,1,2,3,4,5,6,7,8,9,10,11,12,13,14,15
16...31	0,4,8,12,1,5,9,13,2,6,10,14,3,7,11,15
32...47	0,8,4,12,2,10,6,14,1,9,5,13,3,11,7,15

the operator $\lll s_i$ denotes a circular left shift (rotation) by s_i positions. The variable m_{w_i} defines a message word and the variable k_i defines a round constant. The order of accessing the message words is given in Tab.1.

$$r_i = (r_{i\text{-}4} + f_i(r_{i\text{-}1}, r_{i\text{-}2}, r_{i\text{-}3}) + m_{w_i} + k_i) \lll s_i, \quad 0 \le i \le 47. \tag{1}$$

The number of bit positions s_i in a rotation is either $\{3, 7, 11, 19\}$ in the first round, $\{3, 5, 9, 13\}$ in the second round, or it is $\{3, 9, 11, 15\}$ in the third round. The initial values are in hexadecimal notation:

$$(r_{\text{-}4}, r_{\text{-}3}, r_{\text{-}2}, r_{\text{-}1}) = (0\text{x}67452301, 0\text{x}10325476, 0\text{x}98\text{badcfe}, 0\text{xefcdab}89)$$

These initial values are used to initialize the four 32-bit chaining variables (A, B, C, D). After 48 steps, the values $(r_{44}, r_{45}, r_{46}, r_{47})$ are added to the chaining variables (A, B, C, D). If all message blocks have been processed, then the hash value of the input message is determined by the concatenation of the four chaining variables.

2.2   Selecting an Input Difference

In the first step of the Wang *et al.* attack, one determines the difference Δ between the two input messages M and M' (*i.e.* the input difference). In contrast to Dobbertin's attack [Dob98] on MD4, Wang *et al.* do not aim on producing one local collision within MD4 but two. The idea is to have one local collision in the third round that is easily fulfilled and to have another local collision over some steps in the first two rounds. The local collision in the third round determines the input difference (see [Sch06] for further details).

Differential Properties of the *XOR* Function. There are two simple observations that are the foundation for producing the local collision in the third round in the Wang *et al.* attack. The first observation is that an input difference of 2^{31} (mod 2^{32}) implies that only the 31st bit in the message words differ. The second observation is that if two input values of the *XOR* function have a difference of 2^{31}, then this difference is canceled.

Differential Properties of the Update Rule in the Third Round. In the third round of MD4, the function f_i in the update rule (1) is the *XOR* function. We look at step i and assume hereby that there are no differences in the four previous steps. Choosing the message difference to be 2^{31-s_i} in step i, causes the difference after the i-th step to be 2^{31}. In the $(i + 1)$-st step, the difference of 2^{31} propagates through the *XOR* function. In order to cancel it, we choose the difference of $m_{w_{i+1}}$ to be 2^{31} (also -2^{31} would work). Because the difference from the i-th step also goes into the *XOR* function in the $(i + 2)$-nd step, it is clever to set the difference of $m_{w_{i+1}}$ to be $2^{31} + 2^{31-s_{i+1}}$. In this way we cancel the 2^{31} difference in the $(i + 1)$-st step and insert a $+2^{31-s_{i+1}}$ difference that becomes a 2^{31} difference after the rotation by s_{i+1} bit positions. In the $(i + 2)$-nd step, we have on two inputs of the *XOR* function a difference of 2^{31},

Table 2. Propagation of differences in the third round of MD4 according to the update rule $r_i = (r_{i-4} + XOR(r_{i-3}, r_{i-2}, r_{i-1}) + m_{w_i} + k_i) \lll s_i$

Step	Δr_{i-4}	Δr_{i-3}	Δr_{i-2}	Δr_{i-1}	Δr_i	Δm_{w_i}
i	0	0	0	0	2^{31}	2^{31-s_i}
i+1	0	0	0	2^{31}	2^{31}	$2^{31} + 2^{31-s_i}$
i+2	0	0	2^{31}	2^{31}	0	0
i+3	0	2^{31}	2^{31}	0	0	0
i+4	2^{31}	2^{31}	0	0	0	0
i+5	2^{31}	0	0	0	0	2^{31}
i+6	0	0	0	0	0	0

which cancel each other. Hence, the difference after the $(i + 2)$-nd step is zero. The same argument holds for step $i + 3$. In step $i + 4$, the input r_{i+1} of the XOR function has difference 2^{31}, hence it propagates and gets canceled by r_i in the addition. Consequently, in the $(i + 5)$-th step, there is the 2^{31} difference in r_{i+1} value that needs to be canceled. This can be done by inserting the same difference in $m_{w_{i+5}}$. This differential behavior is summarized in Tab. 2. One can choose the starting step i for this local collision. The choice of i determines in which message words the differences are introduced. This in turn determines the length of the differential path that describes the local collision over the steps in the first two rounds. We can also choose the sign of the differences. The choice $i = 35$ leads to $\Delta m_1 = 2^{31}$, $\Delta m_2 = 2^{31} + 2^{28}$ and $\Delta m_{12} = 2^{16}$. As indicated before, we may choose other signs for the differences: the differences $\Delta m_1 = 2^{31}$, $\Delta m_2 = 2^{31} - 2^{28}$ and $\Delta m_{12} = -2^{16}$ also lead to a local collision. In our experiments, this particular choice of i turned out to be the best choice.

2.3 Finding a Differential Path

The second crucial step is to find a so-called differential path that cancels the differences between steps 1 and 24. In the articles of Wang *et al.* such paths were given. However, no insight was provided how these paths were found. It is therefore assumed that the paths were found by hand. This means, that a great deal of intuition by the researchers was needed to determine the paths. The main contribution of this article is the automated search algorithm, which is described in Sect. 4.

2.4 Message Modification

The result of the second step is a differential path and the conditions on the intermediate values that are needed to fulfill the path. These conditions can be translated into equations for the message words that allow to pre-fulfill the conditions. In the third and last step of the attack, one applies different message modification techniques to the message in order to pre-fulfill as many conditions as possible.

Different message modification techniques were introduced by Wang *et al.* There is the single-step message modification technique, which allows to pre-fulfill all conditions that occur in the first round. The second technique is the multi-step message modification and allows to pre-fulfill some conditions in the second round. Other ideas for message modification techniques are the so-called advanced multi-step message modification techniques [WLF+05] and techniques that have been mentioned in [ABB+05]. The number of conditions that cannot be pre-fulfilled determines the overall complexity of the attack.

The main difference between the single-step and the multi-step message modifications is that the singe-step modifications always succeed. This means that all conditions that occur in the first round can be pre-fulfilled whereas this is not the case for the conditions in the second round. Consequently, it is desirable for a path search algorithm to look for paths that have most conditions in the first round of MD4.

3   Notation

This section details the notation that we will use in the remainder of this article. Furthermore we discuss the carry expansion of signed differences and the differential properties of the functions IF and MAJ. The input messages are denoted by $M = (m_0, m_1, ..., m_{15})$ and $M' = (m'_0, m'_1, ..., m'_{15})$. The intermediate steps in MD4 are computed according to (1) and the results are typically represented by the variable r_i.

3.1   Signed Differences

We follow the idea by Wang *et al.* and use signed differences. Because we only use signed differences, we will often refer to them simply as differences.

Definition 1. *The signed difference* Δx *between two 32-bit words* x *and* x' *is defined bitwise by*

$$\Delta x = x' - x = (\delta x_{31}, ..., \delta x_0) \quad with \quad \delta x_j = x'_j - x_j \in \{-1, 0, 1\}, \ 0 \le j \le 31$$

We use the following abbreviation for Δx:

$$\Delta x = \Delta[d_1, d_2, ..., d_w] \quad where \quad d_i = \begin{cases} -j & if \ \delta x_j = -1 \\ j & if \ \delta x_j = 1 \end{cases}$$

Definition 2. *For a given difference* $\Delta[d_1, d_2, ..., d_w]$, *the value* $|d_i|$ *defines a bit position. The value* w *is the Hamming weight of the difference. The value* $\Delta[]$ *denotes the zero difference.*

A nonzero difference $\Delta x = x' - x$ already determines the values of the corresponding bits in x (and therefore in x'):

$$\Delta x = \Delta[d_1, d_2, ..., d_w] \quad \Rightarrow \quad x_{|d_i|} = \begin{cases} 0 & if \ sign(d_i) = 1 \\ 1 & if \ sign(d_i) = -1 \end{cases}$$

The difference Δx also imposes conditions on the value x.

Example 1. *The difference Δx can be represented as follows:*

$$\Delta x = x' - x = \Delta[\text{-27}, 15, \text{-3}]$$
$$= (0,0,0,0,\text{-1},0,0,0,0,0,0,0,0,0,0,0,1,0,0,0,0,0,0,0,0,0,0,0,\text{-1},0,0,0),$$

and it implies that $x_{27} = 1, x_{15} = 0, x_3 = 1$ and $x'_{27} = 0, x'_{15} = 1, x'_3 = 0$.

Remark. *Because we use signed bit differences in our differential analysis, we need to be able to add and rotate signed bit differences throughout each step. For a detailed definition of these operations see [Dau05] or [Sch06]. We only provide one simple case and one example here.*

Lemma 1. *When adding two signed bit differences $\Delta x = \Delta[dx_1]$ and $\Delta y = \Delta[dy_1]$ with hamming weight 1 the following four cases can occur:*

$$\Delta[dx_1] + \Delta[dy_1] = \begin{cases} \Delta[] & if \quad dx_1 = \text{-}dy_1 \\ \Delta[dx_1 + 1] & if \quad dx_1 = dy_1 \ and \ sign(dx_1) = 1 \\ \Delta[\text{-}(|dx_1| + 1)] & if \quad dx_1 = dy_1 \ and \ sign(dx_1) = \text{-}1 \\ \Delta[dx_1, dy_1] & otherwise \end{cases}$$

A difference at position 32 is always discarded. When adding signed bit differences with Hamming weight $w > 1$ a signed carry effect may occur. To rotate a signed bit difference $\Delta x = \Delta[dx_1, dx_2, ..., dx_w]$ each element is rotated as follows:

$$\Delta[dx_i] \lll s = \Delta[dy_i] \ where \ dy_i = \begin{cases} dx_i + s \bmod 32 & if \ sign(dx_i) = 1 \\ \text{-}(|dx_i| + s \bmod 32) & if \ sign(dx_i) = \text{-}1 \end{cases}$$

Example 2. *We look at the sum of $\Delta x = \Delta[31, 27, 16, 15, 4]$ and of $\Delta y = \Delta[31, \text{-}27, 15, \text{-}3]$. Carries at positions $15, 16$ and 31 occur. The carry that comes from position 31 is discarded.*

$$\Delta x + \Delta y = \Delta[31, 27, 16, 15, 4] + \Delta[31, \text{-}27, 15, \text{-}3] = \Delta[17, 4, \text{-}3]$$
$$\Delta x \lll s = \Delta[31, \text{-}27, 15, \text{-}3] \lll 5 = \Delta[20, \text{-}8, 4, \text{-}0]$$

3.2 Carry Expansion of Signed Differences

Because the representation of signed differences is redundant, every (nonzero) element d_i of a signed difference can be expanded as described in the following. Note that differences at position 32 are discarded.

$$\Delta[d_1, ..., d_i, ..., d_w] = \begin{cases} \Delta[d_1, ..., \text{-}d_i, ..., d_w] + \Delta[d_i + 1] & if \ sign(d_i) = 1 \\ \Delta[d_1, ..., \text{-}d_i, ..., d_w] + \Delta[\text{-}(|d_i| + 1)] & if \ sign(d_i) = \text{-}1 \end{cases}$$

This step can be applied recursively on the resulting signed difference, and on the previous signed difference but for a different bit position. We call the number of expansion steps for each element d_i *additional carries*. A specific representation is achieved by imposing conditions on the difference.

Example 3. *In this example the difference $\Delta x = \Delta[-11, 9]$ is expanded, where the maximum number of expansion steps performed in each recursion branch, and thus the number of additional carries, is 2 and where the expanded element is marked with $\overleftarrow{d_i}$:*

$$\Delta x \rightarrow \Delta[-11, \overleftarrow{9}] \rightarrow \Delta[-11, \overleftarrow{10}, \text{-}9] \rightarrow \Delta[-10, \text{-}9]$$
$$\rightarrow \Delta[-\overleftarrow{11}, 10, \text{-}9] \rightarrow \Delta[-12, 11, 10, \text{-}9]$$
$$\rightarrow \Delta[-\overleftarrow{11}, 9] \rightarrow \Delta[-\overleftarrow{12}, 11, 9] \rightarrow \Delta[-13, 12, 11, 9]$$
$$\rightarrow \Delta[-12, \overleftarrow{11}, 9] \rightarrow \Delta[-12, 11, 10, \text{-}9]$$

Hence, all representations for Δx with a maximum carry expansion of two, sorted by their Hamming weight, are:

$$\Delta x = \Delta[-11, 9] = \Delta[-10, \text{-}9]$$
$$= \Delta[-11, 10, \text{-}9] = \Delta[-12, 11, 9]$$
$$= \Delta[-12, 11, 10, \text{-}9] = \Delta[-13, 12, 11, 9]$$

An expanded signed difference can be reduced to an equivalent difference with minimum weight again. However, this is not true if the difference is rotated between expansion and reduction.

Example 4. *This example shows that the weight of an expanded difference cannot be reduced to a difference with equal weight, if the expanded part is rotated over position 31:*

$$\Delta[12] \lll 19 = \Delta[13, \text{-}12] \lll 19 = \Delta[-31, 0] \neq \Delta[31] = \Delta[12] \lll 19$$

3.3   Properties of the Functions *IF* and *MAJ*

In this section we discuss the propagation of signed differences through the functions *IF* and *MAJ*. In order to control the propagation of differences through these functions, we need to impose conditions on the input values. Table 5 (see App. A) shows all cases and conditions that allow to achieve a specific output difference of these functions.

For the *IF* function, the majority of input cases can be manipulated. However, consecutive ones in the input differences have to be avoided if a zero output difference is desired. For the *MAJ* function, we can only influence the output difference if the number of input differences is exactly one. Table 5 shows that the input difference of the *IF* function can be flipped if δx is not zero. Therefore, it can be assumed that in the first two rounds a zero output difference is possible by imposing conditions in most cases.

4   Our Algorithm for the Differential-Path Search

In Sect. 2.2, we have selected the input difference ΔM as $\Delta m_1 = 2^{31} = \Delta[31]$, $\Delta m_2 = 2^{31} - 2^{28} = \Delta[31, \text{-}28]$ and $\Delta m_{12} = -2^{16} = \Delta[-16]$. These differences are introduced in steps 1, 2 and 12 of round one and in steps 19, 20 and 24 of round two. Thus, in order to derive a differential path for MD4, the differences between step 0 and 24 have to cancel each other. The complexity of a brute-force search through all possible paths is too high. To reduce the search space of our algorithm, we have tried to avoid any uncontrolled propagation of differences through the function f_i (see Sect. 3.3) or by carry propagation (see Sect. 3.2). In order to reduce the resulting number of conditions, low weight signed differences are used by default.

The algorithm consists of three major parts which are the *target differences computation* (see Sect. 4.1), the *cancelation search* (see Sect. 4.2), and the *correction step* (see Sect. 4.3). An overview of the algorithm is given in Fig. 4 (see App. B).

In the first part, the target output difference for the function f_i is determined for every step. Therefore, the message differences are computed backward and forward to derive the so-called correction and disturbance differences. They are then combined to define the target differences.

During the cancelation search, all variations of the elements of the target differences are considered. Note, that this is done for every step of MD4. The elements of the target differences need to be canceled by using the properties of the function f_i. To achieve an output difference for f_i at a specific bit position, the input differences Δr_{i-1}, Δr_{i-2} and Δr_{i-3} need to be expanded. Finally, the conditions for each step are derived.

In the correction step, impossible output differences are resolved without searching for a new differential path first. If some contradictions cannot be corrected, additional differences are added to the target differences. These disturbance differences, which we typically derived by hand, distribute the conditions such that a new differential path without contradictions can be found.

4.1 Target Differences Computation

The goal of an algorithm for finding a differential path is to cancel out all differences that are introduced by the message words. Because one of the four state variables is updated in one step, a message difference can be canceled every fourth step. Hence, a message difference introduced in step i can be canceled by introducing an opposite difference in all steps $(i \pm 4k)$. To know where to introduce a difference and to determine its position and sign, the message differences are compute backward and forward (see Fig. 1 left). To reduce the complexity, no propagation through the function f_i or by a carry expansion is considered while deriving the target differences.

Disturbance Differences. Δd_i are simply derived by *forward* computing the message differences:

$$\Delta d_i = \Delta d_{i-4} \lll s_{i-4} + \Delta m_{w_i}$$
$$\text{with } i := \{0, 1, ..., 24\} \text{ and } \Delta d_{-4} = \Delta d_{-3} = \Delta d_{-2} = \Delta d_{-1} = 0$$

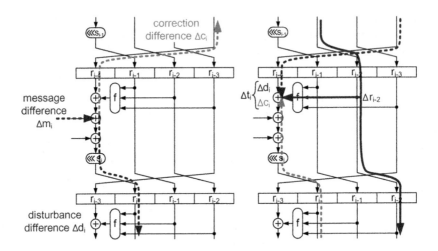

Fig. 1. Left: Forward and backward computation of the message differences to get a target difference for each step. Right: Fulfilling the target difference using the input differences of the function f_i.

Correction Differences. Δc_i are *backward* computed message differences. Using the correction differences it can be determined where to introduce a difference, which in turn can cancel a message difference in a subsequent step:

$$\Delta c_i = (\Delta c_{i+4} + \Delta m_{w_{i+4}}) \ggg s_i$$
$$\text{with } i := \{20, 19, ..., 0\} \text{ and } \Delta c_{24} = \Delta c_{23} = \Delta c_{22} = \Delta c_{21} = 0$$

Target differences are the merged correction and disturbance differences of each step. A target difference Δt_i in step i is the target output difference of the function f_i. This difference is known to cancel a message difference in a previous or later step. The target differences are defined by the sum of the disturbance and correction differences (2). Table 6 shows the target differences of steps $0-24$.

$$\Delta t_i = -(\Delta d_i + \Delta c_i) \tag{2}$$

4.2 Cancelation Search

In this section we describe how to find candidates for differential paths. The main concept is to cancel the target differences in each step using the function f_i. The search is performed recursively over all steps $i = \{0, ..., 24\}$. In order to cancel an element of the target difference we use carry expansions.

Variation of Target Difference Elements. It is not known in advance which elements of the target differences should be canceled in what step. Therefore, in every step i all variations of the elements of the target difference Δt_i have to be considered. These variations are simply called *target variations* in the remainder. If the target difference Δt_i has Hamming weight w then there are 2^w possibilities to cancel elements of the target difference (see Tab. 3).

Carry Expansions. The target variations need to be canceled by the function f_i (see Fig. 1 right). A non-zero output difference of the function f_i is only possible, if a non-zero input difference at the same bit position is available. This is usually not the case. Therefore, the input differences of the function f_i are expanded. Note, that there are again many possibilities to achieve a specific bit position. Each element of the target variation could be canceled by any carry expansion of any input difference of f_i (see Tab. 3). To limit the complexity of the search algorithm, a predefined maximum length for each of the carry expansions (usually 3) is used. Besides limiting the search space, low weight differences reduce the number of conditions as well.

Table 3. This table shows all target variations of $\Delta t_i = \Delta[-29, 20, -17]$ and all carry expansions of $\Delta r_{i-1} = \Delta[19, -17]$, $\Delta r_{i-2} = \Delta[16]$ and $\Delta r_{i-3} = \Delta[14, -7]$ with a maximum length of 2. Each target variation may be canceled by any carry expansion of these inputs of $f_i(r_{i-1}, r_{i-2}, r_{i-3})$.

Δt_i	Δr_{i-1}	Δr_{i-2}	Δr_{i-3}
$\Delta[-29, 20, -17]$	$\Delta[19, -17]$	$\Delta[16]$	$\Delta[14, -7]$
$\Delta[-29, 20\ \ \ \ ]$	$\Delta[18,\ \ 17]$	$\Delta[17, -16]$	$\Delta[15, -14, -7]$
$\Delta[-29,\ \ \ \ -17]$	$\Delta[20, -19, -17]$	$\Delta[18, -17, -16]$	$\Delta[14,\ \ -8,\ \ 7]$
$\Delta[-29\ \ \ \ \ \ \ ]$	$\Delta[19, -18,\ \ 17]$		$\Delta[15, -14,\ \ -8,\ \ 7]$
$\Delta[\ \ \ \ 20, -17]$	$\Delta[20, -19, -18,\ \ 17]$		$\Delta[16, -15, -14, -7]$
$\Delta[\ \ \ \ 20\ \ \ \ ]$	$\Delta[21, -20, -19, -17]$		$\Delta[14,\ \ -9,\ \ 8,\ \ 7]$
$\Delta[\ \ \ \ \ \ \ \ -17]$			
$\Delta[\ \ \ \ \ \ \ \ \ \ \ ]$			

Cancel Possibilities. In this step it is determined which target variation can be achieved by which carry expansion. To achieve one specific target variation, all combinations of the inputs Δr_{i-1}, Δr_{i-2} and Δr_{i-3} of f_i can be tried. However, a target variation can only be met by an input difference, if they share at least the same bit positions. Because most inputs of f_i cannot meet this requirement anyway, the search space can be significantly reduced by considering only combinations that are possible using this principle.

In every step i of the hash function, we first start with the difference Δr_{i-1} of f_i and try to meet all target variations by carry expanding Δr_{i-1}. Some targets will not be met at all, whereas others can be met with several expansions of Δr_{i-1} (see Ex. 5). Note, that it cannot be determined whether a specific output difference of the function f_i is indeed possible until all of its inputs are fixed. Therefore, it is first assumed that the desired target can be canceled and verified in a later step. The input difference with the lowest weight is used by default.

Example 5. *This example shows all cancel possibilities for all variations of the target difference $\Delta t_i = \Delta[-29, 20, -17]$. In this example the expansions of the input $r_{i-1} = \Delta[19, -17]$ are considered. A target variation containing a difference at position 29 cannot be achieved by any input difference listed, whereas the zero target variation can be achieved by all input differences.*

$$\Delta r_{i-1} = \Delta[19, -17] = \Delta[18, 17] = \Delta[20, -19, -17] = \Delta[19, -18, 17]$$
$$= \Delta[21, -20, -19, -17] = \Delta[20, -19, -18, 17]$$

$\Delta t_i = \Delta[-29, \; 20, -17] \quad \Longrightarrow \quad$ *not possible*

$\Delta t_i = \Delta[-29, \; 20 \quad] \quad \Longrightarrow \quad$ *not possible*

$\Delta t_i = \Delta[-29, \qquad -17] \quad \Longrightarrow \quad$ *not possible*

$\Delta t_i = \Delta[-29 \qquad \quad] \quad \Longrightarrow \quad$ *not possible*

$\Delta t_i = \Delta[\qquad 20, -17] \quad \Longrightarrow \quad \Delta[20, -19, -17]$

$\Delta t_i = \Delta[\qquad 20 \quad] \quad \Longrightarrow \quad \Delta[20, -19, -17], \Delta[20, -19, -18, 17]$

$\Delta t_i = \Delta[\qquad \quad -17] \quad \Longrightarrow \quad \Delta[19, -17], \Delta[20, -19, -17], \Delta[21, -20, -19, -17]$

$\Delta t_i = \Delta[\qquad \quad] \quad \Longrightarrow \quad$ *all expansions of* Δr_{i-1}

Already canceled elements of a target difference are removed from the target and the remaining elements are canceled in a later step. All possible target variations are examined recursively. Thus, the message differences are tried to be canceled in *all* steps $i \pm 4k$. After having processed Δr_{i-1}, we continue with Δr_{i-2} and Δr_{i-3}. Note, that Δr_{i-2} and Δr_{i-3} may have already been used to cancel a previous target. Further expansions are only possible if they do not contradict these cancelations. For example, the expansion $\Delta[18, 17] \to \Delta[19, -17]$ is not possible if $\Delta[18]$ has already been used to cancel a target in a previous step.

Deriving the Conditions. The used carry expansions of the input differences finally determine the conditions. Only after we have fixed all three input differences of f_i, we can determine whether a certain output difference is really possible. This is often not the case. However, one of the other cancel possibilities can be tried. In addition, in many cases a previously set condition can contradict with a newly set condition and further expansions need to be tried.

After examining all possible expansions there are usually some contradictions left. A path with at least one contradiction is called an *impossible path*. To reduce the complexity, the search in a branch with too many contradictions is aborted. The result of the cancelation search are a number of paths from step 0 to step 24 that have a zero differences in step 24, but may still have a few contradictions.

4.3   Correction Step

In the correction step, contradictions within impossible paths are corrected. In such impossible paths a specific target difference cannot be met in some step or a zero output difference of the function f_i cannot be achieved. As a consequence, these additional (disturbance) differences induced by the contradictions need to be canceled in some other step.

Correction by Solving Contradictions. To cancel these additional disturbances, they are computed forward and backward through the already determined differential path. As we only need to correct a few new disturbances,

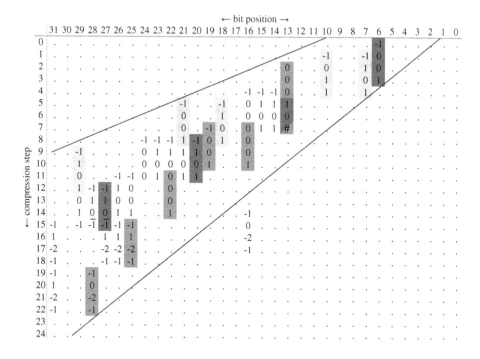

Fig. 2. This figure shows the conditions for a specific differential path. Most differences are rotated with low values. Thus, conditions do not spread (different shades show different rotation values) and so contradictions are more likely to occur although the number of conditions, which is 119, is small. A value of $c = 0$ or $c = 1$ requires $r_{i,j} = c$ for a specific bit j and step i. A negative value c or $\bar{c}$ requires the respective bit to be $r_{i,j} = r_{i-|c|,j}$ or $r_{i,j} \neq r_{i-|\bar{c}|,j}$. The entry marked by # denotes a contradiction.

longer carry expansions can be allowed. However, this does not always work because further contradictions may occur which are even harder to resolve. The reason is, that in the case of MD4 the conditions and differences stick together throughout the whole differential path (see Fig. 2).

Correction by Dispersion Differences. Typically, differences propagate from the least significant bit in the first few steps to the most significant bit in the last steps (see Fig. 2). The reason for this propagation is that the rotation values s_i are very similar for most differences. In order to spread the differences and thus the conditions, *dispersion differences* (Δp_i) are introduced in steps with a high rotation value, *i.e.* $s_3 = 19$. This high rotation allows the dispersion differences to spread within only a few steps. The dispersion differences are then used in the following steps to cancel differences in areas with a low condition density. The dispersion difference $\Delta p_3 = \Delta[6]$, which we determined by hand, leads to a differential path without contradictions (see Fig. 3).

5   Experiments and Results

In our experiments we tried carry expansions with different lengths and different dispersion differences. It turned out that at least in one step, a carry expansion of length three is needed. We further noticed that increasing the maximum length of the carry expansions up to 10 does not lead to less contradictions when using no dispersion differences. During the evaluation of different parameters, we were able to produce over 1000 (similar) differential paths without contradictions so far. One run of our algorithm takes only a couple of minutes. For example, using the disturbance difference $\Delta p_3 = \Delta[6]$ and a maximum carry expansion of 3,

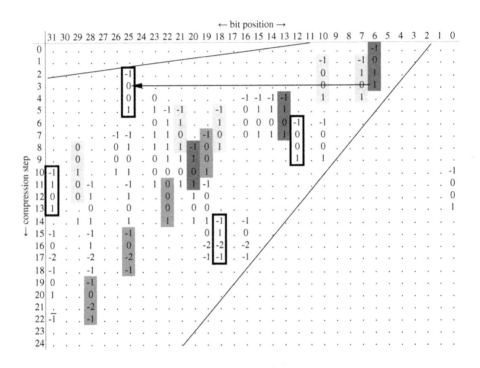

Fig. 3. In our path, differences and conditions are spread by introducing a *dispersion difference* in step $i = 3$ which has a high rotation value ($s_3 = 19$). This dispersion difference causes new conditions, which are marked by a box in this figure. Because of introducing a new difference, the number of conditions is higher (146). However, no contradictions appear.

Table 4. One collision of MD4 using our differential path

M_0	9de70013 4b5611b3 d2ce37bb d3fbfd91 25bb4551 42d059f8 41b1bd57 19ed222e
	4c9c5258 20df2cbf d868c1a8 314acd01 e4aca811 5089a823 bb1912b1 2b61d489
M_0'	9de70013 cb5611b3 42ce37bb d3fbfd91 25bb4551 42d059f8 41b1bd57 19ed222e
	4c9c5258 20df2cbf d868c1a8 314acd01 e4aba811 5089a823 bb1912b1 2b61d489
H_0	877bd941 14da836a 0af87c2e 143a4028

13871 possibilities to cancel the elements of the target differences have been tried. 208 possibilities result in a zero differences in step 24 but have at least 2 contradictions. To correct the contradictions of these paths, 140068 different cancel possibilities to achieve the respective target differences were examined. Finally, 324 paths with no contradiction could be found. The overall number of steps performed was 1964131.

With respect to the message modification, the best of our paths is the one shown in Fig. 3. It has the smallest number of conditions in the second round. Remember that conditions in the first round can be easily pre-fulfilled by the single-step message modification technique. Further, most differences occur in the first few steps of the second round and are thus also easily pre-fulfilled. Our path has 146 conditions with only 22 conditions in round two and 2 conditions in round three. In contrast, the path of Wang *et al.* has 122+2 conditions where 25 conditions occur in round two and 2 conditions occur in round three. The two additional conditions were found by [NSKO05]. Fig. 3 shows all conditions in our path in a graphical manner. Further details about the conditions are provided in App. D and an example of a collision is given in Tab. 4.

6   Conclusions

In this article, we have introduced an algorithm that finds differential paths for the first 24 steps of MD4 in an automated way. Our algorithm is successful: given a difference for the input message, it computes differential paths for MD4. Among the differential paths that we have found so far, there are paths that have fewer conditions in the second round than the path of Wang *et al.* This is an advantage with respect to the message modification techniques; the complexity of a collision attack based on our path is therefore lower.

The techniques that we have used are not very specific for MD4. The forward-backward computation for instance, which is performed in the first part of our algorithm, can be applied in general to determine the target differences. The cancelation search is general in the sense that we try out all carry expansions up to a certain length starting from the simplest one. The use of dispersion differences must be carefully considered for other algorithms where the conditions might be more distributed anyway. In addition, the approach of trying the simplest differences first, and only enlarging the search space if necessary, is also algorithm independent.

Summing up, we have made the first successful step towards an automated search for differential paths, which is the crucial part of Wang *et al.*'s attacks.

Acknowledgements

We would like to thank the anonymous referees and the members of IAIK's Krypto group for their helpful comments.

References

[ABB+05] Daniel Augot, Alex Biryukov, An Braeken, Carlos Cid, Hans Dobbertin, Hakan Englund, Henri Gilbert, Louis Granboulan, Helena Handschuh, Martin Hell, Thomas Johansson, Alexander Maximov, Matthew Parkerand Thomas Pornin, Bart Preneel, Matt Robshaw, and Michael Ward. Ongoing Research Areas in Symmetric Cryptography, January 2005.

[Dau05]   Magnus Daum. *Cryptanalysis of Hash Functions of the MD4-Family*. PhD thesis, Ruhr-Universität Bochum, May 2005.

[Dob98]   Hans Dobbertin. Cryptanalysis of MD4. *Journal of Cryptology*, 11(4):253–271, 1998.

[HPR04]   Philip Hawkes, Michael Paddon, and Gregory G. Rose. Musings on the Wang et al. MD5 Collision. Cryptology ePrint Archive, Report 2004/264, 2004.

[NSKO05] Yusuke Naito, Yu Sasaki, Noboru Kunihiro, and Kazuo Ohta. Improved Collision Attack on MD4. Cryptology ePrint Archive, Report 2005/151, 2005. http://eprint.iacr.org/.

[Sch06]   Martin Schläffer. Cryptanalysis of MD4. Master's thesis, Institute for Applied Information Processing and Communications (IAIK), Graz University of Technology, Inffeldgasse 16a, 8010 Graz, Austria, February 2006.

[WLF+05] Xiaoyun Wang, Xuejia Lai, Dengguo Feng, Hui Chen, and Xiuyuan Yu. Cryptanalysis of the Hash Functions MD4 and RIPEMD. In Ronald Cramer, editor, *Advances in Cryptology – EUROCRYPT 2005: 24th Annual International Conference on the Theory and Applications of Cryptographic Techniques, Aarhus, Denmark, May 22-26, 2005. Proceedings*, volume 3494 of *Lecture Notes in Computer Science*, pages 1–18. Springer, 2005.

[WY05]   Xiaoyun Wang and Hongbo Yu. How to Break MD5 and Other Hash Functions. In Ronald Cramer, editor, *Advances in Cryptology – EUROCRYPT 2005: 24th Annual International Conference on the Theory and Applications of Cryptographic Techniques, Aarhus, Denmark, May 22-26, 2005. Proceedings*, volume 3494 of *Lecture Notes in Computer Science*, pages 19–35. Springer, 2005.

[WYY05a] Xiaoyun Wang, Yiqun Lisa Yin, and Hongbo Yu. Finding Collisions in the Full SHA-1. In Victor Shoup, editor, *Advances in Cryptology - CRYPTO 2005: 25th Annual International Cryptology Conference, Santa Barbara, California, USA, August 14-18, 2005, Proceedings*, volume 3621 of *Lecture Notes in Computer Science*, pages 17–36. Springer, 2005.

[WYY05b] Xiaoyun Wang, Hongbo Yu, and Yiqun Lisa Yin. Efficient Collision Search Attacks on SHA-0. In Victor Shoup, editor, *Advances in Cryptology - CRYPTO 2005: 25th Annual International Cryptology Conference, Santa Barbara, California, USA, August 14-18, 2005, Proceedings*, volume 3621 of *Lecture Notes in Computer Science*, pages 1–16. Springer, 2005.

A   Differential Characteristic of *IF* and *MAJ*

Table 5. Signed differential characteristic of the *IF* and *MAJ* function, with necessary conditions, probability 1 or "-" if the desired output is not possible

$\delta x \delta y \delta z$	$\delta IF = 0$	$\delta IF = 1$	$\delta IF = -1$	$\delta MAJ = 0$	$\delta MAJ = 1$	$\delta MAJ = -1$
0 0 0	1	-	-	1	-	-
0 0 1	$x=1$	$x=0$	-	$x=y$	$x \neq y$	-
0 0 -1	$x=1$	-	$x=0$	$x=y$	-	$x \neq y$
0 1 0	$x=0$	$x=1$	-	$x=z$	$x \neq z$	-
0 1 1	-	1	-	-	1	-
0 1 -1	-	$x=1$	$x=0$	1	-	-
0 -1 0	$x=0$	-	$x=1$	$x=z$	-	$x \neq z$
0 -1 1	-	$x=0$	$x=1$	1	-	-
0 -1 -1	-	-	1	-	-	1
1 0 0	$y=z$	$y=1, z=0$	$y=0, z=1$	$y=z$	$y \neq z$	-
1 0 1	$y=0$	$y=1$	-	-	1	-
1 0 -1	$y=1$	-	$y=0$	1	-	-
1 1 0	$z=1$	$z=0$	-	-	1	-
1 1 1	-	1	-	-	1	-
1 1 -1	1	-	-	-	1	-
1 -1 0	$z=0$	-	$z=1$	1	-	-
1 -1 1	1	-	-	-	1	-
1 -1 -1	-	-	1	-	-	1
-1 0 0	$y=z$	$y=0, z=1$	$y=1, z=0$	$y=z$	-	$y \neq z$
-1 0 1	$y=1$	$y=0$	-	1	-	-
-1 0 -1	$y=0$	-	$y=1$	-	-	1
-1 1 0	$z=0$	$z=1$	-	1	-	-
-1 1 1	-	1	-	-	1	-
-1 1 -1	1	-	-	-	-	1
-1 -1 0	$z=1$	-	$z=0$	-	-	1
-1 -1 1	1	-	-	-	-	1
-1 -1 -1	-	-	1	-	-	1

B   Overview of Our Algorithm

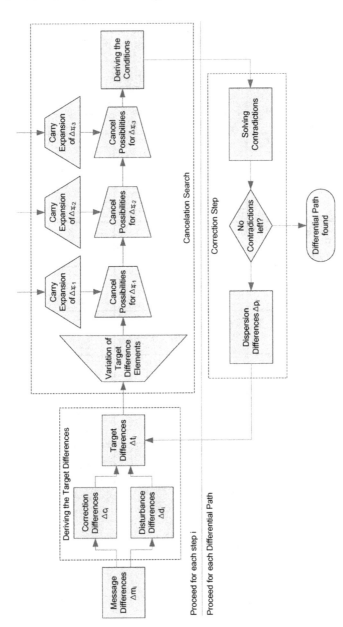

Fig. 4. An overview of the differential path search algorithm. Trapezoids represent expansions and restrictions of the search space.

C   The Target Differences

Table 6. Deriving the target differences Δt_i by forward computation of the disturbance differences Δd_i, and backward computation of the correction differences Δc_i for all message differences of step $0 - 24$. The differences caused by m_{12} of step 12 are highlighted.

step	Δm_{w_i}	s_i	Δd_i	Δc_i	Δt_i
0		3		$\Delta[16, 13, -10, -7]$	$\Delta[-16, -13, 10, 7]$
1	$\Delta[31]$	7	$\Delta[31]$		$\Delta[-31]$
2	$\Delta[31, -28]$	11	$\Delta[31, -28]$		$\Delta[-31, 28]$
3		19		$\Delta[-4]$	$\Delta[4]$
4		3		$\Delta[19, 16, -13, -10]$	$\Delta[-19, -16, 13, 10]$
5		7	$\Delta[6]$		$\Delta[-6]$
6		11	$\Delta[10, -7]$		$\Delta[-10, 7]$
7		19		$\Delta[-23]$	$\Delta[23]$
8		3		$\Delta[22, 19, -16, -13]$	$\Delta[-22, -19, 16, 13]$
9		7	$\Delta[13]$		$\Delta[-13]$
10		11	$\Delta[21, -18]$		$\Delta[-21, 18]$
11		19		$\Delta[-10]$	$\Delta[10]$
12	$\Delta[-16]$	3	$\Delta[-16]$	$\Delta[25, 22, -19]$	$\Delta[-25, -22, 19, 16]$
13		7	$\Delta[20]$		$\Delta[-20]$
14		11	$\Delta[-29, 0]$		$\Delta[29, -0]$
15		19		$\Delta[-29]$	$\Delta[29]$
16		3	$\Delta[-19]$	$\Delta[28, 25, -22]$	$\Delta[-28, -25, 22, 19]$
17		5	$\Delta[27]$		$\Delta[-27]$
18		9	$\Delta[11, -8]$		$\Delta[-11, 8]$
19	$\Delta[-16]$	13	$\Delta[-16]$		$\Delta[16]$
20	$\Delta[31]$	3	$\Delta[31, -22]$	$\Delta[28, -25]$	$\Delta[-31, -28, 25, 22]$
21		5	$\Delta[0]$		$\Delta[-0]$
22		9	$\Delta[20, -17]$		$\Delta[-20, 17]$
23		11	$\Delta[-29]$		$\Delta[29]$
24	$\Delta[31, -28]$	3	$\Delta[31, -28, -25, 2]$		$\Delta[-31, 28, 25, -2]$

D   Detailed Description of Our Path

Table 7. Differential characteristic of our differential path for MD4

Step	r_i	s_i	m_{w_i}	Δm_{w_i}	Δf_i	Δr_i
0	r_0	3	m_0			
1	r_1	7	m_1	$\Delta[31]$		$\Delta[6]$
2	r_2	11	m_2	$\Delta[31,-28]$		$\Delta[10,-7]$
3	r_3	19	m_3		$\Delta[6]$	$\Delta[25]$
4	r_4	3	m_4			
5	r_5	7	m_5			$\Delta[16,-15,-14,-13]$
6	r_6	11	m_6			$\Delta[23,-22,-21,-18]$
7	r_7	19	m_7		$\Delta[23]$	$\Delta[12,10]$
8	r_8	3	m_8		$\Delta[23,-22,16]$	$\Delta[26,-25,19]$
9	r_9	7	m_9			$\Delta[23,-22,-21,-20]$
10	r_{10}	11	m_{10}		$\Delta[-21]$	$\Delta[-29]$
11	r_{11}	19	m_{11}			$\Delta[-31,29,0]$
12	r_{12}	3	m_{12}	$\Delta[-16]$	$\Delta[-22]$	$\Delta[29,-28,-25,22,-20,19]$
13	r_{13}	7	m_{13}		$\Delta[-20]$	
14	r_{14}	11	m_{14}		$\Delta[29]$	
15	r_{15}	19	m_{15}			$\Delta[19,-18,16]$
16	r_{16}	3	m_0		$\Delta[19]$	$\Delta[31,-28,25]$
17	r_{17}	5	m_4			
18	r_{18}	9	m_8			
19	r_{19}	13	m_{12}	$\Delta[-16]$		$\Delta[31]$
20	r_{20}	3	m_1	$\Delta[31]$		$\Delta[-31,28]$
21	r_{21}	5	m_5			
22	r_{22}	9	m_9			
23	r_{23}	13	m_{13}		$\Delta[-31]$	
24	r_{24}	3	m_2	$\Delta[31,-28]$		
...	...					
35	r_{35}	15	m_{12}	$\Delta[-16]$		$\Delta[-31]$
36	r_{36}	3	m_2	$\Delta[31,-28]$	$\Delta[-31]$	$\Delta[-31]$
37	r_{37}	9	m_{10}			
38	r_{38}	11	m_6			
39	r_{39}	15	m_{14}			
40	r_{40}	3	m_1	$\Delta[31]$		

Table 8. Conditions for our differential path

Step	Conditions for r_i
0	$r_{0,6} = r_{-1,6}$
1	$r_{1,6} = 0, r_{1,7} = r_{0,7}, r_{1,10} = r_{0,10}$
2	$r_{2,6} = 1, r_{2,7} = 1, r_{2,10} = 0, r_{2,25} = r_{1,25}$
3	$r_{3,6} = 1, r_{3,7} = 0, r_{3,10} = 0, r_{3,25} = 0$
4	$r_{4,7} = 1, r_{4,10} = 1, r_{4,13} = r_{3,13}, r_{4,14} = r_{3,14}, r_{4,15} = r_{3,15}, r_{4,16} = r_{3,16},$ $r_{4,23} = 0, r_{4,25} = 0$
5	$r_{5,13} = 1, r_{5,14} = 1, r_{5,15} = 1, r_{5,16} = 0, r_{5,18} = r_{4,18}, r_{5,21} = r_{4,21}, r_{5,22} = r_{4,22},$ $r_{5,23} = 1, r_{5,25} = 1$
6	$r_{6,10} = r_{5,10}, r_{6,12} = r_{5,12}, r_{6,13} = 0, r_{6,14} = 0, r_{6,15} = 0, r_{6,16} = 0, r_{6,18} = 1,$ $r_{6,21} = 1, r_{6,22} = 1, r_{6,23} = 0$
7	$r_{7,10} = 0, r_{7,12} = 0, r_{7,13} = 1, r_{7,14} = 1, r_{7,15} = 1, r_{7,16} = 0, r_{7,18} = 0, r_{7,19} = r_{6,19},$ $r_{7,21} = 0, r_{7,22} = 1, r_{7,23} = 1, r_{7,25} = r_{6,25}, r_{7,26} = r_{6,26}$
8	$r_{8,10} = 0, r_{8,12} = 0, r_{8,18} = 1, r_{8,19} = 0, r_{8,20} = r_{7,20}, r_{8,21} = 1, r_{8,22} = 1, r_{8,23} = 1,$ $r_{8,25} = 1, r_{8,26} = 0, r_{8,29} = 0$
9	$r_{9,10} = 1, r_{9,12} = 1, r_{9,19} = 0, r_{9,20} = 1, r_{9,21} = 1, r_{9,22} = 1, r_{9,23} = 0, r_{9,25} = 0,$ $r_{9,26} = 0, r_{9,29} = 0$
10	$r_{10,0} = r_{9,0}, r_{10,19} = 1, r_{10,20} = 0, r_{10,21} = 0, r_{10,22} = 0, r_{10,23} = 0, r_{10,25} = 1,$ $r_{10,26} = 1, r_{10,29} = 1, r_{10,31} = r_{9,31}$
11	$r_{11,0} = 0, r_{11,19} = r_{10,19}, r_{11,20} = 1, r_{11,21} = 1, r_{11,22} = 0, r_{11,23} = 1,$ $r_{11,25} = r_{10,25}, r_{11,28} = r_{10,28}, r_{11,29} = 0, r_{11,31} = 1$
12	$r_{12,0} = 0, r_{12,19} = 0, r_{12,20} = 1, r_{12,22} = 0, r_{12,25} = 1, r_{12,28} = 1, r_{12,29} = 0,$ $r_{12,31} = 0$
13	$r_{13,0} = 1, r_{13,19} = 0, r_{13,20} = 0, r_{13,22} = 0, r_{13,25} = 0, r_{13,28} = 0, r_{13,31} = 1$
14	$r_{14,16} = r_{13,16}, r_{14,18} = r_{13,18}, r_{14,19} = 1, r_{14,20} = 1, r_{14,22} = 1, r_{14,25} = 1,$ $r_{14,28} = 1, r_{14,29} = 1$
15	$r_{15,16} = 0, r_{15,18} = 1, r_{15,19} = 0, r_{15,25} = r_{14,25}, r_{15,28} = r_{14,28}, r_{15,31} = r_{14,31}$
16	$r_{16,16} = r_{14,16}, r_{16,18} = r_{14,18}, r_{16,19} = r_{14,19}, r_{16,25} = 0, r_{16,28} = 1, r_{16,31} = 0$
17	$r_{17,16} = r_{16,16}, r_{17,18} = r_{16,18}, r_{17,19} = r_{16,19}, r_{17,25} = r_{15,25}, r_{17,28} = r_{15,28},$ $r_{17,31} = r_{15,31}$
18	$r_{18,25} = r_{17,25}, r_{18,28} = r_{17,28}, r_{18,31} = r_{17,31}$
19	$r_{19,28} = r_{18,28}, r_{19,31} = 0$
20	$r_{20,28} = 0, r_{20,31} = 1$
21	$r_{21,28} = r_{19,28}$
22	$r_{22,28} = r_{21,28}, r_{22,31} = \overline{r_{18,31}}$
23	
24	

A Study of the MD5 Attacks: Insights and Improvements

John Black[1], Martin Cochran[1], and Trevor Highland[2]

[1] University of Colorado at Boulder, USA
www.cs.colorado.edu/~jrblack, ucsu.colorado.edu/~cochranm
jrblack@cs.colorado.edu, cochranm@cs.colorado.edu
[2] University of Texas at Austin, USA
trevor.highland@gmail.com

Abstract. MD5 is a well-known and widely-used cryptographic hash function. It has received renewed attention from researchers subsequent to the recent announcement of collisions found by Wang et al. [16]. To date, however, the method used by researchers in this work has been fairly difficult to grasp.

In this paper we conduct a study of all attacks on MD5 starting from Wang. We explain the techniques used by her team, give insights on how to improve these techniques, and use these insights to produce an even faster attack on MD5. Additionally, we provide an "MD5 Toolkit" implementing these improvements that we hope will serve as an open-source platform for further research.

Our hope is that a better understanding of these attacks will lead to a better understanding of our current collection of hash functions, what their strengths and weaknesses are, and where we should direct future efforts in order to produce even stronger primitives.

Keywords: Cryptographic Hash Functions, Differential Cryptanalysis, MD5.

1 Introduction

BACKGROUND. MD5 was the last in a succession of cryptographic hash functions designed by Ron Rivest in the early 1990s. It is a widely-used well-known 128-bit iterated hash function, used in various applications including SSL/TLS, IPSec, and many other cryptographic protocols. It is also commonly-used in implementations of timestamping mechanisms, commitment schemes, and integrity-checking applications for online software, distributed filesystems, and random-number generation. It is even used by the Nevada State Gaming Authority to ensure slot-machine ROMs have not been tampered with.

Cryptographic hash functions like MD5 do not have a sound mathematical security definition, but instead rely on the following "intuitive" notions of security: for a hash function h with domain D and range R, we require the following three properties.[1]

---

[1] For a more complete discussion of hash function security definitions, see [12].

M.J.B. Robshaw (Ed.): FSE 2006, LNCS 4047, pp. 262–277, 2006.

Pre-image Resistance: For a given $y \in R$, it should be "computationally infeasible" to find an $x \in D$ such that $h(x) = y$.

Second Pre-image Resistance: For a given $x \in D$, it should be "computationally infeasible" to find a distinct $x' \in D$ such that $h(x) = h(x')$.

Collision Resistance: It should be "computationally infeasible" to find distinct $x, x' \in D$ such that $h(x) = h(x')$.

In all attacks described in this paper, the focus is on violating the last requirement above: that is, we wish to find collisions in MD5.

In 1993 B. den Boer and A. Bosselaers [4] found two messages that collided under MD5 with two different IVs. In 1996 H. Dobbertin [5] published an attack, without details, that found a collision in MD5 with a chosen IV different from MD5's. Finally, at CRYPTO 2004, a team of researchers from the Shandong University in Jinan China, led by Xiaoyun Wang, announced collisions in MD5 as well as collisions in a host of other hash functions including MD4, RIPEMD, and HAVAL-128. Their findings were published at EUROCRYPT in 2005 [15, 16]. The same team presented two papers at the 2005 CRYPTO conference detailing applications of their methods to the hash functions SHA0 and SHA1, with a generated collision for SHA0, and a description on how to obtain collisions in SHA1. Given the variety of hash functions attacked by this team, it seems likely that their approach may prove effective against all cryptographic hashes in the MD family, including all variants of SHA. It therefore seems worthwhile to seek a complete understanding of how this approach works, how it can be improved, and how it can be generalized.

In Wang's short talk at the CRYPTO rump session, few details were given. She presented a brief general overview of the attacks, including the exact differentials for the pairs of colliding message blocks, along with several example collisions and estimations of the time complexity for each attack. In the interim, between her talk and the publication of the team's papers [15, 16], much interest was generated in finding the methods used by the Chinese researchers, and several papers were published on the subject [6, 8, 9]. Unfortunately, some key details of the attacks are omitted from the EUROCRYPT papers, and there are several discrepancies between the analysis done in [6, 9] and the results presented by the Chinese team.

OUR CONTRIBUTIONS. This paper attempts to consolidate and summarize all relevant knowledge of the attacks on MD5 from the works cited above [6, 8, 9, 15, 16], then additionally offer new insights and further improvements to this body of work. Specifically:

- We fully explain the "multi-message modification" technique invented by Wang.
- We offer new insights on how to find other differential paths.
- We use the above insights to demonstrate how to satisfy several more conditions in round 2 of the MD5 computation, thereby significantly speeding up the search for collisions.

- We demonstrate new methods for decreasing the search complexity when finding collisions.
- We provide an "MD5 Toolkit" that uses the above optimizations to produce MD5 collisions faster than any other known implementation; it also serves as a platform for testing further improvements and new ideas.

Along the way, we correct many of the errors made by previous authors in their published analyses, and we use what, we believe, is an improvement in notation. Also, in contrast to the other publications above, we provide full source code implementing our methods as an "MD5 Toolkit." Our hope is that this toolkit will serve as a useful device for researchers wishing to explore further techniques in this line of work. For example, making further code optimizations or search optimizations, adding further conditions, or searching for differential paths in an automated way. The MD5 Toolkit can be found at http://www.cs.colorado.edu/~jrblack/md5toolkit.tar.gz.

Our ultimate goal as a research community is to understand as best we can the way these iterated hash functions work, and the best known attacks against them. Our hope is that the observations offered here, along with the specific improvements we make for MD5 collision-finding, will lead to progress along these lines.

OVERVIEW OF THE PAPER. We begin by covering the notation used throughout the paper. Section 3 reviews the specification of MD5. We then give a high-level overview of the attacks and touch on the motivation and theory behind the attacks in section 4. Then we move on to the details of the attack in section 5.

The remainder of the paper is devoted to detailing our insights and improvements. Specific to MD5, we offer improvements that reduce the best-known time complexity [9] by roughly a factor of three. The methods used by the Chinese team require an expected 2^{37} MD5 computations to find the first block pair of the colliding messages, and an expected 2^{30} MD5 computations to find the second block pair. Klima [9] improved the attack so that an expected 2^{33} and 2^{24} MD5 computations are needed to find the first and second message block pairs, respectively, although Klima did not implement his improved attack for finding the second block pair. Our method improves the attack so that an expected 2^{30} MD5 computations are required to find the first block pair, and we implement Klima's code for finding the second block pair.

The Wang team reported that the example collision they found for the first block took about an hour on an IBM supercomputer, and the second block pair was found in 15 seconds to 5 minutes on the same computer. Our code produces both blocks in an average of 11 minutes on a commodity PC.

LATEST RESULTS. Since the publication of our paper, more progress has been made in attacking MD5. The HashClash project[14] implements Wang's attack using parallel computing resources to produce collisions in under 1 minute. Klima [7] has invented a new technique called "tunneling" also resulting in code that produces collisions in under a minute.

2   Notation

All indices start at 0. This is in contrast to the notation used in the Wang et al. papers, as well as [6, 9]. Thus, for a 4-byte unsigned integer x, the bits are labeled from 0 to 31, with 0 referring to the least significant bit. Let $\{0,1\}^n$ denote the set of all binary strings of length n. For an alphabet Σ, let Σ^* denote the set of all strings with elements from Σ. Let $\Sigma^+ = \Sigma^* - \{\epsilon\}$ where ϵ denotes the empty string. For strings s, t, let $s \parallel t$ denote the concatenation of s and t. For a binary string s let $|s|$ denote the length of s. For a string s where $|s|$ is a multiple of n, let $|s|_n$ denote $|s|/n$. Given binary strings s, t such that $|s| = |t|$, let $s \oplus t$ denote the bitwise XOR of s and t. For a string M such that $|M|$ is a multiple of n, $|M|_n = k$, then we will use the notation $M = (M_0, M_1, M_2, \ldots, M_{k-1})$ such that $|M_0| = |M_1| = |M_2| = \ldots = |M_{k-1}| = n$. We will also use the notation $M = (m_0, m_2, \ldots, m_{k-1})$ such that $|m_0| = |m_2| = \ldots = |m_{k-1}| = n$. This latter notation is used when $n = |m_i| = 32$. The former notation will be used when $n = |M_i| = 512$. We may think of M as a k-tuple if it is convenient (hence the vector notation). Generally, the symbol M will be used for members of $(\{0,1\}^{512})^+$. For a set S of the form $\{A_i : a \le i \le b\}$, we will sometimes denote S as $A_{a:b}$.

XOR DIFFERENTIAL VS. SUBTRACTION DIFFERENTIAL. These methods use a combination of the XOR differential and the subtraction differential, but with an emphasis on the subtraction differential. That is, for two integers $x, x' \in [0, 2^{31} - 1]$, consider the function $\Delta_X(x, x') = x \oplus x'$. This defines the XOR differential for x, x'. Alternatively, define $\Delta_S(x, x')$ as $x' - x \bmod 2^{32}$. This is the subtraction differential. The Chinese authors supply two columns of differentials in their tables of differentials for each step. One column contains the subtraction differential. Another contains what is essentially the XOR differential, but there is extra information included to indicate bit differences. For example, let $\Delta_S(x, x') = 2^2$. There are many possibilities for $\Delta_X(x, x')$ such as these three examples.

- $\Delta_X(x, x') = \text{0x00000004}$ (there is only one bit different between x and x', in index 2)
- $\Delta_X(x, x') = \text{0x0000000c}$ (bit 3 is set in x' but is not set in x, bit 2 is not set in x' but is set in x)
- $\Delta_X(x, x') = \text{0x0000fffc}$ (bit 15 is set in x' but is not set in x, bits 2 through 14 are not set in x' but are set in x

The differential used in [15, 16] captures this type of information by the following notation. Let x be in $[0, 2^{31} - 1]$. Then $x' = x[a_1, a_2, \ldots, a_n, -b_1, -b_2, \ldots, -b_m]$ denotes $x' = x + 2^{a_1} + 2^{a_2} + \cdots + 2^{a_n} - 2^{b_1} - 2^{b_2} \cdots - 2^{b_m} \bmod 2^{32}$. From this information one can compute both $\Delta_X(x, x')$ and $\Delta_S(x, x')$ if and only if for every index i for which x and x' differ $i \in \{a_1, a_2, \ldots, a_n, b_1, b_2, \ldots, b_m\}$. The complete differential tables in the full version of this paper [1] use this specialized differential, but with the above property so that both Δ_X and Δ_S may be computed.

3   The MD5 Algorithm

The following is a brief description of MD5 using the notation that is used to describe the attacks later in this paper. We omit message padding in this description since it has no effect on our attacks. The full specification for MD5 can be found in [11].

MD5 is a hash function in the Merkle-Damgård paradigm [2, 10], where the security of the hash function reduces to the security of its compression function. The MD5 compression function, which we denote as $MD5_c$, accepts as input a 128-bit chaining value CV which we break into four 32-bit values cv_0, cv_1, cv_2, cv_4 and a 512-bit message block M and outputs a 128-bit chaining value CV'. Formally, $MD5_c : \{0,1\}^{128} \times \{0,1\}^{512} \rightarrow \{0,1\}^{128}$. Let $H_0 \in \{0,1\}^{128}$ and let $M = (M_0, M_1, \ldots, M_k)$ for some $k \geq 0$ and $|M_i| \in \{0,1\}^{512}$ for $0 \leq i \leq k$. Then MD5(M) is computed as follows. Let $H_{i+1} = MD5_c(H_i, M_i)$ for $0 \leq i \leq k$. MD5(M) is defined as H_{k+1}.

3.1   The Compression Function $MD5_c$

We now detail the compression function used in MD5. There are 64 intermediate values produced, which we will call step values and denote by Q_i for $0 \leq i < 64$. The step values are computed in the following fashion:

$$T_i \leftarrow \Phi_i(Q_{i-1}, Q_{i-2}, Q_{i-3}) + Q_{i-4} + w_i + y_i$$
$$Q_i \leftarrow Q_{i-1} + (T_i \lll s_i)$$

Where s_i, y_i are step-dependent constants and w_i is the i-th block of the initial message expansion. For $0 \leq i < 64$, $w_i = m_j$ for some $0 \leq j < 16$. The exact message expansion can be found in [11]. By '$x+y$' we mean the addition of x and y modulo 2^{32}, and by '$x \lll y$' we mean the circular left shift of x by y bit positions (similarly, '$x \ggg y$' denotes the circular right shift of x by y bit positions).

The Φ function is defined in the following manner:

$$\begin{aligned}
\Phi_i(x, y, z) &= F(x, y, z) = (x \wedge y) \vee (\neg x \wedge z), & 0 \leq i \leq 15 \\
\Phi_i(x, y, z) &= G(x, y, z) = (x \wedge z) \vee (y \wedge \neg z), & 16 \leq i \leq 31 \\
\Phi_i(x, y, z) &= H(x, y, z) = x \oplus y \oplus z, & 32 \leq i \leq 47 \\
\Phi_i(x, y, z) &= I(x, y, z) = y \oplus (x \vee \neg z), & 48 \leq i \leq 63
\end{aligned}$$

$Q_{-1}, \ldots, Q_{-4}$ are determined by the chaining values to MD5 so that

$$Q_{-4} \leftarrow cv_0, \quad Q_{-3} \leftarrow cv_3, \quad Q_{-2} \leftarrow cv_2, \quad Q_{-1} \leftarrow cv_1$$

The chaining values are initially set to, in big endian byte order,

$$cv_0 \leftarrow \texttt{0x01234567}, \ cv_1 \leftarrow \texttt{0x89abcdef}$$
$$cv_2 \leftarrow \texttt{0xfedcba98}, \ cv_3 \leftarrow \texttt{0x76543210}$$

After all 64 steps are computed, $MD5_c$ computes

$$cv_0' \leftarrow cv_0 + Q_{60}, \ cv_1' \leftarrow cv_1 + Q_{63}, \ cv_2' \leftarrow cv_2 + Q_{62}, \ cv_3' \leftarrow cv_3 + Q_{61}$$

and outputs $CV' \leftarrow cv_0' \parallel cv_1' \parallel cv_2' \parallel cv_3'$.

Because of their importance later, we repeat some of our notation and terminology: for each message *block*, $MD5_c$ has four *rounds*, each of which computes 16 *step values* (for a total of 64).

4   High-Level Overview

Define δ_0 as $(0,0,0,0,2^{31},0,0,0,0,0,0,2^{15},0,0,2^{31},0)$ and δ_1 as $(0,0,0,0,2^{31}, 0,0,0,0,0,0,-2^{15},0,0,2^{31},0)$. Let $M = (M_0, M_1)$ be a 1024-bit string such that $|M_0| = |M_1| = 512$. For any such M let $M_0' = M_0 + \delta_0$, $M_1' = M_1 + \delta_1$ and $M' = (M_0', M_1')$ where addition is done component-wise modulo 2^{32}.

The Wang attacks describe a way of efficiently finding 1024-bit strings M such that $MD5(M) = MD5(M')$. They do this by tracking the differences in the step values during the computation of $MD5(M)$ and $MD5(M')$. Formally, let Q_i denote the output of the i-th round of the MD5 compression function upon input M and let Q_i' denote the output of the i-th round of MD5 upon input M'. Then [16] supplies 128 values (64 for the first block and 64 for the second block) a_i, $0 \le i < 128$ such that if their methods find an M such that $MD5(M) = MD5(M')$, then $Q_i' - Q_i = a_i$ for all Q_i computed during the computation of $MD5_c(M_0)$ and $MD5_c(M_0')$ and $Q_i' - Q_i = a_{i+64}$ for all Q_i computed during the computation of $MD5_c(M_1)$ and $MD5_c(M_1')$. We will call the values $Q_i' - Q_i$ *differentials*. The a_i are the correct or prescribed differentials. Additionally, four extra values are given in [16] that specify the differentials for the intermediate chaining values, or the outputs of $MD5_c(M_0)$ and $MD5_c(M_0')$.

It is not described in [16] or elsewhere how they chose the values for a_i, but in the next subsection we conjecture some ideas on their derivation. Regardless, Wang et al. detail methods for efficiently finding such M by determining conditions on the Q_i such that if those conditions are satisfied then the differentials hold with high probability ([16] mistakenly labels the conditions as 'sufficient'). Very little information is given in [16] as to how the conditions on the Q_i are obtained, but an excellent analysis is given by Hawkes, Paddon, and Rose in [6].

Wang's method for finding an M of the correct form can be described in pseudocode as the following:

Algorithm Find_Collision
while collision_found is false **do**:

1. Use random seeds and deterministic methods to find M which satisfies most conditions on Q_i
2. Compute all Q_i and Q_i' to check to see if differentials are correct
3. **if** (rest_of_differentials_hold) **then** collision_found ← true
   else collision_found ← false

end do
return M

x	y	z	$\Delta x \Rightarrow \Delta F$	$\Delta y \Rightarrow \Delta F$	$\Delta z \Rightarrow \Delta F$	x	y	z	$\Delta x \Rightarrow \Delta G$	$\Delta y \Rightarrow \Delta G$	$\Delta z \Rightarrow \Delta G$
0	0	0			✓	0	0	0	✓		
0	0	1	✓		✓	0	0	1	✓		
0	1	0	✓		✓	0	1	0		✓	✓
0	1	1			✓	0	1	1	✓		✓
1	0	0		✓		1	0	0		✓	✓
1	0	1	✓	✓		1	0	1	✓		✓
1	1	0	✓	✓		1	1	0		✓	
1	1	1		✓		1	1	1	✓		

Fig. 1. Output differences for $F = \Phi_i$, $0 \leq i < 16$ and $G = \Phi_i$, $16 \leq i < 32$

We also note here that the above pseudocode is actually done once for each block of M. First a 512-bit block M_0 is found that satisfies all first-block differentials, then block M_1 is found.

4.1 Finding the Differentials and Conditions

GENERATING MESSAGE DIFFERENTIALS. The derivation of the message and step value differentials used by Wang remains unexplained. We attempt here to conjecture how these were derived, although we stress that this is pure speculation and guesswork.

We begin by noting the following three things:

- The Φ function for round three is just the bitwise XOR of the inputs and is therefore linear - any change in one of the inputs necessarily changes the output in the same bits (formally, for any six 32-bit unsigned integers u, v, w, x, y, z, $H(x \oplus u, y \oplus v, z \oplus w) = H(u, v, w) \oplus H(x, y, z)$). On a related note, as can be seen in figure 1, Φ_i for $0 \leq i < 32$ has some 'absorbing' properties. That is, it is common that bit changes in the input do not change the output.
- The differential is 0 for the last few step values in round 2 and the first few step values for round 3.
- The differential is 2^{31} for almost all step values in rounds 3 and 4.

The full version of this paper contains a detailed description of how this difference in bit 31 is propagated in round three through the introduction of differences via the message words and careful manipulation of certain properties of addition modulo 2^{32} and the function H. Basically, addition of 2^{31} modulo 2^{32} operates the same as XOR. For now, however, let us note that it appears that the message differentials were chosen for this exact reason - propagating a single bit difference through most of round three.

The above analysis leads us to believe the following course of action was used in determining the message and step value differentials:

- Assume that whatever message differences are introduced in the first and second rounds can be absorbed by the Φ_i functions so that there are no differences in the step values used in the first step of round 3.
- Pick message differences so that the difference in bit 31 cascades through the step values. This involves:
  - Picking blocks in the initial message expansion m_a, m_b, m_c, such that $m_a = w_i$, $m_b = w_{i+1}$, $m_c = w_{i+3}$, $32 \leq i < 45$.
  - Let the differential be $m'_b = m_b + 2^{31}$, $m'_c = m_c + 2^{31}$ and $m'_a = m_a + 2^{31-s_i}$ where s_i is the shift value for round i.
- Find a differential path through the first and second rounds, using the message differentials chosen above, so that the difference for the last four step values in round 2 is zero.
- Find sufficient conditions on the step values to guarantee the differential path (the work done in [6] is an excellent resource on this step).
- For the above step try to minimize 2nd round conditions to avoid complicated multi-message modification techniques.

This third to last step is still surrounded in mystery, but one can see that by the properties of the Φ_i functions for rounds 1 and 2 that the task is possible. Although the step update function for MD4 and RIPEMD is different than that of MD5, the Wang et al. attacks [15] on those functions support the above analysis. That is, there is no difference in the step values for the last few steps of round 2 and the message differentials appear to have been chosen to minimize differences in round three by exploiting the linearity of bit 31. Again, we stress that this analysis is guesswork and we eagerly await a full exposition by the authors of [15, 16].

FULFILLING THE CONDITIONS. Conditions on Q_i are conditions on the individual bits of Q_i. For example, for the first block near-collision of MD5 to guarantee the differential they require that the 8-th least significant bit of Q_4 is zero. There are a total of 290 conditions on the round values for the first block attack, and there a total of 310 conditions for step values in the second block.

However, most of these conditions occur in the first and second rounds. This is important because during the first round, one can easily change M so that all the conditions are satisfied because at that point one has complete control over M and any changes do not affect prior computation. The Chinese team denoted these types of changes as "single-message modifications" or "single-step modifications." We will adopt and use the former terminology. Some round two conditions may also be corrected by other methods, which we will refer to as "multi-message modifications," but these methods are considerably more complicated because one has to be sure, because of the initial message expansion, that changes to M do not affect the computation of earlier rounds.

We present efficient methods [9, 16] which satisfy all but 30 conditions for the first block, and all but 24 conditions for the second block. The remaining conditions are satisfied in a probabilistic manner. On the assumption that each condition is satisfied with probability $1/2$, an expected 2^{30} (2^{24}, resp) messages need to be generated before a message M is found which satisfies all the first (second, resp) block conditions. This estimate is actually a tad low, because it does not account for the fact that the conditions on the Q_i are necessary, but not sufficient, for the step differentials to hold, even in the later rounds where the differentials cannot be satisfied deterministically.

5   The Dirty Details

In the full version [1] this section is intended as a detailed step-by-step guide to writing code that implements the MD5 attacks. Here, however, we focus only on some detailed examples of the method known as multi-message modification. It is this method which is perhaps the most unexplained and crucial step of the Chinese methods. We give a general overview of the method and go through details for the first block multi-message modifications based on Klima's paper [9]. In the appendix we go over some new, more complex methods in detail. A more general and comprehensive approach to this technique is presented in [3].

5.1   Multi-message Modification

One of the key ideas in the Chinese papers is that of multi-message modification. This is where after the satisfaction of all first-round conditions has occurred, one may alter several message blocks together to satisfy second round conditions while leaving all first-round conditions satisfied. Despite the importance of these methods for decreasing the time complexity of the attack, the description in [15, 16] is either completely omitted or brief and truncated. We seek here to fully explain the mystery of multi-message modification techniques by covering the general ideas behind the method and then walking through a few examples in detail in the appendix.

GENERAL IDEA. In [16] the method of multi-message modification is given, almost entirely, in a table similar to the following:

			Modify m_i	$a^{new}, b^{new}, c^{new}, d^{new}$
1	m_1	12	$m_1 \leftarrow m_1 + 2^{26}$	d_1^{new}, a_1, b_0, c_0
2	m_2	17	$m_2 \leftarrow ((c_1 - d_1^{new}) \ggg 17) - c_0 - \Phi_2(d_1^{new}, a_1, b_0) - y_2$	c_1, d_1^{new}, a_1, b_0
3	m_3	22	$m_3 \leftarrow ((b_1 - c_1) \ggg 22) - b_0 - \Phi_3(c_1, d_1^{new}, a_1) - y_3$	b_1, c_1, d_1^{new}, a_1
4	m_4	7	$m_4 \leftarrow ((a_2 - b_1) \ggg 7) - a_1 - \Phi_4(b_1, c_1, d_1^{new}) - y_4$	a_2, b_1, c_1, d_1^{new}
5	m_5	12	$m_5 \leftarrow ((d_2 - a_2) \ggg 12) - d_1^{new} - \Phi_5(a_2, b_1, c_1) - y_5$	d_2, a_2, b_1, c_1

The table is a guide to correcting the condition on $Q_{16,31}$, or $a_{5,32}$ in the notation from [16]. The condition is that this bit must be 0. The first column denotes the step number. The second column and third columns denote the message word and the shift value, respectively, used in the computation of the step value. The column under the heading "Modify m_i" details the update needed to correct the step value or message word in that step. The last column lists updates to step variables, if any, after the modification for that step.

How does this all work? Let's walk through the table. Although not shown in the table, the shift value for round 16 is 5. Therefore, the addition of 2^{26} to

m_1 has the net effect of adding 2^{31} to $Q_{16,31}$, which corrects for the condition in question. However, this change to m_1 also changes the value of a step value computed earlier: $Q_1 (= d_1)$. Therefore we must recompute d_1 with the new value of m_1 to obtain d_1^{new} (this is not explicitly shown in the above table, but we will come to this in a bit). The other rows of the table detail how to assimilate the changes in d_1 so that none of the other step values are changed (but the message bits are). Note that we can still change other message bits because in step 16 only one message block has been used to compute more than one step value. Namely, m_1. At the end of the process, $m_1, m_2, m_3, m_4, m_5, Q_1$, and Q_{16} have been changed, but all other step values and message bits remain the same. The change in Q_{16} was to remedy the incorrect condition, and the other changes were necessary to absorb the changes to m_1 and Q_1.

Furthermore, in the paper by Wang et al. discussing their attack on MD4, the table denotes that to update d_1 to d_1^{new} all one needs to do is add 2^{26} shifted by the appropriate amount (in this case 12, so that $d_1^{new} = d_1 + 2^6$). This does not always produce the correct value because shifting and carry expansion do not commute. The safest way to compute d_1^{new} is to just re-do the step value computation. A complete table with the updated computation is given below.

			Modify m_i	$a^{new}, b^{new}, c^{new}, d^{new}$
1	m_1	12	$m_1^{new} \leftarrow m_1 + 2^{26}$	d_1^{new}, a_1, b_0, c_0
			$d_1^{new} \leftarrow a_1 + ((\Phi_1(a_1, b_0, c_0) + d_0 + y_1 + m_1^{new}) \lll 12)$	
2	m_2	17	$m_2^{new} \leftarrow ((c_1 - d_1^{new}) \ggg 17) - c_0 - \Phi_2(d_1^{new}, a_1, b_0) - y_2$	c_1, d_1^{new}, a_1, b_0
3	m_3	22	$m_3^{new} \leftarrow ((b_1 - c_1) \ggg 22) - b_0 - \Phi_3(c_1, d_1^{new}, a_1) - y_3$	b_1, c_1, d_1^{new}, a_1
4	m_4	7	$m_4^{new} \leftarrow ((a_2 - b_1) \ggg 7) - a_1 - \Phi_4(b_1, c_1, d_1^{new}) - y_4$	a_2, b_1, c_1, d_1^{new}
5	m_5	12	$m_5^{new} \leftarrow ((d_2 - a_2) \ggg 12) - d_1^{new} - \Phi_5(a_2, b_1, c_1) - y_5$	d_2, a_2, b_1, c_1

So this is the gist of multi-message modification, but this simple trick does not handle all cases, and unfortunately the details to some of the trickier modifications are not to be found in the Chinese papers. In the appendix we go through an example of a slightly more complex multi-message modification, in addition to attempting to explain the motivation for each step in the method. We hope that by doing so the reader gains a deeper understanding of the (as yet more-or-less unexplained) method.

5.2 1st Block Multi-message Modification

Here we present our methods for finding the first block pair, based on the methods found in [9, 16]. Before detailing our new methods for satisfying three extra conditions, we review and correct the collision-finding pseudocode in Klima's paper [9].

1ST BLOCK COLLISION-FINDING PROGRAM OUTLINE. Klima is able to satisfy four extra conditions from [16] through some clever probabilistic multi-message modifications. The following outline is nearly identical to that which is presented in Klima's full paper [9]. There are a couple of mistakes in Klima's multi-message modification methods as presented in his paper, however. A few of the steps are out of order and some crucial steps are omitted. Here is how the code should look (using our notation with shifted indices):

1. We choose $Q_{2:15}$ fulfilling conditions.
2. We compute $m_{6:15}$: For i going from 6 to 15 do

$$m_i \leftarrow ((Q_i - Q_{i-1}) \ggg s_i) - F(Q_{i-1}, Q_{i-2}, Q_{i-3}) - Q_{i-4} - y_i$$

3. We change Q_{16} until conditions $Q_{16:18}$ are fulfilled. Sometimes this is not possible (because the values of $Q_{12}, Q_{13}, Q_{14}, Q_{15}$ do not allow the conditions on Q_{17} and Q_{18} to hold), and it becomes necessary to change $Q_{2:15}$.

$$Q_{17} \leftarrow Q_{16} + ((G(Q_{16}, Q_{15}, Q_{14}) + Q_{13} + m_6 + y_{17}) \lll s_{17})$$
$$Q_{18} \leftarrow Q_{17} + ((G(Q_{17}, Q_{16}, Q_{15}) + Q_{14} + m_{11} + y_{18}) \lll s_{18})$$

4. All conditions $Q_{2:18}$ are fulfilled now. Moreover, we have free value m_0.
5. We choose Q_{19} arbitrarily, but fulfilling the one condition for it. Then we compute m_0:

$$m_0 \leftarrow ((Q_{19} - Q_{18}) \ggg s_{19}) - G(Q_{18}, Q_{17}, Q_{16}) - Q_{15} - y_{19}$$

6. Compute Q_0 from new value of m_0:

$$Q_0 \leftarrow Q_{-1} + ((F(Q_{-1}, Q_{-2}, Q_{-3}) + Q_{-4} + m_0 + y_0) \lll s_0)$$

7. Compute m_1:

$$m_1 \leftarrow ((Q_{16} - Q_{15}) \ggg s_{16}) - G(Q_{15}, Q_{14}, Q_{13}) - Q_{12} - y_{16}$$

8. Compute Q_1 from new values of m_1, Q_0:

$$Q_1 \leftarrow Q_0 + ((F(Q_0, Q_{-1}, Q_{-2}) + Q_{-3} + m_1 + y_1) \lll s_1)$$

9. Compute $m_{2:5}$: For i going from 2 to 5 do

$$m_i \leftarrow ((Q_i - Q_{i-1}) \ggg s_i) - F(Q_{i-1}, Q_{i-2}, Q_{i-3}) - Q_{i-4} - y_i$$

For step 3, we chose to satisfy the conditions on $Q_{16:18}$ probabilistically, by simply randomly selecting Q_{16} and checking to see whether the other conditions were satisfied. There are only 9 conditions for these three chaining variables, so this can be done quickly. Sometimes no selection of Q_{16} will satisfy the conditions, so in this case our code simply begins anew by randomly selecting another $Q_{2:15}$ such that the first round conditions are satisfied.[2]

After step 9, we continue the computation, checking to see if the remaining conditions are satisfied (each condition is expected to be satisfied with probability near $1/2$, so we expect to iterate over the above pseudocode 2^{30} times before we find a suitable first block pair). If a condition isn't satisfied, then we have to choose a new message. We do this efficiently by iterating over all possible 2^{31} values of Q_{19} in step 5 (simply incrementing Q_{19} after each failed attempt is the

---

[2] We implement this by setting a reasonable upper limit on the number of random selections of Q_{16} which are chosen and tested.

fastest way). If we exhaust all possible values for Q_{19} without finding a suitable message, we return to step 3 and select another value for Q_{16}. In this manner we avoid significant unnecessary computation.

PERFORMANCE. We ran code based on the work done by [13], modified with our extra methods, to find the first block 80 times run on a desktop 3.0 GHz processor. Overall, full two-block collisions were found, on average, in under 5 minutes. This is a dramatic improvement over the timings given by Klima, even after correcting for discrepancies in hardware.

6   A New Method

In our research we found the following optimizing heuristic, which was verified experimentally but not analytically:

> Relying on the fact that the step update function is not very random, we can attempt to identify patterns in the step values which tend to yield solutions. Using this knowledge, we narrow our search space to use only step values which fall within these patterns.

AN EXAMPLE. The methods presented in the appendix provide an analytic method to satisfy conditions on Q_{20} with probability near 15/16 (approximately 94% of the time). Although the analytic solution works, it nearly doubles the computation over the main iterative loop, thus weakening its positive impact on the running time of our collision-finding program. However, we were also able to obtain an equally or perhaps more effective (satisfied conditions around 97% of the time on tests) method for satisfying conditions on Q_{20}.

The pattern is simple. While iterating through values of Q_{19}, as in step 5 of the pseudocode in section 5.2, there are distinct patterns in which values of Q_{19} automatically satisfy conditions on Q_{20}. That is, ignoring our new multi-message modifications for Q_{20}, we tried to identify which values of Q_{19} led to conditions on Q_{20} being satisfied. We found that values of Q_{19} which satisfied conditions on Q_{20} often occurred sequentially in blocks of 128 followed by 128 consecutive values of Q_{19} which *didn't* satisfy conditions on Q_{20}. There were exceptions to this pattern, but the correlation was strong enough to reduce the time complexity. This suggests a new method for satisfying conditions on Q_{20} that requires much less computation:

– While iterating over values of Q_{19}, check to see if the conditions on Q_{20} are satisfied. If not, add 127 to Q_{19} and continue.

Restricting the values of Q_{19} in this manner yields an algorithm for which around 97% of all used values of Q_{19} satisfy conditions on Q_{20}, with the additional benefit that very little overhead is needed compared to the other multi-message modification method.

Several other examples are discussed in the full version of this paper[1].

THE METHOD. We have not attempted to systematically identify and define this approach within our own work. It was merely that through a casual analysis of data patterns observed during coding, we noticed this phenomenon. Nonetheless, it seems that we used the following rough methodology:

- Record values for intermediate step values as well as result (Were all differentials satisfied with these values? If not, how many were satisfied?). Do this for many random choices of M.
- Attempt to find simple patterns in these step values which will yield a good heuristic.

This procedure can have broad or narrow scope. With the above example, we looked at consecutive values of Q_{19} and checked only one condition on Q_{20}. Broadening the scope may yield results, or it may decrease the chances that simple patterns will be easy to find.

The above technique seems possible to automate so that no human interaction is necessary and we believe this is a possible avenue for future research. We suspect that artificial intelligence techniques could be especially useful with this sort of analysis. The main drawback to this method is that one might not be able to easily understand *why* these patterns of data exist. In general, it seems preferable to do as much analysis as possible, but it seems likely that an automated tool to detect these kinds of patterns may be used with great success after analysis becomes too cumbersome or fruitless.

7   The Full Version

We encourage the interested reader to look at our full version [1], which can be found at `http://www.cs.colorado.edu/~jrblack/papers.html`. This version of the paper contains the following additional content:

- Full, complete, updated tables of the 1st and 2nd block conditions and differentials that use our notation.
- Many more details for those wishing to implement the attacks themselves, correcting errors in earlier papers. This includes discussion of the single-message modification technique.
- A detailed step-by-step guide to the 2nd block multi-message modification methods.
- More examples of the technique presented in section 6.
- An in-depth presentation of additional new multi-message modifications for the 1st block.
- A more detailed explanation of the derivation of Wang's differentials.
- Discussion on the performance of various methods.

Acknowledgements

Many thanks to Matt Robshaw and the anonymous FSE reviewers for their comments and to Vincent Rijmen for his work organizing the event and helping coordinate travel. John Black's and Martin Cochran's work was supported by NSF CAREER-0240000 and NSF-0524118. Trevor Highland's work was supported by NSF REU-0244168; his travel to FSE was in part supported by the IACR.

References

1. BLACK, J., COCHRAN, M., AND HIGHLAND, T.   A study of the MD5 attacks: Insights and improvements (full version). Manuscript available at `http://www.cs.colorado.edu/~jrblack/papers.html`.
2. DAMGÅRD, I. A design principle for hash functions. In *CRYPTO* (1989), vol. 435 of *Lecture Notes in Computer Science*, Springer-Verlag, pp. 416–427.
3. DAUM, M. Cryptanalysis of hash functions of the MD4 family. Dissertation available at `http://www.cits.rub.de/imperia/md/content/magnus/dissmd4.pdf`.
4. DEN BOER, B., AND BOSSELAERS, A. Collisions for the compression function of MD5. In *EUROCRYPT* (1993), vol. 765 of *Lecture Notes in Computer Science*, Springer, pp. 293–304.
5. DOBBERTIN, H. Cryptanalysis of MD5 compress. Presented at the rump session of EUROCRYPT '96.
6. HAWKES, P., PADDON, M., AND ROSE, G. G. Musings on the Wang et al. MD5 collision, October 2004. See `http://eprint.iacr.org/2004/264`.
7. KLIMA, V.   Tunnels in hash functions: MD5 collisions within a minute. See `http://eprint.iacr.org/2006/105`.
8. KLIMA, V. Finding MD5 collisions: A toy for a notebook, March 2005.   See `http://eprint.iacr.org/2005/075`.
9. KLIMA, V. Finding MD5 collisions on a notebook PC using multi-message modifications. In *International Scientific Conference Security and Protection of Information* (May 2005).
10. MERKLE, R. C. One way hash functions and DES. In *CRYPTO* (1989), vol. 435 of *Lecture Notes in Computer Science*, Springer-Verlag, pp. 428–446.
11. RIVEST, R. The MD5 message-digest algorithm. *RFC 1321* (April 1992).
12. ROGAWAY, P., AND SHRIMPTON, T. Cryptographic hash-function basics: Definitions, implications and separations for preimage resistance, second-preimage resistance, and collision resistance. In *Fast Software Encryption (FSE 2004)* (2004), vol. 3017 of *Lecture Notes in Computer Science*, Springer, pp. 371–388.
13. STACH, P., AND LIU, V.   MD5 collision generation.   Code available at `http://www.stachliu.com/collisions.html`.
14. STEVENS, M. HashClash. See `http://www.win.tue.nl/hashclash/`.
15. WANG, X., LAI, X., FENG, D., CHEN, H., AND YU, X. Cryptanalysis of the hash functions MD4 and RIPEMD. In *EUROCRYPT* (2005), vol. 3494 of *Lecture Notes in Computer Science*, Springer, pp. 1–18.
16. WANG, X., AND YU, H. How to break MD5 and other hash functions. In *EUROCRYPT* (2005), vol. 3494 of *Lecture Notes in Computer Science*, Springer, pp. 19–35.

A   New Multi-message Modification Methods

We now cover the details of our methods which reduce the overall complexity of the attack to an expected 2^{30} MD5 computations. In subsection 5.1 we covered the basic idea behind multi-message modifications and went over a simple example. However, not all second-round conditions can be handled as easily. In previous papers [15, 16], these more complicated methods are not described at all. Therefore our approach will be to walk through our techniques in detail, attempting to explain our methodology at each step so that the reader gains not only an understanding of our techniques, but hopefully insight into the general technique of multi-message modifications as well.

A.1   New Multi-message Modifications for Correcting $Q_{20,17}$ and $Q_{20,31}$

These modifications take into account all the modifications that Klima has done to correct the conditions on $Q_{0:19}$. These methods satisfy the two conditions on Q_{20} or a_6 (Wang's notation).

OUTLINE OF METHOD. We set up a few conditions on $Q_{16:19}$ so that flipping a couple of bits of Q_{18} and Q_{19} does not affect earlier computations but with high probability satisfies the two conditions on Q_{20}. For example, bit 31 of Q_{20} must be set to 0. Let's say it is 1. Note how Q_{20} is computed:

$$Q_{20} \leftarrow Q_{19} + ((G(Q_{19}, Q_{18}, Q_{17}) + Q_{16} + m_5 + y_{20}) \lll 5)$$

We set conditions on $Q_{17:19}$ so that by default the value of the 26th bit of $G(Q_{19}, Q_{18}, Q_{17})$ is 0, but that if we flip the 26th bits of both Q_{19} and Q_{18} then the value of the 26th bit $G(Q_{19}, Q_{18}, Q_{17})$ changes to 1. To derive such conditions, one has to look at how the function G is computed, but it can easily be verified that if the 26th bits of Q_{18} and Q_{19} are 0, then the 26th bit of $G(Q_{19}, Q_{18}, Q_{17})$ will be zero, and if the 26th bits of Q_{19} and Q_{18} are flipped to 1, then the 26th bit of $G(Q_{19}, Q_{18}, Q_{17})$ will also be flipped. Flipping the 26th bit of $G(Q_{19}, Q_{18}, Q_{17})$ in this manner has the net effect of adding $2^{31} + 2^{26}$ to the value of Q_{20} because we have added 2^{26} to Q_{19}, which occurs twice in the computation of Q_{20} (once in the computation of $G()$ and once by addition to $T_{19} \lll 5$). Adding 2^{31} flips the most significant bit of Q_{20}, like we wanted, and the addition of 2^{26}, which we cannot really avoid, will only re-flip the most significant bit of Q_{20} if the next 5 most significant bits of Q_{20} were originally set, which occurs with probability $1/32$.

At this point the observant reader may ask "Why did we have to flip the 26th bits of both Q_{19} and Q_{18}?". "Why not just flip the 26th bit of Q_{19}?" Here's why: Remember back in Klima's code how m_0 was computed:

$$m_0 \leftarrow ((Q_{19} - Q_{18}) \ggg s_{19}) - G(Q_{18}, Q_{17}, Q_{16}) - Q_{15} - y_{19}$$

If we just changed Q_{19}, we would have to re-compute m_0, which would affect the computation of m_5 a couple of steps later, and Q_{20} would likewise be affected. In fact, changing m_0 by one bit in this way can change m_5 by a bunch of bits, so we must be very careful so we don't have to modify it for our methods to work. By changing both Q_{19} and Q_{18} in the same way, the changes cancel each other out and m_0 is not changed, so long as $G(Q_{18}, Q_{17}, Q_{16})$ is not affected by these changes. It can easily be verified that requiring the condition $Q_{16,26} = 0$ satisfies this goal (1 more condition). It is important to note that these added conditions do not significantly affect the performance of the overall code because they are satisfied in step 3 of Klima's code. Instead of 9 conditions to probabilistically satisfy in step 3, we now have 11, which is still tiny in comparison to the overall runtime.

Okay. So we've sneakily changed Q_{19} and Q_{18} so that m_0 is unaffected and a condition on Q_{20} is fulfilled with high probability. Now we have to fix everything else. We recompute the value of m_{11} from the new value of Q_{18} by the following:

$$m_{11} \leftarrow ((Q_{18} - Q_{17}) \ggg s_{18}) - G(Q_{17}, Q_{16}, Q_{15}) - Q_{14} - y_{18}$$

And we now have to recompute Q_{11} from m_{11}.

$$Q_{11} \leftarrow Q_{10} + ((F(Q_{10}, Q_9, Q_8) - Q_7 - y_{11}) \lll s_{11})$$

Luckily the changes we made to m_{11} don't affect any of the 15 conditions on Q_{11}, so long as bit 2 of Q_{11} is originally set to 0 (so that we don't have to worry about carries). So we add this condition to the list - again, it is fulfilled "for free" by the single-message modification methods presented in the Wang papers (or by the fact that Q_{11} is initially arbitrarily chosen in the Klima paper).

The only thing left to do is to recompute $m_{12:15}$ to absorb the changes in Q_{11}. This can be done without changing any of the other Q variables by simply recomputing $m_{12:15}$ as we did earlier:

$$m_{12} \leftarrow ((Q_{12} - Q_{11}) \ggg s_{12}) - F(Q_{11}, Q_{10}, Q_9) - Q_8 - y_{12}$$
$$m_{13} \leftarrow ((Q_{13} - Q_{12}) \ggg s_{13}) - F(Q_{12}, Q_{11}, Q_{10}) - Q_9 - y_{13}$$
$$m_{14} \leftarrow ((Q_{14} - Q_{13}) \ggg s_{14}) - F(Q_{13}, Q_{12}, Q_{11}) - Q_{10} - y_{14}$$
$$m_{15} \leftarrow ((Q_{15} - Q_{14}) \ggg s_{15}) - F(Q_{14}, Q_{13}, Q_{12}) - Q_{11} - y_{15}$$

That's it. At the end of everything we have changed Q_{19} and Q_{18} so that one condition on Q_{20} has been changed with probability $31/32$, m_{11} and Q_{11} have been changed, but without affecting the conditions on Q_{11}, and m_{12-15} have been changed to absorb the changes in Q_{11} so that no other Q values are affected.

The exact same method can be used to correct the condition on the 17th bit of Q_{20} (just shift all bit values above by 14). There are a total of 8 new conditions that this method requires, but they are all more or less "free."

It is possible that the above methods fail to correct the specified conditions, but the probability that this happens is bounded above by $1/32 + 1/32 = 1/16$.

After each iteration, our code goes back to starting values for $m_{11:15}$, Q_{11}, and Q_{18}, because we need the correct bits of Q_{11} and Q_{18} to be set so that flipping them to satisfy Q_{20} can occur safely.

The Impact of Carries on the Complexity of Collision Attacks on SHA-1*

Florian Mendel**, Norbert Pramstaller,
Christian Rechberger, and Vincent Rijmen

Institute for Applied Information Processing and Communications (IAIK)
Graz University of Technology, Austria
Norbert.Pramstaller@iaik.tugraz.at

Abstract. In this article we present a detailed analysis of the impact of carries on the estimation of the attack complexity for SHA-1. We build up on existing estimates and refine them. We show that the attack complexity is slightly lower than estimated in all published work to date. We point out that it is more accurate to consider probabilities instead of conditions.

1 Introduction

In past years, significant progress has been made in the cryptanalysis of the hash functions MD4, MD5, RIPEMD, SHA-0, and SHA-1 [2,3,5,6,9,10,11,13,14,15]. In 2004 and 2005, Wang *et al.* announced that they had broken the hash functions MD4, MD5, RIPEMD, HAVAL, SHA-0, and SHA-1 [16,17,19,20,21].

SHA-1, a widely used hash function in practice, has attracted most attention over the last years. This year, at the CRYPTO 2005 rump session, Wang *et al.* announced that they have further improved their attack on SHA-1. They updated the attack complexity from 2^{69} to 2^{63} [18].

As it will be explained in Section 2, the attack complexity is mainly determined by the probabilities of so-called 6-step local collisions in a linearized variant of SHA-1. For each local collision, the attacker derives conditions such that the local collision holds for the original SHA-1. Based on the derived conditions the attack complexity is conjectured. The main contribution of this article is that we will show that it is more accurate to look at probabilities instead of estimating the attack complexity based on the number of conditions.

The remainder of this article is structured as following. We start with a short description of the hash function SHA-1 and review the basic attack strategy of Wang *et al.* in Section 2. In Section 3, we perform a detailed analysis of local collisions. Section 3.2 describes how Wang *et al.* derive conditions for local collisions. In Section 3.3 and Section 3.4 we present a more accurate analysis of local collisions and the corresponding probabilities. Based on these results we update the complexity of the collision attack on SHA-1 in Section 3.5. Finally, we present conclusions in Section 4.

---

* The work in this paper has been supported by CRYPTREC.
** This author is supported by the Austrian Science Fund (FWF) project P18138.

2   Collision Attacks on SHA-1

In this section we will review the basic attack strategy for collision attacks on SHA-1. We start with a short description of SHA-1, giving only the details we need later in this article.

2.1   Short Description of SHA-1

The input message is split into 512-bit message blocks (after padding). The compression function is then applied to each of these 512-bit message blocks. The compression function basically consists of two parts: the message expansion and the state update. The message expansion expands the 512-bit input message block into 80 32-bit words W_i that are used in each step of the state update. A single step of the state update is shown in Figure 1.

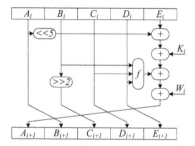

Fig. 1. One step of the state update of SHA-1

As it can be seen in Figure 1, in each step the function f is applied to the inputs B_i, C_i, and D_i. The function f depends on the step number: steps 0 to 19 (round 1) use f_{IF} and steps 40 to 59 (round 3) use f_{MAJ}. f_{XOR} is applied in the remaining steps (round 2 and 4). The functions are defined as:

$$f_{IF}(B, C, D) = B \wedge C \oplus \overline{B} \wedge D \tag{1}$$
$$f_{MAJ}(B, C, D) = B \wedge C \oplus B \wedge D \oplus C \wedge D \tag{2}$$
$$f_{XOR}(B, C, D) = B \oplus C \oplus D \ . \tag{3}$$

For a detailed description of SHA-1 refer to [12].

2.2   The Basic Attack Strategy on SHA-1

In 1998, Chabaud and Joux presented an attack on SHA-0 [3]. They used a linearized variant of SHA-0 to find a characteristic, which we will refer to as *L-characteristic* throughout the remainder of this article. For the linearized variant all modular additions are replaced by XOR and the functions f_{MAJ} and f_{IF} are replaced by f_{XOR}. They observed the following: the probability that the characteristic holds for the original SHA-1 is related to the Hamming weight of the characteristic. In general, the lower the weight the higher the probability.

In 2004 and 2005, Wang *et al.* announced that they have broken the hash functions MD4, MD5, RIPEMD, SHA-0, and SHA-1 [19,20,21]. For the collision attack on SHA-1 they use basically the following strategy, which is also depicted in Figure 2. They search for a low-weight *L-characteristic* that leads to a pseudo collision in the last 60 steps (referred to as P2 in Figure 2). Then by using a nonlinear characteristic (referred to as *NL-characteristic*) in the first 20 steps (referred to as P1 in Figure 2), they are able to turn the pseudo collision into a collision. Furthermore, they improved their attack by searching for an *L-characteristic* that leads to a pseudo-near collision in P2. As before, they turn the pseudo collision into a collision with the *NL-characteristic* and by using two-block messages they construct a collision from the near collision in each block. The fact that it is easier to find a near collision than a collision was observed already by Biham and Chen in [1]. An important property of this attack strategy is that the *NL-characteristic* has no impact on the complexity of the attack since conditions in P1 are fulfilled by using message modification techniques invented by Wang *et al.* Therefore, only the *L-characteristic* determines the attack complexity.

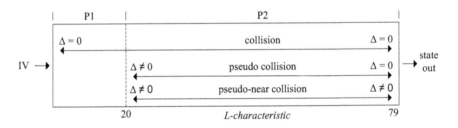

Fig. 2. Attack strategy of Wang *et al.*

The *L-characteristic* consists of overlapped single local collisions as it has been shown in [19]. To determine the attack complexity, Wang *et al.* count the number of conditions for each local collision such that it holds for the original SHA-1. Then they conjecture the attack complexity by assuming that after fulfilling the first 20 steps, random trials are performed to find the colliding messages. The complexity for this random trials is estimated to be $2^{\# \text{ conditions}}$.

Many researchers investigated the *L-characteristic* and tried to find *L-characteristics* with lower weight. A possible approach is to exploit coding theory since finding a low-weight *L-characteristic* in P2 corresponds to finding a low-weight codeword in a linear code describing P2. Results of the coding-theory approach are presented for instance in [8,11,13,15]. In 2005, Jutla and Patthak [7] used a computer aided proof to show that the minimum Hamming weight in the last 60 steps of the SHA-1 message expansion is 25. This low-weight vector is also referred to as the disturbance vector, since it contains the disturbances for the single local collisions. However, Wang *et al.* use a disturbance vector with higher weight (weight = 27). The reason for this is that the vector with higher weight leads to a smaller number of conditions (see [19]). Since the attack complexity is determined by the number of local collisions and the corresponding probabilities (conditions) we will analyze them accurately in the next section.

3   Detailed Analysis of Local Collisions in SHA-1

In the first part of this section, we start with deriving the conditions and corresponding probabilities for all possible local collisions in the *L-characteristic* of SHA-1. We follow the work of Wang *et al.* in [19] to conjecture the overall probability of a collision attack on SHA-1 based on these local collisions. Note that the *L-characteristic* does not include the first 20 steps of SHA-1 and therefore, we only consider the functions f_{XOR} and f_{MAJ} described in Section 2.1. In the second part of this section, we derive a more accurate estimation of the probabilities for local collisions. With this analysis we update the attack complexity of Wang *et al.* presented in [19].

3.1   Notation and Definitions

For the analysis of local collisions we follow the notation given in Table 1. Throughout the remainder of this article we will use signed bit differences. In the following we describe the basic properties of signed bit differences that we require for our analysis. A detailed discussion of signed bit differences can be found in [4, Chapter 4].

We define the sign of a difference in bit position j as

$$w'_j = w_j - w^*_j, \quad \text{where } w_j, w^*_j \in \{0,1\} \text{ and } w'_j \in \{-1,0,+1\} . \quad (4)$$

In particular, if $w'_j = 0$ the difference is zero. The signed bit difference is then defined as $W'_j = w'_j 2^j$. A useful property of signed bit differences is the fact that the difference also includes information about the values of w_j and w^*_j. This is shown in (5).

$$W'_j = \begin{cases} +2^j & \text{if } w_j = 1 \text{ and } w^*_j = 0 \\ 0 & \text{if } w_j = w^*_j \\ -2^j & \text{if } w_j = 0 \text{ and } w^*_j = 1 \end{cases} \quad (5)$$

Table 1. Notation

notation	description
step	the SHA-1 compression function consists of 80 steps, $0 \leq i \leq 79$
round	the SHA-1 compression function consists of 4 rounds $= 4 \times 20$ steps
$W_{i,j}$	bit j of expanded message word in step i, $0 \leq j \leq 31$
w'_j	sign of bit difference in bit position (j mod 32), $w'_j \in \{-1,0,+1\}$
$W'_j = w'_j 2^j$	signed bit difference in bit position (j mod 32), $W'_j \in \{-2^j, 0, +2^j\}$
$W'_{i,j}$	signed bit difference in step i, bit position j
$(j + n \bmod 32)$	bit position j rotated to the left by n positions
$(j - n \bmod 32)$	bit position j rotated to the right by n positions

Table 2. Addition of signed bit differences

A'_j	B'_j	C'_j	S'_j	C'_{j+1}	A'_j	B'_j	C'_j	S'_j	C'_{j+1}
0	0	0	0	0	0	u	v	0	$\frac{1}{2}(u+v)$
0	0	v	$(-1)^{A_j \oplus B_j} v$	$-v(A_j \oplus B_j)$	u	0	v	0	$\frac{1}{2}(u+v)$
0	v	0	$(-1)^{A_j \oplus C_j} v$	$-v(A_j \oplus C_j)$	u	v	0	0	$\frac{1}{2}(u+v)$
v	0	0	$(-1)^{B_j \oplus C_j} v$	$-v(B_j \oplus C_j)$	v	v	$-v$	$(-1)^{A_j \oplus B_j \oplus 1} v$	$(-1)^{A_j \oplus B_j} v$
v	v	v	$(-1)^{A_j \oplus B_j \oplus 1} v$	$(-1)^{A_j \oplus B_j} v$	v	$-v$	v	$(-1)^{A_j \oplus C_j \oplus 1} v$	$(-1)^{A_j \oplus C_j} v$
					$-v$	v	v	$(-1)^{B_j \oplus C_j \oplus 1} v$	$(-1)^{B_j \oplus C_j} v$

Table 3. Differential properties of f_{XOR} and f_{MAJ} for signed bit differences

B'_j	C'_j	D'_j	$f_{XOR}(B'_j, C'_j, D'_j)$	$f_{MAJ}(B'_j, C'_j, D'_j)$
0	0	v	$(-1)^{B_j \oplus C_j} v$	$(B_j \oplus C_j) v$
0	v	0	$(-1)^{B_j \oplus D_j} v$	$(B_j \oplus D_j) v$
v	0	0	$(-1)^{C_j \oplus D_j} v$	$(C_j \oplus D_j) v$

Let us now consider the addition of two signed bit differences. The addition $S = A + B$ is defined as $S_j = A_j \oplus B_j \oplus C_j$ and $C_{j+1} = f_{MAJ}(A_j, B_j, C_j)$ with $C_0 = 0$, where C_{j+1} is the resulting carry of the addition in bit position j. Table 2 lists all possible cases for the output and carry difference of a signed bit addition with $v, u \in \{-1, +1\}$.

To perform the addition of two signed bit differences we can use Table 2 for computing the resulting difference. We know that the output difference is $C'_{j+1} 2^{j+1} + S'_j 2^j$. For instance, if there are two non-zero differences at the input with opposite signs, then both C'_{j+1} and S'_j are zero and hence the output difference is zero. If the differences have the same sign, for instance -2^j and -2^j, the output difference is -2^{j+1}, since $C'_{j+1} = -1$ and $S'_j = 0$.

For our analysis we need the differential properties of f_{XOR} and f_{MAJ} with respect to signed bit differences. In Table 3, we list the cases that occur in a local collision (see Figure 3) where $v \in \{-1, +1\}$. As it can be seen in Table 3, for f_{XOR} the sign of the input difference is flipped with probability $1/2$ depending on the input values. For f_{MAJ} the sign is preserved but the difference propagates with probability $1/2$.

3.2   Considering the Number of Conditions

In [3], Chabaud and Joux showed how the corrections for a single bit disturbance in SHA-0 can be constructed. Since the state update for SHA-0 and SHA-1 is the same, this construction is also valid for SHA-1. Table 4 shows a local collision with signed bit differences for f_{XOR} and f_{MAJ}.

For the local collision defined in Table 4, we can now derive the number of conditions and the corresponding probabilities such that the local collision holds for the original SHA-1. We refer to conditions that contain only expanded message words as *easy* conditions since we can easily fulfill them. Conditions that

Table 4. Local collision (disturbance-corrections) for SHA-1

step	difference			description
		f_{XOR}	f_{MAJ}	
i	$W_i' =$	$+2^j$	$+2^j$	single bit disturbance at bit position j
$i+1$	$W_{i+1}' =$	-2^{j+5}	-2^{j+5}	correction
$i+2$	$W_{i+2}' =$	$\pm 2^j$	-2^j	correction
$i+3$	$W_{i+3}' =$	$\pm 2^{j-2}$	-2^{j-2}	correction
$i+4$	$W_{i+5}' =$	$\pm 2^{j-2}$	-2^{j-2}	correction
$i+5$	$W_{i+8}' =$	-2^{j-2}	-2^{j-2}	correction

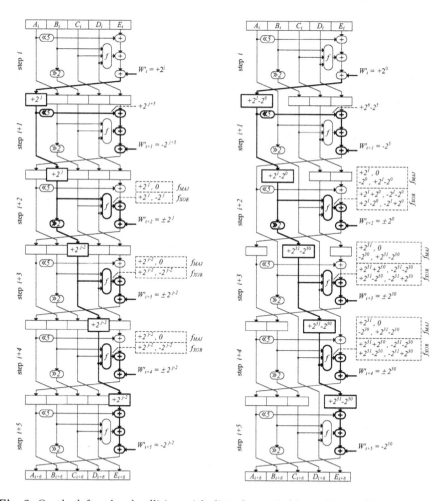

Fig. 3. On the left, a local collision with disturbance in bit position j. No carry occurs in step i. On the right a local collision with disturbance in bit position $j = 0$. In step i a carry occurs. The differences in the dashed rectangles are the possible output differences of f_{XOR} and f_{MAJ}.

include state variables are considered to be *hard* conditions. For the analysis we can assume without loss of generality that the sign of the disturbance is positive, *i.e.* $W_i' = +2^j$. If the disturbance is -2^j, we get the same results by just flipping all the other signs. The propagation of the disturbance and corrections is shown in the left part of Figure 3.

Disturbance in step i. In step i, where the disturbance is introduced, it is required that the disturbance propagates to state variable A_{i+1} without causing a carry in the difference, *i.e.* $A_{i+1}' = W_i' = +2^j$. This occurs with probability $1/2$. If the disturbance is introduced at bit position $j = 31$, it propagates to A_{i+1}' with probability 1.

Correction in step $i+1$. As shown in Figure 3, the difference in state variable A is rotated to the left by 5 positions. Therefore, the correction is $W_{i+1}' = -2^{j+5}$. It follows from Table 2 that if the sign of the correction is the opposite of the sign of the disturbance, then the correction occurs with probability 1. We can ensure the negative sign of the correction with condition CW_{i+1}: $W_{i+1,j+5} \oplus W_{i,j} = 1$. This condition is in W only and we can easily fulfill it.

Correction in step $i + 2$. In this step, we have to consider the modular addition and the function f. As described in Table 3, f_{XOR} flips the sign of the input difference with probability $1/2$. Therefore, for $B_{i+2}' = +2^j$ the output difference of f_{XOR} can be either $+2^j$ or -2^j depending on C_{i+2} and D_{i+2}. Since we cannot easily influence the values of C_{i+2} and D_{i+2} the probability for the correction is $1/2$.

For f_{MAJ} we get the same probability as for f_{XOR} by defining a condition in W only. For the input difference $B_{i+2}' = +2^j$ the possible output difference of f_{MAJ} is either $+2^j$ or 0. This results in a probability of $1/4$. However, if the sign of the correction is negative, then the correction has a probability of $1/2$. This can be ensured by fulfilling condition CW_{i+2}: $W_{i+2,j} \oplus W_{i,j} = 1$.

Correction in step $i + 3$ and $i + 4$. These steps are the same as step $i + 2$ except that the difference $+2^j$ is rotated to the right by 2 positions, *i.e.* $+2^{j-2}$. For f_{XOR} we get a probability of $1/2$ in each step. For f_{MAJ} we also get the probability $1/2$ by fulfilling the following easy conditions in W only: CW_{i+3}: $W_{i+3,j-2} \oplus W_{i,j} = 1$, and CW_{i+4}: $W_{i+4,j-2} \oplus W_{i,j} = 1$.

Correction in step $i + 5$. If all corrections have taken place in the previous steps the signed bit difference is in state variable E. As it can be seen in Figure 3, E_{i+5}' is the same difference as $A_{i+1}' = +2^j$ rotated by 2 to the right, *i.e.* $E_{i+5}' = +2^{j-2}$. We only have to consider the modular addition. As in step $i + 1$, we can fulfill condition CW_{i+5}: $W_{i+5,j-2} \oplus W_{i,j} = 1$ such that the correction has negative sign. Hence, the correction in step $i + 5$ has probability 1.

Local collision with best probability. With the above described probabilities for each step of the local collision we can define a local collision that has the best probability for f_{XOR}. Assume the disturbance is introduced in bit position

Table 5. Probabilities for local collisions in SHA-1

disturbance	probability		easy conditions on W	
	f_{XOR}	f_{MAJ}	f_{XOR}	f_{MAJ}
$j = 1$	2^{-2}	2^{-4}	CW_{i+1}	CW_{i+1}, CW_{i+2}
$j = 26$	2^{-4}	2^{-4}	CW_{i+5}	$CW_{i+2}, CW_{i+3}, CW_{i+4}, CW_{i+5}$
$j = 31$	2^{-3}	2^{-3}	CW_{i+1}, CW_{i+5}	$CW_{i+1}, CW_{i+3}, CW_{i+4}, CW_{i+5}$
$j = 0, 2, \ldots, 25$ $j = 27, \ldots, 30$	2^{-4}	2^{-4}	CW_{i+1}, CW_{i+5}	$CW_{i+1}, CW_{i+2}, CW_{i+3}, CW_{i+4}, CW_{i+5}$

$j = 1$. In step i we have a probability of $1/2$. Since we can easily fulfill condition CW_{i+1} we have a probability of 1 in step $i + 1$. In step $i + 2$ the probability is $1/2$. Now, for steps $i+3$ to $i+5$ the disturbance is rotated to bit position $j = 31$. Since a carry in the difference can be ignored (addition mod 2^{32}), we get a total probability of 2^{-2} for a local collision with disturbance in bit position $j = 1$.

Summary of probabilities of local collisions. Table 5 summarizes the probabilities for all possible local collisions with a single-bit disturbance and lists the easy conditions in W that have to be fulfilled. For the discussion so far we only considered probabilities and easy conditions. However, the probabilities for the modular addition and the functions f_{MAJ} and f_{XOR} can also be described in terms of so-called *hard* conditions. Each single condition is fulfilled with probability $1/2$. Consider for instance f_{MAJ}. The input difference $B_i' = +2^j$ leads to the output difference $+2^j(C_i \oplus D_i)$ (see Table 3). In order to ensure that the difference propagates, we require that $C_i \oplus D_i = 1$. Since we cannot easily influence the values of C_i and D_i, the condition is fulfilled with probability $1/2$. The same can be done for the other cases. For a local collision with disturbance in bit position $j = 1$, we have a probability of 2^{-4}. In other words there are 4 *hard* conditions that we cannot easily fulfill.

With the probabilities listed in Table 5 the complexity of the attack on SHA-1 can be determined. For the description we follow the work of Wang et al. [19]. For the disturbance vector [19, Table 5] we compute the product of all probabilities for each disturbance bit to determine the overall probability and hence the attack complexity.

3.3   Accurate Probability Computation

In Section 3.2, we determined the probabilities of local collisions with disturbances introduced at different bit positions. For the analysis we did not allow carries in step i where the disturbance is introduced. This restriction can actually be relaxed. In the following we will analyze the impact of carries in step i on the probability of local collisions. We will show that the probabilities are actually higher for most bit positions of the disturbance.

Single bit disturbance. We start with a disturbance in bit position $j = 0$. As shown in Table 5 this results in a probability of 2^{-4}. Now consider that a carry occurs in the difference in step i, $i.e.$ the disturbance $W_i' = +2^0$ propagates to $A_{i+1}' = +2^1 - 2^0$. This case is shown on the right hand side in Figure 3.

The carry in step i occurs with probability $1/4$. The difference in bit position $j = 1$ can be seen as a new disturbance that leads to a second local collision with a certain probability. To cancel out the difference $A_{i+1}' = +2^1$ we require that the corrections in the consecutive steps also produce a carry in the difference. As described in Section 3.2, we fulfill condition CW_{i+1} to ensure that $W_{i+1}' = -2^5$. Therefore, the differences cancel out with probability 1 since $(+2^6 - 2^5) + (-2^5) = 0$ (as shown in Table 2, $-2^5 + (-2^5) = -2^6$ and hence $2^6 - 2^6 = 0$). For steps $i+2$ to $i+4$ we first consider f_{XOR}. In step $i+2$ we have a probability of $1/4$ because f_{XOR} flips the sign of a bit difference with probability $1/2$. Since we have two bit differences this results in a probability of $1/4$. The same holds for steps $i+3$ and $i+4$. However, since the disturbance is introduced in bit position $j = 0$, the second difference caused by the carry is rotated to bit position $j = 31$ in step $i+2$. We can ignore carries in this bit position and hence the sign in bit position $j = 31$ has no impact. Therefore, we get a probability of $1/2$ for each step. We can do the same analysis for f_{MAJ}. As already mentioned, f_{MAJ} preserves the sign of the input difference but the difference propagates only with probability $1/2$. Therefore, we cannot exploit bit position $j = 31$—the probability for steps $i+3$ and $i+4$ is $1/4$ each. For step $i+2$ the probability is $1/4$ since CW_{i+2} is fulfilled. In step $i+5$ we have a probability of 1 for f_{XOR} and f_{MAJ} based on the same reasoning as for step $i+1$. With the results of this analysis we can update the probability of Section 3.2. The best probability for f_{XOR} and f_{MAJ} with a disturbance in bit position $j = 0$ is:

$$p(f_{XOR}, j = 0) = 2^{-4} + 2^{-6} = 2^{-3.6781} , \tag{6}$$
$$p(f_{MAJ}, j = 0) = 2^{-4} + 2^{-8} = 2^{-3.9125} . \tag{7}$$

Uncorrectable carries. Let us now consider the case where two carries in step i occur, $i.e.$ $W_i' = +2^0$ propagates to $A_{i+1}' = +2^2 - 2^1 - 2^0$. Two carries occur with probability $1/8$. If we work with the difference in bit position $j = 2$, we encounter the following problem, which we refer to as $uncorrectable$ $carries$. In step $i+2$ the difference is rotated by two positions to the right, $i.e.$ $-2^{31} - 2^{30} + 2^0$. It is not possible to correct the difference $+2^0$ in step $i+3$ anymore since the correction takes place in bit position $j = 30$. For f_{MAJ}, uncorrectable carries for this example take place only in step $i+5$. This is due to the fact that the difference $+2^0$ is blocked by f_{MAJ} with probability $1/2$ in steps $i+2$ to $i+4$. However, in step $i+5$ we cannot correct the difference $+2^0$ since the correction takes place in $j = 30$. Therefore, the probabilities given in (6) and (7) are the best probabilities for both functions with a disturbance in $j = 0$.

If we perform the carry analysis for bit position $j = 1$, we also encounter uncorrectable carries as for the disturbance in $j = 0$. Namely, a carry in step i cannot be corrected anymore in step $i+3$ (step $i+5$ for f_{MAJ}, respectively) and therefore, a carry does not increase the probability for a local collision with disturbance in $j = 1$ for both f_{XOR} and f_{MAJ}. Uncorrectable carries can also occur due to the left rotation by 5 in step $i+1$. A disturbance in $j = 26$ that leads to a carry in step i cannot be corrected anymore in step $i+1$ since the correction W'_{i+1} takes place in bit position $j = 31$ but the carry is rotated to $j = 0$.

Carries that improve the probability of local collisions. After determining the probabilities for $j = 0$ and $j = 1$, we describe now the impact of carry effects for disturbances in bit position $j = 2, \ldots, 31$. Due to uncorrectable carries after bit position $j = 26$ we have to analyze the probability for $j = 2, \ldots, 26$ and $j = 27, \ldots, 31$ separately. We start the explanation for f_{XOR}. For $2 \le j \le 26$ we have the same probability in steps i, $i+2$, $i+3$, and $i+4$, namely the probability that no carry occurs and the probabilities for all possible carries. Note that the probability in steps $i+1$ and $i+5$ is 1 since we fulfill the easy conditions CW_{i+1} and CW_{i+5} (see Section 3.2). For $27 \le j \le 31$ we have the same except that the probability in step $i+2$ is increased by a factor of 2 if the carry in step i reaches bit position $j = 31$. For f_{MAJ} we also assume that the easy conditions in W are fulfilled. Then we get the same probabilities as for f_{XOR} with the difference that for $27 \le j \le 31$ we cannot exploit bit position $j = 31$. In (8) and (9) we give the formulae to compute the accurate probability for a local collision including all carry effects. Probability bounds for (8) and (9) are given in Appendix A. For a disturbance in bit position $j = 3$ the probability for both f_{XOR} and f_{MAJ} is $2^{-3.9068}$ instead of 2^{-4} which is the probability derived by counting conditions.

$$p(f_{XOR}, j) = \begin{cases} 2^{-2} & \text{for } j = 1 \\ 2^{-4} + 2^{-6} & \text{for } j = 0 \\ \sum_{k=1}^{27-j} 2^{-4k} & \text{for } j = 2, \ldots, 26 \\ 2 \cdot 2^{-4 \cdot (32-j)} + \sum_{k=1}^{31-j} 2^{-4k} & \text{for } j = 27, \ldots, 31 \end{cases} \tag{8}$$

$$p(f_{MAJ}, j) = \begin{cases} 2^{-4} & \text{for } j = 1 \\ 2^{-3} & \text{for } j = 31 \\ 2^{-4} + 2^{-8} & \text{for } j = 0 \\ \sum_{k=1}^{27-j} 2^{-4k} & \text{for } j = 2, \ldots, 26 \\ \sum_{k=1}^{32-j} 2^{-4k} & \text{for } j = 27, \ldots, 30 \end{cases} \tag{9}$$

3.4   Disturbances in Consecutive Bit Position

If we have a look at the disturbance vector in [19, Table 5] or [13, Table 7] there occur disturbances in consecutive bit positions, i.e. $W'_i = +2^{j+1} + 2^j$ for f_{XOR}. For the explanation we take the concrete case with disturbance $W'_i = -2^1 + 2^0$, and the five corrections $W'_{i+1} = +2^6 - 2^5$, $W'_{i+2} = +2^1 - 2^0$, $W'_{i+3} = +2^{31} + 2^{30}$, $W'_{i+4} = +2^{31} + 2^{30}$, and $W'_{i+5} = +2^{31} - 2^{30}$. In a straightforward way we can just treat them as separate disturbances and compute the probability based on (8). This results in a probability of

$$p(f_{XOR}, -2^1 + 2^0) = \underbrace{2^{-2}}_{j=1} \cdot \underbrace{(2^{-4} + 2^{-6})}_{j=0} = 2^{-5.678} . \tag{10}$$

Table 6. Update on complexity for collision attack on SHA-1

	[19, Table 9]			our work
disturbance bit position	disturbance index	number of conditions	estimated probability	accurate probability
$j = 1$	$23, 24, 27, 28, 32, 35, 36$	$2 \cdot 7 = 14$	2^{-14}	2^{-14}
$j = 0$	$25, 29, 33$	$4 \cdot 3 = 12$	2^{-12}	$2^{-11.0343}$
$j = 1$	$39, 43, 45, 47, 49$	$4 \cdot 5 = 20$	2^{-20}	2^{-20}
$j = \{2, 3, 4, 5, 7\}$	$65, 68, 71, 73, 74$	$4 \cdot 5 = 20$	2^{-20}	$2^{-5 \cdot 3.9068} = 2^{-19.534}$
		total	2^{-66}	$2^{-64.5683}$

However, by performing a detailed analysis we show that the probability for this case can be improved to $p(f_{XOR}, -2^1 + 2^0) = 2^{-3.678}$ by defining two additional conditions in W only, referred to as CW_i and $CW1_{i+2}$. We assume that the easy conditions described in Section 3.2 are fulfilled. If no carry occurs in step i, both disturbances are corrected with probability 2^{-6}. This follows from Section 3.2. Now consider the case that a carry occurs in step i. Assume that in step i the disturbances have opposite signs, e.g. $W_i' = -2^1 + 2^0$. This can be ensured by fulfilling the new condition CW_i: $W_{i,1} \oplus W_{i,0} = 1$. If a carry occurs in bit position $j = 0$ the difference that propagates to A_{i+1}' is -2^0 since the positive sign of the carry (see Table 2) cancels the negative difference in $j = 1$. This occurs with probability $1/2$. In step $i + 1$ the probability is 1 since CW_{i+1} is fulfilled. In step $i + 2$ we can increase the probability to $1/2$ if the additional condition $CW1_{i+2}$: $W_{i+2,1} \oplus W_{i+2,0} = 1$ is fulfilled. This is based on the same reasoning as for step i. For the remaining steps $i+3$ to $i+4$ we get a probability of $1/2$ for each step. Again, in step $i+5$ we have a probability of 1. Hence we have a total probability of 2^{-4} for the case that a carry occurs in step i. Therefore, the total probability for the disturbance $+2^1 - 2^0$ or $-2^1 + 2^0$ is

$$p(f_{XOR}, -2^1 + 2^0) = \underbrace{2^{-4}}_{\text{carry in } j=0} + \underbrace{2^{-6}}_{\text{no carry in step } i} = 2^{-3.6781} . \qquad (11)$$

Wang et al. use a probability of 2^{-4} for their estimation. For disturbances in other consecutive bit positions the same analysis can be performed. For f_{XOR} the analysis is given in Appendix B.

3.5 Update of Attack Complexity by Wang et al.

With the above analysis we covered all cases of disturbances that occur in the disturbance vector of [19]. Since they count conditions in the last 60 steps of SHA-1 the overall probability can be updated based on (8) and (9). Table 6 lists the comparison with [19, Table 9].

As it can be seen in Table 6 the probability is by a factor of approx. 2.7 higher than estimated in [19]. Note that we did not count the disturbances in step $i = 21$ and step $i = 77$ since some of the conditions are fulfilled due to message modification or truncation. This means that the path of the disturbance is fixed and we cannot exploit any carry effects.

In order to determine the overall probability, we assume that the probabilities of local collisions are independent. To confirm this assumption, we have performed several computer measurements for a few overlapping local collisions. The measurement results match the computed probabilities.

3.6 Importance of Carry Effects

In the case of SHA-1, the improvement of the attack complexity is rather small. This is due to the fact that the disturbance vector is very sparse and the disturbances are introduced in bit positions where we cannot exploit any carry effects due to uncorrectable carries, *e.g.* bit position $j = 1$.

Consider for instance the hash function SHA1-IME [8]. Jutla and Patthak claim to improve the collision resistance of SHA-1 by modifying the existing message expansion with the goal to increase the minimum Hamming weight. By using a computer aided proof they show that the minimum weight in the last 60 steps of the message expansion of SHA1-IME is at least 75. It is clear that the overall complexity increases with a higher weight in the disturbance vector. However, due to the higher weight also the impact of carry effects as shown in this section increases. Therefore, our way of looking at probabilities instead of conditions gives a more accurate complexity estimation.

4   Conclusion and Further Work

In this article we analyzed local collisions and corresponding probabilities in detail. We showed that it is more accurate to consider probabilities instead of conditions for the estimation of the overall attack complexity for collision attacks on SHA-1. This is due to the fact that carry effects increase the probability. Based on the accurate probability computation we updated the complexity of the collision attack on SHA-1 presented by Wang *et al.* Currently we are investigating the impact of our approach on SHA1-IME and local collisions in SHA-256.

Acknowledgements

We would like to thank Christophe De Cannière for fruitful discussions and comments on this article.

References

1. Eli Biham and Rafi Chen. Near-Collisions of SHA-0. In Matthew K. Franklin, editor, *CRYPTO 2004, Santa Barbara, California, USA, August 15-19, 2004, Proceedings*, volume 3152 of *LNCS*, pages 290–305. Springer, 2004.
2. Eli Biham, Rafi Chen, Antoine Joux, Patrick Carribault, Christophe Lemuet, and William Jalby. Collisions of SHA-0 and Reduced SHA-1. In Ronald Cramer, editor, *EUROCRYPT 2005, Aarhus, Denmark, May 22-26, 2005. Proceedings*, volume 3494 of *LNCS*, pages 36–57. Springer, 2005.

3. Florent Chabaud and Antoine Joux. Differential Collisions in SHA-0. In Hugo Krawczyk, editor, *CRYPTO '98, Santa Barbara, California, USA, August 23-27, 1998, Proceedings*, volume 1462, pages 56–71. Springer, 1998.
4. Magnus Daum.   *Cryptanalysis of Hash Functions of the MD4-Family.* PhD thesis, Ruhr Universität Bochum, 2005.   Available at `http://http://www.cits.rub.de/imperia/md/content/magnus/dissmd4.pdf` .
5. Hans Dobbertin. Cryptanalysis of MD4. In Bart Preneel, editor, *Fast Software Encryption, Cambridge, UK, February 21-23, 1996, Proceedings*, volume 1039 of *LNCS*, pages 53–69. Springer, 1996.
6. Hans Dobbertin. Cryptanalysis Of MD4. *Journal of Cryptology*, 11(4):253–271, 1998.
7. Charanjit S. Jutla and Anindya C. Patthak. A Matching Lower Bound on the Minimum Weight of SHA-1 Expansion Code. Cryptology ePrint Archive, Report 2005/266, 2005. `http://eprint.iacr.org/`.
8. Charanjit S. Jutla and Anindya C. Patthak. A Simple and Provably Good Code for SHA Message Expansion. Cryptology ePrint Archive, Report 2005/247, 2005. `http://eprint.iacr.org/`.
9. Vlastimil Klima. Finding MD5 Collisions on a Notebook PC Using Multi-message Modifications, 2005. Preprint, available at `http://eprint.iacr.org/2005/102`.
10. Arjen Lenstra, Xiaoyun Wang, and Benne de Weger. Colliding X.509 Certificates, 2005. Preprint, available online at `http://eprint.iacr.org/2005/067`.
11. Krystian Matusiewicz and Josef Pieprzyk.   Finding good differential patterns for attacks on SHA-1.   Cryptology ePrint Archive, Report 2004/364, 2004. `http://eprint.iacr.org/`.
12. National Institute of Standards and Technology (NIST).   FIPS-180-2: Secure Hash Standard, August 2002.   Available online at `http://www.itl.nist.gov/fipspubs/`.
13. Norbert Pramstaller, Christian Rechberger, and Vincent Rijmen. Exploiting Coding Theory for Collision Attacks on SHA-1. In Nigel P. Smart, editor, *Cryptography and Coding, Cirencester, UK, December 19-21, 2005, Proceedings*, volume 3796 of *LNCS*, pages 78–95. Springer, 2005.
14. Bart Preneel. *Analysis and Design of Cryptographic Hash Functions*. PhD thesis, Katholieke Universiteit Leuven, 1993.
15. Vincent Rijmen and Elisabeth Oswald. Update on SHA-1. In Alfred Menezes, editor, *CT-RSA 2005, San Francisco, CA, USA, February 14-18, 2005, Proceedings*, volume 3376 of *LNCS*, pages 58–71. Springer, 2005.
16. Xiaoyun Wang, Dengguo Feng, Xuejia Lai, and Xiuyuan Yu. Collisions for Hash Functions MD4, MD5, HAVAL-128 and RIPEMD, August 2004. Preprint, available at `http://eprint.iacr.org/2004/199`.
17. Xiaoyun Wang, Xuejia Lai, Dengguo Feng, Hui Chen, and Xiuyuan Yu. Cryptanalysis of the Hash Functions MD4 and RIPEMD. In Ronald Cramer, editor, *EUROCRYPT 2005, Aarhus, Denmark, May 22-26, 2005. Proceedings*, volume 3494 of *LNCS*, pages 1–18. Springer, 2005.
18. Xiaoyun Wang, Andrew Yao, and Frances Yao. New Collision Search for SHA-1, August 2005. Presented at rump session of CRYPTO 2005.
19. Xiaoyun Wang, Yiqun Lisa Yin, and Hongbo Yu. Finding Collisions in the Full SHA-1. In Victor Shoup, editor, *CRYPTO 2005, Santa Barbara, California, USA, August 14-18, 2005, Proceedings*, volume 3621 of *LNCS*, pages 17–36. Springer, 2005.

20. Xiaoyun Wang and Hongbo Yu. How to Break MD5 and Other Hash Functions. In Ronald Cramer, editor, *EUROCRYPT 2005, Aarhus, Denmark, May 22-26, 2005. Proceedings*, volume 3494 of *LNCS*, pages 19–35. Springer, 2005.
21. Xiaoyun Wang, Hongbo Yu, and Yiqun Lisa Yin. Efficient Collision Search Attacks on SHA-0. In Victor Shoup, editor, *CRYPTO 2005, Santa Barbara, California, USA, August 14-18, 2005, Proceedings*, volume 3621 of *LNCS*, pages 1–16. Springer, 2005.

A   Probability Bounds for Single-Bit Disturbances

Based on formulae (8) and (9) in Section 3.3, the probability of f_{XOR} and f_{MAJ} can be bounded as follows. We know that

$$\sum_{k=1}^{27-j} 2^{-4k} = 2^{-4} \frac{1 - 2^{-4(28-j)}}{1 - 2^{-4}} \leq \frac{2^{-4}}{1 - 2^{-4}} = \frac{1}{2^4 - 1} \,,$$

$$\sum_{k=1}^{32-j} 2^{-4k} = 2^{-4} \frac{1 - 2^{-4(33-j)}}{1 - 2^{-4}} \leq \frac{2^{-4}}{1 - 2^{-4}} = \frac{1}{2^4 - 1} \,, \text{ and}$$

$$2 \cdot 2^{-4(32-j)} + \sum_{k=1}^{31-j} 2^{-4k} =$$

$$2^{-4(32-j)+1} + 2^{-4} \frac{1 - 2^{-4(32-j)}}{1 - 2^{-4}} \leq 2^{-3} + \frac{2^{-4}}{1 - 2^{-4}} = \frac{1}{2^3} + \frac{1}{2^4 - 1} \,.$$

Therefore, we get the following bounds on the probability for f_{XOR} and f_{MAJ}:

$$\frac{1}{2^4} \leq p(f_{XOR}, j) \leq \frac{1}{2^4 - 1} \text{ for } j = 2, \ldots, 26 \,, \tag{12}$$

$$\frac{1}{2^4} \leq p(f_{XOR}, j) \leq \frac{1}{2^3} + \frac{1}{2^4 - 1} \text{ for } j = 27, \ldots, 31 \,, \tag{13}$$

$$\frac{1}{2^4} \leq p(f_{MAJ}, j) \leq \frac{1}{2^4 - 1} \text{ for } j = 2, \ldots, 26 \text{ and } j = 27, \ldots, 30 \,, \tag{14}$$

where the lower bound for the probability 2^{-4} is derived by counting conditions. For instance, if we compute the probability for a disturbance in bit position $j = 3$ we get for both f_{XOR} and f_{MAJ} a probability of $2^{-3.9068}$ instead of 2^{-4}.

B   Probabilities for Disturbances in Consecutive Bit Position

Here we give the probabilities for disturbances in consecutive bit positions for f_{XOR}. This is the generalization of the case presented in Section 3.4. Again, we have to consider uncorrectable carries. Uncorrectable carries occur if the disturbances are in bit position $j = 2, 1$ and $j = 27, 26$. In these cases, we get the probability of both disturbances without carry. If $j = 2, 1$, we obtain a probability of $2^{-4}2^{-2} = 2^{-6}$ and $j = 27, 26$ results in $2^{-4}2^{-4} = 2^{-8}$. Let us now consider disturbances in consecutive bit positions from $j = 2, \ldots 25$, *i.e.* the tuples $j = (3, 2), (4, 3), \ldots, (26, 25)$, and from $j = 27, \ldots, 30$, *i.e.* the tuples $j = (28, 27), (29, 28), (30, 29), (31, 30)$. The formulae for all cases are given in (15), where j refers to the right entry of the tuple.

$$p(f_{XOR}, (j+1, j)) = \begin{cases} 2^{-4} + 2^{-6} & \text{for } j = 0 \\ 2^{-4} + 2^{-8} & \text{for } j = 1 \text{ and } j = 26 \\ \sum_{k=1}^{27-j} 2^{-4k} & \text{for } j = 2, \ldots, 25 \\ 2 \cdot 2^{-4(32-j)} + \sum_{k=1}^{31-j} 2^{-4k} & \text{for } j = 27, \ldots, 30 \end{cases} \quad (15)$$

A New Mode of Encryption Providing a Tweakable Strong Pseudo-random Permutation

Debrup Chakraborty and Palash Sarkar

Applied Statistics Unit
Indian Statistical Institute
203 B.T. Road
Kolkata 700108, India
{debrup_r, palash}@isical.ac.in

Abstract. We present PEP, which is a new construction of a tweakable strong pseudo-random permutation. PEP uses a hash-encrypt-hash approach which has been recently used in the construction of HCTR. This approach is different from the encrypt-mask-encrypt approach of constructions such as CMC, EME and EME*. The general hash-encrypt-hash approach was earlier used by Naor-Reingold to provide a generic construction technique for an SPRP (but not a tweakable SPRP). PEP can be seen as the development of the Naor-Reingold approach into a fully specified mode of operation with a concrete security reduction for a tweakable strong pseudo-random permutation. HCTR is also based on the Naor-Reingold approach but its security bound is weaker than PEP. Compared to previous known constructions, PEP is the only known construction of tweakable SPRP which uses a single key, is efficiently parallelizable and can handle an arbitrary number of blocks.

Keywords: mode of operation, tweakable encryption, strong pseudo-random permutation.

1 Introduction

A block cipher is a fundamental primitive in cryptography. A block cipher by itself can encrypt only fixed length strings. Applications in general require encryption of long and arbitrary length strings. A mode of operation of a block cipher is used to extend the domain of applicability from fixed length strings to long and variable length strings. The mode of operation must be secure in the sense that if the underlying block cipher satisfies a certain notion of security, then the extended domain mode of operation also satisfies an appropriate notion of security.

A formal model of security for a block cipher is a pseudo-random permutation [9] which is formalized as a keyed family of permutations. Pseudo-randomness of the permutation family requires a computationally bounded adversary to be unable to distinguish between a random permutation and a permutation picked at random from the family. Strong pseudo-random permutations (SPRPs)

M.J.B. Robshaw (Ed.): FSE 2006, LNCS 4047, pp. 293–309, 2006.

require computational indistinguishability even when the adversary has access to the inverse permutation.

A mode of encryption usually provides two security assurances – privacy and authenticity. For example, OCB [14] is a mode of operation (providing both privacy and authenticity) which extends the domain of a block-cipher to arbitrary strings. An SPRP which can encrypt arbitrary length strings, can be viewed as a mode of operation with a somewhat different goal. Such a mode of operation is length preserving and no tag is produced. Hence, authentication is of limited nature. A change in the ciphertext cannot be detected but the decryption of the tampered ciphertext will result in a plaintext which is indistinguishable from a random string. Additional redundancy in the message introduced by higher level applications might even allow the detection of the tampering. This point is discussed in details by Bellare and Rogaway [1].

Tweakable encryption was introduced by Liskov, Rivest and Wagner [8] which added an extra input called tweak to a block cipher. This allows simplification of several applications. The paper [8] introduced both tweakable PRP and SPRP. In the adversarial model for tweakable SPRP, the adversary queries the encryption (resp. decryption) oracle with a tweak and the plaintext (resp. ciphertext). The adversary is allowed to repeat the tweaks in its queries to the encryption and the decryption oracles. A tweakable SPRP provides a mode of operation having all the advantages of an SPRP with the additional flexibility of having a tweak. The constructions CMC [4], EME [5], EME* [3] and HCTR [17] are proved to be tweakable SPRPs under the assumption that the underlying block cipher is an SPRP. As mentioned in [4], such a primitive is well suited for disk encryption where the tweak can be considered to be the sector address.

Our Contributions: We present a new construction of a tweakable SPRP called PEP (for Polynomial hash-Encrypt-Polynomial hash). PEP uses a block cipher which can encrypt an n-bit string to construct an encryption algorithm which can encrypt mn-bit strings for any $m \geq 1$. It uses two Wegman-Carter [16] style polynomial hashes over the binary field $GF(2^n)$. (Similar hashes have been earlier used in GCM [10] and HCTR.) The new construction is proved to be a tweakable SPRP assuming that the underlying block cipher is an SPRP. Below we mention some interesting features of PEP.

1. PEP is a fully specified mode of operation providing a tweakable SPRP. The security proof for PEP provides a concrete security bound which is the usual quadratic birthday bound earlier obtained for CMC, EME and EME*. The security bound of HCTR is weaker and the security degradation is cubic.

2. The total computation cost of PEP for encrypting an m-block message consists of $m + 5$ block cipher calls for $m \geq 2$ ($m + 3$ for $m = 1$), and $(4m - 6)$ multiplications in $GF(2^n)$. In contrast, CMC requires $2m + 1$ block cipher calls; EME* requires $(2m + m/n + 1)$ block cipher calls; and HCTR requires m block cipher calls and $2(m + 1)$ many $GF(2^n)$ multiplications. The exact comparison of the computation costs between PEP and CMC depends upon several factors such as the implementation platform (hardware or software), availability of suitable co-processors (for software implementation),

availability of parallel encryption blocks (for hardware implementation) and most importantly on the actual block cipher being used *and* the efficiency of its implementation.

3. Currently, PEP fills a gap in the known constructions. It is the only known construction of tweakable SPRP which uses a single key, is efficiently parallelizable and can handle arbitrary number of message blocks. Table 1 in Section 3 provides a detailed comparison of PEP with the other tweakable SPRPs.

Related Constructions: To the best of our knowledge, the first suggestion for constructing an SPRP was made by Naor and Reingold in [11]. They suggested the hash-encrypt-hash approach used in PEP. However, as discussed in [4], the description in [11] is at a top level and also the later work [12] does not fully specify a mode of operation. Perhaps more importantly, the work [12] does not consider *tweakable* SPRP since it predates the introduction of tweakable primitives in [8].

The NR approach was rejected in [4] as not being capable of efficient instantiation. The work in [4] and also the later constructions [5,3] follow a encrypt-mask-encrypt strategy, i.e., there are two layers of encryptions with a layer of masking in between. CMC [4] provides the first efficient, fully specified construction of a tweakable SPRP. Parallel versions of the encrypt-mask-encrypt strategy have been proposed as EME and EME*.

Interestingly, the NR approach made a recent comeback in the HCTR construction. The HCTR construction combines the NR type invertible hash functions with the counter mode of operation. This results in an efficient tweakable SPRP which can handle any message having length $\geq n$ bits. The drawback of HCTR is its weaker security bound and the requirement of having two keys. Currently, PEP can be viewed as the development of the NR approach to the construction of tweakable SPRP.

2   Specification of PEP

We construct the tweakable enciphering scheme PEP from a block cipher $E :$ $\mathcal{K} \times \{0,1\}^n \to \{0,1\}^n$ and call it PEP[E]. The key space of PEP[E] is same as that of the underlying block cipher E and the tweak space is $\mathcal{T} = \{0,1\}^n$. The message space consists of all binary strings of size mn where $m \geq 1$.

Finite Field Arithmetic: An n-bit string can also be viewed as an element in $GF(2^n)$. Thus, we will consider each n-bit string in the specification of PEP as a polynomial over $GF(2)$ modulo a fixed primitive polynomial $\tau(x)$ of degree n. For each n-bit string Z that occur in the description of PEP, we will use $Z(x)$ to denote the corresponding polynomial in $GF(2^n)$. The expressions $p(x)M_1$ (resp. xEN), represent the n-bit string obtained by multiplying the polynomials $p(x)$ and M_1 (resp. x and EN) modulo $\tau(x)$. Also for two n-bit strings Z_1 and Z_2, the expression $Z_1(x)Z_2(x)$ denotes the n-bit string obtained by multiplying $Z_1(x)$

and $Z_2(x)$ modulo $\tau(x)$. Finally, by $R^{-1}(x)$ we will denote the multiplicative inverse of $R(x)$ modulo $\tau(x)$ when $R(x) \neq 0$.

Definition 1. *Let $m \geq 3$ and $p_{m,1}(x), \ldots, p_{m,m}(x)$ be a sequence of polynomials over $GF(2)$ each having degree at most $n-1$. We call such a sequence an* allowed *sequence* with respect to *a primitive polynomial $\tau(x)$ of degree n if the following two conditions hold.*

1. $\bigoplus_{i=1}^{m} p_{m,i}(x) \equiv 0 \bmod \tau(x)$.
2. *For $1 \leq i < j \leq m$, $(p_{m,i}(x) \oplus p_{m,j}(x)) \not\equiv 0 \bmod \tau(x)$.*

From the definition, it is clear that for an allowed sequence to exist, we must have $m \geq 3$. The parameter m in the specification of PEP will represent the number of blocks to be encrypted (or decrypted). Later we show how to easily define such an allowed sequence. Since $\tau(x)$ is fixed, we will simply write "allowed sequence" instead of "allowed sequence with respect to $\tau(x)$".

The notation $\text{bin}(m)$ denotes the n-bit binary representation of the integer m. For example, $\text{bin}(1) = 0^{n-1}1$ and $\text{bin}(2) = 0^{n-2}10$. The specification of PEP consists of three cases: $m = 1$; $m = 2$; and $m \geq 3$. The cases $m = 1$ and $m = 2$ are shown in Figure 2. Figure 3 shows the encryption of a 4-block message. The complete encryption and decryption algorithms are shown in Figure 1.

Remark: For decryption to be possible, we need $R(x)$ to have a multiplicative inverse modulo $\tau(x)$. Since $R(x)$ is a polynomial of degree at most $n - 1$, the only value for which $R(x)$ does not have such an inverse is $R(x) = 0$. For such an R, the protocol is not defined. Note that $R = E_K(T)$. Assuming $E_K()$ to be a random permutation, the probability $R = 0$ is $1/2^n$. Since $n \geq 128$, the probability of getting a T for which the protocol is not defined is negligible.

Basic intuition behind the construction: The basic idea of the construction is to compute a polynomial hash (Wegman-Carter [16]) of the message. (Similar hashes are used in the GCM mode of operation [10] and HCTR.) This hash is XOR-ed with EN to obtain the element MPP whose expression is the following.

$$MPP = EN \oplus \bigoplus_{i=1}^{m} P_i(x)R^{i-1}(x). \tag{1}$$

The mask M_1 is obtained by encrypting MPP. Since we are aiming at an SPRP, we should ensure that each ciphertext bit depends upon all the plaintext bits. To do this, the mask M_1 is "mixed" to the message blocks to obtain the PPP_i's. While doing this we must be careful. During decryption, we will obtain the PPP_i's after the decryption layer. Thus, we should be able to compute MPP from the PPP_i's. To ensure that this can be done, we do two things. First we convert the P_i's to PP_i's by multiplying with R^{i-1}. The second thing is to "distribute" M_1 among the PP_i's while obtaining the PPP_i's so as to ensure that

$$\bigoplus_{i=1}^{m} PPP_i = \bigoplus_{i=1}^{m} PP_i = \bigoplus_{i=1}^{m} P_i(x)R^{i-1}(x) = MPP \oplus EN. \tag{2}$$

This follows from the first property of allowed sequences, namely, $\bigoplus_{i=1}^{m} p_{m,i}(x) \equiv$ $0 \bmod \tau(x)$ and hence MPP can be computed either as $\bigoplus_{i=1}^{m} PP_i \oplus EN$ or as $\bigoplus_{i=1}^{m} PPP_i \oplus EN$. Since allowed sequences exist only for $m \geq 3$, we need to tackle the cases $m = 1$ and $m = 2$ separately. In the case $m = 1$, there is no requirement to "distribute" M_1 among the message blocks while in the case of $m = 2$, this is a bit tricky to do. In the last case, we distribute M_1 as $M_1 \oplus EN$ and $M_1 \oplus EEN$. Note that the XOR of these two elements is $EN \oplus EEN$ (and not 0). However, both EN and EEN can be computed from the tweak and the length of the message and hence $MPP = PP_1 \oplus PP_2 \oplus EN = PPP_1 \oplus PPP_2 \oplus EEN$ and can be computed both during encryption and decryption.

The computation after the encryption layer is similar to the computation before the encryption layer. This is because we are constructing an SPRP and the view from the decryption end will be similar to the view from the encryption end. In the decryption query, we work with $L(x) = R^{-1}(x)$ while computing the polynomial hash. This is required to ensure the consistency of decryption. This leads to decryption requiring one extra inversion operation making it slightly more costlier than encryption.

2.1   Construction of Allowed Sequence of Polynomials

In this section, we provide one construction of allowed sequence. We do not claim this to be the only possibility; there may be others.

Let $\tau(x)$ be a primitive polynomial of degree n. Let $m_{\max}$ be a positive integer such that $\tau(x)$ does not divide any trinomial of degree less than $m_{\max}$. Estimates of $m_{\max}$ have been studied in the context of attacks on the nonlinear combiner model for stream ciphers [7]. (To use a primitive polynomial $\tau(x)$ in such stream ciphers, it is necessary that $\tau(x)$ does not divide a low degree trinomial.) This study indicates that $m_{\max}$ is around $2^{n/3}$. Given $\tau(x)$, there are algorithms for computing sparse multiples of $\tau(x)$. See [15] for a discussion on this issue.

We will be constructing an allowed sequence of length $m_{\max}$ and hence we will not be able to encrypt a message having more than $m_{\max}$ blocks. For $n = 128$, we have $2^{n/3} \approx 2^{42.6}$ and hence $m_{\max}$ is also around $2^{42.6}$. The ability to encrypt messages having at most $2^{42.6}$ many blocks is sufficient for all practical purposes.

Let $m = 3t \leq m_{\max}$, with $t \geq 1$. We define a sequence of polynomials $\tau_{3t,1}(x), \ldots, \tau_{3t,3t}(x)$ in the following manner.

$$\left.\begin{array}{ll} \tau_{3t,i}(x) & = x^i \qquad\qquad\quad \text{for } 1 \leq i \leq 2t; \\ \tau_{3t,2t+i}(x) = x^{2i-1} \oplus x^{2i} & \text{for } 1 \leq i \leq t. \end{array}\right\} \tag{3}$$

Let $3 \leq m \leq m_{\max}$. We define a sequence of polynomials $p_{m,1}(x), \ldots, p_{m,m}(x)$ in the following manner.

$m = 3t$: Define $p_{m,i}(x) = \tau_{m,i}(x)$.

$m = 3t + 1$: Define $p_{m,1} = 1 \oplus x$, $p_{m,2}(x) = x \oplus x^2$, $p_{m,3}(x) = x^2 \oplus x^3$, $p_{m,4}(x) = x^3 \oplus 1$ and $p_{m,4+i}(x) = x^3 \tau_{3(t-1),i}(x)$, for $1 \leq i \leq m$.

Algorithm $E_K^T(P_1, P_2, \ldots, P_m)$	Algorithm $D_K^T(C_1, C_2, \ldots, C_m)$
$R = E_K(T)$;	$R = E_K(T)$;
$EN = E_K(R \oplus \mathrm{bin}(m))$;	$EN = E_K(R \oplus \mathrm{bin}(m))$;
$EEN = E_K(xEN)$;	$EEN = E_K(xEN)$;
if $m == 1$, then	if $m == 1$, then
$\quad PPP_1 = P_1 \oplus EN$;	$\quad CCC_1 = C_1 \oplus xEEN$;
$\quad CCC_1 = E_K(PPP_1)$;	$\quad PPP_1 = D_K(CCC_1)$;
$\quad C_1 = CCC_1 \oplus xEEN$;	$\quad P_1 = PPP_1 \oplus EN$;
$\quad$ return C_1;	$\quad$ return P_1;
endif	endif
	$L(x) = R(x)^{-1}$;
if $m == 2$, then	if $m == 2$, then
$\quad PP_1 = P_1$; $PP_2 = R(x)P_2(x)$;	$\quad CC_1 = C_1$; $CC_2 = L(x)C_2(x)$;
$\quad MPP = PP_1 \oplus PP_2 \oplus EN$;	$\quad MCC = CC_1 \oplus CC_2 \oplus EEN$;
$\quad M_1 = E_K(MPP)$;	$\quad M_2 = E_K(MCC)$;
$\quad PPP_1 = PP_1 \oplus M_1 \oplus EN$;	$\quad CCC_1 = CC_1 \oplus M_2 \oplus EN$;
$\quad PPP_2 = PP_2 \oplus M_1 \oplus EEN$;	$\quad CCC_2 = CC_2 \oplus M_2 \oplus EEN$;
$\quad CCC_1 = E_K(PPP_1)$;	$\quad PPP_1 = D_K(CCC_1)$;
$\quad CCC_2 = E_K(PPP_2)$;	$\quad PPP_2 = D_K(CCC_2)$;
$\quad MCC = CCC_1 \oplus CCC_2 \oplus EN$;	$\quad MPP = PPP_1 \oplus PPP_2 \oplus EEN$;
$\quad M_2 = E_K(MCC)$;	$\quad M_1 = E_K(MPP)$;
$\quad CC_1 = CCC_1 \oplus M_2 \oplus EN$;	$\quad PP_1 = PPP_1 \oplus M_1 \oplus EN$;
$\quad CC_2 = CCC_2 \oplus M_2 \oplus EEN$;	$\quad PP_2 = PPP_2 \oplus M_1 \oplus EEN$;
$\quad C_1 = CC_1$; $C_2 = R(x)CC_2(x)$;	$\quad P_1 = PP_1$; $P_2 = L(x)PP_2(x)$;
$\quad$ return C_1, C_2;	$\quad$ return P_1, P_2;
endif	endif
if $m \geq 3$, then	if $m \geq 3$, then
$\quad R_1 = 1$; $PP_1 = P_1$; $MPP = PP_1$;	$\quad L_1 = 1$; $CC_1 = C_1$; $MCC = CC_1$;
$\quad$ for $i = 2$ to m do	$\quad$ for $i = 2$ to m do
$\quad\quad R_i(x) = R(x)R_{i-1}(x)$;	$\quad\quad L_i(x) = L(x)L_{i-1}(x)$;
$\quad\quad PP_i(x) = R_i(x)P_i(x)$;	$\quad\quad CC_i = L_i(x)C_i(x)$;
$\quad\quad MPP = MPP \oplus PP_i$;	$\quad\quad MCC = MCC \oplus CC_i$;
$\quad$ end for	$\quad$ end for
$\quad MPP = MPP \oplus EN$;	$\quad MCC = MCC \oplus EEN$;
$\quad M_1 = E_K(MPP)$; $MCC = 0^n$;	$\quad M_2 = E_K(MCC)$; $MPP = 0^n$;
$\quad$ for $i = 1$ to m do	$\quad$ for $i = 1$ to m do
$\quad\quad PPP_i = PP_i \oplus p_{m,i}(x)M_1(x)$;	$\quad\quad CCC_i = CC_i \oplus p_{m,i}(x)M_2(x)$;
$\quad\quad CCC_i = E_K(PPP_i)$;	$\quad\quad PPP_i = D_K(CCC_i)$;
$\quad\quad MCC = MCC \oplus CCC_i$;	$\quad\quad MPP = MPP \oplus PPP_i$;
$\quad$ end for	$\quad$ end for
$\quad MCC = MCC \oplus EEN$;	$\quad MPP = MPP \oplus EN$;
$\quad M_2 = E_K(MCC)$;	$\quad M_1 = E_K(MPP)$;
$\quad CC_1 = CCC_1 \oplus p_{m,1}(x)M_2(x)$;	$\quad PP_1 = PPP_1 \oplus p_{m,1}(x)M_1(x)$;
$\quad R_1 = 1$; $C_1 = CC_1$;	$\quad L_1 = 1$; $P_1 = PP_1$;
$\quad$ for $i = 2$ to m do	$\quad$ for $i = 2$ to m do
$\quad\quad CC_i = CCC_i \oplus p_{m,i}(x)M_2(x)$;	$\quad\quad PP_i = PPP_i \oplus p_{m,i}(x)M_1$;
$\quad\quad R_i(x) = R(x)R_{i-1}(x)$;	$\quad\quad L_i(x) = L(x)L_{i-1}(x)$;
$\quad\quad C_i(x) = R_i(x)CC_i(x)$;	$\quad\quad P_i(x) = L_i(x)PP_i(x)$;
$\quad$ end for	$\quad$ end for
$\quad$ return $C_1, C_2, \ldots, C_m$;	$\quad$ return $P_1, P_2, \ldots, P_m$;
endif	endif

Fig. 1. Encryption and Decryption using PEP

$m = 3t + 2$: Define $p_{m,1} = 1 \oplus x$, $p_{m,2}(x) = x \oplus x^2$, $p_{m,3}(x) = x^2 \oplus x^3$, $p_{m,4}(x) = x^3 \oplus x^4$, $p_{m,5}(x) = x^4 \oplus 1$ and $p_{m,5+i}(x) = x^4 \tau_{3(t-1),i}(x)$, for $1 \leq i \leq m$.

From the definition, note that for $m_1 \neq m_2$, we may have $p_{m_1,i}(x) \neq p_{m_2,i}(x)$. Later we show that due to its simple form, multiplication by $p_{m,i}(x)$ is quite efficient.

Proposition 1. *The sequence of polynomials* $p_{m,1}(x), \ldots, p_{m,m}(x)$ *is an allowed sequence.*

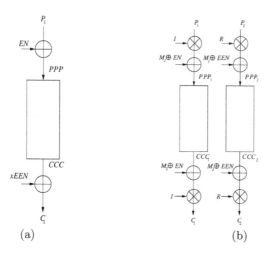

Fig. 2. (a) Enciphering one plaintext block with PEP, (b) Enciphering two plaintext blocks with PEP

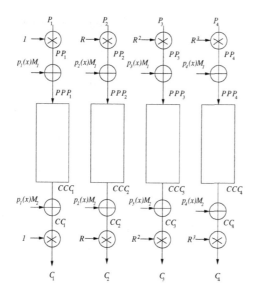

Fig. 3. Enciphering four plaintext blocks with PEP. $R = E_K(T)$, $EN = E_K(R \oplus \text{bin}_n(4))$, $EEN = E_K(xEN)$, $M_1 = E_K(\bigoplus_{i=1}^{4} R^{i-1}(x)PP_i(x) \oplus EN)$, $M_2 = E_K(\bigoplus_{i=1}^{4} CCC_i \oplus EEN)$.

Proof : From the definition of $\tau_{3t,i}(x)$ it is easy to verify that $\bigoplus_{i=1}^{3t} \tau_{3t,i}(x) = 0$. From this it is easy to see that $\bigoplus_{i=1}^{m} p_{m,i}(x) = 0$. This establishes the first condition for allowed sequences.

For the second condition, we must show that $p_{m,i}(x) \oplus p_{m,j}(x) \neq 0 \bmod \tau(x)$. There are three cases to consider.

Both $p_{m,i}(x)$ and $p_{m,j}(x)$ are monomials: By construction the degrees of both $p_{m,i}(x)$ and $p_{m,j}(x)$ are at most $m_{\max} < 2^n - 1$. Hence, by the primitivity of $\tau(x)$ we have the required condition.

Both $p_{m,i}(x)$ and $p_{m,j}(x)$ are 2-nomials: By construction $p_{m,i}(x) = x^{i_1} \oplus x^{i_1+1}$ and $p_{m,j}(x) = x^{j_1} \oplus x^{j_1+1}$. Assume without loss of generality that $i_1 < j_1$. Then, $p_{m,i}(x) \oplus p_{m,j}(x) = x^{i_1}(1 \oplus x)(1 \oplus x^{j_1-i_1})$. Again, the primitivity of $\tau(x)$ ensures the required condition.

One of $p_{m,i}(x)$ and $p_{m,j}(x)$ is a monomial and the other is a 2-nomial: In this case, $p_{m,i}(x) \oplus p_{m,j}(x)$ is a trinomial of degree at most $m_{\max}$. The definition of $m_{\max}$ ensures the required condition.

This completes the proof. □

2.2 Computation of $p_{m,i}(x)M_1(x)$

In the encryption and decryption algorithms, we need to compute $p_{m,i}(x)M_1(x)$ and $p_{m,i}(x)M_2(x)$. We show how these may be efficiently computed. For this it is sufficient to show how to efficiently compute $\tau_{3t,i}(x)M_1$ and $\tau_{3t,i}(x)M_2$.

The polynomials $\tau_{3t,i}(x)$ satisfy the following recurrences:

$$\begin{aligned}
\tau_{3t,i}(x) &= x\tau_{3t,i-1}(x) && \text{for } 2 \leq i \leq 2t; \\
\tau_{3t,2t+1} &= \tau_{3t,1}(x) \oplus \tau_{3t,2}(x); \\
\tau_{3t,i}(x) &= x^2\tau_{3t,i-1}(x) && \text{for } 2t + 2 \leq i \leq 3t.
\end{aligned}$$

Define $M_{1,3t,i} = \tau_{3t,i}(x)M_1$. Then using the above recurrences, we have

$$\begin{aligned}
M_{1,3t,i}(x) &= xM_{1,3t,i-1} && \text{for } 2 \leq i \leq 2t; \\
M_{1,3t,2t+1} &= M_{1,3t,1}(x) \oplus M_{1,3t,2}; \\
M_{1,3t,i}(x) &= x^2 M_{1,3t,i-1} && \text{for } 2t + 2 \leq i \leq 3t.
\end{aligned}$$

Using these recurrences, it is easy to compute all the $\tau_{3t,i}(x)M_1$'s; the requirement is to multiply by either x or x^2 and perform a bitwise XOR. Multiplying by x and x^2 is much more efficient than a general multiplication modulo $\tau(x)$. A similar computation will yield the products $\tau_{3t,i}(x)M_2$.

3 Features of PEP

Here we discuss some of the important features and limitations of PEP.

Message Length: PEP does not produce any ciphertext expansion as it does not produce any tag. The tweak is not considered to be a part of the ciphertext. This is similar to CMC, EME, EME* and HCTR. The current version of PEP can only handle messages whose length is a multiple of n. This is similar to CMC. EME can handle messages of lengths mn, with $1 \leq m \leq n$, while EME* and HCTR can handle messages of lengths $\geq n$. Modification of PEP to handle messages of lengths $\geq n$ is a future task.

There is a (theoretical) restriction on the number of blocks in a particular message to be encrypted by PEP. The number of blocks in any one message to be encrypted by PEP is at most $m_{\max}$. For $n = 128$, this implies that a single message can contain at most around $2^{42.6}$ blocks, which is sufficient for all practical purposes (see Section 2.1).

Single Block Cipher Key: PEP uses the same key for all the block cipher calls. CMC and EME require a single key. On the other hand, EME* requires three keys and HCTR requires two keys. A single block cipher key saves key storage space and key setup costs.

Tweak: Encryption under PEP requires an n-bit tweak. The tweak need not be random, unpredictable or secret. The adversary is allowed to repeat a tweak in queries to the encryption and decryption oracles. The tweak is required for decryption and hence has to be available at both the receiver and the sender ends. This may be achieved by maintaining a shared counter between the two parties or it may be such that it can be understood from the context. An example of the later is the sector address in disk encryption applications.

Online/Offline: An encryption scheme is called online if it can output a stream of ciphertext bits as a stream of plaintext bits arrive. PEP is not online. PEP incorporates the effect of the whole plaintext on each ciphertext bit. Hence, construction of PEP does not allow it to output a single block of ciphertext unless it has seen the total message. We note that no construction of SPRP (tweakable or otherwise) can be online for the same reason as PEP.

Consider the encryption algorithm of PEP for $m \geq 3$. The algorithm consists of three separate **for** loops. The first loop computes the PP_i's and the quantity $\bigoplus_{i=1}^{m} R_i(x)P_i(x)$. The values of the PP_i's need to be stored for use by the second loop. The second loop computes the PPP_i's and the CCC_i's and also the quantity $\bigoplus_{i=1}^{m} CCC_i$. The PPP_i's do not need to be stored but the value of the CCC_i's need to be stored for use by the third loop. The third loop produces the CC_i's and the C_i's which completes the encryption. (Decryption also has a similar structure).

To summarize, the algorithm makes a pass over the P_i's to produce the PP_i's which are stored; makes a pass over the PP_i's to produce the CCC_i's which are also stored; and finally makes a pass over the CCC_i's to produce the C_i's. This makes it a three pass algorithm. Note that the PP_i's can be written over the P_i's and the CCC_i's can be written over the PP_i's. Thus, the intermediate quantities PP_i's and the CCC_i's do not require any extra storage.

HCTR is also a hash-encrypt-hash type construction and requires three passes for reasons similar to that of PEP. The algorithmic descriptions of CMC, EME and EME* as given in the respective papers suggest these algorithms to be three-pass algorithms. On the other hand, a careful consideration of the algorithms show that all of these algorithms can be implemented using two passes over the data. Basically, these algorithms are of the form encrypt-mask-encrypt. The first encryption layer needs to be completed in one pass over the data. Then the mask is computed. The actual masking of the intermediate values and the second encryption layer can be combined in a single pass.

Note that any algorithm requiring more than one pass cannot be online and also at least one set of intermediate variables need to be stored. If we overwrite the P_i's then no extra storage is required for either two or three pass algorithms. On the other hand, if we wish to preserve the P_i's, then the same amount of extra space is required by both two and three pass algorithms. Further, a two pass algorithm is not necessarily more efficient than a three pass algorithm. We have to compare the total amount of computation done by the two algorithms to determine relative efficiency. We consider this issue next.

Computation Cost: PEP performs two polynomial hashes and one layer of block cipher encryption. The total number of block cipher encryptions for an m-block message is 4 if $m = 1$; and is $m + 5$ if $m \geq 2$. The polynomial hashes are used to compute MPP and MCC for $m \geq 2$, the two computations being similar.

To compute MPP we have to compute $\bigoplus_{i=1}^{m} P_i(x)R^{i-1}(x)$. Using Horner's rule this can be computed using $(m - 1)$ multiplications over $GF(2^n)$. But Horner's rule does not compute the values $PP_i = P_i(x)R^{i-1}(x)$. Since the values of PP_i's are also required, using Horner's rule does not help. We have to compute $R^2(x), \ldots, R^{m-1}(x)$ and $P_2(x)R(x), \ldots, P_mR^{m-1}(x)$. These require a total of $2m - 3$ multiplications in $GF(2^n)$. Similarly, a total of $2m - 3$ multiplications are required for computing MCC and the CC_i's. Hence, for $m \geq 2$, PEP requires a total of $4m - 6$ finite field multiplications. In addition, there are the multiplications of the type $p_{m,i}(x)M_1$ and $p_{m,i}(x)M_2$. But as discussed in Section 2.2, these can be computed very efficiently.

If sufficient memory is available, then the values $R^2(x), \ldots, R^{m-1}(x)$ computed during the computation of MPP and the PP_i's can be stored and used during the computation of MCC and the CC_i's. (This can also be combined with the parallel computation strategy discussed below.) This brings down the total number of multiplications to $3m - 4$.

We note that decryption requires the computation of one finite field inverse. The cost of this is amortized over the entire computation and will not reflect on the overall cost if m is moderately large. The main cost will be that for hardware implementation since we will have to implement a finite field inverter requiring more chip area.

Parallelism: The encryption layer is fully parallelizable though the computations of R, EN, EEN, M_1 and M_2 has to be sequential. Thus, for $m \geq 2$, we require at least six parallel encryption rounds irrespective of the number of available block cipher units – five for computing the above quantities and at least one for encrypting the PPP_i's.

The computation of the PP_i's and MPP can be parallelized in the following manner. We illustrate by an example. Suppose there are 4 finite field multipliers available. In the first step, the quantity R^2 is computed. Now consider the following parallel schedule for the four multipliers.

	multiplier 1	multiplier 2	multiplier 3	multiplier 4
Round 1	$(P_2(x) * R(x))$	$(P_3(x) * R^2(x))$	$(R(x) * R^2(x))$	$(R^2(x) * R^2(x))$
Round 2	$(P_4(x) * R^3(x))$	$(P_5(x) * R^4(x))$	$(R^3(x) * R^2(x))$	$(R^4(x) * R^2(x))$
Round 3	$(P_6(x) * R^5(x))$	$(P_7(x) * R^6(x))$	$(R^5(x) * R^2(x))$	$(R^6(x) * R^2(x))$
$\cdots$	$\cdots$	$\cdots$	$\cdots$	$\cdots$

Using this schedule, all the four multipliers can be kept busy in all the rounds (except possibly for the last round and the initial computation of R^2). A similar schedule can be built for computing MCC and the CC_i's. In general using κ many multipliers, all the multiplications can be completed in approximately $\lceil (4m - 6)/\kappa \rceil$ many parallel rounds which is optimal for κ many multipliers.

HCTR uses a polynomial hash which can be evaluated by Horner's rule. On the other hand, there is no straightforward way of parallelizing the polynomial computation *without* increasing the total number of multiplications. The approach described above can be used to obtain parallel implementation of polynomial evaluation and this would approximately double the number of multiplications required in evaluating using Horner's rule.

Provable Security: PEP is provably secure. We state a theorem related to the security of PEP in Section 4 and provide the proof ideas (for the full proof see [2]). The concrete security bound that we obtain for PEP is similar to that obtained for CMC, EME and EME* and is as expected for a mode of operation. Loosely speaking, the theorem shows that the advantage of an adversary in attacking PEP[E] as a tweakable SPRP is bounded above by the advantage of an adversary in attacking E as an SPRP plus an additive factor which is approximately equal to $c\sigma_n^2/2^n$, where c is a constant and σ_n is the total number of blocks (plaintext or ciphertext) provided by the adversary in its queries to the encryption and the decryption oracles. The security bound of HCTR is considerably weaker than the other modes of operations including PEP. For HCTR, the quantity $c\sigma_n^2/2^n$ is replaced by $c\sigma_n^3/2^n$, i.e., there is a cubic security degradation rather than the usual quadratic degradation. One consequence of a weaker security bound is that the secret keys need to changed much earlier compared to the other modes.

Comparison: Table 1 provides a comparison of the various features of PEP and other modes of operations which are SPRPs. We make the following points based on Table 1.

HCTR is the only previously known fully specified tweakable SPRP which is of the hash-encrypt-hash type. The NR construction is too incomplete (and also not tweakable) to permit a proper comparison to other constructions. The main drawback of HCTR is its weaker security bound and the requirement of two keys. It has other good features such as ability to handle all message lengths $\geq n$, and lower computation cost. The encryption layer of HCTR is fully parallel. The two hash layers can be implemented in parallel with computation cost similar to that of PEP.

PEP is currently the only known single key, efficiently parallelizable algorithm which can handle arbitrary number of n-bit blocks. CMC, EME* and HCTR can

Table 1. Comparison of SPRPs using an n-bit block cipher, an n-bit tweak and for m message blocks. (We assume $m \geq 3$.) [BC]: one block cipher invocation; [M]: one $GF(2^n)$ multiplication.

Mode	sec. bnd.	computation cost	keys	msg. len.	passes	enc. layers	parallel?
CMC	$\sigma_n^2/2^n$	$(2m+1)[\text{BC}]$	1	$mn, m \geq 1$	2	2	No
EME	$\sigma_n^2/2^n$	$(2m+2)[\text{BC}]$	1	$mn, 1 \leq m \leq n$	2	2	Yes
EME*	$\sigma_n^2/2^n$	$(2m+\frac{m}{n}+1)[\text{BC}]$	3	$\geq n$	2	2	Yes
HCTR	$\sigma_n^3/2^n$	$m[\text{BC}]$ $+2(m+1)[\text{M}]$	2	$\geq n$	3	1	partial
PEP	$\sigma_n^2/2^n$	$(m+5)[\text{BC}]$ $+(4m-6)[\text{M}]$	1	$mn, m \geq 1$	3	1	Yes

handle such messages, but CMC is strictly sequential; EME* requires one key from $\mathcal{K}$ plus two other n-bit keys; and HCTR requires one key from $\mathcal{K}$ and one n-bit key. On the other hand, EME uses a single key and is fully parallel but can handle at most n many n-bit blocks.

A general comparison of the computation cost of PEP and HCTR with the other modes (CMC, EME and EME*) is difficult. This is because in one case we have less block cipher invocations but $GF(2^n)$ multiplications whereas in the other case we have more block cipher invocations and no $GF(2^n)$ multiplications. Consequently, the comparison depends upon several factors such as:

1. The implementation platform – hardware or software. For software implementation, we need to consider the target architecture and its support for $GF(2^n)$ multiplication. Availability of cryptographic co-processors may provide substantial support for such operation. For hardware implementation, the number of available parallel encryption units and $GF(2^n)$ multiplication units need to be considered.
2. The most important consideration is the design of the actual block cipher being used with the mode of operation *and* its efficient implementation.

Note that PEP approximately trades one block cipher call for four multiplications (HCTR trades one [BC] for two [M]). Implementation results from [10] suggest that using a look-up table it is possible to complete a few $GF(2^n)$ multiplications in the time required for one AES-128 invocation. A more detailed software speed comparison requires efficient implementation of the various modes of operation and can be a topic of future study.

A mode of operation is not designed for use with any particular block cipher. As a side remark, we would like to note that the FIPS 197 specifies AES for protecting "sensitive (unclassified) information". So each government might be having its own block cipher for protecting classified information. The computation cost of PEP vis-a-vis the other modes with respect to such ciphers has to be determined on a case-to-case basis.

The number of passes made by PEP and HCTR is one more than the other modes (though the number of encryption layers is one less). We do not consider this to be a serious problem since all the modes require at least two passes and cannot perform online encryption and decryption. The additional pass of PEP and HCTR by itself does not by itself lead to any efficiency degradation.

4   Security of PEP

4.1   Definitions and Notation

As mentioned earlier, an n-bit block cipher is a function $E : \mathcal{K} \times \{0,1\}^n \to \{0,1\}^n$, where $\mathcal{K} \neq \emptyset$ is called the key space and for any $K \in \mathcal{K}$, $E(K,.)$ is a permutation. We will usually write $E_K()$ instead of $E(K,.)$.

An adversary A is a probabilistic algorithm which has access to some oracles and which outputs either 0 or 1. Oracles are written as superscripts. The notation $A^{\mathcal{O}_1,\mathcal{O}_2} \Rightarrow 1$ denotes the event that the adversary A, interacting with the oracles $\mathcal{O}_1, \mathcal{O}_2$, finally outputs the bit 1.

Let $\mathrm{Perm}(n)$ denote the set of all permutations on $\{0,1\}^n$. We require $E(,)$ to be a strong pseudorandom permutation. The advantage of an adversary in breaking the strong pseudorandomness of $E(,)$ is defined in the following manner.

$$\mathbf{Adv}_E^{\pm \mathrm{prp}}(A) = \Pr\left[K \xleftarrow{\$} \mathcal{K} : A^{E_K(),E_K^{-1}()} \Rightarrow 1\right] -$$
$$\Pr\left[\pi \xleftarrow{\$} \mathrm{Perm}(n) : A^{\pi(),\pi^{-1}()} \Rightarrow 1\right].$$

Formally, a tweakable enciphering scheme is a function $\mathbf{E} : \mathcal{K} \times \mathcal{T} \times \mathcal{M} \to \mathcal{M}$, where $\mathcal{K} \neq \emptyset$ and $\mathcal{T} \neq \emptyset$ are the key space and the tweak space respectively and $\mathcal{M} = \cup_{i \geq 1}\{0,1\}^{ni}$, where n is the length of a message block. We shall often write $\mathbf{E}_K^T(.)$ instead of $\mathbf{E}(K,T,.)$. The inverse of an enciphering scheme is $\mathbf{D} = \mathbf{E}^{-1}$ where $X = \mathbf{D}_K^T(Y)$ if and only if $\mathbf{E}_K^T(X) = Y$.

Let $\mathrm{Perm}^T(\mathcal{M})$ denote the set of all functions $\pi : \mathcal{T} \times \mathcal{M} \to \mathcal{M}$ where $\pi(\mathcal{T},.)$ is a length preserving permutation. Such a $\pi \in \mathrm{Perm}^T(\mathcal{M})$ is called a tweak indexed permutation. For a tweakable enciphering scheme $\mathbf{E} : \mathcal{K} \times \mathcal{T} \times \mathcal{M} \to \mathcal{M}$, we define the advantage an adversary A has in distinguishing $\mathbf{E}$ and its inverse from a random tweak indexed permutation and its inverse in the following manner.

$$\mathbf{Adv}_{\mathbf{E}}^{\pm \widetilde{\mathrm{prp}}}(A) = \Pr\left[K \xleftarrow{\$} \mathcal{K} : A^{\mathbf{E}_K(\cdot,\cdot,\cdot),\mathbf{E}_K^{-1}(\cdot,\cdot,\cdot)} \Rightarrow 1\right] -$$
$$\Pr\left[\pi \xleftarrow{\$} \mathrm{Perm}^T(\mathcal{M}) : A^{\pi(\cdot,\cdot,\cdot),\pi^{-1}(\cdot,\cdot,\cdot)} \Rightarrow 1\right].$$

Pointless queries: An adversary never queries its deciphering oracle with (T,C) if it got C in response to an encipher query (T,M) for some M. Similarly, the adversary never queries its enciphering oracle with (T,M) if it got M as a response to a decipher query of (T,C) for some C. These queries are called *pointless* as the adversary knows what it would get as response for such queries.

Following [5], we define the query complexity σ_n of an adversary as follows. A string X contributes $\max(|X|/n, 1)$ to the query complexity. A tuple of strings $(X_1, X_2, \ldots)$ contributes the sum of the contributions from all oracle queries plus the contribution from the adversary's output. Suppose an adversary makes q queries where the number of n-bit blocks in the ith query is ℓ_i. Then, $\sigma_n = 1 + \sum_{i=1}^{q}(1 + \ell_i) \geq 2q$. Let ρ be a list of resources used by the adversary A and suppose $\mathbf{Adv}_\pi^{\pm\mathrm{xxx}}(A)$ has been defined where π is either a block cipher or a tweakable enciphering scheme. $\mathbf{Adv}_\pi^{\pm\mathrm{xxx}}(\rho)$ denotes the maximal value of $\mathbf{Adv}_\pi^{\pm\mathrm{xxx}}(A)$ over all adversaries A using resources at most ρ. Usual resources of interest are the running time t of the adversary, the number of oracle queries q made by the adversary and the query complexity σ_n ($n \geq 1$).

The notation $\mathrm{PEP}[E]$ denotes a tweakable enciphering scheme, where the n-bit block cipher E is used in the manner specified by PEP. Our purpose is to show that $\mathrm{PEP}[E]$ is secure if E is secure. The notation $\mathrm{PEP}[\mathrm{Perm}(n)]$ denotes a tweakable enciphering scheme obtained by plugging in a random permutation from $\mathrm{Perm}(n)$ into the structure of PEP. For an adversary attacking $\mathrm{PEP}[\mathrm{Perm}(n)]$, we do not put any bound on the running time of the adversary, though we still put a bound on the query complexity σ_n. We show the information theoretic security of $\mathrm{PEP}[\mathrm{Perm}(n)]$ by obtaining an upper bound on $\mathbf{Adv}_{\mathrm{PEP}[\mathrm{Perm}(n)]}^{\pm\widetilde{\mathrm{prp}}}(q, \sigma_n)$. The upper bound is obtained in terms of n and σ_n. For a fixed block cipher E, we bound $\mathbf{Adv}_{\mathrm{PEP}[E]}^{\pm\widetilde{\mathrm{prp}}}(q, \sigma_n, t)$ in terms of $\mathbf{Adv}_E^{\pm\mathrm{prp}}(q, \sigma_n, t')$, where $t' = t + O(\sigma_n)$. We will use the notation $\mathbf{E}_\pi$ as a shorthand for $\mathrm{PEP}[\mathrm{Perm}(n)]$ and $\mathbf{D}_\pi$ will denote the inverse of $\mathbf{E}_\pi$. Thus, the notation $A^{\mathbf{E}_\pi, \mathbf{D}_\pi}$ will denote an adversary interacting with the oracles $\mathbf{E}_\pi$ and $\mathbf{D}_\pi$.

4.2   Statement of Result

The following theorem specifies the security of PEP.

Theorem 1. *Fix n, q and $\sigma_n \geq q$ to be positive integers and an n-bit block cipher $E : \mathcal{K} \times \{0,1\}^n \to \{0,1\}^n$. Then*

$$\mathbf{Adv}_{PEP[Perm(n)]}^{\pm\widetilde{prp}}(q, \sigma_n) \leq \frac{1}{2^{n+1}} \times \left(q^2 + 6(5q + \sigma_n)^2\right) \tag{4}$$

$$\mathbf{Adv}_{PEP[E]}^{\pm\widetilde{prp}}(q, \sigma_n, t) \leq \frac{1}{2^{n+1}} \times \left(q^2 + 6(5q + \sigma_n)^2\right) + \mathbf{Adv}_E^{\pm prp}(q, \sigma_n, t') \tag{5}$$

where $t' = t + O(\sigma_n)$.

Since each query consists of at least one n-bit block, we have $q \leq \sigma_n$ and hence we could write $(q^2 + 6(5q + \sigma_n)^2) \leq c\sigma_n^2$ for some constant c. Upper bounding q by σ_n is proper when σ_n and q are comparable, i.e., when each query consists of a few blocks. On the other hand, if each query consists of a large number of blocks, the bound $q \leq \sigma_n$ is very loose and replacing q by σ_n makes the bound appear worse than what it really is. Hence, we choose to present the bound in terms of both q and σ_n.

The above result and its proof is similar to previous work (see for example [4,5,14]). As mentioned in [4], Equation (5) embodies a standard way to pass from the information theoretic setting to the complexity theoretic setting. Let $\mathbf{E}(.,.,.)$ denote PEP[E]. For any adversary A, we have the following.

$$\mathbf{Adv}^{\pm\widetilde{\mathrm{prp}}}_{\mathrm{PEP}[E]}(A)$$

$$= \Pr\left[K \xleftarrow{\$} \mathcal{K} : A^{\mathbf{E}_K(.,.),\mathbf{E}_K^{-1}(.,.)} \Rightarrow 1\right] - \Pr\left[\pi \xleftarrow{\$} \mathrm{Perm}^{\mathcal{T}}(\mathcal{M}) : A^{\pi(.,.),\pi^{-1}(.,.)} \Rightarrow 1\right]$$

$$= \left(\Pr\left[K \xleftarrow{\$} \mathcal{K} : A^{\mathbf{E}_K(.,.),\mathbf{E}_K^{-1}(.,.)} \Rightarrow 1\right] - \Pr\left[\pi \xleftarrow{\$} \mathrm{Perm}(n) : A^{\mathbf{E}_\pi,\mathbf{D}_\pi} \Rightarrow 1\right]\right) +$$

$$\Pr\left[\pi \xleftarrow{\$} \mathrm{Perm}(n) : A^{\mathbf{E}_\pi,\mathbf{D}_\pi} \Rightarrow 1\right] - \Pr\left[\pi \xleftarrow{\$} \mathrm{Perm}^{\mathcal{T}}(\mathcal{M}) : A^{\pi(.,.),\pi^{-1}(.,.)} \Rightarrow 1\right]$$

$$= X + \mathbf{Adv}^{\pm\widetilde{\mathrm{prp}}}_{\mathrm{PEP}[\mathrm{Perm}(n)]}(A)$$

where

$$X = \left(\Pr\left[K \xleftarrow{\$} \mathcal{K} : A^{\mathbf{E}_K(.,.),\mathbf{E}_K^{-1}(.,.)} \Rightarrow 1\right] - \Pr\left[\pi \xleftarrow{\$} \mathrm{Perm}(n) : A^{\mathbf{E}_\pi,\mathbf{D}_\pi} \Rightarrow 1\right]\right).$$

The quantity X represents an adversary's advantage in distinguishing PEP[E] from PEP[π], where π is a randomly chosen permutation from $\mathrm{Perm}(n)$. Clearly, such an adversary A can also distinguish E from a random permutation and hence $X \leq \mathbf{Adv}^{\pm\mathrm{prp}}_{E}(A)$. This argument shows how (4) is obtained from (5).

We need to consider an adversary's advantage in distinguishing a tweakable enciphering scheme $\mathbf{E}$ from an oracle which simply returns random bit strings. This advantage is defined in the following manner.

$$\mathbf{Adv}^{\pm\mathrm{rnd}}_{\mathrm{MEM}[\mathrm{Perm}(n)]}(A) = \Pr[\pi \xleftarrow{\$} \mathrm{Perm}[n] : A^{\mathbf{E}_\pi,\mathbf{D}_\pi} \Rightarrow 1] - \Pr[A^{\$(.,.),\$(.,.)} \Rightarrow 1]$$

where $\$(.,.)$ returns random bits of length $|M|$. The basic idea of proving (4) is as follows.

$$\mathbf{Adv}^{\pm\widetilde{\mathrm{prp}}}_{\mathrm{PEP}[\mathrm{Perm}(n)]}(A)$$

$$= \Pr\left[\pi \xleftarrow{\$} \mathrm{Perm}(n) : A^{\mathbf{E}_\pi,\mathbf{D}_\pi} \Rightarrow 1\right] - \Pr\left[\pi \xleftarrow{\$} \mathrm{Perm}^{\mathcal{T}}(\mathcal{M}) : A^{\pi(.,.),\pi^{-1}(.,.)} \Rightarrow 1\right]$$

$$= \left(\Pr\left[\pi \xleftarrow{\$} \mathrm{Perm}(n) : A^{\mathbf{E}_\pi,\mathbf{D}_\pi} \Rightarrow 1\right] - \Pr\left[A^{\$(.,.),\$(.,.)} \Rightarrow 1\right]\right) +$$

$$\left(\Pr\left[A^{\$(.,.),\$(.,.)} \Rightarrow 1\right] - \Pr\left[\pi \xleftarrow{\$} \mathrm{Perm}^{\mathcal{T}}(\mathcal{M}) : A^{\pi(.,.),\pi^{-1}(.,.)} \Rightarrow 1\right]\right)$$

$$\leq \mathbf{Adv}^{\pm\mathrm{rnd}}_{\mathrm{PEP}[\mathrm{Perm}(n)]}(A) + \binom{q}{2}\frac{1}{2^n}$$

where q is the number of queries made by the adversary. For a proof of the last inequality see [5].

The main task of the proof now reduces to obtaining an upper bound on $\mathbf{Adv}^{\pm\mathrm{rnd}}_{\mathrm{PEP}[\mathrm{Perm}(n)]}(\sigma_n)$. The complete proof is provided in [2]. The proof uses the standard technique of sequence of games between an adversary and the mode of operation PEP. The proof is similar to the corresponding proofs of CMC [4] and EME [5]. By a sequence of games we show that if in response to any valid query of the adversary, random strings of appropriate lengths are returned then the internal computations of PEP can be performed consistently under the assumption that the block cipher and its inverse are random permutations. The crux of the proof lies in showing that there would seldom be any collisions in the range and domain sets of the block cipher if the adversary queries PEP with valid queries and PEP responds to them by producing random strings. In a later part we remove the randomness associated with the adversary and the game runs on a fixed transcript consisting of the queries and their responses. We show that in such a situation also the internal computations of PEP can be performed consistently.

We later prove (see [2]) that for any adversary making q queries and having query complexity σ_n

$$\mathbf{Adv}_{\text{PEP}[\text{Perm}(n)]}^{\pm\text{rnd}}(q, \sigma_n) \leq \frac{3(5q + \sigma_n)^2}{2^n}.$$

Using this and (6), we obtain

$$\mathbf{Adv}_{\text{PEP}[\text{Perm}(n)]}^{\pm\widetilde{\text{prp}}}(q, \sigma_n) \leq \frac{1}{2^{n+1}} \times \left(q^2 + 6(5q + \sigma_n)^2\right).$$

5 Conclusion

We have presented a new construction for a tweakable SPRP called PEP. Our approach is to use polynomial hash, followed by an encryption layer and again followed by a polynomial hash. This is different from the other constructions of tweakable SPRPs, namely CMC, EME and EME* and is similar to the approach for constructing SPRP (not tweakable SPRP) given in [11]; this approach has also been used in constructing HCTR. PEP offers certain advantages over the known tweakable SPRPs – it is the only know construction which uses a single key, is efficiently parallelizable and can handle an arbitrary number of blocks. We make a detailed comparison between the known constructions of tweakable SPRPs which show that PEP compares quite favorably

Acknowledgement

We would like to thank Peng Wang for pointing out two errors in a previous version of this paper which had specified a different mode of operation.

References

1. M. Bellare and P. Rogaway, Encode-then-encipher encryption: How to exploit nonces or redundancy in plaintexts for efficient cryptography, Advances in Cryptology - Asiacrypt 2000, LNCS 1976, pp. 317-330, Springer, 2000.

2. D. Chakraborty and P. Sarkar, A new mode of encryption providing a tweakable strong pseudorandom permutation, eprint.iacr.org, 2006
3. S. Halevi, EME*. Extending EME to handle arbitrary-length messages with associated data. INDOCRYPT 2004, pp. 315-327, Springer 2004
4. S. Halevi and P. Rogaway. A tweakable enciphering mode, Advances in Cryptology - CRYPTO 2003, LNCS, vol. 2729, pp. 482-499, Springer, 2003.
5. S. Halevi and P. Rogaway. A parallelizable enciphering mode, Topics in Cryptology, CT-RSA 2004, LNCS, vol. 2964, pp. 292-304, Springer, 2004
6. C. S. Jutla: Encryption modes with almost free message integrity. EUROCRYPT 2001: 529-544.
7. S. Maitra, K. C. Gupta and A. Venkateswarlu. Results on multiples of primitive polynomials and their products over $GF(2)$. *Theoretical Computer Science* 341(1-3): 311-343 (2005).
8. M. Liskov, R. L. Rivest and D. Wagner. Tweakable block ciphers. CRYPTO 2002: 31-46.
9. M. Luby and C. Rackoff, How to construct pseudo-random permutations and pseudo-random functions, *SIAM Journal of Computing*, vol. 17, pp. 373-386, 1988.
10. D. A. McGrew and J. Viega. The Security and Performance of the Galois/Counter Mode (GCM) of Operation. Proceedings of Indocrypt 2004, 343-355.
11. M. Naor and O. Reingold. On the construction of pseudo-random permutations: Luby-Rackoff revisited, *J. of Cryptology*, vol 12, pp. 29-66, 1999.
12. M. Naor and O. Reingold. A pseudo-random encryption mode. Manuscript available from www.wisdom.weizmann.ac.il/naor.
13. P. Rogaway. Nonce-based symmetric encryption, Fast Software Encryption (FSE) 2004, LNCS 3017, pp. 348-359, Springer, 2004.
14. P. Rogaway, M. Bellare and J. Black. OCB: A block-cipher mode of operation for efficient authenticated encryption. ACM Conference on Computer and Communication Security 2001: 196-205.
15. D. Wagner. A generalized birthday problem. CRYPTO 2002, LNCS 2442, pp. 288–303, Springer 2002.
16. M. Wegman and L. Carter. New hash functions and their use in authentication and set equality. *Journal of Computer and System Sciences*, vol. 22, 1981, pp. 265–279.
17. P. Wang, D. Feng and W. Wu. HCTR: A variable-input-length enciphering mode. CISC 2005, LNCS 3822, pp. 175–188, 2005.

New Blockcipher Modes of Operation with Beyond the Birthday Bound Security

Tetsu Iwata

Dept. of Computer and Information Sciences,
Ibaraki University
4–12–1 Nakanarusawa, Hitachi, Ibaraki 316-8511, Japan
iwata@cis.ibaraki.ac.jp
http://crypt.cis.ibaraki.ac.jp/

Abstract. In this paper, we define and analyze a new blockcipher mode of operation for encryption, CENC, which stands for Cipher-based ENCryption. CENC has the following advantages: (1) beyond the birthday bound security, (2) security proofs with the standard PRP assumption, (3) highly efficient, (4) single blockcipher key, (5) fully parallelizable, (6) allows precomputation of keystream, and (7) allows random access. CENC is based on the new construction of "from PRPs to PRF conversion," which is of independent interest. Based on CENC and a universal hash-based MAC (Wegman-Carter MAC), we also define a new authenticated-encryption with associated-data scheme, CHM, which stands for CENC with Hash-based MAC. The security of CHM is also beyond the birthday bound.

1 Introduction

A blockcipher mode of operation, or a mode for short, is an algorithm that provides security goals, such as privacy and/or authenticity, based on blockciphers. The mode for privacy is called an encryption mode.

Of many encryption modes, counter (CTR) mode has a number of desirable advantages, and it works as follows. Let E be a blockcipher whose block length is n bits, and let ctr be an n-bit counter. For a plaintext $M = (M_0, \ldots, M_{l-1})$ broken into n-bit blocks, let

$$\begin{cases} C_i \leftarrow M_i \oplus S_i, \text{ where } S_i \leftarrow E_K(\text{ctr} + i) \text{ for } 0 \le i \le l-1, \\ \text{ctr} \leftarrow \text{ctr} + l. \end{cases}$$

The ciphertext is $C = (C_0, \ldots, C_{l-1})$, and $S = (S_0, \ldots, S_{l-1})$ is the keystream.

Starting from [3], provable security (or reduction-based security) is the standard security goal for modes. For encryption modes, we consider the strong security notion of privacy called "indistinguishability from random strings" from [23], which provably implies the more standard notions given in [1]. In this strong notion, the adversary is in the adaptive chosen plaintext attack scenario, and the goal is to distinguish the ciphertext from the random string of the same length (where ctr is not considered part of the ciphertext).

M.J.B. Robshaw (Ed.): FSE 2006, LNCS 4047, pp. 310–327, 2006.

For CTR mode, Bellare, Desai, Jokipii and Rogaway were the first who presented the proof of security [1]. The nonce-based treatment of CTR mode was presented by Rogaway [21]. It was proved that, for any adversary against CTR mode, the success probability is at most $0.5\sigma(\sigma-1)/2^n$ under the assumption that the blockcipher is a secure pseudorandom permutation (PRP), where σ denotes the total ciphertext length in blocks that the adversary obtains. This is the well-known *birthday bound*.

The above analysis is tight. There *is* an adversary that meets the security bound within a constant factor. The adversary simply searches for a collision in the keystream of σ blocks, and guesses the data is the true ciphertext iff there is no collision. It is easy to show that the success probability is at least $0.3\sigma(\sigma-1)/2^n$. This implies that, as long as $E_K(\cdot)$ is a permutation, there is no hope that CTR mode achieves beyond the birthday bound security.

In this paper, we design a new blockcipher mode of operation for encryption. The goals are: (1) beyond the birthday bound security, (2) security proofs with the standard PRP assumption, (3) highly efficient, (4) single blockcipher key, (5) fully parallelizable, (6) allows precomputation of keystream, and (7) allows random access. The original CTR mode achieves all the above goals except for the first one, while we improve the security of CTR mode without breaking its important advantages. As for the security assumption, we do not use the ideal blockcipher model. For efficiency, the number of blockcipher calls is close to CTR mode, and we avoid using any heavy operations, e.g., re-keying.

Now in CTR mode, it is known that if $E_K(\cdot)$ is a secure pseudorandom function (PRF), then for any adversary the success probability 0, well beyond the birthday bound. Thus the natural approach to achieve beyond the birthday bound security is to construct a secure PRF from PRPs and use the PRF in CTR mode, where the security of PRF must be beyond the birthday bound. There are several such constructions [4,10,16,2]. The first construction, due to Bellare, Krovetz, and Rogaway is called data-dependent re-keying [4]. It was proved that the construction achieves beyond the birthday bound security in the ideal blockcipher model. The truncation construction was analyzed by Hall et. al., and they also considered the order construction [10]. Lucks [16] and Bellare and Impagriazzo [2] independently analyzed the construction $G_K(x) = E_K(x\|0) \oplus E_K(x\|1)$, where $x \in \{0,1\}^{n-1}$. Lucks also considered a more generalized construction where d blockciphers are xor'ed to output an n-bit block, and a multiple key version, $G_{K_1,K_2}(x) = E_{K_1}(x) \oplus E_{K_2}(x)$, where $x \in \{0,1\}^n$ [16].

By using these constructions in CTR mode, it is possible to construct encryption modes with beyond the birthday bound. However, there is a significant restriction in efficiency, and/or it breaks several important advantages of the original CTR mode. For example, if the construction from [4] is used, we need the ideal blockcipher model for security proofs and have the efficiency problem for re-keying. The constructions from [10] are not very efficient and the truncation construction has relatively small security improvement. If $G_K(x) = E_K(x\|0) \oplus E_K(x\|1)$ is used, $2l$ blockcipher calls are needed to encrypt l plaintext blocks. We see that the main reason for inefficiency is that the output size of these PRFs is one block (or less).

To achieve beyond the birthday bound security, we first show a new "from PRPs to PRF conversion," where the output size of the new PRF is *larger* than the block size. In particular, our PRF outputs w blocks at a time by using $w+1$ blockcipher calls. The parameter, w, is called a frame width, and one frame is equivalent to nw bits. The frame width, w, can be any fixed positive integer. We prove that the adversary's success probability is at most $w\sigma^3/2^{2n-3} + w\sigma/2^n$, where σ is the total number of blocks that the adversary obtains.

Based on the PRF, we show a new encryption mode with beyond the birthday bound security. The new mode is called CENC, which stands for Cipher-based ENCryption. CENC calls $l + \lceil l/w \rceil$ blockciphers to encrypt l plaintext blocks, and the default value is $w = 2^8$, i.e., we need $l + \lceil l/256 \rceil$ blockcipher calls to encrypt l plaintext blocks. Notice that, with the AES ($n = 128$), one frame corresponds to nw bits, which is $128 \times 256 = 4\text{KBytes}$, and almost all the traffic on the Internet fits in one frame [8]. This implies we need $l+1$ blockcipher calls for these short data, i.e., the cost is *one* blockcipher call per data compared to CTR mode. As for the security, with $w = 2^8$ and the AES, the security bound of CENC is $\hat{\sigma}^3/2^{248} + \hat{\sigma}/2^{121}$, where $\hat{\sigma}$ is (roughly) the total number of blocks that the adversary obtains. The security of CENC is beyond the birthday bound with the standard PRP assumption. Besides, CENC has desirable advantages of CTR mode. It uses single blockcipher key, it is fully parallelizable, allows precomputation of keystream, and random access is possible.

An authenticated-encryption with associated-data scheme, or AEAD scheme, is a scheme for both privacy and authenticity. It takes a plaintext M and a header H, and provides privacy for M and authenticity for both M and H. There are a number of proposals: we have IAPM [13], OCB mode [23], CCM mode [25,12], EAX mode [7], CWC mode [15], GCM mode [19,20], and CCFB mode [17]. Based on CENC and a universal hash-based MAC (Wegman-Carter MAC), we propose a new AEAD scheme called CHM, which stands for CENC with Hash-based MAC. We show that the security of CHM is beyond the birthday bound, which is the first example in literature. The scheme is similar to GCM, achieves higher security with small efficiency loss. It also fixes several undesirable properties of GCM (for example, GCM is not online in the sense that headers must be MACed before starting MAC the ciphertext, and the plaintext length is limited to 64GBytes when used with the AES).

2   Preliminaries

Notation. If x is a string then $|x|$ denotes its length in bits. If x and y are two equal-length strings, then $x \oplus y$ denotes the xor of x and y. If x and y are strings, then $x \| y$ denotes their concatenation. Let $x \leftarrow y$ denote the assignment of y to x. If X is a set, let $x \xleftarrow{R} X$ denote the process of uniformly selecting at random an element from X and assigning it to x. For a positive integer n, $\{0,1\}^n$ is the set of all strings of n bits. For positive integers n and w, $(\{0,1\}^n)^w$ is the set of all strings of nw bits, and $\{0,1\}^*$ is the set of all strings (including the empty string). For positive integers n and m such that $n \leq 2^m - 1$, $[n]_m$ is the m-bit binary representation of n. For a bit string x and a positive integer n such that $|x| \geq n$, $\mathrm{first}(n, x)$ and $\mathrm{last}(n, x)$ denote the first n bits of x and the last n bits of x, respectively. For a positive integer n, 0^n and 1^n denote the n-times repetition of 0 and 1, respectively.

Blockciphers and function families. The blockcipher (permutation family) is a function $E : \mathcal{K} \times \{0,1\}^n \rightarrow \{0,1\}^n$, where, for any $K \in \mathcal{K}$, $E(K, \cdot) = E_K(\cdot)$ is a permutation on $\{0,1\}^n$. The positive integer n is the block length and an n-bit string is called a block. If $\mathcal{K} = \{0,1\}^k$, then k is the key length.

The PRP notion for blockciphers was introduced in [18] and later made concrete in [3]. Let $\mathrm{Perm}(n)$ denote the set of all permutations on $\{0,1\}^n$. This set can be regarded as a blockcipher by considering that each permutation is specified by a unique string. P is a random permutation if $P \xleftarrow{R} \mathrm{Perm}(n)$. An adversary is a probabilistic algorithm (a program) with access to one or more oracles. Let A be an adversary with access to an oracle, either the encryption oracle $E_K(\cdot)$ or a random permutation oracle $P(\cdot)$, and returns a bit. We say A is a *PRP-adversary* for E, and we define

$$\mathbf{Adv}_E^{\mathrm{prp}}(A) \stackrel{\text{def}}{=} \left| \Pr(K \xleftarrow{R} \mathcal{K} : A^{E_K(\cdot)} = 1) - \Pr(P \xleftarrow{R} \mathrm{Perm}(n) : A^{P(\cdot)} = 1) \right|.$$

Similarly, the function family is a function $F : \mathcal{K} \times \{0,1\}^m \rightarrow \{0,1\}^n$, where, for any $K \in \mathcal{K}$, $F(K, \cdot) = F_K(\cdot)$ is a function from $\{0,1\}^m$ to $\{0,1\}^n$. Let $\mathrm{Func}(m, n)$ denote the set of all functions from $\{0,1\}^m$ to $\{0,1\}^n$. This set can be regarded as a function family by considering that each function in $\mathrm{Func}(m, n)$ is specified by a unique string. R is a random function if $R \xleftarrow{R} \mathrm{Func}(m, n)$. Let A be an adversary with access to an oracle, either $F_K(\cdot)$ or a random function oracle $R(\cdot)$, and returns a bit. We say A is a *PRF-adversary* for F, and we define

$$\mathbf{Adv}_F^{\mathrm{prf}}(A) \stackrel{\text{def}}{=} \left| \Pr(K \xleftarrow{R} \mathcal{K} : A^{F_K(\cdot)} = 1) - \Pr(R \xleftarrow{R} \mathrm{Func}(m, n) : A^{R(\cdot)} = 1) \right|.$$

For an adversary A, A's running time is denoted by $time(A)$. The running time is its actual running time (relative to some fixed RAM model of computation) and its description size (relative to some standard encoding of algorithms). The details of the big-O notation for the running time reference depend on the RAM model and the choice of encoding.

The frame, nonce, and counter. The modes described in this paper take a positive integer w as a parameter, and it is called a frame width. For fixed positive integer w (say, $w = 2^8$), a w-block string is called a frame. Throughout this paper, we assume $w \geq 1$. A nonce N is a bit string, where for each pair of key and plaintext, it is used only once. The length of the nonce is denoted by ℓ_{nonce}, and it is at most the block length. We also use an n-bit string called a counter, $\mathtt{ctr}$. This value is initialized based on the value of the nonce, then it is incremented after each blockcipher invocations. The function for increment is

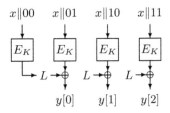

Fig. 1. Example illustration of F. In this example, $w = 3$, $\omega = 1 + \lfloor \log_2 w \rfloor = 2$, and $F : \{0,1\}^k \times \{0,1\}^{n-2} \to (\{0,1\}^n)^3$ where $F_K(x) = (y[0], y[1], y[2])$. Here $x \in \{0,1\}^{n-2}$, $y[0] = L \oplus E_K(x \| 01)$, $y[1] = L \oplus E_K(x \| 10)$, $y[2] = L \oplus E_K(x \| 11)$, where $L = E_K(x \| 00)$.

denoted by $\mathsf{inc}(\cdot)$. It takes an n-bit string x (possibly a counter) and returns the incremented x. We assume $\mathsf{inc}(x) = x + 1 \bmod 2^n$, but other implementations also work, e.g., with LFSRs if $x \neq 0^n$. For $i > 0$, $\mathsf{inc}^i(\mathtt{ctr})$ means $\mathtt{ctr}$ is incremented for i times. Since the value is initialized based on the value of the nonce, there is no need to maintain this value across the massages.

3   The Basic Tool: A New Pseudorandom Function F

In this section, we define a new function family F. It takes two parameters, a blockcipher, and a frame width.

Fix the blockcipher $E : \{0,1\}^k \times \{0,1\}^n \to \{0,1\}^n$, and the frame width w. Define $\omega = 1 + \lfloor \log_2 w \rfloor$, i.e., we need ω bits to represent w. Now we define the function family $F : \{0,1\}^k \times \{0,1\}^{n-\omega} \to (\{0,1\}^n)^w$ as $F_K(x) = (y[0], \ldots, y[w-1])$, where $y[i] = L \oplus E_K(\mathsf{inc}^{i+1}(x \| [0]_\omega))$ for $i = 0, \ldots, w-1$ and $L = E_K(x \| [0]_\omega)$. We call L a mask. See Figure 1 for an example.

We have the following information theoretic result on F.

Theorem 1. *Let $Perm(n)$ and w be the parameters for F. Let A be a PRF-adversary for F making at most q oracle queries. Then*

$$\mathbf{Adv}_F^{\mathrm{prf}}(A) \leq \frac{(w+1)^4 q^3}{2^{2n+1}} + \frac{w(w+1)q}{2^{n+1}}.$$

Notice that w is a constant and the security bound of Theorem 1 is "beyond the birthday bound." Also, if we set $\sigma = qw$ (i.e., the total number of blocks that the adversary obtains) and measure the security bound in terms of σ, we have $\mathbf{Adv}_F^{\mathrm{prf}}(A) \leq w\sigma^3/2^{2n-3} + w\sigma/2^n$, since $1 + w \leq 2w$.

The following definition is useful in proving Theorem 1.

Definition 1. *Let $x = (x_0, \ldots, x_{q-1}) \in (\{0,1\}^{n-\omega})^q$ be an arbitrary $(n - \omega)q$-bit string. We say that "x is distinct," if $x_i \neq x_j$ for $0 \leq i < j \leq q - 1$. Similarly, let $Y = (Y_0, \ldots, Y_{q-1}) \in (\{0,1\}^{nw})^q$ be an arbitrary nqw-bit string, where $Y_i = (y_i[0], \ldots, y_i[w-1]) \in (\{0,1\}^n)^w$ for $0 \leq i \leq q - 1$. We say that "Y is non-zero-distinct," if there is no equal bit strings in $\{0^n, y_i[0], \ldots, y_i[w-1]\}$ for any i s.t. $0 \leq i \leq q - 1$.*

Note that 0^n is included in the definition for "Y is non-zero-distinct." Suppose that $F_K(x_i) = (y_i[0], \ldots, y_i[w-1])$. Then we always have $y_i[j] \neq 0^n$, and we also see that $y_i[j] \neq y_i[j']$ for $j \neq j'$. We allow, for example, $y_i[j] = y_{i'}[j']$ for $i \neq i'$. Intuitively, Definition 1 is the set of possible input-output pairs, and for these pairs the following lemma, which will be used in the proof of Theorem 1, shows that the distribution is close to uniform. This is the crucial observation for the security improvement. There are no collisions in "one frame," but the collision occurs across the frames.

Lemma 1. Let $x = (x_0, \ldots, x_{q-1}) \in (\{0,1\}^{n-w})^q$ and $Y = (Y_0, \ldots, Y_{q-1}) \in (\{0,1\}^{nw})^q$ be arbitrarily fixed bit strings, where x is distinct and Y is non-zero-distinct. Then

$$\frac{p_F}{p_R} \geq 1 - \frac{q^3(w+1)^4}{2^{2n+1}}, \tag{1}$$

where $p_F \stackrel{\text{def}}{=} \Pr(P \stackrel{R}{\leftarrow} Perm(n) : F_P(x_i) = Y_i \text{ for } 0 \leq i \leq q-1)$ and $p_R \stackrel{\text{def}}{=} \Pr(R \stackrel{R}{\leftarrow} Func(n-w, nw) : R(x_i) = Y_i \text{ for } 0 \leq i \leq q-1)$.

The proof is based on the counting argument.

Proof (of Lemma 1). We first count the number of $P \in Perm(n)$ which satisfies $F_P(x_i) = Y_i$ for $0 \leq i \leq q-1$. Let $L_0, \ldots, L_{q-1}$ be n-bit variables. Then the number of $L_0, \ldots, L_{q-1}$ which satisfy $\{L_i, L_i \oplus y_i[0], \ldots, L_i \oplus y_i[w-1]\} \cap \{L_j, L_j \oplus y_j[0], \ldots, L_j \oplus y_j[w-1]\} = \emptyset$ for any $0 \leq i < j \leq q-1$ is at least $\prod_{0 \leq i \leq q-1}(2^n - i(w+1)^2)$, since there are 2^n possibilities for L_0, and once $L_0, \ldots, L_{i-1}$ are fixed, we have at least $2^n - i(w+1)^2$ possibilities for L_i. If we set $L_i = P(x_i\|[0]_w)$, then it is possible to set $P(\text{inc}(x_i\|[0]_w)) = L_i \oplus y_i[0], \ldots, P(\text{inc}^w(x_i\|[0]_w)) = L_i \oplus y_i[w-1]$ uniquely. We have fixed $q(w+1)$ input-output pairs of P, and the remaining $2^n - q(w+1)$ entries can be any value. Therefore, the number of $P \in Perm(n)$ which satisfies $F_P(x_i) = Y_i$ for $0 \leq i \leq q-1$ is at least $(2^n - q(w+1))! \prod_{0 \leq i \leq q-1}(2^n - i(w+1)^2)$.

Then, the left hand side of (1) is at least

$$\frac{(2^n)^{qw}(2^n - q(w+1))! \prod_{0 \leq i \leq q-1}(2^n - i(w+1)^2)}{(2^n)!}$$

$$\geq \prod_{0 \leq i \leq q-1} \frac{1 - \frac{i(w+1)^2}{2^n}}{\left(1 - \frac{i(w+1)}{2^n}\right)\left(1 - \frac{i(w+1)+1}{2^n}\right) \cdots \left(1 - \frac{i(w+1)+w}{2^n}\right)}$$

$$\geq \prod_{0 \leq i \leq q-1}\left(1 - \frac{i(w+1)^2}{2^n}\right)\left(1 + \frac{i(w+1)^2}{2^n} + \frac{w(w+1)}{2^{n+1}}\right). \tag{2}$$

We have used the fact that $(1-\alpha)^{-1} \geq 1 + \alpha$ for $|\alpha| < 1$, and the right hand side of (1) is given by simplifying (2). $\square$

The proof of Theorem 1 uses Lemma 1, and is given in Appendix A.

4   A Relaxed Version F^+

In F, if the input is x, then the mask is always generated with $x\|[0]_w$. In this section, we present a slightly relaxed version of F, called F^+, which removes this restriction. Similarly to F, F^+ takes two parameters, a blockcipher E : $\{0,1\}^k \times \{0,1\}^n \to \{0,1\}^n$, and a frame width w.

Now the function family F^+ : $\{0,1\}^k \times \{0,1\}^n \to (\{0,1\}^n)^w$ is defined as $F_K^+(x) = (y[0], \dots, y[w-1])$, where $y[i] = L \oplus E_K(\text{inc}^{i+1}(x))$ for $i = 0, \dots, w-1$ and $L = E_K(x)$.

Observe that F^+ takes n-bit x as input, and the mask is generated with x. Also, it is not hard to show that F^+ is a good PRF as long as there is no collision in the input to E.

Let A be an adversary that makes at most q oracle queries and let $x_i \in \{0,1\}^n$ denote A's i-th query. Define $X_i = \{x_i, \text{inc}(x_i), \text{inc}^2(x_i), \dots, \text{inc}^w(x_i)\}$, i.e., X_i is the set of input to E in the i-th query. We say that A is $input\text{-}respecting$ if $X_i \cap X_j = \emptyset$ for any $0 \le i < j \le q-1$, regardless of oracle responses and regardless of A's internal coins.

We have the following information theoretic result on F^+.

Corollary 1. *Let $Perm(n)$ and w be the parameters for F^+. Let A be a PRF-adversary for F^+ making at most q oracle queries, where A is input-respecting. Then*

$$\mathbf{Adv}_{F^+}^{\text{prf}}(A) \le \frac{(w+1)^4 q^3}{2^{2n+1}} + \frac{w(w+1)q}{2^{n+1}}.$$

The proof is almost the same as that of Theorem 1, and omitted.

5   CENC: Cipher-Based ENCryption

In this section, we propose a new (nonce-based) encryption scheme, CENC. It takes three parameters, a blockcipher, a nonce length, and a frame width.

Fix the blockcipher E : $\{0,1\}^k \times \{0,1\}^n \to \{0,1\}^n$, the nonce length ℓ_{nonce} and the frame width w, where $1 \le \ell_{\text{nonce}} < n$. CENC consists of two algorithms, the encryption algorithm (CENC.Enc) and the decryption algorithm (CENC.Dec). Both algorithms internally use the keystream generation algorithm (CENC.KSGen). These algorithms are defined in Figure 2. A picture illustrating CENC.KSGen is given in Figure 3.

CENC.Enc has the following syntax. CENC.Enc : Key $\times$ Nonce $\times$ Plaintext $\to$ Ciphertext, where Key is $\{0,1\}^k$, Nonce is $\{0,1\}^{\ell_{\text{nonce}}}$, and Plaintext and Ciphertext are $\{M \in \{0,1\}^* \mid |M| \le n2^{\ell_{\text{max}}}\}$, i.e., the set of bit strings at most ℓ_{max} blocks, where ℓ_{max} is the largest integer satisfying $\ell_{\text{max}} \le w(2^{n-\ell_{\text{nonce}}} - 1)/(w+1)$. It takes the key K, the nonce N, and the plaintext M to return the ciphertext C. We write $C \leftarrow \text{CENC.Enc}_K(N, M)$. The decryption algorithm CENC.Dec : Key $\times$ Nonce $\times$ Ciphertext $\to$ Plaintext takes K, N, C to return M. We write $M \leftarrow \text{CENC.Dec}_K(N, C)$. For any K, N, and M, we have $M \leftarrow \text{CENC.Dec}_K(N, \text{CENC.Enc}_K(N, M))$.

Algorithm CENC.Enc$_K$(N, M)	Algorithm CENC.KSGen$_K$(ctr, l)
100 ctr ← ($N\|0^{n-\ell_{\text{nonce}}}$)	300 for j ← 0 to $\lceil l/w \rceil - 1$ do
101 l ← $\lceil |M|/n \rceil$	301 L ← E_K(ctr)
102 S ← CENC.KSGen$_K$(ctr, l)	302 ctr ← inc(ctr)
103 C ← $M \oplus$ first($|M|, S$)	303 for i ← 0 to $w - 1$ do
104 return C	304 S_{wj+i} ← E_K(ctr) $\oplus L$

Algorithm CENC.Dec$_K$(N, C)	305 ctr ← inc(ctr)
200 ctr ← ($N\|0^{n-\ell_{\text{nonce}}}$)	306 if $wj + i = l - 1$ then
201 l ← $\lceil |C|/n \rceil$	307 S ← ($S_0\|S_1\| \cdots \|S_{l-1}$)
202 S ← CENC.KSGen$_K$(ctr, l)	308 return S
203 M ← $C \oplus$ first($|C|, S$)	
204 return M	

Fig. 2. Definition of the encryption algorithm CENC.Enc (left top), the decryption algorithm CENC.Dec (left bottom), and the keystream generation algorithm CENC.KSGen (right), which is used in both encryption and decryption

CENC.Enc and CENC.Dec call CENC.SKGen to generate the keystream of required length, where the length is in blocks. The encryption (resp. decryption) is just the xor of the plaintext (resp. ciphertext) and the keystream.

The keystream generation algorithm, CENC.KSGen, takes K, the initial counter value ctr, and a non-negative integer l. The output is a keystream S, where the length of S is l blocks. We write S ← CENC.KSGen$_K$(ctr, l).

In CENC.KSGen, we first generate an n-bit mask, L. $\lceil l/w \rceil$ is the number of frames, incomplete frame counts as one frame. We see that $\lceil l/w \rceil$ masks are generated in line 301. For each mask, w blocks of the keystream are generated in line 304 (except for the last frame, as the last frame may have fewer than w blocks). If l blocks of keystream are generated in line 306, the resulting S is returned in line 308. Observe that the blockcipher is invoked for $l + \lceil l/w \rceil$ times, since we generate $\lceil l/w \rceil$ masks and we have l blocks of keystream, where each block of keystream requires one blockcipher invocation.

Discussion and default parameters. CENC takes the blockcipher $E : \{0,1\}^k \times \{0,1\}^n \to \{0,1\}^n$, the nonce length ℓ_{nonce} ($1 \le \ell_{\text{nonce}} < n$) and the frame width w, as the parameters. With these parameters, CENC can encrypt at most $2^{\ell_{\text{nonce}}}$ plaintexts, and the maximum length of the plaintext is $\ell_{\max}$ blocks. Note that $\ell_{\max}$ is derived by solving $\ell_{\max} + \lceil \ell_{\max}/w \rceil \le 2^{n-\ell_{\text{nonce}}}$ in $\ell_{\max}$, and in general, the bound on $\ell_{\max}$ is $\ell_{\max} \le 2^{n-\ell_{\text{nonce}}-1}$ since $\lceil \ell_{\max}/w \rceil \le \ell_{\max}$. As we will present in Section 6, the security bound of CENC is $(w+1)^4 \hat{\sigma}^3/w^3 2^{2n+1} + (w+1)\hat{\sigma}/2^{n+1}$, where $\hat{\sigma}$ is (roughly) the total number of blocks processed by one key.

Our default parameters are, E is any blockcipher such that $n \ge 128$, $\ell_{\text{nonce}} = n/2$, and $w = 2^8 = 256$. For example, if we use the AES, CENC can encrypt at most 2^{64} plaintexts, the maximum length of the plaintext is 2^{63} blocks (2^{37}GBytes), and the security bound is $\hat{\sigma}^3/2^{248} + \hat{\sigma}/2^{121}$ (we used $(w+1)^4/w^3 < 261 < 2^9$), thus $\hat{\sigma}$ should be sufficiently smaller that 2^{82} blocks (2^{56}GBytes).

$N\|0^{n-\ell_{\mathrm{nonce}}}$

↓

ctr

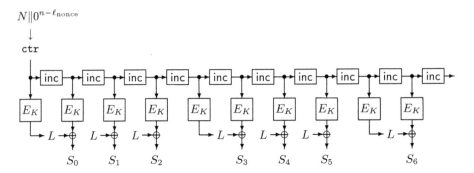

Fig. 3. Illustration of the keystream generation algorithm. This example uses $w = 3$ and outputs $l = 7$ blocks of keystream $S = (S_0, \ldots, S_6)$. This S is used in both encryption and decryption. The mask L is updated after generating w blocks of keystream. The counter ctr is incremented for $l + \lceil l/w \rceil = 10$ times, and there are 10 blockcipher invocations.

The frame width, w, should be large enough so that we can implement CENC efficiently. On the other hand, it affects the security bound. We chose $w = 2^8 = 256$, which implies 256 blocks of keystream are generated with 257 blockcipher invocations, thus the cost is about 0.4% compared to CTR mode. We see that the efficiency loss is very small in both software and hardware. Also, the security bound is low enough with this value of w. We do not recommend $w > 2^8$ (when $n = 128$) because of the security loss.

64-bit blockciphers. We do not claim that CENC is generally useful for $n = 64$, since there are restrictions on the nonce length (thus the number of plaintexts), and the plaintext length.

For example, if we use Triple-DES and $(\ell_{\mathrm{nonce}}, w) = (32, 256)$, CENC can encrypt at most 2^{32} plaintexts, and the maximum length of the plaintext is 2^{31} blocks (16GBytes), which may not be enough for general applications (still, it is comparable to CTR mode). In this case, the security bound is $\hat{\sigma}^3/2^{120} + \hat{\sigma}/2^{57}$, which implies $\hat{\sigma}$ should be sufficiently smaller that 2^{40} blocks (2^{13}GBytes).

The limitations of the nonce length and the plaintext length can be removed if we use a counter (instead of a nonce) that is maintained across the plaintexts. This "counter version of CENC" is more suitable for 64-bit blockciphers.

6   Security of CENC

CENC is a symmetric encryption scheme. Before showing the security results on CENC, we first formally define what we mean by symmetric encryption schemes, and what we mean by such schemes to be secure.

Symmetric encryption schemes. A (nonce-based) symmetric encryption scheme is a pair of algorithms $S\mathcal{E} = (\mathcal{E}, \mathcal{D})$ where $\mathcal{E}$ is a deterministic encryption algorithm $\mathcal{E}$: Key × Nonce × Plaintext → Ciphertext and $\mathcal{D}$ is a deterministic decryption algorithm $\mathcal{D}$: Key × Nonce × Ciphertext → Plaintext. The key space Key is a set of keys, and is a nonempty set having a distribution (the uniform distribution when the set is finite). The nonce space Nonce, the plaintext space Plaintext, and the ciphertext space Ciphertext are nonempty sets of strings. We write $\mathcal{E}_K(N, M)$ for $\mathcal{E}(K, N, M)$ and $\mathcal{D}_K(N, C)$ for $\mathcal{D}(K, N, C)$. We require that $\mathcal{D}_K(N, \mathcal{E}_K(N, M)) = M$ for all $K \in$ Key, $N \in$ Nonce and $M \in$ Plaintext.

Nonce-respecting adversary. Let A be an adversary with access to an encryption oracle $\mathcal{E}_K(\cdot, \cdot)$. This oracle, on input (N, M), returns $C \leftarrow \mathcal{E}_K(N, M)$. Let $(N_0, M_0), \ldots, (N_{q-1}, M_{q-1})$ denote its oracle queries. The adversary is said to be nonce-respecting if $N_0, \ldots, N_{q-1}$ are always distinct, regardless of oracle responses and regardless of A's internal coins.

Privacy of symmetric encryption schemes. We adopt the strong notion of privacy for nonce-based encryption schemes from [23]. This notion, which we call indistinguishability from random strings, provably implies the more standard notions given in [1].

Let A be an adversary with access to an oracle, either the encryption oracle $\mathcal{E}_K(\cdot, \cdot)$ or $\mathcal{R}(\cdot, \cdot)$, and returns a bit. The $\mathcal{R}(\cdot, \cdot)$ oracle, on input (N, M), returns a random string of length $|\mathcal{E}_K(N, M)|$. We say that A is a PRIV-adversary for $\mathcal{SE}$. We assume that any PRIV-adversary is nonce-respecting. The advantage of PRIV-adversary A for $\mathcal{SE} = (\mathcal{E}, \mathcal{D})$ having key space Key is

$$\mathbf{Adv}_{\mathcal{SE}}^{\mathrm{priv}}(A) \stackrel{\mathrm{def}}{=} \left| \Pr(K \stackrel{R}{\leftarrow} \mathsf{Key} : A^{\mathcal{E}_K(\cdot, \cdot)} = 1) - \Pr(A^{\mathcal{R}(\cdot, \cdot)} = 1) \right|.$$

Security results on CENC. Let A be a nonce-respecting PRIV-adversary for CENC, and assume that A makes at most q oracle queries, and the total length of these queries is at most σ blocks, where "the total length of queries" is defined as follows: if A makes q queries $(N_0, M_0), \ldots, (N_{q-1}, M_{q-1})$, then the total length of queries is $\sigma = \lceil |M_0|/n \rceil + \cdots + \lceil |M_{q-1}|/n \rceil$, i.e, the total number of blocks of plaintexts. We have the following information theoretic result.

Theorem 2. *Let $Perm(n)$, ℓ_{nonce}, and w be the parameters for CENC. Let A be a nonce-respecting PRIV-adversary for CENC making at most q oracle queries, and the total length of these queries is at most σ blocks. Then*

$$\mathbf{Adv}_{\mathrm{CENC}}^{\mathrm{priv}}(A) \leq \frac{(w+1)^4 \hat{\sigma}^3}{w^3 2^{2n+1}} + \frac{(w+1)\hat{\sigma}}{2^{n+1}}, \tag{3}$$

where $\hat{\sigma} = \sigma + qw$.

If we use the rough inequality of $w + 1 \leq 2w$, then we have the simpler form, $\mathbf{Adv}_{\mathrm{CENC}}^{\mathrm{priv}}(A) \leq w\hat{\sigma}^3/2^{2n-3} + w\hat{\sigma}/2^n$.

The proof of Theorem 2 is based on the contradiction argument. If there exists a nonce-respecting PRIV-adversary A such that $\mathbf{Adv}_{\mathrm{CENC}}^{\mathrm{priv}}(A)$ is larger than the right hand side of (3), then we can construct an input-respecting PRF-adversary B for F^+ which contradicts Corollary 1. The proof is given in Appendix B.

Given Theorem 2, we have the following complexity theoretic result.

Corollary 2. *Let* $E : \{0,1\}^k \times \{0,1\}^n \rightarrow \{0,1\}^n$, ℓ_{nonce}, *and* w *be the parameters for CENC. Let* A *be a nonce-respecting PRIV-adversary for CENC making at most* q *oracle queries, and the total length of these queries is at most* σ *blocks. Then there is a PRP-adversary* B *for* E *making at most* $(w+1)\hat{\sigma}/w$ *oracle queries, time*$(B) = time(A) + O(n\hat{\sigma}w)$, *and* $\mathbf{Adv}^{\text{prp}}_E(B) \geq \mathbf{Adv}^{\text{priv}}_{\text{CENC}}(A) - w\hat{\sigma}^3/2^{2n-3} - w\hat{\sigma}/2^n$, *where* $\hat{\sigma} = \sigma + qw$.

The proof of Corollary 2 is given in [11].

7   CHM: CENC with Hash-Based MAC

In this section, we present a new (nonce-based) authenticated-encryption with associated-data (AEAD) scheme, CHM. It takes six parameters, a blockcipher, a nonce length, a tag length, a frame width, and two constants.

Fix the blockcipher $E : \{0,1\}^k \times \{0,1\}^n \rightarrow \{0,1\}^n$, the nonce length ℓ_{nonce}, the tag length τ, the frame width w, and two n-bit constants const_0 and const_1. We require that $1 \leq \ell_{\text{nonce}} < n$, $1 \leq \tau \leq n$, $\text{const}_0 \neq \text{const}_1$, and $\text{first}(1, \text{const}_0) = \text{first}(1, \text{const}_1) = 1$ (the most significant bits of const_0 and const_1 are both 1).

CHM consists of two algorithms, the encryption algorithm (CHM.Enc) and the decryption algorithm (CHM.Dec). These algorithms are defined in Figure 4. Both algorithms use the keystream generation algorithm (CHM.KSGen) and a hash function (CHM.Hash). CHM.KSGen is equivalent to CENC.KSGen defined in Figure 2, and the hash function CHM.Hash is defined in Figure 5.

The syntax of the encryption algorithm is CHM.Enc : Key × Nonce × Header × Plaintext → Ciphertext × Tag, where the key space Key is $\{0,1\}^k$, the nonce space Nonce is $\{0,1\}^{\ell_{\text{nonce}}}$, and the header space Header is $\{0,1\}^*$. The plaintext space Plaintext and ciphertext space Ciphertext are $\{M \in \{0,1\}^* \mid |M| \leq n2^{\ell_{\text{max}}}\}$, where ℓ_{max} is the largest integer satisfying $\ell_{\text{max}} \leq w(2^{n-\ell_{\text{nonce}}-1}-1)/(w+1)-1$. The tag space Tag is $\{0,1\}^\tau$. It takes the key K, the nonce N, the header H, and the plaintext M to return the ciphertext C and the tag T. We write $(C,T) \leftarrow \text{CHM.Enc}_K(N, H, M)$. The decryption algorithm CHM.Dec : Key × Nonce × Header × Ciphertext × Tag → Plaintext $\cup$ {reject} takes K, N, H, C and T to return M or a special symbol reject. We write $M \leftarrow \text{CHM.Dec}_K(N, H, C, T)$ or reject $\leftarrow \text{CHM.Dec}_K(N, H, C, T)$.

CHM is the natural combination of CENC and a universal hash function-based MAC (Wegman-Carter MAC). As a universal hash function, we chose the standard polynomial-based hash, since it is efficient in both software and hardware, and it is well studied. The multiplication is done in the finite field $\text{GF}(2^n)$ using a canonical polynomial to represent field elements. The suggested canonical polynomial is the lexicographically first polynomial among the irreducible polynomials of degree n that have a minimum number of nonzero coefficients. For $n = 128$ the indicated polynomial is $x^{128} + x^7 + x^2 + x + 1$.

Discussion and default parameters. CHM takes six parameters, the blockcipher $E : \{0,1\}^k \times \{0,1\}^n \rightarrow \{0,1\}^n$, the nonce length ℓ_{nonce}, the tag length τ, the

Algorithm CHM.Enc$_K(N, H, M)$	Algorithm CHM.Dec$_K(N, H, C, T)$
100 $S_0 \leftarrow E_K(\text{const}_0)$	200 $S_0 \leftarrow E_K(\text{const}_0)$
101 $S_1 \leftarrow E_K(\text{const}_1)$	201 $S_1 \leftarrow E_K(\text{const}_1)$
102 $l \leftarrow \lceil |M|/n \rceil$	202 $l \leftarrow \lceil |C|/n \rceil$
103 $\mathbf{ctr} \leftarrow (0\|N\|0^{n-\ell_{\text{nonce}}-1})$	203 $\mathbf{ctr} \leftarrow (0\|N\|0^{n-\ell_{\text{nonce}}-1})$
104 $S \leftarrow \text{CHM.KSGen}_K(\mathbf{ctr}, l+1)$	204 $S \leftarrow \text{CHM.KSGen}_K(\mathbf{ctr}, l+1)$
105 $S_2 \leftarrow \text{first}(n, S)$	205 $S_2 \leftarrow \text{first}(n, S)$
106 $S_3 \leftarrow \text{last}(nl, S)$	206 $\text{Hash}_0 \leftarrow \text{CHM.Hash}_{S_0}(C)$
107 $C \leftarrow M \oplus \text{first}(|M|, S_3)$	207 $\text{Hash}_1 \leftarrow \text{CHM.Hash}_{S_1}(H)$
108 $\text{Hash}_0 \leftarrow \text{CHM.Hash}_{S_0}(C)$	208 $T' \leftarrow \text{Hash}_0 \oplus \text{Hash}_1 \oplus S_2$
109 $\text{Hash}_1 \leftarrow \text{CHM.Hash}_{S_1}(H)$	209 $T' \leftarrow \text{first}(\tau, T')$
110 $T \leftarrow \text{Hash}_0 \oplus \text{Hash}_1 \oplus S_2$	210 if $T' \neq T$ then return reject
111 $T \leftarrow \text{first}(\tau, T)$	211 $S_3 \leftarrow \text{last}(nl, S)$
112 return (C, T)	212 $M \leftarrow C \oplus \text{first}(|C|, S_3)$
	213 return M

Fig. 4. Definition of the encryption algorithm CHM.Enc (left), and the decryption algorithm CHM.Dec (right). CHM.KSGen is equivalent to CENC.KSGen in Figure 2, and CHM.Hash is defined in Figure 5.

Algorithm CHM.Hash$_S(M)$
100 $M \leftarrow M\|10^{n-1-(|M| \bmod n)}$
101 $l \leftarrow |M|/n$
102 $\text{Hash} \leftarrow 0^n$
103 for $i \leftarrow 0$ to $l-1$ do
104      $\text{Hash} \leftarrow (\text{Hash} \oplus M_i) \cdot S$
105 return Hash

Fig. 5. Definition of CHM.Hash : $\{0,1\}^n \times \{0,1\}^* \rightarrow \{0,1\}^n$. M_i is the i-th block of $M\|10^{n-1-(|M| \bmod n)}$, i.e., $(M_0, \ldots, M_{l-1}) = M\|10^{n-1-(|M| \bmod n)}$. Multiplication in line 104 is in GF(2^n).

frame width w, and two n-bit constants const_0 and const_1. With these parameters, CHM can encrypt at most $2^{\ell_{\text{nonce}}}$ plaintext-header pairs, and the maximum length of the plaintext is $\ell_{\max}$ blocks ($\ell_{\max}$ is derived by solving $\ell_{\max}+1+\lceil(\ell_{\max}+1)/w\rceil \leq 2^{n-\ell_{\text{nonce}}-1}$ in $\ell_{\max}$). As we will present in Section 8, the security bound of CHM is $(w+1)^3\tilde{\sigma}^2/w^2 2^{2n-3} + (w+1)^4\tilde{\sigma}^3/w^3 2^{2n+1} + 1/2^n + (w+1)\tilde{\sigma}/2^{n+1}$ for privacy, and $(w+1)^3\tilde{\sigma}^2/w^2 2^{2n-3} + (w+1)^4\tilde{\sigma}^3/w^3 2^{2n+1} + 1/2^n + (w+1)\tilde{\sigma}/2^{n+1} + (1 + H_{\max} + M_{\max})/2^\tau$ for authenticity, where $\tilde{\sigma}$ is (roughly) the total number of blocks processed by one key, $M_{\max}$ is the maximum block length of plaintexts, and $H_{\max}$ is the maximum block length of headers.

Our default parameters are, E is any blockcipher such that $n \geq 128$, $\ell_{\text{nonce}} = n/2 - 1$, $\tau \geq 96$, $w = 2^8 = 256$, $\text{const}_0 = 1^{n-1}\|0$ and $\text{const}_1 = 1^n$.

With these parameters, if we use the AES, CHM can encrypt at most 2^{63} plaintexts-header pairs, and the maximum length of the plaintext is 2^{63} blocks (2^{37}GBytes), and the security bounds are $\tilde{\sigma}^3/2^{242} + \tilde{\sigma}/2^{120}$ for privacy, and

$\tilde{\sigma}^3/2^{242} + \tilde{\sigma}/2^{120} + (1 + H_{\max} + M_{\max})/2^\tau$ for authenticity. This implies $\tilde{\sigma}$ should be sufficiently smaller that 2^{80} blocks (2^{54}GBytes), and $H_{\max}$ and $M_{\max}$ should be small enough so that $(1 + H_{\max} + M_{\max})/2^\tau$ is low enough.

8   Security of CHM

CHM is an authenticated-encryption with associated-data (AEAD) scheme. Before showing the security results on CHM, we first formally define what we mean by AEAD schemes, and what we mean by such schemes to be secure.

AEAD schemes. A (nonce-based) authenticated-encryption with associated-data (AEAD) scheme is a pair of algorithms $\mathcal{AE} = (\mathcal{E}, \mathcal{D})$ where $\mathcal{E}$ is a deterministic encryption algorithm $\mathcal{E} : \mathsf{Key} \times \mathsf{Nonce} \times \mathsf{Header} \times \mathsf{Plaintext} \to \mathsf{Ciphertext} \times \mathsf{Tag}$ and $\mathcal{D}$ is a deterministic decryption algorithm $\mathcal{D} : \mathsf{Key} \times \mathsf{Nonce} \times \mathsf{Header} \times \mathsf{Ciphertext} \times \mathsf{Tag} \to \mathsf{Plaintext} \cup \{\mathsf{reject}\}$. The key space Key is a set of keys. The nonce space Nonce and the header space Header (also called the space of associated data), the plaintext space $\mathsf{Plaintext}$ and the ciphertext space $\mathsf{Ciphertext}$ are nonempty sets of strings. (We note that there is a more general treatment where $\mathsf{Ciphertext}$ and Tag are not separated. See [7]. We separate them for simplicity.) We write $\mathcal{E}_K(N, H, M)$ for $\mathcal{E}(K, N, H, M)$ and $\mathcal{D}_K(N, H, C, T)$ for $\mathcal{D}(K, N, H, C, T)$. We require that $\mathcal{D}_K(N, H, \mathcal{E}_K(N, H, M)) = M$ for all $K \in \mathsf{Key}$, $N \in \mathsf{Nonce}$, $H \in \mathsf{Header}$ and $M \in \mathsf{Plaintext}$.

Privacy of AEAD schemes. We follow the security notion from [7]. Let A be an adversary with access to an oracle, either the encryption oracle $\mathcal{E}_K(\cdot, \cdot, \cdot)$ or $\mathcal{R}(\cdot, \cdot, \cdot)$, and returns a bit. The $\mathcal{R}(\cdot, \cdot, \cdot)$ oracle, on input (N, H, M), returns a random string of length $|\mathcal{E}_K(N, H, M)|$. We say that A is a PRIV-adversary for $\mathcal{AE}$. We assume that any PRIV-adversary is nonce-respecting (i.e., if $(N_0, H_0, M_0), \ldots, (N_{q-1}, H_{q-1}, M_{q-1})$ is A's oracle queries, $N_0, \ldots, N_{q-1}$ are always distinct, regardless of oracle responses and regardless of A's internal coins). The advantage of PRIV-adversary A for AEAD scheme $\mathcal{AE} = (\mathcal{E}, \mathcal{D})$ having key space Key is

$$\mathbf{Adv}_{\mathcal{AE}}^{\mathrm{priv}}(A) \overset{\mathrm{def}}{=} \left| \Pr(K \overset{R}{\leftarrow} \mathsf{Key} : A^{\mathcal{E}_K(\cdot, \cdot, \cdot)} = 1) - \Pr(A^{\mathcal{R}(\cdot, \cdot, \cdot)} = 1) \right|.$$

Authenticity of AEAD schemes. A notion of authenticity of ciphertext for AEAD schemes was formalized in [23,22] following [14,6,5]. This time, let A be an adversary with access to an encryption oracle $\mathcal{E}_K(\cdot, \cdot, \cdot)$ and returns a tuple, (N, H, C, T). This tuple is called a forgery attempt. We say that A is an AUTH-adversary for $\mathcal{AE}$. We assume that any AUTH-adversary is nonce-respecting. (The condition is understood to apply only to the adversary's encryption oracle. Thus a nonce used in an encryption-oracle query may be used in a forgery attempt.) We say A forges if A returns (N, H, C, T) such that $\mathcal{D}_K(N, H, C, T) \neq$ reject but A did not make a query (N, H, M) to $\mathcal{E}_K(\cdot, \cdot, \cdot)$ that resulted in a response (C, T). That is, adversary A may never return a forgery attempt

(N, H, C, T) such that the encryption oracle previously returned (C, T) in response to a query (N, H, M). Then the advantage of AUTH-adversary A for AEAD scheme $\mathcal{AE} = (\mathcal{E}, \mathcal{D})$ having key space Key is

$$\mathbf{Adv}_{\mathcal{AE}}^{\mathrm{auth}}(A) \stackrel{\mathrm{def}}{=} \Pr(K \stackrel{R}{\leftarrow} \mathsf{Key} : A^{\mathcal{E}_K(\cdot, \cdot, \cdot)} \text{ forges}).$$

Privacy results on CHM. Let A be a nonce-respecting PRIV-adversary for CHM, and assume that A makes at most q oracle queries, and the total plaintext length of these queries is at most σ blocks, where "the total plaintext length of queries" is defined as follows: if A makes queries $(N_0, H_0, M_0), \ldots, (N_{q-1}, H_{q-1}, M_{q-1})$, then $\sigma = \lceil |M_0|/n \rceil + \cdots + \lceil |M_{q-1}|/n \rceil$, i.e., the total number of blocks of plaintexts. We have the following information theoretic result.

Theorem 3. *Let $Perm(n)$, ℓ_{nonce}, τ, w, $\mathbf{const}_0$ and $\mathbf{const}_1$ be the parameters for CHM. Let A be a nonce-respecting PRIV-adversary making at most q oracle queries, and the total plaintext length of these queries is at most σ blocks. Then*

$$\mathbf{Adv}_{\mathrm{CHM}}^{\mathrm{priv}}(A) \leq \frac{(w+1)^3 \tilde{\sigma}^2}{w^2 2^{2n-3}} + \frac{(w+1)^4 \tilde{\sigma}^3}{w^3 2^{2n+1}} + \frac{1}{2^n} + \frac{(w+1)\tilde{\sigma}}{2^{n+1}}, \qquad (4)$$

where $\tilde{\sigma} = \sigma + q(w+1)$.

Note that there is no restriction on the header length. If we use $w + 1 \leq 2w$, we have the simpler form, $\mathbf{Adv}_{\mathrm{CHM}}^{\mathrm{priv}}(A) \leq w\tilde{\sigma}^2/2^{2n-6} + w\tilde{\sigma}^3/2^{2n-3} + 1/2^n + w\tilde{\sigma}/2^n$.

The proof of Theorem 3 is given in [11]. From Theorem 3, we have the following complexity theoretic result.

Corollary 3. *Let $E : \{0,1\}^k \times \{0,1\}^n \to \{0,1\}^n$, ℓ_{nonce}, τ, w, $\mathbf{const}_0$ and $\mathbf{const}_1$ be the parameters for CHM. Let A be a nonce-respecting PRIV-adversary making at most q oracle queries, and the total plaintext length of these queries is at most σ blocks. Then there is a PRP-adversary B for E making at most $(w+1)\tilde{\sigma}/w$ oracle queries, $time(B) = time(A) + O(n\tilde{\sigma}w)$, and $\mathbf{Adv}_E^{\mathrm{prp}}(B) \geq \mathbf{Adv}_{\mathrm{CHM}}^{\mathrm{priv}}(A) - w\tilde{\sigma}^2/2^{2n-6} - w\tilde{\sigma}^3/2^{2n-3} - 1/2^n - w\tilde{\sigma}/2^n$, where $\tilde{\sigma} = \sigma + q(w+1)$.*

The proof of Corollary 3 is given in [11].

Authenticity results on CHM. Let A be an AUTH-adversary for CHM, and assume that A makes at most q oracle queries (including the final forgery attempt), the total plaintext length of these queries is at most σ blocks, the maximum plaintext length of these queries is at most M_{max} blocks, and the maximum header length of these queries is at most H_{max} blocks. Here, if A makes queries $(N_0, H_0, M_0), \ldots, (N_{q-2}, H_{q-2}, M_{q-2})$, and returns the forgery attempt (N^*, H^*, C^*, T^*), then σ, M_{max} and H_{max} are defined as

$$\begin{cases} \sigma \stackrel{\mathrm{def}}{=} \lceil |M_0|/n \rceil + \cdots + \lceil |M_{q-2}|/n \rceil + \lceil |C^*|/n \rceil, \\ M_{\mathrm{max}} \stackrel{\mathrm{def}}{=} \max\{\lceil |M_0|/n \rceil, \ldots, \lceil |M_{q-2}|/n \rceil, \lceil |C^*|/n \rceil\}, \\ H_{\mathrm{max}} \stackrel{\mathrm{def}}{=} \max\{\lceil |H_0|/n \rceil, \ldots, \lceil |H_{q-2}|/n \rceil, \lceil |H^*|/n \rceil\}. \end{cases}$$

We say A's query resource is $(q, \sigma, M_{\mathrm{max}}, H_{\mathrm{max}})$. We have the following information theoretic result.

Theorem 4. *Let* $Perm(n)$, ℓ_{nonce}, τ, w, const_0 *and* const_1 *be the parameters for CHM. Let A be a nonce-respecting AUTH-adversary whose query resource is* $(q, \sigma, M_{\max}, H_{\max})$. *Then* $\mathbf{Adv}^{\text{auth}}_{\text{CHM}}(A)$ *is at most*

$$\frac{(w+1)^3 \tilde{\sigma}^2}{w^2 2^{2n-3}} + \frac{(w+1)^4 \tilde{\sigma}^3}{w^3 2^{2n+1}} + \frac{1}{2^n} + \frac{(w+1)\tilde{\sigma}}{2^{n+1}} + \frac{1 + H_{\max} + M_{\max}}{2^\tau}, \quad (5)$$

where $\tilde{\sigma} = \sigma + q(w+1)$.

If we use $w + 1 \le 2w$, we have the simpler form, $\mathbf{Adv}^{\text{auth}}_{\text{CHM}}(A) \le w\tilde{\sigma}^2/2^{2n-6} + w\tilde{\sigma}^3/2^{2n-3} + 1/2^n + w\tilde{\sigma}/2^n + (1 + H_{\max} + M_{\max})/2^\tau$.

The proof of Theorem 4 is given in [11]. From Theorem 4, we have the following complexity theoretic result.

Corollary 4. *Let* $E : \{0,1\}^k \times \{0,1\}^n \to \{0,1\}^n$, ℓ_{nonce}, τ, w, const_0, *and* const_1 *be the parameters for CHM. Let A be a nonce-respecting AUTH-adversary whose query resource is* $(q, \sigma, M_{\max}, H_{\max})$. *Then there is a PRP-adversary B for E making at most* $(w+1)\tilde{\sigma}/w$ *oracle queries,* $time(B) = time(A) + O(n\tilde{\sigma}w)$, *and* $\mathbf{Adv}^{\text{prp}}_E(B) \ge \mathbf{Adv}^{\text{auth}}_{\text{CHM}}(A) - w\tilde{\sigma}^2/2^{2n-6} - w\tilde{\sigma}^3/2^{2n-3} - 1/2^n + w\tilde{\sigma}/2^n - (1 + H_{\max} + M_{\max})/2^\tau$, *where* $\tilde{\sigma} = \sigma + q(w+1)$.

9   Discussions

Counter-based versions. CENC and CHM use a nonce, and it is natural to consider their counter-based versions. Call them CENC-C and CHM-C, respectively. They use an n-bit counter maintained across the plaintexts (usually by the sender). The drawback is the difficulty of implementation and it is relatively harder to use them properly, which is the reason why we have concentrated on the nonce-based schemes. The advantage of CENC-C and CHM-C is that, the nonce length and the maximum plaintext length restrictions are removed, while the security is unchanged (further, non-adaptive version of PRP is enough for the security proofs). The restrictions only come from the security bound (instead of the schemes). Thus, if carefully implemented and properly used, these counter versions are suitable especially for 64-bit blockciphers

Tightness of the security bounds. For CTR mode, the security bound is tight up to a constant factor. However, for CENC and CHM (and the PRF F in Section 3), we do not know the tightness of our security bounds. The tightness is an open question. For example, if we take CENC, the bound is $O(w\hat{\sigma}^3/2^{2n} + w\hat{\sigma}/2^n)$. The question is the existence of an adversary A that breaks the privacy of CENC with about $\hat{\sigma} = 2^{82}$ data (without breaking the pseudorandomness of the AES), *or* the proof that the security is better than the above. We conjecture that the bound of CENC can be improved to $O(w\hat{\sigma}/2^n)$, possibly by using the technique from [2][1].

---

[1] However, it is not possible to check the details of the proof of [2], since only a sketch is given.

Acknowledgement

The author would like to thank Kazumaro Aoki, Fumihiko Sano, and Akashi Satoh for useful comments.

References

1. M. Bellare, A. Desai, E. Jokipii, and P. Rogaway. A concrete security treatment of symmetric encryption. Proceedings of *The 38th Annual Symposium on Foundations of Computer Science, FOCS '97,* pp. 394–405, IEEE, 1997.
2. M. Bellare, and R. Impagliazzo. A tool for obtaining tighter security analyses of pseudorandom function based constructions, with application to PRP → PRF convention. *Cryptology ePrint Archive,* Report 1999/024, Available at http://eprint.iacr.org/, 1999.
3. M. Bellare, J. Kilian, and P. Rogaway. The security of the cipher block chaining message authentication code. *JCSS,* vol. 61, no. 3, pp. 362–399, 2000. Earlier version in *Advances in Cryptology—CRYPTO '94,* LNCS 839, pp. 341–358, Springer-Verlag, 1994.
4. M. Bellare, T. Krovetz, and P. Rogaway. Luby-Rackoff backwards: Increasing security by making block ciphers non-invertible. *Advances in Cryptology—EUROCRYPT '98,* LNCS 1403, pp. 266–280, Springer-Verlag, 1998.
5. M. Bellare, and C. Namprempre. Authenticated encryption: Relations among notions and analysis of the generic composition paradigm. *Advances in Cryptology—ASIACRYPT 2000,* LNCS 1976, pp. 531–545, Springer-Verlag, 2000.
6. M. Bellare, and P. Rogaway. Encode-then-encipher encryption: How to exploit nonces or redundancy in plaintexts for efficient cryptography. *Advances in Cryptology—ASIACRYPT 2000,* LNCS 1976, pp. 317–330, Springer-Verlag, 2000.
7. M. Bellare, P. Rogaway, and D. Wagner. The EAX mode of operation. *Fast Software Encryption, FSE 2004,* LNCS 3017, pp. 389–407, Springer-Verlag, 2004.
8. K. Claffy, G. Miller, and K. Thompson. The nature of the beast: Recent traffic measurements from an Internet backbone. Proceedings of *INET '98.* Available at http://www.caida.org/outreach/papers/1998/Inet98.
9. D. Delov, C. Dwork, and M. Naor. Non-malleable cryptography. *SIAM J. Comput.,* vol. 30, no. 2, pp. 391–437, 2000.
10. C. Hall, D. Wagner, J. Kelsey, and B. Schneier. Building PRFs from PRPs. *Advances in Cryptology—CRYPTO '98,* LNCS 1462, pp. 370–389, Springer-Verlag, 1998.
11. T. Iwata. New blockcipher modes of operation with beyond the birthday bound security. Full version of this paper. Available from the author, 2006.
12. J. Jonsson. On the Security of CTR + CBC-MAC. *Selected Areas in Cryptography, 9th Annual Workshop (SAC 2002),* LNCS 2595, pp. 76–93. Springer-Verlag, 2002.
13. C.S. Jutla. Encryption modes with almost free message integrity. *Advances in Cryptology—EUROCRYPT 2001,* LNCS 2045, pp. 529–544, Springer-Verlag, 2001.
14. J. Katz, and M. Yung. Unforgeable encryption and chosen ciphertext secure modes of operation. *Fast Software Encryption, FSE 2000,* LNCS 1978, pp. 284–299, Springer-Verlag, 2000.
15. T. Kohno, J. Viega, and D. Whiting. CWC: A high-performance conventional authenticated encryption mode. *Fast Software Encryption, FSE 2004,* LNCS 3017, pp. 408–426, Springer-Verlag, 2004.

16. S. Lucks. The sum of PRPs is a secure PRF. *Advances in Cryptology—EUROCRYPT 2000*, LNCS 1807, pp. 470–484, Springer-Verlag, 2000.
17. S. Lucks. The two-pass authenticated encryption faster than generic composition. *Fast Software Encryption, FSE 2005*, LNCS 3557, pp. 284–298, Springer-Verlag, 2005.
18. M. Luby and C. Rackoff. How to construct pseudorandom permutations from pseudorandom functions. *SIAM J. Comput.*, vol. 17, no. 2, pp. 373–386, 1988.
19. D. McGrew, and J. Viega. The Galois/Counter mode of operation (GCM). Submission to NIST. Available at http://csrc.nist.gov/CryptoToolkit/modes/, 2004.
20. D. McGrew, and J. Viega. The security and performance of Galois/Counter mode of operation. *Progress in Cryptology—INDOCRYPT 2004*, LNCS 3348, pp. 343–355, Springer-Verlag, 2004.
21. P. Rogaway. Nonce-based symmetric encryption. *Fast Software Encryption, FSE 2004*, LNCS 3017, pp. 348–358, Springer-Verlag, 2004.
22. P. Rogaway. Authenticated-encryption with associated-data. *Proceedings of the ACM Conference on Computer and Communications Security, ACM CCS 2002*, pp. 98–107, ACM, 2002.
23. P. Rogaway, M. Bellare, J. Black, and T. Krovetz. OCB: a block-cipher mode of operation for efficient authenticated encryption. *ACM Trans. on Information System Security (TISSEC)*, vol. 6, no. 3, pp. 365–403, 2003. Earlier version in *Proceedings of the eighth ACM Conference on Computer and Communications Security, ACM CCS 2001*, pp. 196–205, ACM, 2001.
24. M.N. Wegman, and J.L. Carter. New hash functions and their use in authentication and set equality. *JCSS*, vol. 22, pp. 256–279, 1981.
25. D. Whiting, R. Housley, and N. Ferguson. Counter with CBC-MAC (CCM). Submission to NIST. Available at http://csrc.nist.gov/CryptoToolkit/modes/, 2002.

A   Proof of Theorem 1

Proof (of Theorem 1). Without loss of generality, we assume that A makes exactly q oracle queries and A does not repeat an oracle query. Also, since A is computationally unbounded, we assume that A is deterministic. Now we can regard A as a function $f_A : (\{0,1\}^{nw})^q \to \{0,1\}$. To see this, let $Y = (Y_0, \ldots, Y_{q-1})$ be an arbitrary nqw-bit string, where each Y_i is nw bits. The first query, x_0, is determined by A. If we return Y_{i-1} as the answer for x_{i-1}, the next query x_i is determined, and finally, if we return Y_{q-1} as the answer for x_{q-1}, the output of A, either 0 or 1, is determined. Therefore, the output of A and the q queries, $x_0, \ldots, x_{q-1}$, are all determined by fixing Y. Note that for any Y, the corresponding sequence of queries $x = (x_0, \ldots, x_{q-1})$ is distinct. Let $\mathbf{v}_{\text{one}} = \{Y \in (\{0,1\}^{nw})^q \mid f_A(Y) = 1\}$, and $\mathbf{v}_{\text{dist}} = \{Y \in (\{0,1\}^{nw})^q \mid Y$ is non-zero-distinct$\}$. Observe that $|\mathbf{v}_{\text{dist}}| = ((2^n - 1)(2^n - 2) \cdots (2^n - w))^q \geq 2^{nwq}(1 - qw(w+1)/2^{n+1})$, and therefore, we have

$$|\mathbf{v}_{\text{one}} \cap \mathbf{v}_{\text{dist}}| \geq |\mathbf{v}_{\text{one}}| - 2^{nwq}qw(w+1)/2^{n+1}. \tag{6}$$

Let $P_R \overset{\text{def}}{=} \Pr(R \overset{R}{\leftarrow} \text{Func}(n - \omega, nw) : A^{R(\cdot)} = 1)$. Then we have

$$P_R = \sum_{Y \in \mathbf{v}_{\text{one}}} p_R = \frac{|\mathbf{v}_{\text{one}}|}{(2^{nw})^q}. \tag{7}$$

PRF-adversary B
If A **makes a query** (N_i, M_i)**:**
100    $\text{ctr} \leftarrow (N_i \| 0^{n-\ell_{\text{nonce}}})$
101    $l \leftarrow \lceil |M_i|/n \rceil$
102    $S \leftarrow \text{CENC.KSGen.Sim}(\text{ctr}, l)$
103    $C_i \leftarrow M_i \oplus \text{first}(|M_i|, S)$
104   **return** C_i
If A **returns** b**:**
200   **output** b

Algorithm CENC.KSGen.Sim(ctr, l)
300   **for** $j \leftarrow 0$ **to** $\lceil l/w \rceil - 1$ **do**
301        $Y_j \leftarrow \mathcal{O}(\text{ctr})$
302        $\text{ctr} \leftarrow \text{inc}^{w+1}(\text{ctr})$
303    $Y \leftarrow (Y_0, \ldots, Y_{\lceil l/w \rceil - 1})$
304    $Y \leftarrow \text{first}(nl, Y)$
305   **return** Y

Fig. 6. The PRF-adversary B for F^+ based on the PRIV-adversary A for CENC

On the other hand, let $P_F \overset{\text{def}}{=} \Pr(P \overset{R}{\leftarrow} \text{Perm}(n) : A^{F_P(\cdot)} = 1)$. Then

$$P_F = \sum_{Y \in \mathbf{v}_{\text{one}}} p_F \geq \sum_{Y \in (\mathbf{v}_{\text{one}} \cap \mathbf{v}_{\text{dist}})} p_F \geq \left(1 - \frac{q^3(w+1)^4}{2^{2n+1}}\right) \sum_{Y \in (\mathbf{Y}_{\text{one}} \cap \mathbf{Y}_{\text{dist}})} \frac{1}{(2^{nw})^q}$$

where the last inequality follows from Lemma 1. Then P_F is at least

$$\left(1 - \frac{q^3(w+1)^4}{2^{2n+1}}\right) \frac{|\mathbf{v}_{\text{one}} \cap \mathbf{v}_{\text{dist}}|}{(2^{nw})^q} \geq \left(1 - \frac{q^3(w+1)^4}{2^{2n+1}}\right) \left(P_R - \frac{qw(w+1)}{2^{n+1}}\right)$$

from (6) and (7). Now, we have $P_F \geq P_R - q^3(w+1)^4/2^{2n+1} - qw(w+1)/2^{n+1}$, and by applying the same argument to $1 - P_F$ and $1 - P_R$, we have $1 - P_F \geq 1 - P_R - q^3(w+1)^4/2^{2n+1} - qw(w+1)/2^{n+1}$. $\qquad \square$

B   Proof of Theorem 2

Proof (of Theorem 2). Suppose for a contradiction that $\mathbf{Adv}_{\text{CENC}}^{\text{priv}}(A)$ is larger than the right hand side of (3). Let the oracle $\mathcal{O}$ be either $F_P^+(\cdot)$ or $R(\cdot) \in \text{Func}(n, nw)$. Consider the PRF-adversary B for F^+ in Figure 6, where B uses the nonce-respecting PRIV-adversary A for CENC as a subroutine.

We see that if $\mathcal{O}$ is $F_P^+(\cdot)$, then B gives A a perfect simulation of CENC.Enc, since $F_P^+(\cdot)$ corresponds to "one frame" of CENC.KSGen, and therefore the outputs of CENC.KSGen.Sim(ctr, l) and CENC.KSGen$_P(\text{ctr}, l)$ are the same. This implies $\Pr(P \overset{R}{\leftarrow} \text{Perm}(n) : B^{F_P^+(\cdot)} = 1) = \Pr(P \overset{R}{\leftarrow} \text{Perm}(n) : A^{\text{CENC.Enc}_P(\cdot, \cdot)} = 1)$. Also, it is easy to check that B is input-respecting. On the other hand, if $\mathcal{O}$ is $R(\cdot)$, then B gives A a perfect simulation of $\mathcal{R}$. That is, $\Pr(R \overset{R}{\leftarrow} \text{Func}(n, nw) : B^{R(\cdot)} = 1) = \Pr(A^{\mathcal{R}(\cdot, \cdot)} = 1)$. Therefore, we have $\mathbf{Adv}_{F^+}^{\text{prf}}(B) = \mathbf{Adv}_{\text{CENC}}^{\text{priv}}(A)$.

Suppose that the queries made by A are $(N_0, M_0), \ldots, (N_{q-1}, M_{q-1})$. If we let $l_i = \lceil |M_i|/n \rceil$, then B makes $\lceil l_0/w \rceil + \cdots + \lceil l_{q-1}/w \rceil$ queries, which is at most $(l_0 + \cdots + l_{q-1})/w + q \leq \sigma/w + q = \hat{\sigma}/w$ queries. Note that this holds regardless of the value of $l_0, \ldots, l_{q-1}$. From the assumption for a contradiction, $\mathbf{Adv}_{\text{CENC}}^{\text{priv}}(A)$ is larger than the right hand side of (3), which implies $\mathbf{Adv}_{F^+}^{\text{prf}}(B) > (w+1)^4 \hat{\sigma}^3 / w^3 2^{2n+1} + (w+1)\hat{\sigma}/2^{n+1}$. This contradicts Corollary 1. $\qquad \square$

The Ideal-Cipher Model, Revisited: An Uninstantiable Blockcipher-Based Hash Function

John Black

Dept. of Computer Science, University of Colorado, Boulder CO 80309, USA
jrblack@cs.colorado.edu
www.cs.colorado.edu/~jrblack

Abstract. The Ideal-Cipher Model of a blockcipher is a well-known and widely-used model dating back to Shannon [25] and has seen frequent use in proving the security of various cryptographic objects and protocols. But very little discussion has transpired regarding the meaning of proofs conducted in this model or regarding the model's validity. In this paper, we briefly discuss the implications of proofs done in the ideal-cipher model, then show some limitations of the model analogous to recent work regarding the Random-Oracle Model [2]. In particular, we extend work by Canetti, Goldreich and Halevi [5], and a recent simplification by Maurer, Renner, and Holenstein [15], to exhibit a blockcipher-based hash function that is provably-secure in the ideal-cipher model but trivially insecure when instantiated by any blockcipher.

Keywords: Ideal-Cipher Model, Information-Theoretic Cryptography, Random-Oracle Model, Uninstantiability.

1 Introduction

THE STANDARD MODEL. Before we can prove the security of a cryptographic system or object, we must specify what model we are using. The most common model used in modern cryptography is the so-called "standard model." Here we use no special mathematical objects such as infinite random strings or random oracles [2], and we abstract our communications system typically as a reliable but insecure channel. We have not been able to achieve most common cryptographic goals in the standard model without making additional complexity-theoretic hardness assumptions, because we still have no proof that any of our standard cryptographic building blocks have computational lower bounds. The common assumptions are typically that factoring the product of large primes is hard, or that discrete log is intractible in certain sufficiently large groups, or that AES is a good pseudo-random permutation (PRP) [16]. The standard model is usually well-accepted in our community despite the fact that proofs done in this model rest upon unproven assumptions and that already much relevant real-world context has been abstracted away (timing, power consumption, error

M.J.B. Robshaw (Ed.): FSE 2006, LNCS 4047, pp. 328–340, 2006.

messages, and other real-world effects are typically not included as part of the model in spite of the demonstrated fact they are often relevant to security).

THE RANDOM-ORACLE MODEL. When proofs in the standard model are unappealing or are provably impossible (eg, see [19]), we often resort to proofs using an alternative model. By far the best-known is the "Random-Oracle Model." The random-oracle model was used for some time before being formalized by Bellare and Rogaway [2], and continues to see widespread use today (there are more than a hundred instances; for a few examples see [11,18,2,21,24]). In the random-oracle model we have a public random function, accessible to all parties, which typically accepts any string from $\{0,1\}^*$ and outputs n bits. For each element in its domain, the corresponding n-bit output is uniform and independent from all other outputs. Proofs conducted in the random-oracle model often admit schemes which are provably-secure and more efficient than schemes which have been proven secure in the standard model, and for this reason the random-oracle model has been widely-adopted.

Of course random oracles do not exist in practice, and if the schemes proven secure in the random-oracle model are going to be put into use, we must choose some object to implement the random oracle. This step is called "instantiation." Most often, random oracles are instantiated with cryptographic hash functions such as SHA-1 [20]. The following question then arises: now that we have instantiated our random oracle with a concrete function, what security guarantees do we have? Does our proof in the random-oracle model have any bearing on the security of the instantiated system?

For quite some time there has been concern in our community that the random-oracle model should be treated with suspicion, and proofs in the standard model should be preferred. As a recent example, the main selling point of the Cramer-Shoup cryptosystem [7] is that it is provably-secure in the standard model and still practical (and, as with most proofs in the standard model, comes with an assumption: the Decisional Diffie-Hellman assumption [4]). Further doubt has been recently cast on the random-oracle model due to a string of results exhibiting schemes which are provably-secure in the random-oracle model but are completely insecure when instantiated by *any* hash function [5,15,6,1]. Schemes of this type are called "uninstantiable."

It has been noted [2] that proofs done in the random-oracle model *do* guarantee one thing: if the adversary treats the instantiated random oracle as a black box, promising not to think about its inner workings, promising not to exploit any unnatural behavior related to the fact that we have instantiated with some algorithm that has a compact description, then the proof remains valid in the standard model. Of course there is no guarantee that real adversaries would abide by such restrictions, and indeed they would be remiss if they did. Nonetheless, no scheme has thus far been proven secure in the random-oracle model and then broken once instantiated, unless this was the goal from the start.

THE IDEAL-CIPHER MODEL. Blockciphers are a common building block for cryptographic protocols. In the standard model the associated assumption for blockciphers is that they are "pseudo-random permutations" (PRPs). By this we mean (informally) that an n-bit blockcipher under a secret randomly-chosen key is computationally indistinguishable from a randomly-chosen n-bit permutation. Proofs conducted using this assumption typically give reductions showing that if an adversary breaks some scheme, then there exists an associated adversary that can efficiently distinguish the underlying blockcipher from random.

There are countless examples where the PRP assumption in the standard model is sufficient, but there are also plenty of cases where we cannot get a proof to go through. In certain cases it can be shown that blockcipher-based schemes we believe to be secure cannot have a proof of security using only the PRP assumption in the standard model [26]. In this case we are faced with either abandoning attempts at a proof, or using an alternate model.

The blockcipher analog for the random-oracle model is variously called the "Shannon Model," the "Black-Box Model," or the "Ideal-Cipher Model." We will prefer the latter name in this paper.

Though not as widely-used as the random-oracle model, the ideal-cipher model dates back to Shannon [25] and has been used in a variety of settings (see, for example [27,17,10,13,9,3,12]). In the ideal-cipher model we think of a blockcipher E with k-bit key and n-bit blocksize as being chosen uniformly from the set of all possible blockciphers of this form. For each key, there are $2^n!$ permutations, and since any permutation may be assigned to a given key, there are $(2^n!)^{2^k}$ possible blockciphers. When we instantiate our black box, it becomes some particular blockcipher. AES with a 128-bit key is one choice from the nearly $2^{2^{263}}$ blockciphers we could have chosen (though in the spirit of Kolmogorov complexity and in line with the main result of this paper, we should note that the vast majority of these blockciphers will not have an efficient and compact C implementation).

The ideal-cipher model is analogous to the random-oracle model with three notable exceptions:

- The ideal cipher has a permutivity requirement that random oracles obviously do not.
- Adversaries interacting with an ideal-cipher oracle are typically given access to both the cipher and its inverse.
- The blocksize n of the ideal cipher is typically fixed a priori. This means that an ideal cipher is a finite object while the random oracle is an infinite one.

The ideal-cipher model has been used in a variety of settings, and like the random-oracle model, some researchers question the wisdom of its use. The argument is completely analogous: if a scheme is proved secure in the ideal-cipher model, what exactly are we guaranteed once the ideal cipher is instantiated by a real blockcipher? And if the answer is essentially "not much," then what is the value of such proofs? A common argument against the ideal-cipher model is that most real-world blockciphers have distinguishing patterns which would exist with exceedingly small probability in a collection of random permutations. The key complementation property of DES is a typical example of this [16]. Although no such properties are currently known for AES, some blockcipher experts who are comfortable with the assumption that AES is a good PRP are reluctant to model AES as ideal because of practical concerns: the AES key schedule, for instance, is quite simple and it perhaps contains related-key properties we have not yet discovered.

As compensation to the adversary for his respecting the blockcipher as a black box, we often endow him with limitless computational resources. In this respect, many proofs done in the ideal-cipher model are information theoretic. This too is unrealistic, but here we are giving the *adversary* more power rather than enhancing the *objects* themselves. Nonetheless, it is saying something about the strength of our model that it allows us to achieve information-theoretic security.

The main result of this paper is to exhibit a blockcipher-based hash function that is secure in the ideal-cipher model against information-theoretic adversaries but which is trivially insecure once instantiated with any blockcipher. In order to state this result more clearly, we take a short detour to review blockcipher-based hash functions.

BLOCKCIPHER-BASED HASH FUNCTIONS. One area of recently-renewed interest involves constructing hash functions from blockciphers. This approach, dating back at least to Rabin [23], uses some blockcipher E with an n-bit key and an n-bit blocksize, and builds a compression function from it. Iterating this function then hopefully produces a collision-resistant hash function. Preneel, Govaerts, and Vandewalle [22] conducted a systematic study of a class of 64 blockcipher-based hash functions. They focused on compression functions of the form $f(h_{i-1}, m_i) = E_a(b) \oplus c$ where $a, b, c \in \{h_{i-1}, m_i, h_{i-1} \oplus m_i, v\}$ for some fixed constant v. We can now hash any $M \in (\{0,1\}^n)^+$ by writing $M = M_1 \cdots M_\ell$ and then setting h_0 to some constant (typically 0^n) and setting $h_i = f(h_{i-1}, m)$. We return h_ℓ as the digest. The PGV analysis consisted of testing a series of attacks on each of these iterated hash functions. Black, Rogaway and Shrimpton [3] considered these same 64 constructions exhibiting either an attack or a proof of security (in the ideal-cipher model) for each. They determined that 20 of the 64 schemes were provably collision-resistant up to the birthday bound. For one example, see Figure 1.

Although a proof of security for a blockcipher-based hash function in the *standard model* would be prefered, it has been shown that the PRP assumption is insufficient for building a collision-resistant hash function [26]. Indeed, one can easily imagine a blockcipher $\widetilde{E} : \{0,1\}^n \times \{0,1\}^n \rightarrow \{0,1\}^n$ that is a good PRP, but which fails when used in the MMO construction of Figure 1. For example, let blockcipher $E : \{0,1\}^n \times \{0,1\}^n \rightarrow \{0,1\}^n$ be a good PRP and consider blockcipher $\widetilde{E}$ defined as follows:

$$
\widetilde{E}(K, X) = \begin{cases} K & \text{if } X = K \\ E(K, K) & \text{if } X = E^{-1}(K, K) \\ E(K, X) & \text{otherwise} \end{cases}
$$

So $\widetilde{E}$ is the same blockcipher as E with one change: we now have the invariant that $E(K, K) = K$ for all $K \in \{0,1\}^n$. Clearly $\widetilde{E}$ is a good PRP since E was: for a *randomly*-chosen key K, $\widetilde{E}(K, \cdot)$ is computationally indistinguishable from

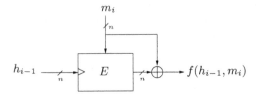

Fig. 1. The Matyas-Meyer-Oseas (MMO) compression function [14], called H_1 in [3]. $E: \{0,1\}^n \times \{0,1\}^n \to \{0,1\}^n$ is a block cipher; the hatch mark denotes the location of the key. Iterating this compression function results in a provably-secure blockcipher-based hash function in the ideal-cipher model.

a random permutation. However, using $\widetilde{E}$ in MMO would be inadvisable: it is trivial to find collisions. Specifically, let H be MMO built on $\widetilde{E}$ with $h_0 = 0^n$. Then $H(a \parallel \widetilde{E}(0,a) \oplus a) = 0^n$ for all $a \in \{0,1\}^n$.

MAIN RESULT. Given the recent string of results calling into question the validity of the random-oracle model, it is natural to ask if there are similar results which can be shown for the ideal-cipher model. Specifically, is it possible to exhibit some cryptographic scheme which is provably secure in the ideal-cipher model and yet breaks when instantiated with *any* blockcipher? Given that ideal ciphers are finite objects whereas random oracles are infinite objects, this fact might lead one to ask whether results for the random-oracle model (in particular uninstantiability results) might break down in the ideal-cipher setting given that ideal ciphers can be described with a finite string. We will show that the answer to the above question is "yes": we exhibit a blockcipher-based hash function which is provably collision-resistant in the ideal-cipher model and for which it is trivial to find collisions once the ideal cipher has been instantiated.

We follow the approaches of [5,15], adapting them to blockciphers and hash functions, and moving into the concrete (rather than asymptotic) setting. The main idea is to create a blockcipher-based hash function $\widetilde{H}$ that acts normally on most inputs, but acts insecurely when given a description of its oracle as an input. In the latter case, $\widetilde{H}$ tests the oracle description embedded in its input against the oracle it already has by submitting some number of test values. If the oracles agree on all values, $\widetilde{H}$ outputs a user-specified value which was also given in the input. The difficulty here is showing that $\widetilde{H}$ remains secure even when behaving this way, and the crucial point is that there a far more possible ideal ciphers with specified input-output pairs than there are encodings to represent them. All of this is formalized and rigorously proven in Section 3.

RELATED WORK. Virtually no discussion of the ideal-cipher model has transpired prior to this work. As already mentioned, much relevant work has appeared in the analogous random-oracle setting. Random oracles were used implicitly at least 18 years ago by Fiat and Shamir in their seminal work on identification schemes [11]. Bellare and Rogaway formalized the notion and argued that the model afforded a path to efficient protocols; as examples, they gave efficient non-malleable and chosen-ciphertext-secure encryption schemes, a signature scheme secure against adaptive chosen-message attack, and an efficient zero-knowledge proof protocol [2]. Canetti, Goldreich, and Halevi gave the first uninstantiable protocol for the random-oracle model: they exhibited a signature scheme which was provably-secure in the random-oracle model but which acted insecurely (gave up its key) when instantiated [5]. Their proof is quite complex, involving techniques similar to Micali's CS-proofs [18]. The same authors later extended their result to show that there exists a signature scheme, limited to short messages, which is also uninstantiable [6]. Maurer, Renner, and Holenstein generalized the results of [5]; they introduced a generalization of indistinguishability called "indifferentiability" which captures the notion of shared random objects (like random oracles) [15]. They state general theorems which imply the result of [5] as a special case, and give an explicit simplified proof of that result. Their proof is very much in the spirit of classical Kolmogorov complexity theory as is ours in the present paper. Nielsen [19] exhibited a protocol that had a simple solution in the random-oracle model, but which had no provable instantiation in the standard model. Bellare, Boldyreva, and Palacio exhibited the first "natural" scheme, a hybrid encryption scheme, secure in the random-oracle model but uninstantiable [1]. Dent adapted techniques from [5] to show an uninstantiable signature scheme in the generic group model (generic groups are finite objects like ideal ciphers) [8].

2   Definitions

BASIC NOTIONS. Let $\kappa, n \geq 1$ be numbers. A *blockcipher* is a map $E \colon \{0,1\}^\kappa \times \{0,1\}^n \to \{0,1\}^n$ where, for each $k \in \{0,1\}^\kappa$, the function $E_k(\cdot) = E(k, \cdot)$ is a permutation on $\{0,1\}^n$. Parameter n is called the *blocksize* of E, and n will be understood to be this quantity throughout the paper. If E is a blockcipher then E^{-1} is its inverse, where $E_k^{-1}(y)$ is the string x such that $E_k(x) = y$. Let $\mathrm{Bloc}(\kappa, n)$ be the set of all block ciphers $E \colon \{0,1\}^\kappa \times \{0,1\}^n \to \{0,1\}^n$. Choosing a random element of $\mathrm{Bloc}(\kappa, n)$ means that for each $k \in \{0,1\}^\kappa$ one chooses a random permutation $E_k(\cdot)$.

A (blockcipher-based) *hash function* is a map $H \colon \mathrm{Bloc}(\kappa, n) \times D \to R$ where $\kappa, n, c \geq 1$, $D \subseteq \{0,1\}^*$, and $R = \{0,1\}^c$. The function H must be given by a program that, given M, computes $H^E(M) = H(E, M)$ using an E-oracle. Hash function $f \colon \mathrm{Bloc}(\kappa, n) \times D \to R$ is a *compression function* if $D = \{0,1\}^a \times \{0,1\}^b$ for some $a, b \geq 1$ where $a + b \geq c$. Fix $h_0 \in \{0,1\}^a$. The *iterated hash* of compression function $f \colon \mathrm{Bloc}(\kappa, n) \times (\{0,1\}^a \times \{0,1\}^b) \to \{0,1\}^a$ is the hash function $H \colon \mathrm{Bloc}(\kappa, n) \times (\{0,1\}^b)^* \to \{0,1\}^a$ defined by $H^E(m_1 \cdots m_\ell) = h_\ell$ where $h_i = f^E(h_{i-1}, m_i)$. Set $H^E(\varepsilon) = h_0$. We often omit the superscript E to f and H.

We write $x \xleftarrow{\$} S$ for the experiment of choosing a random element from the finite set S and calling it x. An *adversary* is an algorithm with access to one or more oracles. We write these as superscripts. The notation $|x|$ denotes the size of the string x, in bits, and the notation $x[i \ldots j]$ denotes the substring of string x starting at the i-th bit of x and terminating at the j-th bit, inclusive. All bits are numbered starting from 1, and ascending left-to-right. Finally, $x \parallel y$ denotes the concatenation of strings x and y.

COLLISION RESISTANCE. To quantify the collision resistance of a blockcipher-based hash function H we instantiate the blockcipher by a randomly chosen $E \in \text{Bloc}(\kappa, n)$. An adversary A is given oracles for $E(\cdot, \cdot)$ and $E^{-1}(\cdot, \cdot)$ and wants to find a *collision* for H^E—that is, M, M' where $M \neq M'$ but $H^E(M) = H^E(M')$. We look at the number of queries that the adversary makes and compare this with the probability of finding a collision.

Definition 1 (Collision Resistance). Let H be a blockcipher-based hash function, $H : \text{Bloc}(\kappa, n) \times D \to R$, and let A be an adversary. Then the advantage of A in finding collisions in H is the real number

$$\mathbf{Adv}_H^{\text{coll}}(A) = \Pr\left[E \xleftarrow{\$} \text{Bloc}(\kappa, n); (M, M') \xleftarrow{\$} A^{E, E^{-1}} : \right.$$

$$\left. M \neq M' \wedge H^E(M) = H^E(M')\right]$$

For $q \geq 1$ we write $\mathbf{Adv}_H^{\text{coll}}(q) = \max_A\{\mathbf{Adv}_H^{\text{coll}}(A)\}$ where the maximum is taken over all adversaries that ask at most q oracle queries (ie, E-queries $+ E^{-1}$ queries).

3   An Uninstantiable Blockcipher-Based Hash Function

In [3] we find 20 blockcipher-based hash function constructions that are provably secure in the ideal-cipher model. Specifically, it is shown that $\mathbf{Adv}_H^{\text{coll}}(q) = \Theta(q^2/2^n)$ for 20 blockcipher-based hash functions H. This bound is about the best we can hope for: a truly random function would have the same bound due to the birthday phenomenon.

The proofs in [3] are carried out in the ideal-cipher model and the adversaries are information theoretic. In this section we will show that any scheme H from this set can be transformed into a related scheme $\widetilde{H}$ such that $\widetilde{H}$ is uninstantiable. We first outline the method and then give the details.

MAIN IDEA. Our goal is to produce an uninstantiable blockcipher-based hash function. We will do this by transforming some scheme which is provably secure in the ideal-cipher model. For concreteness, select any of the 20 secure schemes from [3] and call it H.

We will describe a related blockcipher-based hash function $\widetilde{H}$ which is uninstantiable. The idea has its roots in Kolmogorov complexity. We adapt the approach of Maurer, Renner, and Holenstein [15]; when $\widetilde{H}$ processes input M, it first decomposes M into three parts: $M = (\pi, c, v)$ where the details of this decomposition are left for later. The first parameter, π is considered to be the encoding of a Universal Turing Machine (UTM), encoded in some well-defined manner. The second parameter $c \in \{0,1\}^\sigma$ is a counter that is ignored by $\widetilde{H}$, and the final parameter $v \in \{0,1\}^n$ is the value that the adversary would like to have output by $\widetilde{H}$.

Now $\widetilde{H}$ uses its blockcipher oracle $\mathcal{O}$ to compute $\mathcal{O}(i, 0^n)$ for all $1 \leq i \leq |\pi|$. (Why we choose this range will become apparent in the proof below.) It also computes $\pi(i, 0^n)$ for the same set of i-values. If $\mathcal{O}(i, 0^n) = \pi(i, 0^n)$ for all $1 \leq i \leq |\pi|$, $\widetilde{H}$ outputs v. If not, $\widetilde{H}$ outputs $H(M)$.

Now consider two cases: in the first case, the oracle to $\widetilde{H}$ was an ideal cipher I. This means that it is highly unlikely there is a sufficiently-short Turing-machine encoding, π, such that $\pi(\cdot, 0^n)$ would correctly match I on all $|\pi|$ points, and therefore it is extremely unlikely that we would have $I(i, 0^n) = \pi(i, 0^n)$ for all $1 \leq i \leq |\pi|$. This means that in all likelihood $\widetilde{H}$ would output $H(M)$, and we know this construction is provably collision resistant. And so in this case $\mathbf{Adv}_{\widetilde{H}}^{\mathrm{coll}}(q)$ is $\Theta(q^2/2^n)$ by [3].

Now consider the case where the oracle to $\widetilde{H}$ is some blockcipher E; in other words we have instantiated oracle $\mathcal{O}$ with blockcipher E. There therefore exists some Turing machine π that implements E. Therefore an adversary may simply output two queries $M_1 = (\pi \parallel 0^\sigma \parallel v)$ and $M_2 = (\pi \parallel 1^\sigma \parallel v)$ for any fixed string $v \in \{0,1\}^n$ he desires. Since $\widetilde{H}$ will discover that $E(i, 0^n) = \pi(i, 0^n)$ for all $1 \leq i \leq |\pi|$, it will output v for both queries, and this adversary will have trivially found a collision.

Note that things could not be worse for $\widetilde{H}$, in fact: not only can we find collisions, but we can find preimages for any output value, second preimages for any output value, and 2^σ inputs which collide on any chosen value.

A DETAILED DESCRIPTION. We now proceed to formalize and prove correct the informal discussion just given. Throughout the remainder of this section, n will denote the blocksize of our blockciphers.

Definition 2. *Blockcipher E is said to be k-efficient if it can be implemented as a Turing machine never requiring more than k steps to produce its output.*

For example, all modern blockciphers are 2^{20}-efficient. For the remainder of this section, k is assumed to be some fixed value. We next exhibit an uninstantiable blockcipher-based hash function. Here, by "uninstantiable" we mean that a given hash function H has $\mathbf{Adv}_H^{\mathrm{coll}}(q) = O(q^2/2^n)$, and is therefore secure in the ideal-cipher model, but any instantiation of its blockcipher oracle with a blockcipher E results in a trivially insecure hash function.

For the remainder of this section we will let H denote some blockcipher-based hash function which is known to be secure in the ideal-cipher model (such as MMO, in Figure 1). We now give the algorithm $\widetilde{H}$ which is an uninstantiable variant of H, then we prove its various properties.

```
Algorithm H̃(M)
10    if |M| ≤ n + σ then return H^O(M)
20    v ← M[|M| − n + 1 ... |M|]
21    π ← M[1 ... |M| − n − σ]
30    if ¬TuringValid(π) then return H^O(M)
40    for i ← 1 to |π|
41        Run π on input (i, 0^n) for at most k steps
42        if π does not output n bits then return H^O(M)
43        if π(i, 0^n) ≠ O(i, 0^n) then return H^O(M)
50    return v
```

Fig. 2. An uninstantiable variant of the provably-secure blockcipher-based hash function H. If the input encodes a valid UTM, we evaluate $|\pi|$ values on this UTM and check against our oracle $\mathcal{O}$. If they match, we simply output v, the last n bits of M. There are σ bits of M which are ignored in order to help the attacker produce 2^σ colliding inputs with digest v. The UTM π is run for at most k steps, where k is a fixed parameter of the scheme.

Algorithm $\widetilde{H}$ accepts messages M from the domain $(\{0, 1\}^n)^+$ and outputs n bits. As usual, the domain could be extended to $M \in \{0, 1\}^*$ with an unambiguous padding rule. We fix two parameters to the algorithm: H, the provably-secure blockcipher-based hash function just mentioned, and a counter-size $\sigma > 0$. We assume the domain of H has been extended to $\{0, 1\}^*$ so we can dispense with concerns about message sizes in our construction of $\widetilde{H}$. We further fix some binary encoding scheme for Universal Turing Machines (UTMs) such that any UTM can be encoded into a binary string. Furthermore, we assume there is an efficient function TuringValid that returns true when given a string π that is a valid UTM encoding under our fixed convention. Finally, we let $\mathcal{O}$ denote the blockcipher-oracle which is used by $\widetilde{H}$. The algorithm to compute $\widetilde{H}^{\mathcal{O}}(M)$ is given in Figure 2.

We are now faced with arguing that $\widetilde{H}$ is uninstantiable. First notice that $\widetilde{H}$ is efficient: we assume that oracle calls are constant-time, so therefore $H^{\mathcal{O}}$ runs in time linear in the length of the input M. Since we run π for at most k steps, the whole algorithm runs in time $O(k|\pi|) = O(|M|)$.

Theorem 1. [$\widetilde{H}$ is uninstantiable] Fix some provably-secure blockcipher-based hash function H and some $\sigma > 0$. Let $\mathcal{O}$ be a blockcipher oracle. Then function $\widetilde{H}$ as described above is uninstantiable.

Proof. There are two things we must prove: first, that $\widetilde{H}$ is secure in the ideal-cipher model. That is, $\mathbf{Adv}^{\mathrm{coll}}_{\widetilde{H}^{\mathcal{O}}}(q) = O(q^2/2^n)$. Second, that $\widetilde{H}^E$ is insecure for any efficient blockcipher E.

We begin by showing $\widetilde{H}^{\mathcal{O}}$ is secure when $\mathcal{O}$ is modeled by an ideal cipher. Fix q and suppose adversary A makes q oracle queries to $\mathcal{O}$. (Throughout the proof, we will assume $q \leq 2^{n/2}$ since q-values in excess of this render the bound vacuous.) At the end of this process, A must output a pair of distinct messages M and M' in the hope that $\widetilde{H}^{\mathcal{O}}(M) = \widetilde{H}^{\mathcal{O}}(M')$. The probability that he succeeds is the advantage we wish to bound.

There are two types of collisions A may construct given the outputs from his q queries to $\mathcal{O}$. The first collision is event C_1: there exist two distinct messages M_1, M_2 such that they collide under the original hash function (ie, $H(M_1) = H(M_2)$). We have selected H such that $\Pr[C_1] = O(q^2/2^n)$. The other type of collision A might construct given his q oracle-query outputs results in event C_2 which we describe next.

Extract π as in line 21, and observe the **for** loop at lines 40 through 43. If at any time, π does not output n bits, or if the n bits it does output do not agree with $\mathcal{O}$, we relegate the computation to H. Therefore we are concerned with the condition that π correctly computes the $|\pi|$ values required by the test on line 43. If $\pi(i, 0^n) = \mathcal{O}(i, 0^n)$ for all $1 \leq i \leq |\pi|$, we say π is a "qualifying" program. We define event C_2 as true if there exists a qualifying program with length at most q bits. If C_2 occurs, A will certainly have set v (computed at line 20) to a colliding value, and so we therefore wish to bound $\Pr[C_2]$.

Adversary A has made q queries to $\mathcal{O}$ and would like to now encode some qualifying program π into M, with $|\pi| \leq q$. To this end, there are two possibilities: (1) A outputs a program π where C_2 is guaranteed because he has queried $\mathcal{O}$ at all points from 1 to $|\pi|$ and there was a qualifying program, or (2) A outputs a program π where there exists some point j with $1 \leq j \leq |\pi|$ that A did not query, yet C_2 occurs by chance. In the second case, $\widetilde{H}$ will ask $\pi(j)$ and the probability over choices of $\mathcal{O}$ that $\pi(j, 0^n) = \mathcal{O}(j, 0^n)$ is $1/2^n$. Therefore in this case $\Pr[C_2] \leq 1/2^n$.

We therefore concern ourselves with the first case, where C_2 occurs because A has queried $\mathcal{O}(\cdot, 0^n)$ at all points from 1 to $|\pi|$. The encoding scheme is of course fixed a priori. Therefore $\Pr[C_2]$ is computed over choices of $\mathcal{O}$. Let Q_ℓ be the event that there exists a qualifying program of size ℓ. So $C_2 = Q_1 \vee \cdots \vee Q_q$. For fixed ℓ there are at most 2^ℓ possible Turing-Valid encodings π with $|\pi| = \ell$. We evaluate, at line 43, $\mathcal{O}(i, 0^n)$ for $1 \leq i \leq |\pi|$. Since we are iterating on the key value for $\mathcal{O}$, there is no permutivity, and therefore outputs will be uniform on $\{0, 1\}^n$. This means that, for a fixed i, the probability that $\pi(i, 0^n) = \mathcal{O}(i, 0^n)$ is 2^{-n}. The probability this will happen ℓ times is therefore $2^{-n\ell}$, and given there are 2^ℓ possible encodings, we see

$$\Pr_{\mathcal{O}}[Q_\ell] \leq 2^\ell/2^{n\ell} = 1/2^{\ell(n-1)} \leq 1/2^{n-1}.$$

So the chance of finding a qualifying program within q queries is

$$\Pr_{\mathcal{O}}[C_2] = \Pr_{\mathcal{O}}[Q_1 \vee \cdots \vee Q_q] \leq \sum_{\ell=1}^{q} 1/2^{n-1} = \frac{q}{2^{n-1}}.$$

Finally, the chance that A can find any collision in q queries is bounded by $\Pr[C_1 \vee C_2] \leq \Pr[C_1] + \Pr[C_2] = O(q^2/2^n) + q/2^{n-1} = O(q^2/2^n)$, as required.

The second case is quite straightforward. We wish to show that $\widetilde{H}^E$ is insecure for any efficient blockcipher E. Since E *does* have a concise Turing-Valid encoding π, we may simply write two messages

$$M = \pi \parallel 0^\sigma \parallel 0^n \qquad \text{and} \qquad M' = \pi \parallel 1^\sigma \parallel 0^n.$$

Since the oracle to $\widetilde{H}$ is E, and since π agrees with E on *every* point, the **if** condition in line 43 will never hold and we will return $v = 0^n$ for each message. Thus we have $\widetilde{H}^E(M) = \widetilde{H}^E(M') = 0^n$, yielding a collision with zero oracle queries required.

PREIMAGE, SECOND PREIMAGE, AND MULTICOLLISIONS. Since the instantiated form of $\widetilde{H}$ allows us full control over the output, we can clearly find 2^σ preimages for any digest of our choice, and we can similarly find $2^\sigma - 1$ second preimages for any given value. Similarly, we can find multicollisions for any value, and 2^σ collisions for each of the 2^n possible outputs. In this sense, $\widetilde{H}^E$ is much worse than just failing to be collision resistant: it fails to have any security properties at all.

On a technical note, the alert reader will notice that we at no time defined what "insecure" means for a blockcipher-based hash function that has been instantiated. This is because all concrete hash functions are "insecure" if security requires the nonexistence of any efficient program that outputs a colliding pair of inputs! (Since collisions must exist for any non-injective map f, there exists a program that simply outputs a colliding pair for any given f.) Nonetheless, there exists an intuitive notion of security for fixed functions like SHA-1, and clearly the instantiated version of hash function $\widetilde{H}$ is insecure in this sense.

ARTIFICIALITY. Like all other uninstantiable schemes, $\widetilde{H}$ is quite artificial. It is uninstantiable only because it was designed to be, and upon inspection no one would use such a scheme. It remains to be seen whether there is a more natural construction (where "natural" is necessarily subjective). Thus far, as in the random-oracle model analog, no scheme proven secure in the ideal-cipher model has been broken after instantiation, unless that was the goal from the start.

4   Conclusion and Open Questions

Although the scheme just presented is quite unnatural, it does arouse suspicion as to the wisdom of blindly using the ideal-cipher model in proofs of security. More evidence to support this suspicion could be provided by showing that H^{AES} is insecure for a hash scheme H from [3] that is provably-secure in the ideal-cipher model. Such an attack would necessarily exploit specific features of AES, but since AES is generally thought to be well-designed, it would add fuel to the fire.

Probably the short-signature results of [6] could be extended to this setting, but a more interesting question is whether there exists a "natural" scheme that is provably-secure in the ideal-cipher model but uninstantiable. Hash functions probably are not the right place to look for these, but there are many other objects whose proofs rely on the ideal-cipher model that might provide settings where natural examples of uninstantiable schemes could be constructed.

Acknowledgements

We would like to thank Zully Ramzan, Phillip Rogaway and Thomas Shrimpton for their comments and suggestions. John Black's work was supported by NSF CAREER-0240000 and a gift from the Boettcher Foundation.

References

1. BELLARE, M., BOLDYREVA, A., AND PALACIO, A. An uninstantiable random-oracle-model scheme for a hybrid-encryption problem. In *Advances in Cryptology – EUROCRYPT '04* (2004), vol. 3027 of *Lecture Notes in Computer Science*, Springer-Verlag, pp. 171–188.
2. BELLARE, M., AND ROGAWAY, P. Random oracles are practical: A paradigm for designing efficient protocols. In *Proceedings of the 1st ACM Conference on Computer and Communications Security* (1993), pp. 62–73.
3. BLACK, J., ROGAWAY, P., AND SHRIMPTON, T. Black-box analysis of the block-cipher-based hash-function constructions from PGV. In *Advances in Cryptology – CRYPTO '02* (2002), vol. 2442 of *Lecture Notes in Computer Science*, Springer-Verlag.
4. BONEH, D. The decision Diffie-Hellman problem. In *Proceedings of the Third Algorithmic Number Theory Symposium* (1998), vol. 1423 of *Lecture Notes in Computer Science*, Springer-Verlag, pp. 48–63.
5. CANETTI, R., GOLDREICH, O., AND HALEVI, S. The random oracle methodology, revisited. In *Proceedings of the 30th ACM Symposium on the Theory of Computing* (1998), ACM Press, pp. 209–218.
6. CANETTI, R., GOLDREICH, O., AND HALEVI, S. On the random-oracle methodology as applied to length-restricted signature schemes. In *First Theory of Cryptography Conference* (2004), vol. 2951 of *Lecture Notes in Computer Science*, Springer-Verlag, pp. 40–57.
7. CRAMER, R., AND SHOUP, V. A practical public key cryptosystem provably secure against adaptive chosen ciphertext attack. In *Advances in Cryptology – CRYPTO '93* (1998), Lecture Notes in Computer Science, Springer-Verlag, pp. 13–25.
8. DENT, A. Adapting the weaknesses of the random oracle model to the generic group model. In *Advances in Cryptology – ASIACRYPT '02* (2002), vol. 2501 of *Lecture Notes in Computer Science*, Springer-Verlag, pp. 100–109.
9. DESAI, A. The security of all-or-nothing encryption: Protecting against exhaustive key search. In *Advances in Cryptology – CRYPTO '00* (2000), vol. 1880 of *Lecture Notes in Computer Science*, Springer-Verlag.
10. EVEN, S., AND MANSOUR, Y. A construction of a cipher from a single pseudorandom permutation. In *Advances in Cryptology – ASIACRYPT '91* (1992), vol. 739 of *Lecture Notes in Computer Science*, Springer-Verlag, pp. 210–224.

11. FIAT, A., AND SHAMIR, A. How to prove yourself: Practical solutions to identification and signature problems. In *Advances in Cryptology – CRYPTO '86* (1986), Lecture Notes in Computer Science, Springer-Verlag, pp. 186–194.

12. JAULMES, E., JOUX, A., AND VALETTE, F. On the security of randomized CBC-MAC beyond the birthday paradox limit: A new construction. In *Fast Software Encryption (FSE 2002)* (2002), vol. 2365 of *Lecture Notes in Computer Science*, Springer-Verlag, pp. 237–251.

13. KILIAN, J., AND ROGAWAY, P. How to protect DES against exhaustive key search (An analysis of DESX). *Journal of Cryptology 14*, 1 (2001), 17–35.

14. MATYAS, S., MEYER, C., AND OSEAS, J. Generating strong one-way functions with cryptographic algorithms. *IBM Technical Disclosure Bulletin 27*, 10a (1985), 5658–5659.

15. MAURER, U. M., RENNER, R., AND HOLENSTEIN, C. Indifferentiability, impossibility results on reductions, and applications to the random oracle methodology. In *First Theory of Cryptography Conference* (2004), vol. 2951 of *Lecture Notes in Computer Science*, Springer-Verlag, pp. 21–39.

16. MENEZES, A., VAN OORSCHOT, P., AND VANSTONE, S. *Handbook of Applied Cryptography*. CRC Press, 1996.

17. MERKLE, R. One way hash functions and DES. In *Advances in Cryptology – CRYPTO '89* (1990), G. Brassard, Ed., vol. 435 of *Lecture Notes in Computer Science*, Springer-Verlag.

18. MICALI, S. CS-proofs. In *Proceedings of IEEE Foundations of Computing* (1994), pp. 436–453.

19. NIELSEN, J. B. Separating random oracle proofs from complexity theoretic proofs: The non-committing encryption case. In *Advances in Cryptology – CRYPTO '02* (2002), vol. 2442 of *Lecture Notes in Computer Science*, Springer-Verlag, pp. 111–126.

20. NIST. Secure hash standard (FIPS 180-1). 1995.

21. POINTCHEVAL, D., AND STERN, J. Security proofs for signature schemes. In *Advances in Cryptology – EUROCRYPT '96* (1996), Lecture Notes in Computer Science, Springer-Verlag, pp. 387–398.

22. PRENEEL, B., GOVAERTS, R., AND VANDEWALLE, J. Hash functions based on block ciphers: A synthetic approach. In *Advances in Cryptology – CRYPTO '93* (1994), Lecture Notes in Computer Science, Springer-Verlag, pp. 368–378.

23. RABIN, M. Digitalized signatures. In *Foundations of Secure Computation* (1978), R. DeMillo, D. Dobkin, A. Jones, and R. Lipton, Eds., Academic Press, pp. 155–168.

24. SCHNORR, C.-P. Efficient signature generation by smart cards. *Journal of Cryptology 4*, 3 (1991), 161–174.

25. SHANNON, C. Communication theory of secrecy systems. *Bell Systems Technical Journal 28*, 4 (1949), 656–715.

26. SIMON, D. Finding collsions on a one-way street: Can secure hash functions be based on general assumptions? In *Advances in Cryptology – EUROCRYPT '98* (1998), Lecture Notes in Computer Science, Springer-Verlag, pp. 334–345.

27. WINTERNITZ, R. A secure one-way hash function built from DES. In *Proceedings of the IEEE Symposium on Information Security and Privacy* (1984), IEEE Press, pp. 88–90.

How Far Can We Go on the x64 Processors?

Mitsuru Matsui

Information Technology R&D Center
Mitsubishi Electric Corporation
5-1-1 Ofuna Kamakura Kanagawa, Japan
Matsui.Mitsuru@ab.MitsubishiElectric.co.jp

Abstract. This paper studies the state-of-the-art software optimization methodology for symmetric cryptographic primitives on the new 64-bit x64 processors, AMD Athlon64 (AMD64) and Intel Pentium 4 (EM64T). We fully utilize newly introduced 64-bit registers and instructions for extracting maximal performance of target primitives. Our program of AES with 128-bit key runs in 170 cycles/block on Athlon 64, which is, as far as we know, the fastest implementation of AES on a PC processor.

Also we implemented a "bitsliced" AES and Camellia for the first time, both of which achieved very good performance. A bitslice implementation is important from the viewpoint of a countermeasure against cache timing attacks because it does not require lookup tables with a key-dependent address. We also analyze performance of SHA256/512 and Whirlpool hash functions and show that SHA512 can run faster than SHA256 on Athlon 64. This paper exhibits an undocumented fact that 64-bit *right* shifts and 64-bit rotations are extremely slow on Pentium 4, which often leads to serious and unavoidable performance penalties in programming encryption primitives on this processor.

Keywords: Fast Software Encryption, x64 Processors, Bitslice.

1 Introduction

This paper explores instruction-level software optimization techniques for the new 64-bit x64 architecture on Athlon 64 and Pentium 4 processors. Of course a 64-bit programming is not a new topic; Alpha, PA-RISC, Sparc, etc. have been already studied in many literatures, but the new x64 architecture is extremely important and promising in the sense that it is a superset of the currently dominant x86 architecture and Microsoft finally launched 64-bit Windows running on the x64 processors in the PC market.

Interestingly and ironically, the x64 architecture was initially designed and published by AMD under the name AMD64, and later followed by Intel under the name EM64T. EM64T is binary compatible with AMD64, but the internal hardware design of Pentium 4 is completely different from that of Athlon 64. Intel pursues higher clock frequency with a deep pipeline and AMD seeks for higher superscalability and lower memory latency with an on-chip I/O controller (HyperTransport). Which of Intel and AMD is fast has been always a controversial issue.

M.J.B. Robshaw (Ed.): FSE 2006, LNCS 4047, pp. 341–358, 2006.

From programmer's side, the x64 structure is what was long awaited; one of its biggest impacts is liberation from the "register starvation", which was x86 programmer's nightmare. This paradigm shift could be compared with the liberation from the "segment wall" of a 16-bit program in 1980's.

Our interest in the x64 architecture is two-fold. First we would like to see to what extent performance gain is expected on this architecture for symmetric cryptographic primitives, and then compare AMD with Intel, targeting at Athlon 64 and Pentium 4 processors. To do this, we started at looking at instruction-level performance. AMD has published a reliable list of an instruction latency and throughput of 64-bit instructions [20], but no similar information is currently available for Pentium 4. Our own experiments exhibit that some 64-bit instructions of Pentium 4 are unexpectedly slow, which can lead to serious and unavoidable performance penalties in programming encryption primitives.

Our first target primitive is AES [6]. We show that Athlon 64 is very suitable for an AES program mainly because it can issue up to two memory instructions in parallel. This boosts encryption speed of AES and, as a result, our program runs at the speed of 170 cycles/block, which is the fastest record of AES on a PC processor. Pentium 4 has a higher clock frequency than Athlon 64 in general, but still Athlon 64 seems to outperform Pentium 4 for AES.

The next target is Camellia [1]. the structure of Camellia should fit to 64-bit processors in nature. However, due to a long dependency chain, it was not easy to obtain high performance. We here propose a two-block parallel encryption, which can be used in a non-feedback mode such as the CTR mode, and show that its encryption speed on Athlon 64 reaches 175 cycles/block, thanks to the doubled number of registers.

We next develop a bitslice implementation of AES and Camellia. The bitslice technique for speeding up an encryption algorithm was introduced by Biham [4], which was remarkably successful for DES key search on 64-bit processors such as Alpha. We revisit this implementation from the viewpoint of a countermeasure against cache attacks [17].

A bitsliced cipher can achieve a good performance if the number of registers is many and a register size is long, which fully meets the x64 architecture. We carefully optimized the S-boxes of AES and Camellia in bit-level, and succeeded in obtaining very good and performance. Note that a bitslice implementation of AES was also discussed by Rudra et al. [18], but our paper for the first time reports a measured performance of bitsliced AES and Camellia implemented on a real processor.

Finally we study software performance of hash functions SHA256/512 and Whirlpool. Our x64 implementation results show that SHA512 can be faster than SHA256 on Athlon 64, and also faster than Whirlpool on both of Athlon 64 and Pentium 4 processors, unlike the results on 32-bit Pentium shown in [13][14].

Table 1 shows our reference machines and environments. Throughout this paper, we refer to Pentium 4 with Prescott core as Pentium 4; it is a current dominant core of Intel's EMT64 architecture.

Table 1. Our reference machines and environments

Processor Name	AMD Athlon 64 3500+	Intel Pentium 4 HT
Core Name	Winchester	Prescott
Other Processor Info		Stepping 4 Revision 14
Clock Frequency	2.2GHz	3.6GHz
Cache (Code/Data)	64KB / 64KB	12Kμops / 16KB
Memory	1GB	1GB
Operation System	Windows XP 64-bit Edition	
Compiler	Microsoft Visual Studio 2005 beta	

2   The x64 Architecture

2.1   x86 vs x64

The x64 (or x86-64) is the first 64-bit processor architecture that is a superset of the x86 architecture. It was initially proposed and implemented by AMD and, in an ironic twist of processor history, later adopted by Intel under the name EM64T. Most of the extended features of the x64 architecture are what PC programmers long awaited:

1. The size of general registers is extended to 64 bits. The 32-bit `eax` register, for instance, is now lower half of the 64-bit `rax` register.
2. Additional eight general registers `r8-r15` and eight xmm registers `xmm8-xmm15` are introduced.
3. Almost all x86 instructions now accept 64-bit operands, including rotate shift instructions.

In particular the register increase has liberated PC programmers from the nightmare of register starvation. It is highly expected that these benefits open up new possibilities of fast and efficient cryptographic applications in near future. On the other side, using these extended features may cause the following new penalties, which can be serious in some cases:

1. An instruction requires an additional prefix byte in using a 64-bit operand or a new register. An increase in instruction length reduces decoding rate.
2. A 64-bit instruction is not always as efficient as its corresponding 32-bit instruction. Performance of an instruction might vary in 32-bit mode and 64-bit mode.

How fast a specific instruction runs is an issue of processor hardware design, not instruction set design. We will see that the second penalty above can be serious for Intel Pentium 4 processor and show that a great care must be taken when we implement a cryptographic algorithm on this processor.

2.2 Athlon 64 vs Pentium 4

Athlon 64 is a 3-way superscalar processor with 12 pipeline stages. It can decode and execute up to three instructions per cycle. Its three ALU's and three AGU's work independently and simultaneously, and moreover up to two 64-bit read/write instructions can access data cache each cycle in any combination of reads and writes. Hence, for example, an n-time repetition of the following (highly practical) code, which consists of five micro operations, works in n cycles, that is, 5μops/cycle.

```
xor   rax,TABLE1[rsi*8]      ; 64-bit load and 64-bit ALU
xor   rbx,TABLE2[rsi*8]      ; 64-bit load and 64-bit ALU
add   rsi,1                  ;                 64-bit ALU
```

Almost all of the 64-bit instructions of Athlon 64 runs in the same performance as its corresponding 32-bit x86 instructions, which is why Athlon 64 is often called a genuine 64-bit processor. The internal architecture of this processor is well-documented by AMD [20] and relatively easy to understand. A document written by Hans de Vries [21] is also helpful for understanding the architecture of Athlon 64. Working on this processor is generally less frustrating than exploring on Pentium 4.

One drawback of Athlon 64 is that it can fetch only 16 bytes of instructions from instruction cache per cycle. This means that the decoding stage can be still a bottleneck of performance, unlike Pentium 4. It is hence critically important for programmers to reduce an average instruction length for obtaining maximal performance on this processor.

A prominent feature of the Pentium 4 processor family is that instructions are cached after decoded, and hence the decoding capability is not a performance limiter any more, as long as a critical loop is covered by the trace cache (instruction cache) entirely.

There exist two different core architectures in the Pentium 4 family, of which we treat a newer one, the Prescott core, in this paper. Prescott has a deep 31-stage pipeline and achieves high clock frequency. At the time of writing, the highest frequency of the Athlon 64 family is 2.8GHz, while that of the Pentium 4 family is 3.8GHz, 36% faster than fastest Athlon 64.

Intel has not published pipeline architecture details of the Prescott core, nor documented information about how EM64T instructions are handled in its pipeline stages. To optimize a program on Pentium 4, we have to refer to not only Intel's document of the 32-bit architecture IA-32 [10], which is often erroneous, but also resources outside Intel such as Agner [8] Kartunov [12].

As far as we know, Prescott can run continuously three micro operations per cycle in an average (some resource says four, but we are not sure), which is less than Athlon 64. Moreover, many instructions of Prescott have a longer latency and/or a lower throughput than those of Athlon 64. This is a clear consequence of the high clock frequency of the Prescott core. Table 2 shows a brief comparative summary of these processors.

Table 2. A simple comparison between Athlon 64 and Pentium 4

	Athlon 64	Pentium 4
Current highest clock frequency	2.8GHz	3.8GHz (good)
Decoding bottleneck	possible	mostly no (good)
Average instruction latency	low (good)	sometimes high
Maximal continuous execution rate	5μops/cycle (good)	3μops/cycle

Which of AMD and Intel is faster is always a controversial issue. We will show in this paper that in many cryptographic algorithms Athlon 64 outperforms Pentium 4 on the 64-bit platform, except a code using xmm instructions (SSE2), even if we take into consideration Pentium 4's faster clock frequency.

2.3   Instruction Latency and Throughput

Table 3 shows a list of an instruction latency (left) and throughput (right) of some of x64 instructions. We derived all the data in the list experimentally. Specifically, we measured the number of execution cycles of a code that consists of 100-1000 repetitions of a target instruction. Some fractional values on Pentium 4 are approximate. A latency of an instruction is n, when its result can be used in n cycles after the instruction has been issued. A throughput is the maximal number of the same instructions that can run continuously in parallel per cycle.

We believe that this table is of independent interest. It is quite surprising that 64-bit right shifts `shr` and 64-bit rotations `ror`,`rol` are extremely slow on Pentium 4. We do not know what was behind in this decision in Intel. Clearly this is a bad news for programmers of cryptographic algorithms.

Table 3. A list of an instruction latency and throughput of Pentium 4 and Athlon 64

Processor	Pentium 4 (EM64T)		Athlon 64 (AMD64)	
Operand Size	32	64	32	64
`mov reg,[mem]`	4, 1	4, 1	3, 2	3, 2
`mov reg,reg`	1, 3	1, 3	1, 3	1, 3
`movzx reg,reg8L`	1, 3	1, 3	1, 3	1, 3
`movzx reg,reg8H`	2, 4/3	-	1, 3	-
`add reg,reg`	1, 2.88	1, 2.88	1, 3	1, 3
`sub reg,reg`	1, 2.88	1, 2.88	1, 3	1, 3
`adc reg,reg`	10, 2/5	10, 2/5	1, 5/2	1, 5/2
`sbb reg,reg`	10, 2/5	10, 2/5	1, 5/2	1, 5/2
`xor/and/or reg,reg`	1, 7/4	1, 7/4	1, 3	1, 3
`not reg`	1, 7/4	1, 7/4	1, 3	1, 3
`shr reg,imm`	1, 7/4	7, 1	1, 3	1, 3
`shl reg,imm`	1, 7/4	1, 7/4	1, 3	1, 3
`ror/rol reg,imm`	1, 1	7, 1/7	1, 3	1, 3

Table 3. (*continued*)

Processor		Pentium 4 (EM64T)	Athlon 64 (AMD64)
Operand Size		128	128
movdqa	xmm,[mem]	-, 1	-, 1
movdqa	xmm,xmm	7, 1	2, 1
movd xmm,reg + movd reg,xmm		13, -	14, -
paddb/paddw/paddd xmm,xmm		2, 1/2	2, 1
paddq	xmm,xmm	5, 2/5	2, 1
pxor/pand/por	xmm,xmm	2, 1/2	2, 1
psllw/pslld/psllq xmm,xmm		2, 1/2	2, 1
pslldq	xmm,xmm	4, 1/2	2, 1
psrlw/psrld/psrlq xmm,xmm		2, 1/2	2, 1
psrldq	xmm,xmm	4, 1/2	2, 1

Throughout this paper, we assume that a memory read/write is one μop each, and hence that xor reg,[mem] and xor [mem],reg consist of two μops and three μops, respectively. An exact μop break-down rule has not been published.

After our creating this table, we found an independent (but not formally published) result obtained by Granlund [9]. Table 3 contains several results that are not covered in [9]. Our results mostly agree with Granlund's for Athlon 64, but look slightly different for some instructions of Pentium 4. It is known that Pentium 4 Prescott has many variations (stepping, revision), which can lead to subtly different instruction timings. Since Intel has not published detailed information on the hardware design of Prescott, it is difficult to derive the precise timing information.

3   AES

First we discuss a fast implementation of AES on the x64 architecture. See [6] for the detailed specification of the AES algorithm. In [13], an x86 code of the "basic component" of AES, which corresponds to Subbytes+Shiftrows+Mixcolumns, was proposed for Pentium 4 processors. Code 1 shows the proposed code with a modification for the x64 platform. One round of AES, except the final round, can be implemented with four additional xor instructions, which corresponds to AddRoundKey, and four-time repetition of the basic component:

```
movzx    esi,al                    ; first address
mov/xor  reg32_1,table1[rsi*4]     ; first table lookup
movzx    esi,ah                    ; second address
mov/xor  reg32_2,table2[rsi*4]     ; second table lookup
shr      eax,16
movzx    esi,al                    ; third address
mov/xor  reg32_3,table3[rsi*4]     ; third table lookup
movzx    esi,ah                    ; fourth address
mov/xor  reg32_4,table4[rsi*4]     ; fourth table lookup
```

Code 1. An x64 implementation of the basic component of AES

Alternatively, the last two lines can be also written in the following form, which was recommended for Pentium 4 with Prescott core, where using ah is a bit expensive:

```
shr       eax,8                    ; fourth address
mov/xor   reg32_4, table4[rax*4]   ; fourth table lookup
```

Code 2. An alternative implementation of the basic component (part)

Since each of the four tables occupies 1KB and we need additional four tables for the final round, a total of 8KB data memory is needed for the entire AES tables. This implementation is highly optimized and well scheduled, and hence also works on 64-bit environments excellently.

Readers might be tempted to write the following code instead, which looks more "genuine" 64-bit style:

```
movzx     rsi,al
mov/xor   reg32_1,table1[rsi*4]
movzx     rsi,ah                   ; no such instruction
mov/xor   reg32_2,table2[rsi*4]
shr       rax,16                   ; slow on Pentium 4
movzx     rsi,al
mov/xor   reg32_3,table3[rsi*4]
movzx     rsi,ah                   ; no such instruction
mov/xor   reg32_4,table4[rsi*4]
```

Code 3. This code does not work

Code 3 does not work since a higher 8-bit partial register such as ah can be used only in x86 code (the third and eighth lines), which is one of the small number of exceptional instructions that do not have an extended 64-bit form. If we change movzx rsi,ah into the original form movzx esi,ah, then code 3 works as expected, but is still slow on Pentium 4 because shr rax,16 is a 64-bit right shift instruction.

Moreover, since the number of higher 8-bit registers are still limited to four (ah,bh,ch,dh) in the x64 environment, we have to assign eax,ebx,ecx,edx to reg32_1,...,reg32_4, which are used as address registers in the next round, to minimize the number of instructions, but this is impossible without saving/restoring at least one input register in each round. In the x86 environment, we had to access temporary memory for this due to register starvation, but in the x64, we can use a new register instead, which slightly improves performance.

In summary, we should keep an x86 style, using new registers for temporary memory in implementing AES on x64 environments. Table 4 summarizes our implementation results of the AES algorithm with 128-bit key on Athlon 64 and Pentium 4 processors, where the right most column shows the best known result on 32-bit Pentium 4:

Our program runs very fast on the Athlon 64 processor. As far as we know, this is the fastest AES implementation ever made on a PC processor; faster

Table 4. Our implementation results of AES with 128-bit key

Processor	Athlon 64 64-bit	Pentium 4 64-bit	Pentium 4 32-bit [13]
cycles/block	170	256	284
cycles/byte	10.6	16.0	17.8
instructions/cycle	2.74	1.81	-
μops/cycle	3.53	2.34	-

than Pentium 4 even if we take into consideration higher clock frequency of the Pentium 4 processor. This is mainly because Athlon 64 can execute two memory load instructions with 3 latency cycles in parallel. The number of memory reads for one block encryption of AES is 4 (for plaintext loads) + 11 × 4 (for subkey loads) + 16 × 10 (for table lookups) = 208, which means that Pentium 4 takes at least 208 cycles/block for one block encryption.

Considering an instruction latency of Athlon 64, the theoretical limit of AES performance on this processor seems around 16 cycles/round = 160 cycles/block. Our result is hence reaching closely this limit.

4   Camellia

The next example of our implementation is another 128-bit block cipher Camellia [1]. Recently Camellia has been adopted in the NESSIE project [16], Japan's CRYPTREC project [5] and also the ISO/IEC 18033-3 standard [11].

Camellia supports three key sizes; 128 bits, 192 bits and 256 bits as AES, where we treat the 128-bit key version. The basic structure of Camellia is Feistel type, consisting eighteen rounds with additional four small FL functions. Figures 1 and 2 show the F-function and the FL-function, respectively.

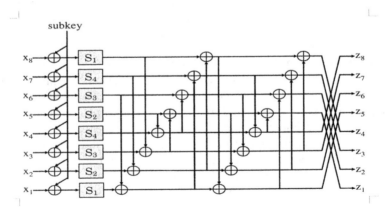

Fig. 1. The F-function of Camellia

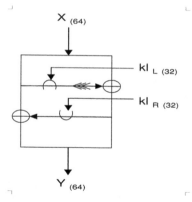

Fig. 2. The FL-function of Camellia

The F-function has a typical SP-structure of eight-byte width, which is quite suitable for 64-bit processors. Code 4 shows an implementation of (half of) the F-function of Camellia with the minimum number of x64 instructions, which should get the benefits of 64-bit registers:

```
movzx    esi,al                  ; first address
xor      rbx,table1[rsi*4]       ; first table lookup
movzx    esi,ah                  ; second address
xor      rbx,table2[rsi*4]       ; second table lookup
shr      rax,16
movzx    esi,al                  ; third address
xor      rbx,table3[rsi*4]       ; third table lookup
movzx    esi,ah                  ; fourth address
xor      rbx,table4[rsi*4]       ; fourth table lookup
shr      rax,16
...                              ; fifth to eighth
```

Code 4. An x64 implementation of the F-function of Camellia

In practice, however, this code does not run very fast because of a long dependency chain; especially the xor chain hinders parallel execution of multiple instructions. Although we can obtain some performance improvement by introducing some intermediate variables to cut the dependency chain, the resultant performance gain is limited.

On the other side, it should be noted that code 4 is using only three registers. Therefore it is possible to compute two blocks in parallel without register starvation, which is expected to boost the performance. This parallel computation method can be applied to a non-feedback mode of operation, such as a counter mode. Our optimized code for encrypting two blocks of Camellia in parallel, where they are interleaved in every half round, runs in 175 cycles/block on Athlon 64; this performance is almost the same as that of AES.

Table 5. Performance of our two-block parallel program of Camellia with 128-bit key

Processor	Athlon 64	Pentium 4
cycles/block	175	457
cycles/byte	10.9	28.6
instructions/cycle	2.46	0.94
μops/cycle	3.28	1.26

On the other hand, the performance on Pentium 4 is still poor because of a long latency of the 64-bit right shift instructions. Probably a code without using 64-bit instructions could be faster on Pentium 4. Table 5 summarizes our implementation results of (two-block parallel) Camellia. Further optimization efforts on Pentium 4 are now ongoing.

Code 5 is our implementation of the FL-function on 64-bit registers without dividing a 64-bit input into two 32-bit halves. The required number of instructions of this tricky code is a bit smaller than that of a straightforward method.

```
mov   rcx,[key]      ; load 64 bits (k1L+k1R)
and   rcx,rax        ; rax = input
shr   rcx,32
rol   ecx,1
xor   rax,rcx
mov   ecx,[key]      ; load 32 bits again (k1R)
or    ecx,eax
shl   rcx,32
xor   rax,rcx        ; rax = output
```

Code 5. An x64 implementation of the FL-function of Camellia

5   Bitslice Implementation

This section discusses a bitslice implementation of AES and Camellia. The bitslice implementation was initially proposed by Biham [4], which makes an n-block parallel computation possible, where n is a block size and one (software) instruction corresponds to n simultaneous one-bit (hardware) operations, by regarding the i-th bit of register j as the j-th bit of the i-th block.

In general, this implementation can be faster than an ordinary implementation when the following conditions are met:

- The bit-level complexity of the target algorithm is small.
- The number of registers of the target processor is many.
- The size of registers of the target processor is long.

The bitslice implementation was successful for DES [4] and MISTY [15] on the Alpha processor, since these algorithms are small in hardware and Alpha has thiry-two 64-bit general registers.

It is obvious that there is no hope of gaining performance using the bitslice technique on x86 processors, which have only eight 32-bit registers. However we have now sixteen 64-bit registers and it is now an interesting topic to see to what extent the x64 architecture contributes to fast encryption. Note that 128-bit xmm registers are of no use, due to a poor latency and throughput of SSE2 instructions.

Moreover, a program written in the bitslice method does not use any table lookups with a key-dependent address. This means that a bitsliced code is safe against implementation attacks such as cache timing attacks [17]. As far as we know, this is the first paper that describes bitslice implementations of AES and Camellia on an actual processor.

Clearly the most critical part of the bitslice program of these cipher algorithms is a design of the 8x8 S-boxes. To minimize the number of instructions of each of the S-boxes, which is composed of an inversion function over $GF(2^8)$ and a linear transformation for either of AES and Camellia, we should look at its hardware implementation, not software, due to the nature of the bitslice implementation.

Satoh et. al [19] proposed to design an inversion circuit of $GF(2^8)$ using circuits of $GF(2)$ in hardware by recursively applying circuits of a subfield of index two. We further considered an optimality of linear transformations that are required before and after the inversion function to design the entire S-box structure, and reached the following basis of $GF(2^8)$ over $GF(2)$ for a bitslice S-box with a small number of instructions:

$$(1, \beta^5, \beta, \beta^6, \alpha, \beta^5\alpha, \beta\alpha, \beta^6\alpha),$$

where $\alpha^8 + \alpha^6 + \alpha^5 + \alpha^3 + 1 = 0$, and $\beta = \alpha^6 + \alpha^5 + \alpha^3 + \alpha^2 \in GF(2^4)$.

Table 6 shows the number of x64 instructions required for implementing the S-box of AES and Camellia, respectively. The "Before inversion"/"After inversion" column shows the number of register-register logical instructions required for the linear transformation before/after the Galois field inversion, respectively. The two numbers of the "Inversion on $GF(2^8)$" column shows the number of register-register logical instructions and register-memory load/store instructions. In the inversion part, all of the fifteen 64-bit general registers are used except the stack register, and additional five 64-bit temporary memory areas are needed. Appendix B shows a source code of our implementation of this AES S-box.

Using this S-box implementation, we made the entire bitslice programs of AES and Camellia. Table 7 shows the resultant performance of our codes.

The speed shown in this table is slower than that of the ordinary implementation method shown in the previous sections, but is still in a very practical level.

Table 6. The number of x64 instructions of the S-box of AES and Camellia

	Before inversion	Inversion on $GF(2^8)$	After inversion	Total
AES S-box	12	156(reg) + 21(mem)	16	205
Camellia S-box	12	156(reg) + 21(mem)	14	203

Table 7. Our implementation results of bitsliced AES and Camellia with 128-bit key

Algorithm	AES		Camellia	
Processor	Athlon 64	Pentium 4	Athlon 64	Pentium 4
cycles/block	250	418	243	415
cycles/byte	15.6	26.1	15.2	25.9
instructions/cycle	2.75	1.66	2.74	1.61
μops/cycle	3.20	1.93	2.99	1.75

Note that in the bitslice implementation Camellia is slightly faster than the bitsliced AES. This is mainly because Camellia has a fewer number of S-boxes (144) than AES (160), though Camellia has an additional FL-function. Also the performance of Athlon 64 is excellent again, since three logical instructions can run on Athlon in parallel, but only two on Pentium 4.

6   Hash Functions: SHA256/512 and Whirlpool

This section briefly shows our implementations of three recent hash functions SHA256, SHA512 [7] and Whirlpool [2][3], all of which are now under consideration for an inclusion in the new version of the ISO/IEC 10118 standard. Note that the message block size of SHA256 and Whirlpool is 64 bytes, while SHA512 has a 128-byte message block.

Table 8 summarizes our performance results of SHA256, where the first column presents an ordinary implementation using general registers, and the second column shows a four-block parallel implementation (in the sense of [13][14]). It is seen that the x64 code of Athlon 64 establishes an excellent superscalability, 2.88 instructions/cycle, which is very close to its structural limit, 3 instructions/cycle. On the other hand, Pentium 4 is faster than Athlon 64 on the xmm code, considering Pentium 4's faster clock frequency.

Table 8. Our implementation results of SHA256

Processor	Athlon 64		Pentium 4	
Instructions	x64 (1b)	xmm (4b)	x64 (1b)	xmm (4b)
cycles/block	1173	1154	1600	1235
cycles/byte	18.3	18.0	25.0	19.3
instruction/cycle	2.88	1.15	2.11	1.08
μops/cycle	3.16	1.22	2.31	1.14

Table 9 illustrates our implementation results of SHA512, where the second column shows a two-block parallel code using xmm instructions. Also the right two columns are previous results shown on [13]. A remarkable fact is that SHA512 runs faster than SHA256 on Athlon 64 because an 64-bit instruction runs in the same latency/throughput as its corresponding 32-bit instruction on this processor. On the other side, Pentium 4 is very slow due to a long latency of 64-bit rotate operations, which are unavoidable in programming SHA512.

Table 9. Our implementation results of SHA512

Processor	Athlon 64		Pentium 4		Pentium 4 [13]	
Instructions	x64 (1b)	xmm (2b)	x64 (1b)	xmm (2b)	mmx (2b)	xmm (4b)
cycles/block	1480	2941	3900	3059	5294	3111
cycles/byte	11.6	23.0	30.5	23.9	41.4	24.3
instruction/cycle	2.85	1.15	1.08	1.10	-	-
μops/cycle	3.17	1.21	1.20	1.14	-	-

Table 10. Our implementation results of Whirlpool

Processor	Athlon 64	Pentium 4	Pentium 4 [13]
Instructions	x64	x64	mmx
cycles/block	1537	2800	2319
cycles/byte	24.0	43.8	36.2
instruction/cycle	2.27	1.24	-
μops/cycle	3.08	1.69	-

Our final example is Whirlpool. Our experimental results presented in Table 10 shows that Whirlpool is not faster than SHA512 on either of Athlon 64 and Pentium 4 in 64-bit environments.

7   Concluding Remarks

This paper explored the state-of-the-art implementation techniques for speeding up symmetric primitives on the x64 architecture. In many cases Athlon 64 attains better performance than Pentium 4 EM64T, even if Pentium 4's higher clock frequency is taken into consideration. Probably the slow 64-bit right shifts and 64-bit rotations of Pentium 4 will be (should be) redesigned in the next core architecture for EM64T.

We also showed the first bitslice implementation of AES and Camellia on these processors and demonstrated that our program achieved very good performance. We believe that a bitslice implementation has a significant and practical impact from the viewpoint of resistance from cache timing attacks. For interested readers, we summarize the coding style we adopted and how we measured clock cycles of our programs in appendix A, and list an assembly language source code of a bitsliced S-box of AES in appendix B.

References

[1] K. Aoki, T. Ichikawa, M. Kanda, M. Matsui, S. Moriai, J. Nakajima, T. Tokita: "The 128-Bit Block Cipher Camellia", IEICE Trans. Fundamentals, Vol.E85-A, No.1, pp.11-24, 2002.
[2] P. Barreto, V. Rijmen: "The Whirlpool Hashing Function", Proceedings of First Open NESSIE Workshop, Heverlee, Belgium, 2000.

[3] P. Barreto: "The Whirlpool Hash Function", http://planeta.terra.com.br/informatica/paulobarreto/WhirlpoolPage.html

[4] E. Biham: "A Fast New DES Implementation in Software", Proceedings of Fast Software Workshop FSE'97, Lecture Notes in Computer Science, Vol.1267, pp.260-272, Springer-Verlag, 1997.

[5] Cryptography Research and Evaluation Committees: The CRYPTREC Homepage http://www.cryptrec.org/

[6] Federal Information Processing Standards Publication 197, "Advanced Encryption Standard (AES)", NIST, 2001.

[7] Federal Information Processing Standards Publication 180-2, "Secure Hash Standard", NIST, 2002.

[8] A. Fog: "How To Optimize for Pentium Family Processors", Available at http://www.agner.org/assem/

[9] T. Granlund: "Instruction latencies and throughput for AMD and Intel x86 Processors", Available at http://swox.com/doc/x86-timing.pdf

[10] IA-32 Intel Architecture Optimization Reference Manual, Order Number 248966-011, http://developer.intel.ru/download/design/Pentium4/manuals/24896611.pdf

[11] ISO/IEC 18033-3, "Information technology - Security techniques - Encryption algorithms - Part3: Block ciphers", 2005.

[12] Victor Kartunov: "Prescott: The Last of the Mohicans? (Pentium 4: from Willamette to Prescott)" http://www.xbitlabs.com/articles/cpu/display/netburst-1.html

[13] M. Matsui, S. Fukuda: "How to Maximize Software Performance of Symmetric Primitives on Pentium III and 4 Processors", Proceedings of Fast Software Workshop FSE2005, Lecture Notes in Computer Science, Vol.3357, pp.398-412, Springer-Verlag, 2005.

[14] J. Nakajima, M. Matsui: "Performance Analysis and Parallel Implementation of Dedicated Hash Functions on Pentium III", IEICE Trans. Fundamentals, Vol.E86-A, No.1, pp.54-63, 2003.

[15] J. Nakajima, M. Matsui: "Fast Software Implementations of MISTY1 on Alpha Processors", IEICE Trans. Fundamentals, Vol.E82-A, No.1, pp.107-116, 1999.

[16] New European Schemes for Signatures, Integrity, and Encryption (NESSIE), https://www.cosic.esat.kuleuven.ac.be/nessie/

[17] D. A. Osvik, A. Shamir, E. Tromer: "Full AES key extraction in 65 milliseconds using cache attacks" Crypto 2005 rump session.

[18] A. Rudra, P. Dubey, C. Jutla, V. Kummar, J. Rao, P. Rohatgi: "Efficient Rijndael Encryption Implementation with Composite Field Arithmetic", Proceedings of CHES 2001, Lecture Notes in Computer Science, Vol.2162, pp.171-184, Springer-Verlag, 2001.

[19] A. Satoh, S. Morioka, K. Takano, S. Munetoh: "A Compact Rijndael Hardware Architecture with S-Box Optimization", Proceedings of Asiacrypt 2001, Lecture Notes in Computer Science, Vol.2248, pp.239-254, Springer-Verlag, 2001.

[20] Software Optimization Guide for AMD64 Processors, Publication 25112, http://www.amd.com/us-en/assets/content_type/white_papers_and_tech_docs/25112.PDF

[21] Hans de Vries: "Understanding the detailed Architecture of AMD's 64 bit Core", http://chip-architect.com/news/2003_09_21_Detailed_Architecture_of_AMDs_64bit_Core.html

Appendix A: Coding Style and How to Measure Cycles

The coding style of our programs is basically the same as that in [13]. Our programs are thread-safe; that is, we did not use any static memory area except read-only constant tables. Also we did not write any key dependent code such as a self-modifying trick.

Our assembly codes have the following interface that is callable from C language, and we assume that the subkey has been given in the third argument. In the x64 environments, the arguments are usually passed through registers, not through stack, but this calling convention does not affect performance of encryption functions seriously if `block` is sufficiently large.

We also assume that all addresses are appropriately aligned, at least on a 16-byte boundary to reduce possible mis-alignment penalties.

```
Function( uchar *plaintext, uchar *ciphertext, uint *subkey, int block )
```

The method for measuring a speed of `Function` that we adopted is that using the `cpuid+rdtsc` instruction sequence, which is common in x86 processors, as shown below:

```
xor     eax,eax
cpuid                       ; pipeline flush
rdtsc                       ; read time stamp
mov     CLK1,eax            ; current time
xor     eax,eax
cpuid

Function(..., int block)

xor     eax,eax
cpuid                       ; pipeline flush
rdtsc                       ; read time stamp
mov     CLK2,eax            ; current time
xor     eax,eax
cpuid
```

Code 6. A code sequence for measuring a speed of `Function`

We first ran the code above and recorded `CLK2-CLK1`. Then we removed `Function` from the code, ran again the code and recorded `CLK2-CLK1`. Since the second record is an overhead of the measurement itself, we subtracted the second record from the first record, then divded it by `block` and adopted the resultant value as "cycles/block". In practice, we made the measurement 100 times, of which we removed exceptional cases due to, for instance, an interruption caused by an operation system, and took an average on the remaning cases.

Strictly speaking the `rdtsc` instruction returns 64-bit clock tics to `edx` and `eax`, but we used only lower 32 bits, because if an overflow of `eax` took place during the measurement, it could be removed as an exceptional case.

Appendix B: Source Program of Bilsliced AES Sbox

This appendix shows a source code of our bitslice implementation of the Sbox of AES, which is written in x64 assembly language with Microsoft MASM syntax. The complete program is described as a macro with eight register inputs and eight register outputs. We wrote several instructions in a single line for saving space.

```
;**************************************************
;*  Bitslice Implementation of Sbox of AES       *
;*  Using x64 Instructions (AMD64 / EM64T)       *
;*                                               *
;*  Input  (rax,rbx,rcx,rdx,rbp,r8,r9,r10)       *
;*  Output (rbx,rdx,rax,r15,rbp,rcx,r10,r9)      *
;*                                               *
;*  205 Instructions (184 logical 21 memory)     *
;*   40 Temporary Memory Bytes                   *
;*                                               *
;*  (C) Mitsuru Matsui  2005,2006                *
;**************************************************

SBOX  MACRO
  InBasisChange   rax,rbx,rcx,rdx,rbp,r8,r9,r10
  Inv_GF256       r10,rbp,rbx,rcx,rdx,r8,r9,rax,r11,r12,r13,r14,r15,rsi,rdi
  OutBasisChange  r10,rbp,rbx,rcx,rdx,r8,r9,rax,r15
ENDM


;**************************
;*  InBasisChange:  (12)  *
;**************************

InBasisChange  MACRO  g0,g1,g2,g3,g4,g5,g6,g7
     xor  g6,g5    xor  g6,g1    xor  g5,g4    xor  g7,g5    xor  g4,g3
     xor  g4,g0    xor  g0,g2    xor  g7,g0    xor  g3,g2    xor  g2,g6
     xor  g3,g1    xor  g6,g4
ENDM

;**************************
;*  OutBasisChange:  (16)  *
;**************************

OutBasisChange  MACRO  g0,g1,g2,g3,g4,g5,g6,g7,g8
     xor  g1,g3    xor  g1,g5    xor  g1,g0    mov  g8,g1    xor  g8,g2
     xor  g1,g4    xor  g2,g6    xor  g6,g1    xor  g1,g7    xor  g7,g2
     xor  g2,g3    xor  g3,g5    xor  g3,g0    xor  g0,g4    xor  g0,g7
     xor  g4,g5
     ;We can skip the follwing four NOTs by modifying subkey in advance.
     ;not  g6      ;not  g0      ;not  g7      ;not  g4
ENDM
```

```
;**************************************************************
;*  Mul_GF4:  Input x0-x1,y0-y1  Output x0-x1  Temp t0  (8)  *
;**************************************************************

Mul_GF4  MACRO   x0,x1,y0,y1,t0
    mov  t0,x1    xor  x1,x0    and  x0,y0    and  t0,y1
    xor  y0,y1    and  x1,y0    xor  x1,x0    xor  x0,t0
ENDM


;**********************************************
;*  Inv_GF4:  Input x0,x1  Output x0,x1  (2)  *
;**********************************************

Inv_GF4  MACRO   x0,x1
    and  x0,x1    not  x0
ENDM


;********************************************************************
;*  Mul_GF16:  Input x0-x3,y0-y3  Output x0-x3  Temp t0-t3  (35)  *
;********************************************************************

Mul_GF16  MACRO   x0,x1,x2,x3,y0,y1,y2,y3,t0,t1,t2,t3
    mov  t0,x2    mov  t1,x3    mov  t2,y3
    mov  t3,t0    and  t3,t2    and  t0,y2    and  t2,t1    and  t1,y2
    xor  t1,t3    xor  t0,t1    xor  t1,t2
    xor  x2,x0    xor  x3,x1    xor  y2,y0    xor  y3,y1

    Mul_GF4   x2,x3,y2,y3,t3
    Mul_GF4   x0,x1,y0,y1,t3

    xor  x2,x0    xor  x3,x1    xor  x0,t1    xor  x1,t0
ENDM


;************************************************************
;*  Inv_GF16:  Input x0-x3  Output x0-x3  Temp t0-t3  (34)  *
;************************************************************

Inv_GF16  MACRO   x0,x1,x2,x3,t0,t1,t2,t3
    mov  t0,x0    mov  t1,x1    xor  t0,x2    xor  t1,x3    mov  t2,t0

    Mul_GF4   x0,x1,t0,t1,t3

    xor  x0,x3    xor  x1,x2

    Inv_GF4   x0,x1

    mov  t0,x0

    Mul_GF4   x2,x3,t0,x1,t3
```

```
    Mul_GF4   x0,x1,t2,t1,t3
ENDM

;**********************************************************************
;*  Inv_GF256:  Input x0-x7  Output x0-x7  Temp t0-t3,s0-s2  (177)  *
;**********************************************************************

Inv_GF256 MACRO   x0,x1,x2,x3,x4,x5,x6,x7,t0,t1,t2,t3,s0,s1,s2
    mov   t0,x0    mov   t1,x1    mov   t2,x2    mov   t3,x3    xor   t0,x4
    xor   t1,x5    xor   t2,x6    xor   t3,x7

    mov   [rsp+0],t0    mov   [rsp+8],t1
    mov   [rsp+16],t2   mov   [rsp+24],t3
    mov   [rsp+32],x7

    Mul_GF16   x0,x1,x2,x3,t0,t1,t2,t3,s0,s1,s2,x7

    mov   x7,[rsp+32]

    xor   x0,x4    xor   x1,x4    xor   x2,x4    xor   x3,x4    xor   x1,x5
    xor   x3,x5    xor   x2,x6    xor   x2,x7    xor   x3,x7

    Inv_GF16   x0,x1,x2,x3,t0,t1,t2,t3

    mov   t0,[rsp+0]    mov   t1,[rsp+8]
    mov   t2,[rsp+16]   mov   t3,[rsp+24]
    mov   [rsp+0],x0    mov   [rsp+8],x1
    mov   [rsp+16],x2   mov   [rsp+24],x3

    Mul_GF16   x0,x1,x2,x3,t0,t1,t2,t3,s0,s1,s2,x7

    mov   t0,[rsp+0]    mov   t1,[rsp+8]
    mov   t2,[rsp+16]   mov   t3,[rsp+24]
    mov   [rsp+24],x3   mov   x7,[rsp+32]

    Mul_GF16   x4,x5,x6,x7,t0,t1,t2,t3,s0,s1,s2,x3

    mov   x3,[rsp+24]
ENDM
```

Computing the Algebraic Immunity Efficiently

Frédéric Didier and Jean-Pierre Tillich

Projet CODES, INRIA Rocquencourt, Domaine de Voluceau,
78153 Le Chesnay cédex
{frederic.didier, jean-pierre.tillich}@inria.fr

Abstract. The purpose of algebraic attacks on stream and block ciphers is to recover the secret key by solving an overdefined system of multivariate algebraic equations. They become very efficient if this system is of low degree. In particular, they have been used to break stream ciphers immune to all previously known attacks. This kind of attack tends to work when certain Boolean functions used in the ciphering process have either low degree annihilators or low degree multiples. It is therefore important to be able to check this criterion for Boolean functions. We provide in this article an algorithm of complexity $O\left(m^d\right)$ (for fixed d) which is able to prove that a given Boolean function in m variables has no annihilator nor multiple of degree less than or equal to d. This complexity is essentially optimal. We also provide a more practical algorithm for the same task, which we believe to have the same complexity. This last algorithm is also able to output a basis of annihilators or multiples when they exist.

Keywords: Algebraic attacks, Algebraic immunity, Stream ciphers, Boolean functions, Annihilator, Low degree multiple.

1 Introduction

Algebraic attacks have proved to be a powerful class of attacks which might threaten both block and stream ciphers [CM03, Cou03, CP02, CDG05, Arm04]. The idea is to set up an algebraic system of equations verified by the key bits and to try to solve it. For instance, this kind of approach can be quite effective [CM03] on stream ciphers which consist of a linear pseudo-random generator hidden with non-linear combining functions acting on the outputs of the generator to produce the final output. For such an attack to work, it is crucial that the combining functions have low degree multiples or low degree annihilators. The reason for this is that it ensures that the algebraic system of equations verified by the secret key is also of small degree, which is in general essential for being able to solve it. This raises the fundamental issue of determining whether or not a given function has non-trivial low degree multiples or annihilators [Car04, MPC04, DGM04, BP05, DMS05]. The smallest degree for which this happens is called the *algebraic immunity* of the function.

Here we are going to address this issue for Boolean functions. Note that this is the main case of interest in this setting and that the algorithms presented here can be generalized to fields of larger size. This problem has already been

M.J.B. Robshaw (Ed.): FSE 2006, LNCS 4047, pp. 359–374, 2006.

considered in the literature. In [MPC04] three algorithms are proposed to decide whether or not a Boolean function in m variables has a non-trivial annihilator of degree at most d. Note that it is readily checked that finding a multiple of degree d for f amounts to finding an annihilator of degree d for $1 + f$. Therefore it is sufficient for computing the algebraic immunity to be able to test whether or not a Boolean function has annihilators of a certain degree. All three algorithms have complexities of order $O\left(\binom{m}{d}^3\right)$. Gröbner bases have also been suggested to perform this task (see [FA03]) but it is still unclear if they perform better than the aforementioned algorithms.

Recall that a Boolean function g of degree d in m variables is an annihilator of a Boolean function f with support of size N iff the $k \overset{\text{def}}{=} \binom{m}{0} + \binom{m}{1} + \cdots + \binom{m}{d}$ coefficients of the monomials of g verify a system of N linear equations (the equations express the fact that g has to be equal to zero at points for which f is equal to 1). We will suppose here that computing a value of f is in $O(1)$ and that it is the only thing we can do. This implies that the complexity of checking whether or not a Boolean function has an annihilator of degree at most d is at least of order $\Omega(k)$ since we need to check the value of f on at least k points.

We are going to present here a new algorithm (namely Algorithm 1 in this paper) which is able to prove the non-existence of annihilators of a certain degree efficiently. More precisely, we prove that for fixed d, the expected running time for our algorithm to prove that there is no non-trivial annihilator of maximum degree d is of order $O(k)$. This algorithm might fail to prove such a property for certain Boolean functions which have actually annihilators of this kind, we prove however that the proportion of such functions is negligible. In view of the previous lower bound, this algorithm is essentially optimal.

We also present another algorithm, namely Algorithm 2, which computes a basis for the annihilators of degree $\leq d$. We conjecture that the average running time of this algorithm is not worse than the average running time of Algorithm 1 when the latter succeeds in proving that there is no annihilator (which would then be of order $O(k)$), but we are only able to prove that its average running time is in this case of order $O(k(\log m)^2)$. Several remarks can be made here.

- It should be noted that the case of small d and rather large m is definitely interesting in cryptography. We wish to emphasize that it implies that we are far from checking all the entries of f to perform such a task. For instance, we are able to test that a random function in 64 variables has no non-trivial annihilator of degree 5 in a few minutes.
- It should be stressed here that our proof does not rely on any assumption about the linear system of equations which arises when a degree d annihilator is sought after. In particular, we do not assume that it behaves like a random system of linear equations, but we rely on a general result about the probability of not being able to recover erasures for linear codes for which the generalized weights distribution is known [Did05].
- The complexity of Algorithm 2 is not the same when there is an annihilator of the specified degree. We note that in this case, if we want to be sure that the output is indeed an annihilator, then the complexity of such an

algorithm is necessarily at least of order the number of entries of f. We present in Section 4 a way to modify this algorithm to make it faster in this case. Its output is in general only a basis of a space which contains the space of annihilators we look after. This algorithm and Algorithm 2 itself have been implemented. In our experiments, we found no example for which the output was not the space of annihilators itself.

2   Algebraic Immunity of a Boolean Function

In this section we recall basic facts about Boolean functions and algebraic immunity. We also introduce the tools we will need to analyze the complexity of our algorithms. In all this paper, we consider the binary vector space $\mathcal{B}(m)$ of m-variable Boolean functions, that is the space of functions from $\{0,1\}^m$ to $\{0,1\}$. It will be convenient to view $\{0,1\}$ as the field over two elements, what we denote by $\mathbf{F}_2$. It is well known that such a function f can be written in an unique way as an m-variable polynomial over $\mathbf{F}_2$ where the degree in each variable is at most 1 using the Algebraic Normal Form (ANF) :

$$f(x_0, \ldots, x_{m-1}) = \sum_{u \in \mathbf{F}_2^m} \alpha_u \left(\Pi_{i=0}^{m-1} x_i^{u_i} \right) \text{ where } \alpha_u \in \mathbf{F}_2, u = (u_0, \ldots, u_{m-1}).$$

By *monomial*, we mean in what follows, a polynomial of the form $\Pi_{i=0}^{m-1} x_i^{u_i}$ with $(u_0, \ldots, u_{m-1}) \in \mathbf{F}_2^m$. The *degree* of f is the maximum weight of the u's for which $\alpha_u \neq 0$. By listing the images of a Boolean function f over all possible values of the variables, that is $(f(x))_{x \in \mathbf{F}_2^m}$ (with some particular order over $\mathbf{F}_2^m$) we can also view it as a binary word of length $n \overset{\text{def}}{=} 2^m$. The *weight* of a Boolean function f is denoted by $|f|$ and is equal to $\sum_{x \in \mathbf{F}_2^m} f(x)$ (the sum being performed over the integers). We also denote in the same way the (Hamming) weight of a binary 2^m-tuple. A *balanced* Boolean function is a function with weight equal to half its length, that is $n/2$.

Dealing with algebraic immunity, we will be interested in the subspace $\mathcal{B}(d, m)$ of $\mathcal{B}(m)$ formed by all Boolean functions of degree $\leq d$. Note that the set of monomials of degree $\leq d$ forms a basis of $\mathcal{B}(d, m)$, we call it the *monomial basis*. By counting the number of such monomials we obtain that the dimension k of $\mathcal{B}(d, m)$ is given by $k = \sum_{i=0}^{d} \binom{m}{i}$.

As mentioned in the introduction, the algebraic immunity quantifies the immunity of a cryptosystem to some recent algebraic attacks. These attacks try to break a cryptosystem by solving an algebraic system involving the key bits. The equations involved often depend on a Boolean function f used in the ciphering process. The idea is that f can be replaced by its annihilators or multiples to obtain a new system of lower degree. The complexity of the attack depends on the degree of this system which is nothing but the algebraic immunity of f. Considering the pointwise product of two m-variable Boolean functions f and g, $\forall x \in \mathbf{F}_2^m, \quad f.g(x) = f(x)g(x)$, a function g is an *annihilator* of f if and only if $f.g$ is equal to 0, and a function g is a *multiple* of f if and only if there exists a Boolean function h such that $f.h$ is equal to g. Note that in the latter case, since the existence of such a function is equivalent to the fact that the set of zeros of f is included in the set of zeros of g, we also have $f.g = g$, which is equivalent to $(f + 1)g = 0$, i.e. g is an annihilator of $1 + f$. By definition, the algebraic immunity of f is the smallest degree d such that f admits a non-trivial annihilator or multiple of degree d. By the previous remark, this is also the smallest degree of a non-trivial annihilator of f or $1 + f$. In this paper, we will be interested in computing efficiently the algebraic immunity of a Boolean function. For achieving this aim, it is clearly sufficient to be able to find efficiently the smallest degree of a non-trivial annihilator of a Boolean function. Note that a Boolean function f admits a non-trivial annihilator of degree $\leq d$ iff the following set of equations with unknowns $\beta_u \in \mathbf{F}_2$, where $u = (u_0, \ldots, u_{m-1})$ ranges over all binary m-tuples of weight $\leq d$, has a non-zero solution

$$\left(\sum_{|u| \leq d} \beta_u x_0^{u_0} \ldots x_{m-1}^{u_{m-1}} = 0 \right)_{x=(x_0,\ldots,x_{m-1}) \in \mathbf{F}_2^m : f(x)=1} \tag{1}$$

This just expresses the fact that a Boolean function g of degree $\leq d$ (that is a function $g(x_0, \ldots, x_{m-1}) = \sum_{u=(u_0,\ldots,u_{m-1}) \in \mathbf{F}_2^m : |u| \leq d} \beta_u x_0^{u_0} \ldots x_{m-1}^{u_{m-1}}$) annihilates f if and only if for all points x at which f evaluates to 1, g is equal to 0. Our task will be to solve this linear system efficiently.

To estimate the complexity of the algorithms given in the following sections, we will use this result

Theorem 1. *Let λ be the fraction of m-variable Boolean functions of weight w with a non-trivial annihilator of degree $\leq d$, $k = \binom{m}{0} + \binom{m}{1} + \cdots + \binom{m}{d}$ and $n = 2^m$. Then $\lambda \leq \exp\left[\frac{n}{2^d}\left(\frac{k}{n}(d\ln 2 + 3) + \ln\left(1 - \frac{w}{n}\right)\right)\right]$.*

Proof. See [Did05].   □

3   Proving the Non-existence of an Annihilator Efficiently

We provide in this section an efficient algorithm which proves that a given Boolean function has no non-trivial annihilator up to a given degree. We hasten to say that we are going to present an improved version of this algorithm in the next section which will also be able to output a basis of the annihilator space up to a certain degree. The purpose of the algorithm presented here is that it can be analyzed rigorously and has average complexity of order $O(m^d)$ to prove that an m-variable Boolean function has no non-trivial annihilator of degree $\leq d$.

This algorithm makes heavily use of Gaussian elimination in its basic subroutines. What we call *lazy Gaussian elimination* in what follows simply consists in solving system (1) by inserting one by one the equations in a matrix that is kept reduced all the time.

When our algorithm looks for annihilators of degree $\leq d$ for an m-variable Boolean function f, it will apply lazy Gaussian elimination on subfunctions of f and will look for annihilators of lower degree. More precisely, it will apply lazy Gaussian elimination for other values of d and m, that is degrees d' in the range $1 \leq d' \leq d$ and number of variables m' of the form $2d + 1 + \lceil \log m \rceil$. We need a bound on the probability that an m'-variable Boolean function has no non-trivial annihilator in $\mathcal{B}(d', m')$. We will focus on *random balanced Boolean function* which is one of the most significant random model in a cryptographic context. The analysis carried over here can of course be applied to more general probabilistic models.

Lemma 1. *Let f be a random balanced Boolean function in m variables. Consider an m'-variable Boolean function f' obtained by fixing $m - m'$ variables of f. Let $m' = 2d + 1 + \lceil \log m \rceil$ and $1 \leq d' \leq d$, then as m tends to infinity, the probability that there is a non-trivial degree d' annihilator for f' is upper bounded by $e^{-2^{2d-d'}(1+o(1))m}$.*

Proof. See appendix, Section A.

Moreover, we also need to give an upper bound on the expected running time of the previous lazy Gaussian elimination. It is quite likely that it is of order $O(k^3)$, but we are only able to prove the slightly weaker statement :

Lemma 2. *The expected running time of lazy Gaussian elimination applied to the f' of the previous lemma is upper bounded by $O(d'k'^3)$, where $k' = \binom{m'}{0} + \cdots + \binom{m'}{d'}$.*

Proof. See appendix, Section B.

Our algorithm is based on the the classical $(u, u+v)$ decomposition of a Boolean function. Given a Boolean function f of degree r in algebraic normal form, one can write $f(x_0, \ldots, x_{m-1}) = u(x_0, \ldots, x_{m-2}) + x_{m-1}v(x_0, \ldots, x_{m-2})$ where u and v are Boolean functions in $m - 1$ variables. Note that for $x_{m-1} = 0$ we have $f(x_0, \ldots, x_{m-1}) = u(x_0, \ldots, x_{m-2})$ whereas for $x_{m-1} = 1$ we have $f(x_0, \ldots, x_{m-1}) = u(x_0, \ldots, x_{m-2}) + v(x_0, \ldots, x_{m-2})$. In other words, u is the restriction of f to the space $x_{m-1} = 0$ and $u + v$ is the restriction of f for $x_{m-1} = 1$. Moreover, the degree of u is at most r and the degree of v is at most $r - 1$.

Now, if there exists a non-trivial annihilator g for f of degree $\leq d$, then by decomposing g in $(u', u' + v')$ we either have:
- $u' \neq 0$ and $u'.u = 0$. This yields a non-trivial annihilator of u of degree $\leq d$.
- u' is zero but not v'. Then $v'.(u + v) = 0$ and we get a non-trivial annihilator of $u + v$ of degree $\leq d - 1$.

This simple remark is the underlying idea of our algorithm : to check that f has no non-trivial annihilator of degree $\leq d$ we check that neither u (or what is the same the restriction of f to $x_{m-1} = 0$) has a non-trivial annihilator of degree $\leq d$ nor $u + v$ (i.e. the restriction of f to $x_{m-1} = 1$) has a non-trivial annihilator of degree $\leq d - 1$. We perform this task recursively by decomposing u and $u+v$ further, up to the time the number of variables in the decomposition is equal to $2d + 1 + \lceil \log m \rceil$. This is illustrated by Figure 1. The first number in each couple represents the maximum annihilator degree to consider. The second

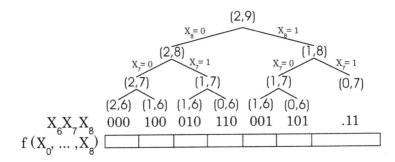

Fig. 1. Recursive decomposition of a 9-variable Boolean function used to find an annihilator of degree smaller than or equal to 2. The decomposition is performed up to 6-variable subfunctions. The bottom squares represent the domains of the subfunctions which are leaves of the decomposition.

number corresponds to the number of variables in the corresponding subfunction. We will continue the decomposition up to a certain depth as indicated above. Note that the decomposition may also stop because the degree of the annihilator to consider is zero.

It is important to notice the following fact

Fact 1. *The first level of decomposition of a Boolean function $f(x_0, \ldots, x_{m-1})$ corresponds for the left part to the restriction of f to $x_{m-1} = 0$, whereas the right part corresponds to the restriction of f to $x_{m-1} = 1$. More generally, consider a subfunction f' at the l-th level of the decomposition. It is associated to a path $p = (p_1, \ldots, p_l) \in \mathbf{F}_2^l$ starting from the root (that is f) by choosing either left or right children at each level. Here the value of p_i is 0 if we go from level $i-1$ to level i by choosing the left children and 1 otherwise. Notice that f' is the restriction of f to $x_{m-1} = p_1, \ldots, x_{m-l} = p_l$. Moreover, if we look for an annihilator of degree d for f, we search for an annihilator of degree $d - |p|$ for f'.*

More formally, our algorithm is the following

Algorithm 1. (Immunity verification) The input is an m-variable Boolean function f and a parameter d. The output is Yes if it can prove that there is no non-trivial annihilator of degree $\leq d$, No otherwise.

1. [decomposition] Recursively decompose the variable space of f according to d and this up to subfunctions in $2d + 1 + \lceil \log m \rceil$ variables. For each leaf, execute step 2.
2. [Subfunction verification] Use lazy Gaussian elimination with the correct degree to compute the annihilator space of the considered subfunction. If there is a non-trivial annihilator then go directly to step 4, otherwise continue.
3. [Immune] Output Yes: f has no non-trivial annihilator of degree $\leq d$.
4. [Unknown] Output No: we cannot prove the immunity of f.

Since in most cases of interest in cryptography, d is small, we can continue the decomposition quite far and still have with high probability that each subfunction does not admit any annihilator of the required degree. The point is that verifying with a classical algorithm that there is no annihilator for each subfunction is much faster than verifying it for the whole function. The real gain comes actually from the fact that the decomposition often ends at subfunctions for which we just have to check that the constant function 1 is not an annihilator. For this, we just have to find a corresponding value of the subfunction which is equal to 1. This can be done in $O(1)$ expected time. Hence, the more we decompose f, the faster we get but the more chance we have to output "No", since the probability that at least one of the subfunctions admits an annihilator increases. Choosing to decompose the functions up to a number of variables equal to $2d + 1 + \lceil \log m \rceil$ is a good tradeoff. For this choice we are able to prove the following

Theorem 2. *Let f be a random balanced m-variable Boolean function. Let d be fixed, then as m tends to infinity, Algorithm 2 runs in $O(k) = O\left(m^d\right)$ expected time and proves that there is no non-trivial annihilator up to degree d, except for a proportion of functions which is at most $O\left(e^{-2^d m(1+o(1))}\right)$.*

Proof. See Appendix C.

4   An Efficient Algorithm for Finding Annihilators

It the previous section, we considered a fixed depth decomposition, but the depth of decomposition may vary. For a given function f, the best decomposition for our problem is actually the smallest decomposition such that each subfunction has no non-trivial annihilator of the required degree.

By using a specific order on the monomials it is possible to work on this best decomposition. In practice, this new algorithm (Algorithm 2) will perform really well as shown by the running time presented later and will even be able to output an annihilator when it exists. However, analyzing directly Algorithm 2 seems to be more involved. We can only get a good upper bound on its complexity when there is a decomposition for which Algorithm 1 outputs Yes, and even in this case, the bound is slightly larger than the $O(k)$ bound of Algorithm 1.

4.1   The Algorithm

We describe here the new algorithm. Its relation with Algorithm 1 is not clear at first sight but will be explained in the next subsection. We will heavily use a specific order on the monomials and on elements of $\mathbf{F}_2^m$. This order is induced by the integer order on the set of integers $[\![0, 2^m - 1]\!]$ by viewing an m-tuple x (or a monomial X) as a binary representation of an integer. More precisely an element $(x_0, \ldots, x_{m-1}) \in \mathbf{F}_2^m$ or the corresponding monomial $X_0^{x_0} \ldots X_{m-1}^{x_{m-1}}$ are both associated to the integer $\sum_{i=0}^{m-1} x_i 2^i$. This identification allows us to compare monomials with points in $\mathbf{F}_2^m$ and to speak about intervals. For instance for $a, b \in \mathbf{F}_2^m$ with $a < b$ we define $[a, b) \stackrel{\text{def}}{=} \{y \in \mathbf{F}_2^m : a \leq y < b\}$. We also index from 0 to $k - 1$ the m-variable monomials of degree $\leq d$ arranged in increasing order.

We are now ready to describe Algorithm 2. The idea behind it is just to compute incrementally (for an x varying in $\mathbf{F}_2^m$) an annihilator basis of f restricted to the first x points.

Algorithm 2. (Incremental algorithm) The input is a Boolean function f in m variables and a number d. The output is a basis of the subspace of annihilators for f of degree $\leq d$.

1. [Initialization] Set the basis stack S to the constant function 1. Set the current monomial index i to 0. Initialize x to the first value in $\mathbf{F}_2^m$ such that $f(x) = 1$.
2. [Add basis element?] While the monomial of index $(i + 1)$ is smaller than or equal to x, push it onto the top of S and increment i.
3. [Remove basis element?] If any, find the element of S closest to the top that evaluates to 1 on x. XOR it with all the other elements of S that evaluate to 1 on x and remove it from S.
4. [Skip to next monomial?] If the annihilator space is trivial, increment i, push the monomial of index i onto the top of S and set x to be this monomial.
5. [Loop?] If possible, increment x until $f(x) = 1$ and go back to step 2.
6. [end] Execute a last time step 2 and return the current basis stored in S.

The correctness of this algorithm follows from the two following lemmas

Lemma 3. *Given a monomial $X_0^{x_0} \ldots X_{m-1}^{x_{m-1}}$ of degree d, the associated monomial function is zero on all entries strictly smaller than $x = (x_0, \ldots, x_{m-1})$. Moreover, this function is equal to 1 on the interval $[x, x')$ where x' is the first point in $\mathbf{F}_2^m$ greater than x of weight $\leq d$.*

Proof. The aforementioned monomial function evaluates to 1 on an m-tuple $y = (y_0, \ldots, y_{m-1})$ iff for all i such that $x_i = 1$ we also have $y_i = 1$. In such a case, we necessarily have $\sum_{i=0}^{m-1} y_i 2^i \geq \sum_{i=0}^{m-1} x_i 2^i$ which concludes the first part of the proof. Now, let us define i_0 to be the minimum of the set $\{i, x_i \neq 0\}$. Denote by x' the first point greater than x of weight less than or equal to d. The second assertion follows from the fact that all y' strictly between x and x' coincide with x for all positions greater than or equal to i_0. This implies that the aforementioned monomial evaluates to 1 at such y's. $\qquad\square$

Lemma 4. *Let $A_{<x}(d)$, respectively $A_{\leq x}(d)$ be the set of m-variable Boolean functions g generated by the monomials of degree $\leq d$ which are smaller than or equal to x that satisfy $f(y)g(y) = 0$ for all entries $y \in \mathbf{F}_2^m$ smaller than x, resp. smaller than or equal to x. We use the same notation when x is a monomial by identifying x with the corresponding point in $\mathbf{F}_2^m$. For a value x chosen by the algorithm, we denote by x^- its previously chosen value. Then*

- *The set S obtained after completing step 2 is a basis for $A_{<x}(d)$.*
- *The set S obtained after completing step 3 is a basis of $A_{\leq x}(d)$.*

- If a monomial X has been added in step 4, then just after completing step 4, $A_{<X}(d) = \{0, X\}$.
- For step 5, if no monomial has been added in step 4 just before, then after completing step 5 we have that $f(x) = f(x^-) = 1$ and $f(y) = 0$ for all y such that $x^- < y < x$. If a monomial X has been added in step 4 just before, we have after step 5 has been completed, that $A_{<X}(d) = \{0, X\}$.

Proof. The key point is that by Lemma 3 the value of a Boolean function on x depends only on the monomials in its ANF smaller than or equal to x.

Let us check by induction on the number of times we perform step 2 that S is a basis of $A_{<x}(d)$ after step 2 has been completed. This is obviously true the first time we perform step 2. The assertions on step 3,4,5 are also clearly true the first time they are performed. Consider now a further step 2 for which all assertions about the previous steps 2,3,4,5 hold. There are two possibilities depending on whether the previous step 4 has added or not a monomial to S.

Case 1 : Assume that step 4 has added a monomial X.
Then $A_{<X}(d) = \{0, X\}$. All monomials added in step 2 are precisely the monomials X' of degree $\leq d$ such that $x \geq X' > X$. Note that for all these X' we have $X'(y)f(y) = 0$ for all $y < x$. This is also true for monomial X. Therefore all elements in S belong to $A_{<x}(d)$. All the elements in S are monomials and are necessarily independent. Consider now an element g in $A_{<x}(d)$. Let us first note that the ANF of g contains no monomial $< X$, otherwise the Boolean function formed by the part of the ANF of g involving only monomials $< X$ would belong to $A_{<X}(d)$, which is impossible. g is then necessarily a sum of monomials in S since we have put in S all monomials of degree $\leq d$ between X and x. S is therefore a basis of $A_{<x}(d)$.

Case 2 : Assume that step 4 has added no monomial. In this case, we have added to S all monomials between x^- (exclusive) and x (inclusive) of degree $\leq d$. Note that $f(y) = 0$ for all y such that $x^- < y < x$. From this, it is easy to check that all elements in S belong to $A_{<x}(d)$. They are also clearly independent. As before, consider a $g \in A_{<x}(d)$. Write $g = g_{\leq x^-} + g_{>x^-}$, where the ANF of $g_{\leq x^-}$ consists in the monomials of the ANF of g which are smaller than or equal to x^-. Notice from Lemma 3 that $g_{\leq x^-}$ belongs to $A_{\leq x^-}(d)$. It is therefore generated by elements in S. $g_{>x^-}$ is also clearly generated by the monomials added in step 2. S is therefore a basis of $A_{<x}(d)$.

The assertion on step 3 follows at once from the assertion on step 2, and the assertions about step 4 and 5 are straightforward. $\qquad\square$

Remarks

1. Step 4 might seem odd at first sight, because it could be omitted without changing the correctness of the algorithm. However it is an essential step for having a low complexity algorithm. The point is that it avoids checking a lot of values of x satisfying $f(x) = 1$ which are useless for the algorithm.
2. In step 3, taking the element of S closest to the top of the stack is a heuristic which really helps in being fast. Define the trailing monomial of a Boolean function as the smallest monomial appearing in its ANF, the closer we are

to the top of the stack, the larger the trailing monomial will be. Taking elements the closest possible to the top of the stack will therefore tend to reduce the number of monomials in its ANF form. XOR'ing such a function to the other elements of the stack will therefore be more efficient.

3. Observe that Algorithm 2 necessarily outputs the empty-set if f evaluates to 1 in at least one point in any interval $[X_1, X_2)$ where X_1 and X_2 are two consecutive monomials of degree $\leq d$ (the monomials being ordered as explained above). This result comes from the second part of Lemma 3 and implies that such a function has no non-trivial annihilator in $\mathcal{B}(d, m)$. In particular, this applies to the inverse m-variable majority function which is defined by $f(x) = 1$ iff $|x| \leq \lfloor m/2 \rfloor$. This function is equal to 1 on all monomials of degree $\leq \lfloor m/2 \rfloor$ and has therefore no non-trivial annihilator of degree $\leq \lfloor m/2 \rfloor$.

4.2   Complexity Issues

Let us now explain the relationship between Algorithm 1 and Algorithm 2 when there exists a decomposition for which the output of Algorithm 1 is "YES I can prove that there are no annihilator of degree $\leq d$". Indeed let Algorithm 1* be a modified version of Algorithm 1 where we use Algorithm 2 on subfunctions instead of lazy Gaussian elimination. Then we have

Proposition 1. *The complexity of Algorithm 2 is upper-bounded by the small-est[1] complexity of Algorithm 1* when it outputs* YES.

Proof. Assume that we apply Algorithm 1* to a certain decomposition for which the corresponding output is YES. Consider the first restriction of f for which we look for an annihilator of degree $\leq d$. Notice that this subfunction is the restriction of f on $[0, a)$, where the integer associated to a is equal to $2^{m'}$, m' being the number of variables on which the restriction depends. Algorithm 1* coincides with Algorithm 2 on this subfunction and outputs that there are no annihilator. Notice now that the second subfunction considered by Algorithm 1* is a restriction of f over an interval of the form $[a, b)$, and so on for all the other subfunctions which correspond to restrictions of f. Note that these intervals are consecutive and that they are examined by Algorithm 1* in a consecutive order. Since Algorithm 1* notices that every subfunction has no annihilator of the corresponding degree, Algorithm 2 also reduces in applying itself to each subfunction separately. The reduction is done implicitly because in this case an element in S will have only non zero coordinates on monomials in the current subfunction interval. The proposition follows.   □

However, the problem is that this does not show that Algorithm 2 is as efficient as Algorithm 1. If we use Algorithm 2 instead of lazy Gaussian elimination on the subfunctions in Algorithm 1, we are only able to use the weak bound

---

[1] Recall that this means that the minimum is taken over all decompositions of f for which all subfunctions have no non-trivial annihilators.

$E(m', d') \leq n'k'^2$ instead of the bound of Lemma 2 ($E(m', d')$ representing the expected running time of Algorithm 2 on the corresponding subfunction). But we still have $E(m', 0) = O(1)$ and we obtain

Proposition 2. *The complexity of Algorithm 1* applied to the decomposition of the previous section is $O\left(k(\log m)^2\right)$.*

Proof. Let us first bound the expected complexity of Algorithm 2 when applied to a subfunction f' on m' variables to find annihilators of degree $\leq d'$. When $d' \geq 1$ we will simply use the worst case complexity where we consider all n' points and for each of them, step 3 and 4 run in k'^2. Thus we have $E(m', d') = O(n'k'^2)$. In the case $d' = 0$, because of step 4, we will stop as soon as we get a point where $f' = 1$ and we have the same expected running time as the lazy Gaussian algorithm, that is $E(m', 0) = O(1)$.

In the end, with the new $E(m', d') = O(n'k'^2)$ we obtain a complexity in $O(k(\log m)^2)$ for Algorithm 1* which is only slightly larger than the bound of complexity of Algorithm 1. □

4.3   Benchmarks

We have implemented a version of Algorithm 2 in C language running on a Pentium 4 at 2.6GHz with 1Gb memory. The (d, m) entry in the following tables means that we calculated the annihilator subspace in $\mathcal{B}(d, m)$ of a balanced m-variable Boolean function.

First of all, we ran some computations to find high degree annihilators. The results are presented in Figure 2. Notice that even for such large values of d corresponding to m, Algorithm 2 performs much better than lazy Gaussian elimination. Notice that the memory usage is better too, allowing to deal with cases where lazy Gaussian elimination ran out of memory.

d,m	2,6	3,8	4,10	5,12	6,14	7,16	8,18	9,20
k	22	93	386	1586	6476	26333	106762	431910
Lazy G	0s	0s	0s	0.1s	5s	2m30s	oom	oom
Algo 2	0s	0s	0s	0.01s	0.5s	20s	15m	12h

Fig. 2. Running time to check a high degree immunity for random balanced functions. The abbreviation oom means Out Of Memory. The lowest degree of a non-trivial annihilator of all the tested functions appeared to be the maximum possible, that is $\lceil \frac{m}{2} \rceil$ for an m-variable balanced Boolean function.

After this, we ran Algorithm 2 for small values of d and large values of m. The table in figure 3 clearly reflects the theoretical complexity in $O(k)$.

We have also checked the performances of our algorithm when there is an annihilator. The results are displayed in Figure 4. To make sure that the balanced Boolean function in m variables had an annihilator of a specified degree d, we took a random g in $\mathcal{B}(m)$ of degree d and weight w greater than 2^{m-1}. Then

d,m	4,32	5,32	6,32	7,32	3,64	4,64	5,64	2,128	3,128	2,256
k	4.10^4	2.10^5	1.10^6	4.10^6	4.10^4	6.10^5	8.10^6	8.10^3	3.10^5	3.10^4
Algo 2	1s	5s	30s	2m40s	1s	32s	8m	0.1s	32s	0.3s

Fig. 3. Running time to check low degree immunity of functions with a large number of variables. The functions were chosen randomly among balanced function and no non-trivial annihilators were found.

d,m	2,30	3,30	4,30	5,30	6,30
k	466	4526	3.10^4	2.10^5	8.10^5
Algo 2	13m	1h	3h45	-	-
Algo 2*	1s	1s	4s	31s	4m34s

Fig. 4. Running times for Algorithm 2 and 2* when $m = 30$ and a non-trivial annihilator of degree d exists

we chose $w - 2^{m-1}$ random values of g equal to 1 and made them equal to 0 to obtain a balanced function f. The f obtained this way has obviously $1 + g$ as an annihilator of degree d. We chose $m = 30$ for our experiments. In this case, Algorithm 2 has to check all 2^{30} points and this can be really long. It is possible to modify the algorithm to obtain a new algorithm (Algorithm 2*) which runs much faster by avoiding to consider all points of f. Since we know that there is an annihilator, the first modification is to skip to the next monomial in step 4 as soon as the annihilator space is of dimension 1. To speed up the process even more, another modification is to consider (at random) only a fraction of all the points where f is equal to 1 (one half for example). However, with this last modification, the output of the new algorithm is just a candidate set for the annihilator subspace : it always contains the right annihilator space, but it could be much larger. In spite of this fact, we found an annihilator space of dimension 1 at the end of all our experiments.

5   Conclusion

This article presents two algorithms of low complexity which allow to detect whether or not a given Boolean function has low degree multiples or low degree annihilators. They allow to check this kind of property even for Boolean functions with a large number of variables. They can be used to build cryptographically strong functions by devising such functions with respect to other criterions (non-linearity, being balanced, resiliency, ...) and by checking afterwards that such functions are also immune against algebraic attacks.

The analysis of the complexity of Algorithm 2 is probably somehow pessimistic and it is likely that it can be improved. It should also be interesting to improve this algorithm when we apply it to a Boolean function admitting a non-trivial annihilator of the specified degree.

References

[Arm04]   Frederik Armknecht. On the existence of low-degree equations for algebraic attacks. 2004. http://eprint.iacr.org/2004/185/.

[BP05]   An Braeken and Bart Preneel. On the algebraic immunity of symmetric Boolean functions. 2005. http://eprint.iacr.org/2005/245/.

[Car04]   Claude Carlet. Improving the algebraic immunity of resilient and non-linear functions and constructing bent functions. 2004. http://eprint.iacr.org/2004/276/.

[CDG05]   Nicolas Courtois, Blandine Debraize, and Eric Garrido. On exact algebraic [non-]immunity of S-boxes based on power functions. Cryptology ePrint Archive, Report 2005/203, 2005. http://eprint.iacr.org/2005/203.

[CM03]   Nicolas Courtois and Willi Meier. Algebraic attacks on stream ciphers with linear feedback. *Advances in Cryptology – EUROCRYPT 2003*, LNCS 2656:346–359, 2003.

[Cou03]   Nicolas Courtois. Fast algebraic attacks on stream ciphers with linear feedback. In *Advances in Cryptology-CRYPTO 2003*, volume 2729 of *LNCS*, pages 176–194. Springer Verlag, 2003.

[CP02]   Nicolas Courtois and Josef Pieprzyk. Cryptanalysis of block ciphers with overdefined systems of equations. *Asiacrypt 2002*, LNCS 2501, 2002. http://eprint.iacr.org/2002/044.

[DGM04]   Deepak Kumar Dalai, Kishan Chand Gupta, and Subhamoy Maitra. Results on algebraic immunity for cryptographically significant Boolean functions. In Anne Canteaut and Kapalee Viswanathan, editors, *INDOCRYPT*, volume 3348 of *Lecture Notes in Computer Science*, pages 92–106. Springer, 2004.

[Did05]   Frédéric Didier. A new bound on the block error probability after decoding over the erasure channel. *Submited to IEEE IT*, July 2005. http://www-rocq.inria.fr/codes/Frederic.Didier/papers/Didier05.pdf.

[DMS05]   Deepak Kumar Dalai, Subhamoy Maitra, and Sumanta Sarkar. Basic theory in construction of Boolean functions with maximum possible annihilator immunity. 2005. http://eprint.iacr.org/2005/229/.

[FA03]   J.-C. Faugère and G. Ars. An algebraic cryptanalysis of nonlinear filter generator using Gröbner bases. *Rapport de Recherche INRIA*, 4739, 2003.

[MPC04]   Willi Meier, Enes Pasalic, and Claude Carlet. Algebraic attacks and decomposition of Boolean functions. *Lecture Notes in Computer Science*, 3027:474–491, April 2004.

A   Proof of Lemma 1

Let $n' = 2^{m'}$, $k' = \binom{m'}{0} + \binom{m'}{1} + \cdots + \binom{m'}{d'}$ and $n = 2^m$. Note that $n' \geq 2^{2d+1}m$. We start the proof by noticing that the probability that the weight of f' is w is given by

$$\frac{\binom{n-n'}{n/2-w}\binom{n'}{w}}{\binom{n}{n/2}}.$$

By Stirling's formula this probability can be shown to be of the form $O(e^{n'(h(w/n')-1)})$, with $h(x) = -x \ln x - (1-x)\ln(1-x)$. Using Theorem 1 we obtain that the probability that f' has an annihilator of degree $\leq d$ is of the form

$$\sum_{w=o}^{n'} O\left(e^{2^{2d+1} m \left[h(w/n') - 1 + \frac{1}{2^{d'}} \left(\ln(1 - w/n') + \frac{k'(d' \ln 2 + 3)}{n'} \right) \right]} \right).$$

It is straightforward to check that $\frac{k'}{n'} = o(1)$ as m' tends to infinity. This implies that the $\frac{k'(d' \ln 2 + 3)}{n'}$ term in the exponent tends to zero. Moreover, we have a linear (in m) number of terms in the sum. Hence, we can upperbound it by its largest term times an $e^{mo(1)}$ factor. Putting all these facts together, we obtain that the probability that f' has an annihilator of degree d' is upperbounded by an expression of the form

$$e^{2^{2d+1} m \sup_{x \in [0,1]} g(x, d')(1 + o(1))},$$

where $g(x, d') \overset{\text{def}}{=} h(x) - 1 + \frac{1}{2^{d'}} \ln(1 - x)$. It can be checked that for $d' \geq 1$ we have that $\sup_{x \in [0,1]} g(x, d') \leq -\frac{1}{2^{d'+1}}$. The upper-bound given in Lemma 1 follows directly from this last remark.

B   Proof of Lemma 2

Let $n' = 2^{m'}$. First of all, the maximum complexity of lazy Gaussian elimination is $O(n'k'^2)$. Moreover, by using Theorem 1, the processing of $w_0 = k'(\lceil d' \ln 2 \rceil + 5)$ equations in lazy Gaussian elimination fails to prove the immunity of f' in $O(w_0 k'^2)$ with a probability smaller than

$$e^{\frac{n'}{2^{d'}} \left(\frac{k'}{n'}(d' \ln(2) + 3) + \ln\left(1 - \frac{w_0}{n'}\right) \right)} \leq e^{-\frac{k'}{2^{d'}-1}(1 + o(1))}$$

We can split the expected running time E in two. One part for the case where the Hamming weight of f' is greater than w_0 and one part where this is not the case. We get

$$E \leq \left[E_{w_0} + O(w_0 k'^2) + e^{-\frac{k'}{2^{d'}-1}(1 + o(1))} O(n' k'^2) \right] + P(|f'| < w_0) O(n' k'^2)$$

where E_{w_0} is the expected number of function evaluations we have to perform before actually finding w_0 points where f' is equal to 1. Notice that in this setting $k' = O(m'^{d'})$ and $e^{-\frac{k'}{2^{d'}-1}(1 + o(1))} n' = O(1)$. The probability that we have to check i points to find w_0 points for which an m-variable balanced Boolean function evaluates to 1 is given by

$$\binom{i-1}{w_0 - 1} \frac{\binom{n-i}{n/2 - w_0}}{\binom{n}{n/2}} = \binom{i-1}{w_0 - 1} \frac{(n)!(n/2)!(n/2)!}{(n-i)!(n/2 - w_0)!(n/2 + w_0 - i)!}$$

We can upper-bound this expression by

$$\binom{i-1}{w_0 - 1} \frac{n^i}{2^i (n-i)^i} \leq \frac{\binom{i-1}{w_0 - 1}}{2^i} \frac{1}{(1 - \frac{n'}{n})^{n'}}$$

Therefore we have for the expected number of function evaluations

$$E_{w_0} = \frac{1}{(1 - \frac{n'}{n})^{n'}} \sum_{i=w_0}^{n'} \binom{i-1}{w_0-1} \frac{i}{2^i} = O(w_0)$$

In the same way, we have

$$P(|f'| < w_0) \le \sum_{i=0}^{w_0-1} \frac{\binom{n'}{i}\binom{n-n'}{n/2-i}}{\binom{n}{n/2}} \le w_0 \binom{n'}{w_0} \frac{\binom{n-n'}{\frac{n-n'}{2}}}{\binom{n}{n/2}} \le \frac{w_0 n'^{w_0}}{2^{n'}} \frac{1}{(1 - \frac{n'}{n})^{n'}}$$

So $P(|f'| < w_0)n' = O(1)$ which concludes the proof.

C   Proof of Theorem 2

The proof of this theorem starts by first evaluating $N(m, d, m', d')$ which is the number of times we obtain an m'-variable subfunction for which we have to determine whether or not it has a degree $\le d'$ annihilator.

Lemma 5. *We have* $N(m, d, m', d') = \binom{m-m'}{d-d'}$.

Proof. We use Fact 1 (and also the notation introduced there) and notice that such subfunctions correspond to paths $p = (p_1, p_2, \ldots, p_{m-m'}) \in \mathbf{F}_2^{m-m'}$ for which $|p| = d - d'$. The number of paths of this kind is obviously $\binom{m-m'}{d-d'}$.   □

Let $E(m', d')$ be the expected running time of checking whether or not the m'-variable subfunction has a non-trivial annihilator of degree $\le d'$. The crux of obtaining a low-complexity algorithm is that

Lemma 6. *For any* m', $E(m', 0) = O(1)$.

Proof. The idea is that in this case we just have to check whether or not the constant function 1 is an annihilator for the subfunction. This can be done by finding a point of the subfunction which evaluates to 1. Let $n = 2^m$ and $n' = 2^{m'}$. The probability that we have to check i points to find a point for which an m-variable balanced Boolean function evaluates to 1 is equal to $\frac{\binom{n-i}{n/2-1}}{\binom{n}{n/2}}$. Therefore the expected time is equal to

$$\sum_{i=1}^{n'} i \frac{\binom{n-i}{n/2-1}}{\binom{n}{n/2}} + n' \frac{\binom{n-n'}{n/2}}{\binom{n}{n/2}}.$$

The last term corresponds to the probability that the subfunction is the all-zero function and tends to 0 as n' tends to infinity. For the sum, we notice that $\frac{\binom{n-i}{n/2-1}}{\binom{n-i-1}{n/2-1}} = \frac{n/2-i+1}{n-i} \le \frac{1}{2}$ and therefore the expected time is smaller than

$$\sum_{i=1}^{n'} i \left(\frac{1}{2}\right)^i + o(1) = O(1).$$ □

We finish the proof of the first assertion in Theorem 2 by noticing that the expected running time E of our algorithm is equal to

$$\sum_{i=1}^{d} N(m, d, 2d + 1 + \lceil \log m \rceil, i) E(2d + 1 + \lceil \log m \rceil, i) + E_0$$

where :

- $2d + 1 + \lceil \log m \rceil$ is the number of variables at which we stop the decomposition.
- E_0 is the expected time spent in checking the subfunctions for which $d' = 0$.

From Lemma 6, E_0 depends on the number of subfunctions for which $d' = 0$. It corresponds to paths of weight exactly d in the decomposition tree and we get

$$E_0 = O\left(\binom{m - (2d + 1 + \lceil \log m \rceil)}{d} \right).$$

We use Lemma 2 to bound the $E(2d+1+\lceil \log m \rceil, i)$'s (and by noticing that the k'''s involved in this lemma are of the form $O\left((2d + 1 + \log m)^i\right)$ and we obtain that

$$E = O\left(\binom{m - 2d - 1 - \lceil \log m \rceil}{d} + \sum_{i=1}^{d} \binom{m - 2d - 1 - \lceil \log m \rceil}{d - i} i(2d + 1 + \log m)^{3i} \right)$$

$$= O\left(m^d + \sum_{i=1}^{d} m^{d-i} i(2d + 1 + \log m)^{3i} \right)$$

$$= O(m^d) = O(k).$$

It should be noted that it is really the time spent in proving that subfunctions have not the constant 1 as annihilator which is responsible for the $O(k)$ term in the average complexity. The time spent in proving that the other subfunctions have annihilators of larger degree is negligible.

To prove that there is only a negligible part of functions for which Algorithm 1 answers "No", we proceed as follows. The probability P_1 that Algorithm 2 exits for values of d' greater than 0 can be bounded by using lemmas 5 and 1 by

$$P_1 \leq \sum_{d'=1}^{d} \binom{m - 2d - 1 - \lceil \log m \rceil}{d - d'} e^{-2^{2d - d'}(1 + o(1))m}$$

$$\leq O(m^d) e^{-2^{d+1}(1 + o(1))m} + e^{-2^d(1 + o(1))m}$$

$$\leq e^{-2^d(1 + o(1))m}$$

The probability P_2 that Algorithm 1 exits for values of d' equal to 0 is clearly upper-bounded by the number of subfunctions for which d' equals 0 times the probability that such a subfunction evaluates to 0 on a domain of size $2^{2d+1+\lceil \log m \rceil}$. This last probability is clearly of the form $O\left(e^{-2^{2d+1}m(\ln 2 + o(1))}\right)$. Summing P_1 and P_2 yields an upper-bound for the probability we wish to bound and leads to the second assertion in Theorem 2.

Upper Bounds on Algebraic Immunity of Boolean Power Functions

(Extended Abstract)

Yassir Nawaz[1], Guang Gong[1], and Kishan Chand Gupta[2]

[1] Department of Electrical and Computer Engineering
University of Waterloo
Waterloo, ON, N2L 3G1, Canada
[2] Centre for Applied Cryptographic Research
University of Waterloo
Waterloo, ON, N2L 3G1, Canada
ynawaz@engmail.uwaterloo.ca, G.Gong@ece.uwaterloo.ca,
kgupta@math.uwaterloo.ca

Abstract. Algebraic attacks have received a lot of attention in studying security of symmetric ciphers. The function used in a symmetric cipher should have high algebraic immunity ($\mathcal{AI}$) to resist algebraic attacks. In this paper we are interested in finding $\mathcal{AI}$ of Boolean power functions. We give an upper bound on the $\mathcal{AI}$ of any Boolean power function and a formula to find its corresponding low degree multiples. We prove that the upper bound on the $\mathcal{AI}$ for Boolean power functions with Inverse, Kasami and Niho exponents are $\lfloor \sqrt{n} \rfloor + \lceil \frac{n}{\lfloor \sqrt{n} \rfloor} \rceil - 2$, $\lfloor \sqrt{n} \rfloor + \lceil \frac{n}{\lfloor \sqrt{n} \rfloor} \rceil$ and $\lfloor \sqrt{n} \rfloor + \lceil \frac{n}{\lfloor \sqrt{n} \rfloor} \rceil$ respectively. We also generalize this idea to Boolean polynomial functions. All existing algorithms to determine $\mathcal{AI}$ and corresponding low degree multiples become too complex if the function has more than 25 variables. In our approach no algorithm is required. The $\mathcal{AI}$ and low degree multiples can be obtained directly from the given formula.

Keywords: Algebraic attacks, Algebraic immunity, Inverse exponent, Kasami exponent, Polynomial functions, Power functions, Niho exponent.

1   Introduction

The idea behind the algebraic attacks is to express the cipher as a system of multivariate equations whose solution gives the secret key. The complexity of the attack depends on the degree of these equations. Therefore the existence of low degree equations, to express a cipher, is crucial for algebraic attacks.

The algebraic attacks on stream ciphers composed of LFSR(s) and a nonlinear combining function f were proposed by Courtois and Meier in [8, 9]. The authors presented several scenarios under which low degree equations exist for ciphers using a combining function f with small number of inputs. These low degree equations are obtained by producing low degree multiples of f, i.e., by

M.J.B. Robshaw (Ed.): FSE 2006, LNCS 4047, pp. 375–389, 2006.
© International Association for Cryptologic Research 2006

multiplying f with a low degree function g such that fg is of low degree. In [26] Meier, Pasalic and Carlet reduced the scenarios (given in [8, 9]) under which low degree equations may exist to two and showed that existence of low degree equations is equivalent to the existence of low degree annihilators of f or $f + 1$.

Krause and Armknecht extended algebraic attacks to combiners with memory in [2]. They proved that algebraic equations always exist for such combiners and also gave an upper bound on the degree of such equations in terms of their input size and memory. In [10] Courtois further extended these attacks to combiners with memory and several outputs and provided an upper bound on the degree of equations for such combiners in terms of the size of their input, output and memory. An improvement on the algebraic attacks called fast algebraic attacks was presented in [7]. These attacks have been further examined in [3, 21].

The first algebraic attack on a block cipher was discussed in [30]. In [11] Courtois and Pieprzyk showed that AES can be attacked by solving a system of quadratic equations. This is possible because the only nonlinear component in AES, i.e., the S-box, can be expressed as a system of quadratic Boolean equations. They also introduced a definition of the resistance of the S-box to algebraic attacks based on the number and type of equations that describe the S-box. Low degree equations that describe the S-box are again essential for the low complexity of the algebraic attack. Cheon and Lee [6] used this definition to determine the resistance of S-boxes (based on Gold, Kasami and inverse exponents) against algebraic attacks. However their results have been disputed by Courtois et al. in [12]. Another algebraic attack on AES was given in [27, 28].

Since the existence of low degree equations for simple combiners, combiners with memory, and S-boxes is important for algebraic attacks, Armknecht combined the three cases in [1]. He showed that finding low degree equations for simple combiners, combiners with memory, and S-boxes can be reduced to the same problem of finding low degree annihilators.

In other direction there is increasing interest in the construction of Boolean functions with highest $\mathcal{AI}$. So far there are only three known constructions [14, 5, 15] that can acheive maximum possible $\mathcal{AI}$ $\lceil \frac{n}{2} \rceil$, where n is the number of inputs to the function. But the constructed function lacks certain cryptographic properties making it unsuitable to be used in a cryptosystem.

Except [27] all other techniques for finding the low degree equations have been developed from the theory of Boolean functions. Even, functions and S-boxes designed over finite field $\mathbb{F}_{2^n}, n > 2$ are analyzed according to this approach. We can refer to them as polynomial functions or mappings. For example the filter function $f : \mathbb{F}_{2^{16}} \to \mathbb{F}_2$ used in the stream cipher SFINKS [4] is a component of the inverse mapping in $\mathbb{F}_{2^{16}}$. S-box used in AES [13] and stream cipher SNOW [22] consists of inverse mapping in $\mathbb{F}_{2^8}$. A power mapping from $\mathbb{F}_{2^n}$ to $\mathbb{F}_{2^n}$ can be decomposed into n component functions, from $\mathbb{F}_{2^n}$ to $\mathbb{F}_2$, called Boolean power functions. In other words a Boolean power function is a monomial trace function which will be introduced later.

In this paper we use the theory of polynomial functions to analyze the $\mathcal{AI}$ of Boolean power function. This approach allows us to obtain meaningful results that are very difficult to obtain from the theory of Boolean functions. For example we derive upper bounds on the $\mathcal{AI}$ of many Boolean power functions that are much lower than the optimal upper bound presented in literature. We show that $\mathcal{AI}$ of functions based on inverse, Kasami and Niho exponents decreases drastically as n increases. More over the existing algorithms to determine the $\mathcal{AI}$ (and finding low degree equations) of functions are very slow and are not practical for $n > 25$. Our approach has no such limitation. The $\mathcal{AI}$ of any Boolean power function can be obtained directly from the formula regardless of the value of n. Similarly the low degree equations required for the algebraic attack are also obtained directly from the formula.

We believe that besides determining the $\mathcal{AI}$ of polynomial functions, the approach presented in this paper can be used to analyze and design polynomial functions with high $\mathcal{AI}$ along with other cryptographic properties.

2   Definitions and Preliminaries

In this section we provide the necessary preliminary material required in the later sections.

2.1   Polynomial Functions

Let $\mathbb{F}_2$ be the finite field of two elements. We consider the domain of an n-variable Boolean function to be the vector space $(\mathbb{F}_2^n, +)$ over $\mathbb{F}_2$, where $+$ is used to denote the addition operator over both $\mathbb{F}_2$ and the vector space $\mathbb{F}_2^m$.

The Hamming weight of an integer i is the number of nonzero coefficients in the binary representation of i and is denoted by $H(i)$.

For a binary string, λ consecutive ones (1's) preceded by zero and followed by zero is called a run of ones of length λ. We are only interested in the total number of runs of ones in a given binary string and not their length λ. Furthermore we consider our runs of ones to be cyclic. For example 1100011110011111 has two (not three) cyclic runs of ones.

Any n variable Boolean function $h: \mathbb{F}_2^n \rightarrow \mathbb{F}_2$, can be uniquely represented as a multivariate polynomial over $\mathbb{F}_2$, called the *algebraic normal form*,

$$h(x_1, \ldots, x_n) = a_0 + \sum_{1 \leq i \leq n} a_i x_i + \sum_{1 \leq i < j \leq n} a_{i,j} x_i x_j + \ldots + a_{1,2,\ldots,n} x_1 x_2 \ldots x_n,$$

where the coefficients $a_0, a_i, a_{i,j}, \ldots, a_{1,2,\ldots,n} \in \mathbb{F}_2$. The degree of the Boolean function h, denoted by $deg(h)$, is the same as the degree of the multivariate polynomial.

An (n, m) S-box (or vectorial function) is a map $F: \mathbb{F}_2^n \rightarrow \mathbb{F}_2^m$ and has component functions $f_1, \cdots, f_m$.

Let $\mathbb{F}_{2^n}$ be the finite field with 2^n elements. A Trace function $Tr: \mathbb{F}_{2^n} \rightarrow \mathbb{F}_{2^m}$, is given by [24, page 51]

$$Tr_m^n(x) = \sum_{i=0}^{n/m-1} x^{2^{mi}}, x \in \mathbb{F}_{2^n}.$$

A Cyclotomic coset C_s modulo $2^n - 1$ is defined as [25, page 104]

$$C_s = \{s, s \cdot 2, \cdots, s \cdot 2^{n_s-1}\},$$

where n_s is the smallest positive integer such that $s \equiv s2^{n_s} \pmod{2^n - 1}$. The subscript s is chosen as the smallest integer in C_s, and s is called the coset leader of C_s. For example the cyclotomic cosets modulo 15 are:

$$C_0 = \{0\}, C_1 = \{1, 2, 4, 8\}, C_3 = \{3, 6, 12, 9\}, C_5 = \{5, 10\}, C_7 = \{7, 14, 13, 11\},$$

where $\{0,1,3,5,7\}$ are coset leaders modulo 15.

Any non-zero polynomial function $f: \mathbb{F}_{2^n} \to \mathbb{F}_2$, can be represented as a sum of trace functions [20, page 178]:

$$f(x) = \sum_{k \in \Gamma(n)} Tr_1^{n_k}(A_k x^k) + A_{2^n-1} x^{2^n-1}, A_k \in \mathbb{F}_{2^{n_k}}, A_{2^n-1} \in \mathbb{F}_2,$$

where $\Gamma(n)$ is the set consisting of all coset leaders modulo $2^n - 1$, n_k is the size of the coset C_k, and $Tr_1^{n_k}(x)$ is the trace function from $\mathbb{F}_{2^{n_k}} \to \mathbb{F}_2$.

If $f(x)$ is balanced , we have [20]

$$f(x) = \sum_{k \in \Gamma(n)} Tr_1^{n_k}(A_k x^k), A_k \in \mathbb{F}_{2^{n_k}}, x \in \mathbb{F}_{2^n}. \tag{1}$$

The algebraic degree of f, denoted by $deg(f)$, is given by the largest w such that $A_k \neq 0$ and $H(k) = w$. There is a natural correspondence between Boolean functions h and polynomial functions f [20, page 334]. Let $\{\alpha_0, \ldots, \alpha_{n-1}\}$ be a basis for $\mathbb{F}_{2^n}$, then this correspondence is given by:

$$h(x_0, \ldots, x_{n-1}) = f(\alpha_0 x_0 + \ldots \alpha_{n-1} x_{n-1}).$$

A monomial or single trace term function f is a function that can be represented by a single trace term, $f(x) = Tr_1^n(\beta x^t)$ where $\beta \in \mathbb{F}_{2^n}$ and t is the coset leader of C_t.

2.2 Algebraic Immunity $\mathcal{AI}$

A Boolean function f is said to admit an annihilating function g, if $f * g = 0$. In [26] $\mathcal{AI}$ of f, denoted by $\mathcal{AI}(f)$, is defined as the minimum value of d such that f or $f + 1$ admits an annihilating function of degree d.

Proposition 1 of [26] states that existence of the relations of form $f * g = h$, where g and h have degree at most d, means the existence of annihilating function g' of degree at most d (as $f*(g+h) = 0$). Therefore if for f we can find a function g such that degree of $f * g$ is d then we can say that $\mathcal{AI}(f) \leq d$.

Fact 1. [8, Theorem 6.0.1] *Let f be any Boolean function with n inputs. Then there is a Boolean function $g \neq 0$ of degree at most $\lceil \frac{n}{2} \rceil$ such that $f * g$ is of degree at most $\lceil \frac{n}{2} \rceil$.*

Fact 1 shows that the upper bound on the $\mathcal{AI}$ of any Boolean function is $\lceil \frac{n}{2} \rceil$. To establish upper bound on the $\mathcal{AI}$ of a polynomial function f we will try to find multipliers g such that the degree of $f * g$ is less than $\lceil \frac{n}{2} \rceil$.

2.3 Monomial Trace Functions and Power Mappings

Monomial Trace functions are represented by a single trace term in polynomial form. There are several compelling reasons to study the $\mathcal{AI}$ of monomial trace functions. Any polynomial function can be expressed as a sum of monomial trace functions. Therefore, for a constant multiplier g, the $\mathcal{AI}$ of any function f is upper bounded by the maximum $\mathcal{AI}$ of any monomial trace function in its polynomial representation. This bound may not always be tight but in certain cases it can reveal the weakness of a function against algebraic attacks. Another important class of functions are the Boolean power functions that can be represented as monomial trace functions. These functions are of interest as they can be used as combining or filtering functions in stream ciphers [4]. These functions can easily be constructed from a power mapping in a finite field. Power mappings can be represented as $F : x \rightarrow x^a$ in $\mathbb{F}_{2^n}$ and are classified based on exponent a. Some famous exponents that have been studied for use in S-boxes are Inverse, Gold, Kasami, Welch and Niho [18, 23, 16, 17]. These mappings can easily be decomposed into Boolean power functions or monomial trace functions:

Let $\{\alpha_0, ..., \alpha_{n-1}\}$ and $\{\beta_0, ..., \beta_{n-1}\}$ be the dual basis [25, page 117] of $\mathbb{F}_{2^n}$. Then an S-box based on power mapping ($F : x \rightarrow x^a$) can be represented as $F(x) = \sum_{j=0}^{n-1} Tr_1^n(\beta_j x^{2^i t})\alpha_j$, $x \neq 0$ and its component functions can be represented as monomial trace functions of the form $f_j(x) = Tr_1^n(\beta_j x^{2^i t})$, where $a \in C_t$. It is conventional to represent monomial trace functions in the form $Tr_1^n(\beta x^t)$ where t is a coset leader of C_t. Note that we can write $a = 2^{-i}t$ for some i. So for any exponent a we can write the monomial trace function in the standard forms as

$$f(x) = Tr_1^n(\beta x^a) = Tr_1^n(\beta x^{2^{-i}t}) = Tr_1^n(\beta^{2^i} x^t),$$

since $Tr_1^n(x) = Tr_1^n(x^2)$. Next we find the $\mathcal{AI}$ of monomial trace functions.

3 Algebraic Immunity of Monomial Trace Functions

We provide the following proposition that will be used to derive upper bound on the $\mathcal{AI}$ of monomial trace functions.

Proposition 1. *Let* $f(x) = Tr_1^n(\beta x^t)$ *and* $g(x) = Tr_1^m(\gamma x^r)$ *be monomial trace functions, where* $x \in \mathbb{F}_{2^n}$, t *and* r *are the coset leaders of cosets* C_t *and* C_r. *The sizes of the cosets* C_t *and* C_r *are* n *and* m *respectively,* $m|n$, *and* $\beta \in \mathbb{F}_{2^n}$, $\gamma \in \mathbb{F}_{2^m}$. *Then*

$$deg(f(x)g(x)) = \max_{0 \leq i < m} H(r + t2^{-i})$$

Proof. Note both $f(x)$ and $g(x)$ are n variable Boolean functions. From the definition of trace function we can write (see also [19])

$$f(x)g(x) = \sum_{j=0}^{n-1}(\beta x^t)^{2^j} \sum_{l=0}^{m-1}(\gamma x^r)^{2^l} = \sum_{j=0}^{n-1}\sum_{l=0}^{m-1} \beta^{2^j} \gamma^{2^l} x^{t2^j + r2^l}$$

$$= \sum_{k=0}^{m-1} Tr_1^n(\gamma\beta^{2^k} x^{r+t2^k}),$$

where the algebraic degree of $f(x)g(x)$ is given by the largest w such that $\gamma\beta^{2^k} \neq 0$ and $H(r + t2^k) = w$. Let $k = m - i$, we have $t2^k \equiv t2^{m-i} \equiv t2^{-i} \bmod 2^n - 1$. Therefore

$$deg(f(x)g(x)) = \max_{0 \le k < m} H(r+t2^k) = \max_{0 \le i < m} H(r+t2^{-i}) \qquad \Box$$

Proposition 1 shows that we only need to add r to the members of coset C_t, and the highest hamming weight of the resulting integers gives the maximum possible degree of $f(x)g(x)$.

In the following Theorem we derive an upper bound on the $\mathcal{AI}$ of monomial trace functions based on a property of the exponent t, i.e., the number of runs of 1's in the binary representation of t.

Theorem 1. Let $l = \lfloor \sqrt{n} \rfloor$, $k = n - \lfloor \frac{n}{l} \rfloor l$ and $f(x) = Tr_1^n(\beta x^t)$, where $\beta \in \mathbb{F}_{2^n}$ and t is the coset leader of C_t. Let $g(x) = Tr_1^m(x^r)$, where

$$m = \begin{cases} l, \ k = 0; \\ n, \ 0 < k < l \end{cases}, \ and \ r = \begin{cases} 1 + \sum_{i=1}^{\frac{n}{l}-1} 2^{il} & , k = 0 \\ 1 + 2^k + \sum_{i=1}^{\lfloor \frac{n}{l} \rfloor -1} 2^{il+k} & , 0 < k < l \end{cases}.$$

Then

$$deg(f(x)g(x)) \le ul + \left\lceil \frac{n}{l} \right\rceil - 1, \qquad (2)$$

where u is the number of runs of 1s in the binary representation of t.

To prove Theorem 1 we need the following two lemmas.

Lemma 1. r is a coset leader modulo $2^n - 1$.

Proof. The above can be established by examining the binary representation of r.
    Case: $k = 0$

Therefore r is the coset leader modulo $2^n - 1$.

Case: $0 < k < l$

$$
\begin{array}{ll}
r & = \overbrace{00\cdots01}^{l}\overbrace{00\cdots01}\ldots\ldots\overbrace{00\cdots01}^{l}\overbrace{00\cdots01}^{k} \\
2r & = 00\cdots10\,00\cdots10\ldots\ldots 00\cdots10\,00\cdots10 > r
\end{array}
$$

$$
2^l r = \overbrace{00\cdots01}^{l}\overbrace{00\cdots01}\ldots\ldots\overbrace{00\cdots01}^{k}\overbrace{00\cdots01}^{l} > r
$$

$$
2^n r = \overbrace{00\cdots01}^{l}\overbrace{00\cdots01}\ldots\ldots\overbrace{00\cdots01}^{l}\overbrace{00\cdots01}^{k} = r
$$

Therefore r is the coset leader modulo $2^n - 1$.   □

Note: For $k = 0, |C_r| = l$ and for $0 < k < l, |C_r| = n$, where $|C_r|$ is the size of the coset C_r.

In Lemma 1 ⌢ is used to indicate the size of a segment in bits. From here onwards we will use ⌢ to represent the size of the segment as before and ⌣ to represent the number of 1's in the segment

Lemma 2. $H(r + t2^{-i}) \leq ul + \lceil \frac{n}{l} \rceil - 1, \ 0 \leq i \leq m - 1$ and $r, t, u,$ and l are as defined in Theorem 1.

Proof. Consider the binary representations of r and t. In Lemma 1, r consists of $\lfloor \frac{n}{l} \rfloor$ identical l bit segments when $k = 0$. If $k \neq 0$, then k least significant bits of r form an additional segment. All $\lceil \frac{n}{l} \rceil$ segments have hamming weight 1. We can segment t in the same way as r however these segments may or may not be identical. We will represent a segment of r and t as r' and t' respectively. Now let us consider the addition of r' and t'. Initially we will restrict ourselves to the case where the binary representation of t' has at most one run. Now consider all possible transitions in t' with and without carry.

Case 1: $1 \rightarrow 0$ transition

$$
\begin{array}{ll}
t' = \overbrace{111\cdots10\cdots00}^{j} & t' = \overbrace{111\cdots10\cdots00}^{j} \quad \leftarrow carry \\
r' = 000\cdots00\cdots01 & r' = 000\cdots00\cdots01 \\
+\,-\,-\,-\,-\,-\,-\,- & +\,-\,-\,-\,-\,-\,-\,- \\
111\cdots10\cdots01 & 111\cdots10\cdots10
\end{array}
$$

$$H(r' + t') = j + 1, j < l - 1 \qquad\qquad H(r' + t') = j + 1, j < l - 1$$

$$t' = \overbrace{111\cdots11}^{j}\overset{\curvearrowright}{0}$$
$$r' = 000\cdots001$$
$$+ \, -\,-\,-\,-\,-\,-\,-$$
$$111\cdots111$$

$$H(r'+t') = j+1, j = l-1$$

$$carry \leftarrow \quad \overbrace{111\cdots11}^{j}\overset{\curvearrowright}{0} \quad \leftarrow carry$$
$$t' \quad = 111\cdots110$$
$$r' \quad = 000\cdots001$$
$$+ \, -\,-\,-\,-\,-\,-\,-$$
$$000\cdots000$$

$$H(r'+t') = 0, j = l-1$$

Case 2: $0 \to 1$ transition

$$t' = \overbrace{000\cdots0}^{j}1\cdots11$$
$$r' = 000\cdots00\cdots01$$
$$+ \, -\,-\,-\,-\,-\,-\,-$$
$$000\cdots10\cdots00$$

$$H(r'+t') = 1, j < l$$

$$t' = \overbrace{000\cdots0}^{j}1\cdots11 \quad \leftarrow carry$$
$$r' = 000\cdots00\cdots01$$
$$+ \, -\,-\,-\,-\,-\,-\,-$$
$$000\cdots10\cdots01$$

$$H(r'+t') = 2, j < l$$

Case 3: No transition

$$t' = 000\cdots00$$
$$r' = 000\cdots01$$
$$+ \, -\,-\,-\,-\,-$$
$$000\cdots01$$

$$H(r'+t') = 1$$

$$\leftarrow carry$$
$$t' = 000\cdots00$$
$$r' = 000\cdots01$$
$$+ \, -\,-\,-\,-\,-$$
$$000\cdots10$$

$$H(r'+t') = 1$$

$$carry \leftarrow$$
$$t' \quad = 111\cdots11$$
$$r' \quad = 000\cdots01$$
$$+ \, -\,-\,-\,-\,-$$
$$000\cdots00$$

$$H(r'+t') = 0$$

$$carry \leftarrow \qquad \leftarrow carry$$
$$t' \quad = 111\cdots11$$
$$r' \quad = 000\cdots01$$
$$+ \, -\,-\,-\,-\,-$$
$$000\cdots01$$

$$H(r'+t') = 1$$

From the above cases it is clear that $H(r'+t')$ achieves maximum value l for $1 \to 0$ transition (no carry case). For $0 \to 1$ transition the maximum value is 2 and when there is no transition it is 1.

Now consider the following complete binary representation of r and t.

$$t = \overbrace{*\,*\cdots*\,*}^{l}\ldots\ldots\overbrace{1110\cdots00}^{l}\ldots\ldots\overbrace{*\,*\cdots*\,*}^{l}$$
$$r = 0\,0\cdots0\,1\ldots\ldots0000\cdots01\ldots\ldots0\,0\cdots0\,1$$

where $*$ can be either 1 or 0.

First we assume that each segment t' has at most one run. To get the maximum possible value of $H(r+t2^{-i})$ each segment with a $1 \to 0$ transition must contribute l number of 1's to the sum. Since there are u number of $1 \to 0$ transitions this adds up to ul number of 1's. For u number of $1 \to 0$ transitions there are at most u segments with $0 \to 1$ transitions, each contributing a maximum of 2 number of 1's to the sum, i.e., $2u$ number of 1's. All the remaining $\lceil \frac{n}{l} \rceil - 2u$ segments with no transitions can contribute at most single 1 to the sum. So the total contribution is $ul + 2u + \lceil \frac{n}{l} \rceil - 2u = ul + \lceil \frac{n}{l} \rceil$. Now consider the first segment from right to left that contains a $0 \to 1$ transition (right most segment with a $0 \to 1$ transition). For this segment to contribute 2 number of 1's to the sum it must receive a carry, otherwise it will contribute a single 1 (see case 2). Since the right most segment never receives a carry, the only way a carry can be generated between the right most segment and the right most segment with a $0 \to 1$ transition is due to the presence of a segment with all 1's (see all cases without carry). However this all 1's segment contributes zero number of 1's to the sum. Therefore we subtract 1 from $ul + \lceil \frac{n}{l} \rceil$. Therefore the maximum possible value of $H(r + t2^{-i})$ is $ul + \lceil \frac{n}{l} \rceil - 1$.

Suppose a segment t' has more than one runs. Then it must contain a $1 \to 0$ transition. In our analysis of segments with single runs, we assumed that each segment with a $1 \to 0$ transition must contribute l number of 1's to the sum. As the size of the segment is l the contribution of a segment with more than one runs must be less than or equal to the contribution of the segment with one run. Therefore $H(r + t2^{-i})$ is upper bounded by $ul + \lceil \frac{n}{l} \rceil - 1$ □

Proof of Theorem 1.
The assertion follows directly from Lemma 1, Lemma 2 and Proposition 1.   □

Remark 1. The significance of Theorem 1 is two fold. It gives the upper bound on the $\mathcal{AI}$ of f, i.e., $ul + \lceil \frac{n}{l} \rceil - 1$ and it also gives the low degree multiplier g $(deg(g) = \lceil \frac{n}{l} \rceil)$. In Theorem 1 we give only one multiplier however we can get $2^m - 1$ distinct non zero multipliers by taking $g(x) = Tr_1^m(\beta x^r)$, where $\beta \in \mathbb{F}_{2^m}$. Note only m of them are linearly independent. Also note that these multipliers can be computed directly from the formula in Theorem 1 with almost no effort.

Remark 2. From Fact 1 we know that $\mathcal{AI}$ of any function is at most $\lceil \frac{n}{2} \rceil$. Let v be the degree of $f(x)$, then, to obtain a meaningful upper bound on $\mathcal{AI}$, $deg(f(x)g(x)) \leq \min \left(v, \lceil \frac{n}{2} \rceil \right)$. So we have the following condition on u,

$$u \leq \min \left(\frac{v - \lceil \frac{n}{l} \rceil + 1}{l}, \frac{\lceil \frac{n}{2} \rceil - \lceil \frac{n}{l} \rceil + 1}{l} \right). \tag{3}$$

For many cryptogaphically useful power mappings, u is very small. For example, $u = 1$ for inverse, and $u = 2$ for Kasami, Gold, Welch and Niho. Therefore Theorem 1 can give very useful bounds for these mappings. In fact in most cases using the proof technique of Theorem 1 and exploiting the specific binary form of each exponent, we can further improve this bound. Since functions with Gold and Welch exponents have very small degrees (2 and 3 respectively) we will only consider inverse, Kasami and Niho exponents in this paper.

4   Inverse Exponent

Inverse mappings $x \rightarrow x^{-1}$ in $\mathbb{F}_{2^n}$ can be decomposed in to n monomial trace functions of the form $Tr_1^n(\beta x^{-1})$. The degree of the monomial trace functions with inverse exponents is $n - 1$. The inverse exponent consists of a single run of 1's. From Theorem 1 its $\mathcal{AI}$ is upper bounded by $l + \lceil \frac{n}{l} \rceil - 1$. However in Theorem 2 we show that for inverse function this bound can be improved to $l + \lceil \frac{n}{l} \rceil - 2$.

Lemma 3. Let $t = 2^{n-1} - 1$. Then $H(r + t2^{-i}) \leq l + \lceil \frac{n}{l} \rceil - 2$, $0 \leq i \leq m - 1$ and r is defined as in Theorem 1.

Proof. Consider the binary representation of r and t.

$$
\begin{array}{l}
\overbrace{\qquad}^{l} \qquad\qquad \overbrace{\qquad}^{l}\ \overbrace{\qquad}^{l} \\
t = 01\cdots11\,11\cdots11\ldots\ldots11\cdots11\,11\cdots11 \\
r = 00\cdots01\,00\cdots01\ldots\ldots00\cdots01\,00\cdots01 \\
+\ -\ -\ -\ -\ -\ -\ -\ -\ -\ -\ -\ -\ -\ - \\
\underbrace{10\cdots01}\,\underbrace{00\cdots01}\ldots\ldots\underbrace{00\cdots01}\,\underbrace{00\cdots00} \\
\quad\ 2 \qquad\quad 1 \qquad\qquad\quad 1 \qquad\quad\ 0
\end{array}
$$

$H(r + t) = \lceil \frac{n}{l} \rceil$.
We can see that $H(r + t2^{-i})$ is maximized when $i = l - 2$

$$
\begin{array}{l}
\qquad\qquad \overbrace{\qquad}^{l} \qquad\qquad\quad \overbrace{\qquad}^{l}\ \overbrace{\qquad}^{l} \\
t2^{-(l-2)} = 11\cdots01\,11\cdots11\ldots\ldots11\cdots11\,11\cdots11 \\
r \qquad\quad = 00\cdots01\,00\cdots01\ldots\ldots00\cdots01\,00\cdots01 \\
+\ -\ -\ -\ -\ -\ -\ -\ -\ -\ -\ -\ -\ -\ -\ -\ -\ - \\
\underbrace{11\cdots11}\,\underbrace{00\cdots01}\ldots\ldots\underbrace{00\cdots01}\,\underbrace{00\cdots00} \\
\quad\ l \qquad\qquad 1 \qquad\qquad\quad 1 \qquad\quad\ 0
\end{array}
$$

$H(r + t2^{-(l-2)}) = l + \lceil \frac{n}{l} \rceil - 2$.

Therefore $H(r + t2^{-i}) \leq l + \lceil \frac{n}{l} \rceil - 2$, $0 \leq i \leq n - 1$.   □

Theorem 2. Let $f(x) = Tr_1^n(\beta x^{-1})$ and $g(x) = Tr_1^m(x^r)$. Then

$$deg(f(x)g(x)) = l + \left\lceil \frac{n}{l} \right\rceil - 2 \tag{4}$$

where β, m, r and l are the same as defined in Theorem 1.

Proof. From Lemma 3, $t = 2^{n-1} - 1$. Since $Tr_1^n(x) = Tr_1^n(x^2)$, we have

$$f(x) = Tr_1^n(\beta x^{-1}) = Tr_1^n(\beta x^{2t}) = Tr_1^n(\beta^{2^{n-1}} x^t). \tag{5}$$

From Lemma 1, r is a coset leader and from Lemma 3, Eqn.(5) and Proposition 1, $deg(f(x)g(x)) \leq l + \lceil \frac{n}{l} \rceil - 2$.   □

This bound on $\mathcal{AI}$ is much less than theoretical optimal value $\lceil \frac{n}{2} \rceil$ for higher values of n (see Table 1).

5 Kasami Exponents

An n variable monomial trace function with Kasami exponent, $f(x)$, can be defined as $f(x) = Tr_1^n(\beta x^e)$, where $e = 2^{2s} - 2^s + 1$, $\gcd(n, s)=1$ and $1 \leq s \leq \frac{n}{2}$ [23]. Note that $f(x)$ is not balanced for all values of n. The algebraic degree of $f(x)$ is $s + 1$. We will only consider Kasami exponents that give the highest algebraic degree. The Kasami exponent consists of 2 runs of 1's in its binary representation. From Theorem 1 its $\mathcal{AI}$ is upper bounded by $2l + \lceil \frac{n}{l} \rceil - 1$. However in Theorem 3 we show that this bound can be improved to $l + \lceil \frac{n}{l} \rceil$.

Theorem 3. Let $f(x) = Tr_1^n(\beta x^e)$, where e is the Kasami exponent that gives highest algebraic degree, i.e., $\frac{n+1}{2}$ for n odd, $\frac{n}{2}$ for $n \equiv 0 \bmod 4$ and $\frac{n}{2} - 1$ for $n \equiv 2 \bmod 4$. Then

$$deg(f(x)g(x)) \leq l + \left\lceil \frac{n}{l} \right\rceil$$

where β, l and $g(x)$ are defined in Theorem 1.

Proof. Let $t = e2^{-s}$, then

$$f(x) = Tr_1^n(\beta x^e) = Tr_1^n(\beta x^{t2^s}) = Tr_1^n(\beta^{2^{n-s}} x^t), \tag{6}$$

since $Tr_1^n(x) = Tr_1^n(x^2)$. The assertion follows directly from Lemma 5 (Proof is given in appendix), Lemma 1, Eqn.(6) and Proposition 1.   □

Remark 3. Though in Theorem 3 bound on $\mathcal{AI}$ is proved for Kasami exponent that gives the highest algebraic degree, it is easy to prove that this bound holds for any Kasami exponent. The proof is given in [29]. This bound is much lower than the optimal bound $\lceil \frac{n}{2} \rceil$ for large n (see Table 1).

6 Niho Exponent

An n variable monomial trace function with Niho exponent [17,6], $f(x)$, can be defined as $f(x) = Tr_1^n(\beta x^e)$, where $e = 2^s + 2^{\frac{s}{2}} - 1$, $n = 2s + 1$ when s is even and $e = 2^s + 2^{\frac{3s+1}{2}} - 1$, $n = 2s + 1$ when s is odd. The degree of Niho function in n variables is $\frac{n+3}{4}$ for $n \equiv 1 \bmod 4$ and $\frac{n+1}{2}$ for $n \equiv 3 \bmod 4$.

The Niho exponent consists of 2 runs of 1's in its binary representation. From Theorem 1 its $\mathcal{AI}$ is upper bounded by $2l + \lceil \frac{n}{l} \rceil - 1$. However Theorem 4 show that this bound can be improved to $l + \lceil \frac{n}{l} \rceil$. The proof of the theorem is very similar to the proof of Kasami case, so we provide Theorem 4 without proof (proof is given in [29]).

Theorem 4. Let $f(x) = Tr_1^n(\beta x^e)$, where e is a Niho exponent. Then

$$deg(f(x)g(x)) \leq l + \left\lceil \frac{n}{l} \right\rceil$$

where β, l and $g(x)$ are defined in Theorem 1.

Table 1 shows how the upper bound on the $\mathcal{AI}$ of monomial trace functions with Inverse, Kasami and Niho exponents decreases as n increases. *Also note that in [4], $\mathcal{AI}$ of inverse function for $n = 16$ is given as 6 which is confirmed by our bound. This shows that our bound is tight for $n = 16$.*

Table 1. $\mathcal{AI}$ bounds for Inverse, Kasami and Niho functions

f	n	$deg(f)$	Bound from Fact 1 ($\lceil \frac{n}{2} \rceil$)	Our bound on $\mathcal{AI}$
Inverse	16	15	8	6
	36	35	18	10
	100	99	50	18
Kasami	16	8	8	8
	36	18	18	12
	100	50	50	20
Niho	15	8	8	8
	35	18	18	12
	99	50	50	20

7   Generalization to Polynomial Functions

Any balanced polynomial function $f(x)$ can be represented by Eqn.(1) which is reproduced here as

$$f(x) = \sum_{k \in \Gamma(n)} Tr_1^{n_k}(A_k x^k), A_k \in \mathbb{F}_{2^{n_k}}, x \in \mathbb{F}_{2^n},$$

which is simply a sum of monomial trace functions. If we only consider monomial trace functions as multipliers the result in Theorem 1 can be generalized to all balanced polynomial functions. Let u_k be the number of runs of 1's in the binary representation of k and $u = \max_{k \in \Gamma(n)}\{u_k\}$ such that $A_k \neq 0$. Let $g(x)$ be a monomial trace function defined in Theorem 1. Then

$$deg(f(x)g(x)) \le ul + \left\lceil \frac{n}{l} \right\rceil - 1.$$

To obtain a meaningful bound on $\mathcal{AI}$ of f, u must satisfy Eqn.(3). The above result implies that $\mathcal{AI}$ of f is upper bounded by the maximum $\mathcal{AI}$ of the single trace functions in the polynomial representation of f.

8   Conclusions

In this paper we use the theory of polynomial functions to provide an upper bound on $\mathcal{AI}$ of Boolean power functions. The low degree multiples are also obtained directly from the formula for any n. This is particularly useful as there are no existing algoritms to find the $\mathcal{AI}$ of a function with more than 25 variables. We improve the $\mathcal{AI}$ bound on inverse, Kasami and Niho functions and show that their $\mathcal{AI}$ is very low. We also generalize our results to polynomial functions.

References

1. F. Armknecht, On the Existence of Low-degree Equations for Algebraic Attacks, *Cryptology ePrint Archive, Report 2004/185*, http://eprint.iacr.org/, 2004.
2. F. Armknecht, Algebraic Attacks on Combiners with Memory, *Advances in Cryptology-CRYPTO 2003*, LNCS 2729, pp. 162-176, Springer-Verlag, 2003.

3. F. Armknecht, Improving Fast Algebraic Attacks *Fast Software Encryption 2004*, LNCS 3017, pp. 65-82, Springer-Verlag, 2003.
4. A. Braeken, J. Lano, N. Mentens, B. Preneel and I. Verbauwhede, SFINKS: A Synchronous Stream Cipher for Restricted Hardware Environments, *eSTREAM Project report 2005/026*, Available at http://www.ecrypt.eu.org/stream/.
5. A. Braeken and B. Preneel, On the Algebraic Immunity of Symmetric Boolean Functions, *To appear in Indocrypt 2005*.
6. J. Cheon and D. Lee, Resistance of S-Boxes Against Algebraic Attacks, *Fast Software Encryption 2004*, LNCS 3017, pp. 83-94, Springer-Verlag, 2004.
7. N. Courtois, Fast Algebraic Attacks on Stream Ciphers with Linear Feedback, *Advances in Cryptology-CRYPTO 2003*, LNCS 2729, pp. 176-194, Springer-Verlag, 2003.
8. N. Courtois and W. Meier, Algebraic Attacks on Stream Ciphers with Linear Feedback, *Advances in Cryptology-EUROCRYPT 2003*, LNCS 2656, pp. 346-359, Springer-Verlag, 2003.
9. N. Courtois and W. Meier, Algebraic Attacks on Stream Ciphers with Linear Feedback, Extended version of [8], available at http://cryptosystem.net/stream
10. N. Courtois, Algebraic Attacks on Combiners with Memory and Several Outputs, *ICISC 2004*, LNCS 3506, pp. 3-20, Springer-Verlag, 2004.
11. N. Courtois and Pieprzyk J., Cryptanalysis of Block Ciphers with Overdefined Systems of Equations, *Advances in Cryptology-ASIACRYPT 2002*, LNCS 2501. Springer-Verlag, 2002.
12. N. Courtois, B. Debraize and E. Garrido, On Exact Algebraic [Non]Immunity of S-boxes Based on Power Functions, *Cryptology ePrint Archive, Report 2005/203*, http://eprint.iacr.org/, 2005.
13. J. Daemen, and V.Rijmen, *The Design of Rijndael*, Springer-Verlag, 2002.
14. D. K. Dalai, K. C. Gupta and S. Maitra, Cryptographically Significant Boolean Functions: Construction and Analysis in Terms of Algebraic Immunity. *Fast Software Encryption 2005*, LNCS 3557, pp. 98-111, Springer-Verlag, 2005.
15. D. K. Dalai, S. Maitra and S. Sarkar, Basic Theory in Construction of Boolean Functions with Maximum Possible Annihilator Immunity. *To appear in Designs, Codes and Cryptography*.
16. H. Dobbertin, Almost Perfect Nonlinear Power Functions on $GF(2^n)$: The Welch Case. *IEEE Transactions on Information Theory*, Vol. 45, No. 4, pp. 1271-1275, 1999.
17. H. Dobbertin, Almost Perfect Nonlinear Power Functions on $GF(2^n)$: The Niho Case. *Information and Computation*, Vol. 151, pp. 57-72, 1998.
18. R. Gold, Maximal Recursive Sequences with 3 valued cross-correlation function, *IEEE Transactions on Information Theory*, Vol. 14, pp. 154-156, 1968.
19. S. W. Golomb and G. Gong, Hyper-Cyclotomic Algebra, *Sequences and their Applications, SETA'01*, Discrete Mathematics and Theoretical Computer Science, Springer, 2001, pp. 154-165. CORR 2001-33.
20. S. W. Golomb, and G. Gong, *Signal Design for Good Correlation: For Wireless Communication, Cryptography, and Radar*, Cambridge University Press, ISBN 0521821045, 2005.
21. P. Hawkes, G. Rose, Rewriting Variables: The Complexity of Fast Algebraic Attacks on Stream Ciphers, *Advances in Cryptology-CRYPTO 2004*, volume LNCS 3152, pp. 390-406, Springer-Verlag, 2004.
22. P. Ekdahl, and T. Johansson, SNOW-A New Version of the Stream Cipher SNOW, *Selected Areas in Cryptography, 2002*, LNCS 2595, pp. 47-61, Springer-Verlag 2003.

23. T. Kasami, The Weight Enumerators for Several Classes of Subcodes of the Second Order Binary Reed-Muller Codes, *Infor. Contr.*, Vol. 18, pp. 369-394, 1971.
24. R. Lidl and H. Niederreiter, *Introduction to Finite Fields and their Applications.* Cambridge University Press, 1994.
25. F. J. MacWilliams and N. J. A. Sloane, *The Theory of Error Correcting Codes.* North Holland, 1986.
26. W. Meier, E. Pasalic, and C. Carlet, Algebraic Attacks and Decomposition of Boolean Functions, *Advances in Cryptology EUROCRYPT 2004*, LNCS 3027, pp.474-491, Springer-Verlag, 2004.
27. S. Murphy and M. Robshaw, Essential Algebraic Structure within AES, *Advances in Cryptology Crypto 2002*, LNCS 2442, pp.1-16, Springer-Verlag, 2002.
28. S. Murphy and M. Robshaw, Comments on the Security of the AES and the XSL Technique, *Electronic Letters*, Vol. 39, pp. 26-38, 2003.
29. Y. Nawaz, G. Gong and K. Gupta, Upper Bounds on Algebraic Immunity of Boolean Power Functions, *Preprint*.
30. I. Schaumuller-Bichl, Cryptanalysis of the Data Encryption Standard by the Method of Formal Coding, *Advances in Cryptology EUROCRYPT-1982*, LNCS 149, pp.235-255, Springer-Verlag, 1983.

A   Appendix

In order to study the binary representation of e we need the following result.

Lemma 4. *Let $n > 4$ be any integer and $f(x) = Tr_1^n(\beta x^e)$, where $e = 2^{2s} - 2^s + 1$, $\gcd(n, s)=1$ and s is the highest value less than $\frac{n}{2}$. Then*

1. *If $n \equiv 0 \mod 4$, then $\gcd(n, s)=1$ where $s = \frac{n}{2} - 1$*
2. *If $n \equiv 2 \mod 4$, then $\gcd(n, s)=1$, where $s = \frac{n}{2} - 2$*
3. *If n is odd $\gcd(n, s)=1$, where $s = \frac{n-1}{2}$*

Proof. Let n be even and 2 divides $\frac{n}{2}$. Let $k > 1$ be an integer such that k divides both n and $\frac{n}{2} - 1$. As $\frac{n}{2}$ is even, $\frac{n}{2} - 1$ is odd, and so divisor k must be odd. Since k divides n, and k is odd, it must also divide $\frac{n}{2}$ and hence it can not divide $\frac{n}{2} - 1$. Hence the contadiction. So $\gcd(n, \frac{n}{2} - 1)=1$ and hence $s = \frac{n}{2} - 1$. The proves of the other two cases are similar.   □

Lemma 5. *Let $t = e2^{-s}$ where e and s are defined in Lemma 4. Then $H(r + t2^{-i}) \le l + \lceil \frac{n}{l} \rceil$, $0 \le i \le m - 1$ and r is defined in Theorem 1.*

Proof. From Lemma 4 the binary representation of t is :

$$t = 2^{\frac{n-1}{2}} + 2^{\frac{n+1}{2}} - 1 = \overbrace{00\cdots01}^{s}0\,1\overbrace{011\cdots11}^{s}, \qquad n \text{ is odd.}$$

$$t = 2^{\frac{n}{2}-1} + 2^{\frac{n}{2}+1} - 1 = \overbrace{00\cdots01}^{s}00\overbrace{0011\cdots11}^{s}, \quad n \equiv 0 \mod 4.$$

$$t = 2^{\frac{n}{2}-2} + 2^{\frac{n}{2}+2} - 1 = \overbrace{00\cdots01}^{s}0000\overbrace{11\cdots11}^{s}, \; n \equiv 2 \mod 4.$$

Let us consider the addition of r and t. We only consider the case where $2 \mid \frac{n}{2}$. The other two cases are similar. The binary representation of t has 2 runs of 1's out of which one run consists of single 1. This 1 can only contribute a single 1 in any segment of $r + t$. Now we can see that $H(r + t2^{-i})$ is maximized when $i = l - 1$.

$$
\begin{array}{l}
t2^{-(l-1)} = \overbrace{11\cdots10}^{l}\,\overbrace{00\cdots00}\ldots\ldots 0\cdots10\,\overbrace{0001\cdots11}^{l}\ldots\ldots \overbrace{11\cdots11}^{l} \\
r \qquad\quad = 00\cdots01\,00\cdots01\ldots\ldots 0\cdots001\,000\cdots01\ldots\ldots 00\cdots01 \\
+\ \text{-----------------------------} \\
\qquad\qquad \underbrace{11\cdots11}_{l}\,\underbrace{00\cdots01}_{1}\ldots\ldots \underbrace{0\cdots101}_{2}\,\underbrace{010\cdots01}_{2}\ldots\ldots \underbrace{00\cdots00}_{0}
\end{array}
$$

$H(r + t2^{-(l-1)}) = l + \lceil \frac{n}{l} \rceil$.

Therefore $H(r + t2^{-i}) \le l + \lceil \frac{n}{l} \rceil,\ 0 \le i \le n - 1$ $\qquad\qquad\square$

Chosen-Ciphertext Attacks Against MOSQUITO

Antoine Joux[1,3] and Frédéric Muller[2]

[1] DGA

[2] HSBC-France
Frederic.Muller@m4x.org

[3] Université de Versailles-Saint-Quentin, France
Antoine.Joux@m4x.org

Abstract. Self-Synchronizing Stream Ciphers (SSSC) are a particular class of symmetric encryption algorithms, such that the resynchronization is automatic, in case of error during the transmission of the ciphertext.

In this paper, we extend the scope of chosen-ciphertext attacks against SSSC. Previous work in this area include the cryptanalysis of dedicated constructions, like KNOT, HBB or SSS. We go further to break the last standing dedicated design of SSSC, *i.e.* the ECRYPT proposal MOSQUITO. Our attack costs about 2^{70} computation steps, while a 96-bit security level was expected. It also applies to $\Gamma\Upsilon$ (an ancestor of MOSQUITO) therefore the only secure remaining SSSC are block-cipher-based constructions.

1 Introduction

Symmetric encryption algorithms are generally split in two parts : stream ciphers and block ciphers. On the one hand, stream ciphers manipulate the plaintext by **short packets** of data (for instance bit per bit), using a **time-dependent transform**. Typically the output of a PRNG (Pseudo-Random Number Generator) is XORed to the plaintext. On the other hand, block ciphers manipulate the plaintext by **larger packets** of data (typically 128 bits for AES [17]) using a **fixed transform**.

Self-Synchronizing Stream Ciphers (SSSC) are a special primitive : they are often considered as a simple subclass of stream ciphers, but there are also some similarities with block ciphers. Their main property is that, when some error occurs in the transmission of the ciphertext, the decryption algorithm eventually corrects it, after a short sequence of incorrectly decrypted bits. Hence a SSSC achieves the features of an encryption algorithm and resynchronization after transmission errors in one single primitive. They are suitable in situations where encryption is needed, but no additional bandwidth is possible for error-correction (see Maurer's paper for a nice survey on the use of SSSC [14]). In practice, few SSSC's are actually used and it is not clear that such algorithms will be important in the future [3]. However, from a theoretical point of view, it is a very challenging subject, because no dedicated SSSC has yet been built, that resists

M.J.B. Robshaw (Ed.): FSE 2006, LNCS 4047, pp. 390–404, 2006.

all known attacks. Some block-cipher-based constructions are possible, but it would be nice to have a dedicated solution, that is both secure and efficient.

To guarantee the automatic resynchronization, it is a requirement that the encryption of the i-th bit of the plaintext depends only on the key and a **small part of the previous ciphertext bits**. We denote by K the secret key, $\{P_i\}_{i \geq 0}$ the plaintext bits and $\{C_i\}_{i \geq 0}$ the ciphertext bits. Typically, a SSSC is such that :

$$C_i = P_i \oplus F(K, C_{i-1}, \ldots, C_{i-T})$$

for some function F and some integer T which is called the **memory** of the SSSC. It is clear that such an encryption scheme is invertible. Besides, if an error occurs in the ciphertext transmission, the decryption algorithm automatically resynchronizes after transmitting T correct ciphertext bits. It is possible to realize a SSSC with a dedicated design, or with a block cipher in an appropriate mode, like the Cipher FeedBack (CFB) mode [9].

From a bitwise point of view, SSSC operate as stream ciphers, since every plaintext bit can be encrypted separately using the time-dependent transform $x \longrightarrow x \oplus F(K, C_{i-1}, \ldots, C_{i-T})$. On the other hand, looking at the ciphertext, a fixed-transform is applied to each T-bit block, which is more similar to a block cipher. The difference is that a block cipher is an invertible mapping on n-bit inputs, while the F function is a n-to-1 mapping. ¿From the designer's point of view, a dedicated SSSC is often looked at as a special mode of operation for stream ciphers with ciphertext feedback (see HBB [20] among others), while cryptanalysis methods are often related to the field of block cipher.

First, we review the usual design methods for SSSC. Secondly, we review the existing attacks against dedicated SSSC, like KNOT, HBB or SSS. Finally, we extend the scope of these attacks, in order to break the only standing dedicated design, MOSQUITO. The complexity of our attack is 2^{70} steps, while the expected security level was 96 bits. To summarize, we observe that only block-cipher-based constructions remain secure in this area.

2   Design Methods for SSSC

Following the terminology introduced previously, all SSSC operate by

$$C_i = P_i \oplus F(K, C_{i-1}, \ldots, C_{i-T})$$

The difference between the designs lie in the way F is built and in the value of T. Figure 1 presents the general description of a SSSC.

2.1   Block-Cipher-Based Constructions

A typical solution is to start from a block cipher E_K that operates on n bit inputs, using a secret key K. Then, one builds a SSSC with memory of n bits by :

$$F(K, C_{i-1}, \ldots, C_{i-n}) = E_K(C_{i-1}, \ldots, C_{i-n}) \& 1;$$

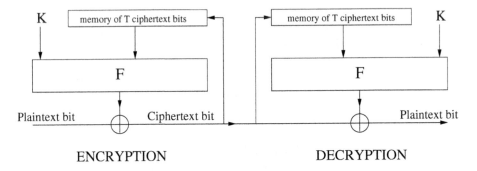

ENCRYPTION   DECRYPTION

Fig. 1. General description of a SSSC

In other terms, E is applied and only the Least Significant Bit (LSB) of the output is kept. This is called the Cipher FeedBack (CFB) mode, with 1-bit feedback [9]. More generally, the CFB mode can be extended to t-bit feedback for any t between 1 and n (see Figure 2), but only the 1-bit version is self-synchronizing. The CFB mode with 1-bit feedback is very inefficient, as one full application of E must be processed to encrypt one plaintext bit. However, there exists more efficient alternatives, like the OCFB mode of operation [1].

There is no generic attack against CFB or OCFB, provided the underlying block cipher is secure. Preneel *et al.* pointed out some possible attacks when one reduces the number of rounds of the block cipher to improve the efficiency of the CFB mode [18]. Actually, there exists a security proof for the CFB mode with n-bit feedback, against chosen plaintext attacks [10]. It is also widely believed that the CFB mode with t-bit feedback is secure for any t, although no generic security proof has yet been published.

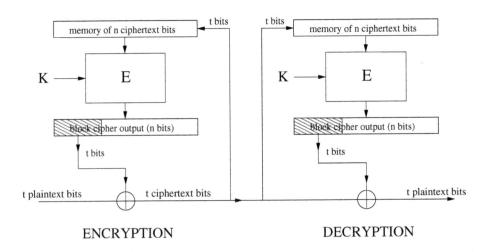

ENCRYPTION   DECRYPTION

Fig. 2. The CFB mode of operation with t-bit feedback

2.2   Dedicated Constructions

Dedicated constructions of SSSC were first introduced by Maurer [14]. Later, Daemen *et al.* reconsidered the design of dedicated SSSC, from a practical perspective [5]. They pointed out that it was not very efficient to recompute the whole function F for each new ciphertext bit introduced. Therefore, they suggested to split the design of the SSSC into two parts :

- An **updatable part** Q which is generally a register with an internal state of size m bits. The state of the register at time i, denoted Q_i, should depend on the last T ciphertext bits, and possibly the key :

$$Q_i = G(K, C_{i-1}, \ldots, C_{i-T})$$

  Then **an update function** is specified, in order to compute efficiently Q_{i+1} from Q_i and C_i. The function G is never actually computed in practice, since the register Q is generally initialized with a m-bit constant, and then the update function is applied as many times as necessary. Note that the memory T and the register length m are not necessary equal, however it is necessary that $m \geq T$, in order to store enough information in the register.
- An **output filter** f which takes the state of the register Q_i and computes the output bit, :

$$F(K, C_{i-1}, \ldots, C_{i-T}) = f(K, Q_i)$$

This filter f often looks like a "light" block cipher.

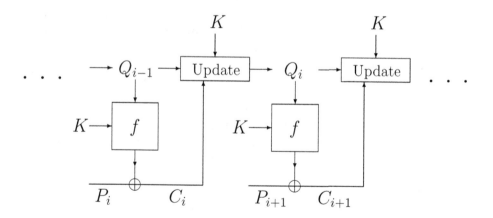

Fig. 3. General Framework of Dedicated SSSC

Figure 3 represents the general framework of such constructions. To guarantee the self-synchronization, it is necessary that the "old" ciphertext bits are "forgotten" after T updates. This constraint is often satisfied using a shift register-oriented design (this is the case for SSS and for the KNOT-MOSQUITO family).

Concerning the role of the secret key, at least f or G must use K as input, but it is rare that both do. Block-cipher-based constructions are the limit case of this framework, with a very simple (unkeyed) shift register as the updatable part, and a complicated (keyed) block cipher as the output filter. Dedicated constructions try to reach a better trade-off with a more complicated (non-linear) updatable part, but a lighter output filter.

2.3   The KNOT-MOSQUITO Family

An interesting family of dedicated SSSC is the "KNOT-MOSQUITO family". KNOT was first proposed as an example of efficient dedicated design by Daemen et al. in their paper of 1992 that dealt with SSSC in a more general perspective [5]. In 1995, Daemen discovered a statistical imbalance in the output of KNOT [4], which was further investigated later [12]. This weakness of the output filter motivated a switch to a tweaked version of KNOT called $\Gamma\Upsilon$ which was proposed in Daemen's PhD thesis [4].

In 2003, a new weakness of KNOT was pointed out by Joux and Muller, which allows to recover the secret key with complexity of 2^{62} steps [12]. Their idea is to apply methods from differential cryptanalysis in order to detect internal collisions in the updatable register. This attack does not apply against $\Gamma\Upsilon$. Recently, as part of the eSTREAM competition for stream ciphers [8] launched by the European project ECRYPT, Daemen and Kitsos proposed a new self-synchronizing stream cipher called MOSQUITO [6]. It is a close variant of $\Gamma\Upsilon$ which was designed to avoid the security problems of KNOT.

All algorithms in this family follow the framework introduced in Section 2.2, with a memory of $T = 96$ bits and an updatable register Q of size $m = 128$ bits, which is non-linear and key-dependent. The type of register used in this family are also called **Conditional Complementing Shift Registers** (CCSR). See [5] or [6] for more details. The output filter is an unkeyed iterative construction, which gradually reduces the state from 128 bits to 1 bit, after 8 rounds[1]. The difference between the 3 algorithms in this family (KNOT, $\Gamma\Upsilon$ and MOSQUITO) lies in the way Q is updated, and in the details of the 8 rounds of the filter. All algorithms in the family are designed to use of **a 96-bit secret key**.

To summarize, the KNOT-MOSQUITO family is an interesting family of dedicated SSSC, since it is not a "tweaked" mode of operation of a stream cipher, like SSS or HBB. Therefore, it is very interesting from a research perspective. However, all the algorithms of the family are subject to differential attacks, as pointed out in Section 4.

2.4   SSS

SSS (Self-Synchronizing Sober) is a new dedicated SSSC, submitted to the eS-TREAM competition [19] by Rose et al. It belongs to the SOBER family of

---

[1] The authors mention 7 rounds, but it depends on whether the final XOR is counted as an 8-th round or not.

stream ciphers [11]. This family uses Linear Feedback Shift Registers (LFSR) operating in $GF(2^n)$ with $n = 8, 16$ or 32. An output filter is applied to some cells from the LFSR, in order to extract a pseudo-random output. This generally relies on n-bit instructions as well, in order to obtain a software-efficient design. Additional functionalities (like authenticated encryption or self-synchronization) have been suggested in tweaked versions of the SOBER family. The general idea is to add an auxiliary input in the LFSR to introduce either the plaintext (for the integrity mode) or the ciphertext (for the self-synchronizing mode).

While few cryptanalysis results are known against the algorithms from the SOBER family in encryption mode, it is well known that tweaking a stream cipher in order to add integrity or self-synchronization completely modifies the cryptanalysis scenario [16]. Indeed, an attacker potentially gains the ability to control the content of the LFSR. A devastating attack against the integrity mode of SOBER-128 was described in 2004 by Watanabe et al. [24].

SSS was also broken by Daemen, Lano and Preneel, shortly after its publication [7]. They described a chosen ciphertext attack which allows to retrieve a key-dependent secret table. According to the designers of SSS, such attacks fall outside the threat model, but Daemen et al. argued that chosen ciphertext attacks are practical and that they are the standard way to evaluate SSSC.

2.5 HBB

HBB (Hiji-bij-bij) is a software-oriented stream cipher, proposed in 2003 by Sarkar [20]. It is a new construction, where the usual LFSRs have been replaced by cellular automata. In addition, an output filter operating on a 128-bit output is used. This filter has some similarities with a block cipher design (use of S-boxes and linear diffusion layers). The Basic (B) mode of operation of HBB is a traditional stream cipher, for which some attacks faster than brute-force have been published in 2005 [13,15].

In addition, a Self-Synchronizing (SS) mode of operation for HBB was also proposed by Sarkar. It is based on a slight modification of the cipher, where the cellular automata is filled with ciphertext bits, instead of being evaluating autonomously. While the primitives are unchanged, this modification completely changes the cryptanalysis scenario. Joux and Muller showed a devastating key-recovery attack against this SS mode, which requires only 2^{12} bits of chosen ciphertext [13]. Basically, they exploited the weak differential properties of the output filter.

2.6 Other Proposals

Another proposition of dedicated SSSC was made by Arnault and Berger [2], as part of their work on Feedback with Carry Shift Registers (FCSR). Their proposal was later broken [25] using a chosen ciphertext attack.

Like for SSS and HBB, it appears that building a SSSC by tweaking a conventional stream cipher is not a good idea. Many devastating attacks have been published : as soon as one considers chosen ciphertext attacks, the cryptanalyst no longer looks at the SSSC as a stream cipher, but instead he analyzes the G function directly (see Section 2.2). The properties of G can be considered under chosen input attacks : either differential properties are used (case of HBB or KNOT) or the possibility to guess and identify individually some portions of the key (case of SSS).

3   The Cryptanalysis Framework Against SSSC

3.1   Chosen Ciphertext Attacks Against SSSC

We first observe that the "natural way" to cryptanalyze a SSSC is by considering chosen ciphertext attacks. This is natural from a theoretical point of view, but also from a practical point of view, as already pointed out by Daemen *et al.* [7].

Theoretically, we are comparing dedicated SSSC with block ciphers in CFB mode. Block ciphers are built to resist both chosen plaintext and chosen ciphertext attacks, so it would be unfair to compare two algorithms that do not take into account the same attack scenarios. In addition, the existence of chosen ciphertext attacks generally reveals design weaknesses that could later be extend to much more realistic scenarios.

Moreover, there are still some scenarios where an attacker could have access to a decryption oracle. This is not necessarily much more difficult than accessing an encryption oracle. For instance, one could consider an active attacker that modifies the communication channel (in order to obtain the ciphertext he wants) and then observes the result of the decryption.

3.2   Chosen Plaintext Attacks Against SSSC

In some attack scenarios against SSSC, chosen ciphertext attacks can even be turned into chosen plaintext attacks. Assume that the attacker needs to obtain a chosen ciphertext sequence equal to $(C_1, \ldots, C_i)$ in order to attack a SSSC. He can achieve it with chosen plaintext only, assuming an **adaptive chosen plaintext attack**.

In such a scenario, we assume that the attacker can reset the encryption algorithm to its initial state, at any point[2]. Then, he tries both value of the bit P_1 and resets the algorithm if the value of C_1 is not what he wants. Then, the process is repeated as long as necessary. On average, this requires about $i/2$ resets, where i is the length of the needed ciphertext sequence, $(C_1, \ldots, C_i)$.

This gives an example of a classical scenario, where there exists a bridge between chosen plaintext and chosen ciphertext attacks against SSSC. In the following, we focus on chosen ciphertext attacks.

---

[2] If he has access to several copies of the algorithm using the same key, a similar attack applies. The idea is to throw away a copy of the algorithm and use a new copy, instead of doing the reset.

4   Differential Attacks Against MOSQUITO

4.1   A Short Overview of the Design

Describing in details the MOSQUITO stream cipher is a long task, since a large number of equations should be stated. We invite the reader to check the details in the original specification [6]. Here, we only give a short overview of the design.

The register Q

As mentioned in Section 2.3, MOSQUITO uses a non-linear shift register as the updatable part : Q has a length of 128 bits. Its content is noted in the later as $(x_1, \ldots, x_{128})$. At time i, the content of Q is updated by introducing the i-th ciphertext bit, and also the secret key. More precisely, each bit x_j is updated by applying a simple boolean function $h_j : \{0,1\}^4 \longrightarrow \{0,1\}$, whose inputs are chosen among $(x_1, \ldots, x_{j-1})$, the key bits, and the introduced ciphertext bit.

One can observe that the propagation goes always in the direction of increasing indexes. This guarantess that the influence of "old" ciphertext bits eventually vanishes, which makes the resynchronization possible.

Actually, one could also express directly the content of register Q at time i has a function of the last 96 ciphertext bits and the key. However, this would lead to a very complicated expression. It is much clearer to describe Q by the update equations (*i.e.* the h_j). See [6] for the expression of all these equations.

The output filter f

After the update of register Q, the $(i+1)$-th ciphertext bit is obtained by XORing the $(i+1)$-th plaintext bit to the output of the filter f. This filter is a fixed, unkeyed transform applied to the state of the register Q. Therefore it is a boolean function from 128 bits to 1 bits.

In order to be computed efficiently, f can be written as the composition of 8 simple transforms (also called rounds) applied to internal states of decreasing size. For instance, the first round, noted φ_1, is applied to the content of Q (*i.e.* 128 bits), but its output size is only 53 bits. Therefore φ_1 is a transform from $\{0,1\}^{128}$ to $\{0,1\}^{53}$. After the 8 rounds, the final output is simply one bit. Like for the update of the register Q, each round can be represented by a small set of boolean functions from 4 bits to 1 bit. The reader should refer to [6] for the expression of all these equations.

4.2   Overview of the New Attack

Our attack is similar to a differential cryptanalysis of a block cipher : we find differential characteristics for both parts of the cipher[3] :

- First, we find a differential characteristic for the output filter f. In the case of MOSQUITO, f is an unkeyed 8-round transform. We are looking for a 128-bit difference Δ such that

---

[3] Regarding the mathematical tools, there are also relations with linear cryptanalysis since we are interested in small statistical deviations called bias, as it will appear later.

$$f(Q_i \oplus \Delta) \oplus f(Q_i)$$

is equal to 0 with probability $p = 0.5 \cdot (1+\varepsilon)$ and $|\varepsilon|$ as large as possible. In the case of KNOT, Joux and Muller [12] exploited $\Delta = 0$ which implied that $\varepsilon = 1$. Such collisions on Q could be obtained from two distinct ciphertexts (*i.e.* the function G was not injective). In the newer algorithms of the family ($\Gamma\Upsilon$ and MOSQUITO), this attack has been countered by making G injective. Therefore, the difference $\Delta = 0$ is not reachable, unless the ciphertexts are equal. We **extend the scope of Joux-Muller's attack to non-zero differences**.

– Secondly, we describe how to build two ciphertexts such that the two values of the state Q_i differ exactly by the previous difference Δ. This step will require to guess some portion of the key. We consider ε^{-2} such pairs, in order to detect an imbalance on the outputs of the filter f.

4.3   Differential Characteristic for f

The filter f of MOSQUITO looks like a block cipher with 8 simple rounds, applied successively. Our analysis focuses on the differential properties of the first round transform, φ_1. Since its output size is smaller than its input size, it is not injective. Hence, we can expect to find an input difference Δ, such that, after the first round

$$\varphi_1(x) = \varphi_1(x \oplus \Delta)$$

with good probability. If such a collision occurs after applying φ_1, this will imply the equality of the outputs of f, since no new input is introduced in the 7 following rounds. As mentioned previously, each output bit of φ_1 is computed using a simple boolean function, written as :

$$\tau : (a, b, c, d) \longrightarrow a \oplus b \oplus c \cdot (d \oplus 1) \oplus 1$$

applied to 4 among the input bits of φ_1, noted $x = (x_1, \ldots, x_{128})$. Since $4 \cdot 53 = 212 < 2 \cdot 128 = 256$, we know that at least $256 - 212 = 44$ input bits are processed only once by the function τ. A quick analysis shows that this observation concerns :

$$(x_1, \ldots x_{17}), (x_{54}, \ldots, x_{60}), (x_{71}, \ldots, x_{75}), (x_{114}, \ldots, x_{128})$$

which are all used only once in φ_1. Some of them are only used as 3-rd or 4-th input bit of τ, so if we flip them, the output of τ may be unchanged. This observation concerns the set of bits :

$$S = \{x_1, \ldots x_{17}, x_{71}, \ldots, x_{75}\}$$

If we flip exactly one bit in this set S, the output of φ_1 is unchanged with probability 0.5. Consequently, the output of f is also unchanged with probability 0.75 (even when φ_1 is changed, the output bit can still be the same with probability 0.5). Hence, we found **differential characteristics on F with bias $\varepsilon = 0.5$ and such that the input difference is non-zero**.

4.4   Advanced Differences

An advanced method consists in flipping two well-chosen input bits of φ_1. As an example, consider the output bit number 29 and 33 of φ_1. They can be computed by

$$x_{81} \oplus x_{65} \oplus x_{66} \cdot (x_{48} \oplus 1) \oplus 1$$
$$x_{80} \oplus x_{66} \oplus x_{65} \cdot (x_{49} \oplus 1) \oplus 1$$

The bits x_{65} and x_{66} are used nowhere else in φ_1. If an attacker flips these two bits simultaneously, there is a probability 0.25 that the output of φ_1 is unaffected (the condition is that $x_{48} = x_{49} = 0$). Hence the outputs of F are equal with probability 0.625, and we have a differential characteristic, with bias $\varepsilon = 0.25$ and such that the input difference is still non-zero (only the bits x_{65} and x_{66} are flipped). There exists other strategies to flip 2 or more bits, while keeping a reasonable bias. However, the difference we have just described will turn out to be the most useful.

4.5   Analysis of the Updatable Part

Our goal is now to build two ciphertexts sequences that map to two register states Q_i and Q_i' such that $\Delta = Q_i \oplus Q_i'$ is one of the "useful" differences identified in the previous section. We may need to repeat this process ε^{-2} times in order to actually observe the predicted bias. Finding 2 appropriate ciphertext sequences can be done using **an exhaustive search on an appropriate portion of the key**.

4.6   Using the Difference on x_{17} Only

We first describe an attack that targets the difference Δ_{17}, which is defined as difference equal 0 on every input bit of f, except x_{17}. At first glance, one could envisage to work with the difference Δ_1 which has the same differential behavior regarding f, however, the resulting complexity of the attack would be worse, as it will appear below.

We focus on the updatable register, to obtain register states that differ from Δ_{17} The state at time i is expressed as :

$$Q_i = G(K, C_{i-1}, \ldots, C_{i-96})$$

We denote the key bits by $K = (k_1, \ldots, k_{96})$. The crucial observation, that we refer to as **observation $\mathcal{O}$** is that, for $1 \leq t \leq 88$, the value of bit x_t at time i depends only on the key bits $k_1, \ldots, k_t$ and the ciphertext bits $C_{i-1}, \ldots, C_{i-t}$.

The basic idea of our attack is to **guess only the 79 least significant bits of the key**. This guess splits the register Q into a left part for which we always know the internal values (thanks to observation $\mathcal{O}$) and a right part which generally remains unknown.

At any time, we can **control the differential behavior on the left part**. The only way we can have information about the right part is by letting the "natural" propagation from left to right in Q bring a zero-difference on the right part (resynchronization effect). We combine these ideas in the following. More precisely, our attack proceeds in three steps, represented in Figure 4 :

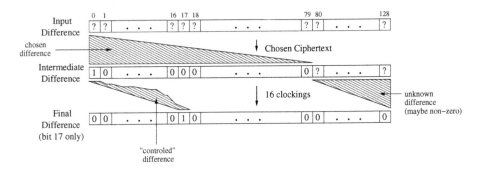

Fig. 4. The attack against MOSQUITO using Δ_{17}

- First, we guess the key bits $k_1, \ldots, k_{79}$. Because of the observation $\mathcal{O}$, we can easily determine the value of $V = (x_1, \ldots, x_{79})$ for any introduced ciphertext.
- Secondly, we pick at random two ciphertext sequences that yield the same V, except that the bit x_1 is flipped. This step is easy to achieve in practice : start from an arbitrary ciphertext sequence and compute the corresponding value of V. Then, flip bit x_1 and clock 79 times backward the register Q (see Figure 3). So we find which ciphertext sequence should be introduced in order to reach this modified state.
- Introduce 16 additional ciphertext bits and clock 16 times the register. The difference will always propagate from position 1 to position 17 during these 16 clockings. But, observation $\mathcal{O}$ guarantees that there is no difference afterward on bits $x_1, \ldots, x_{16}$. Moreover, we can **control the difference during these 16 extra clockings**, in order for the difference not to propagate further than position 17.

There are 2 ways to "control" the difference during these 16 extra clockings. The first method consists in specifying a differential path and writing down the boolean equations for this propagation to be satisfied. Here, about 20 such conditions are needed, which remains reasonable in practice. A second method is to test random ciphertexts, until a "good" one is found. Two tricks make this idea quite efficient : First, we use "early-abort" in order to quickly eliminate the "bad" candidates. Secondly, we observe that this propagation is almost independent on our guess of key bits, $k_{18}, \ldots, k_{79}$, so we can tell if a ciphertext is good or not, before guessing the whole key.

Both methods have been tested on a standard PC and are very efficient. The bottleneck of this attack is the 2-nd step, where we need to repeat 2^{79} the execution of 79 clockings of Q. For comparison, the basic step in an exhaustive search requires 96 clockings of Q. Overall, we need at least $\varepsilon^{-2} = 4$ such plaintext pairs, in order to detect the predicted bias. So we estimate the complexity to about 2^{81} steps (compared to 2^{96} for an exhaustive search).

4.7   Using the Difference on x_{65} and x_{66}

The idea of this attack is essentially the same as the previous one, except that we guess only 66 key bits. Therefore, we can predict only the state of bit number 1 to 66 of the state Q_i. Then, we use 64 extra clockings and we hope to control the difference, during these advances in order to obtain finally a difference on cells number 65 and 66 only.

   The approach based on specifying a differential path and writing the corresponding boolean conditions seems very painful. Therefore, we adopted a purely statistical approach : we tested random ciphertext sequences and used "early-abort" : as soon as the difference starts to spread to several bits, we discard the tested value.

   Our implementation is very efficient and, for a given key guess, finds instantly two satisfying ciphertext sequences, *i.e.* such that the final difference is only located on the bits 65 and 66 of the state of Q. This needs to be repeated about ε^{-2} times, so the complexity of our attack is about

$$2^{66} \times 4^2 = 2^{70}$$

steps, compared to 2^{96} steps for an exhaustive search.

4.8   Some Comments

There are several interesting comments to make about these two attacks. First the specific tricks needed to control the difference by specifying some set of sufficient boolean conditions are related to the techniques developed against hash functions like MD5 or SHA-1 by Wang *et al.* [21,22,23]. There could be further improvements to the cryptanalysis of the KNOT-MOSQUITO family, by looking in that area of research.

   Secondly, our attacks also applies against $\Gamma\Upsilon$, since the underlying primitives (register Q and filter f) are the same. Concerning KNOT, there are slight differences in the primitives, but a similar attack should also apply. However the result by Joux and Muller, exploiting internal collisions for Q_i are slightly more efficient [12]. Actually, our attack can be viewed a generalization of this previous result.

   We used 4 pairs of ciphertext sequences in Section 4.6 and 8 pairs of ciphertexts in Section 4.7. These are optimistic figures. It is well known that a small security margin is generally needed above ε^{-2} to actually detect a bias ε, with a small false alarm probability. It is not necessary that this probability is 2^{-96} here. Indeed, we have already guessed a large portion of the key, so 20 or 30 ciphertext pairs should probably be sufficient. Application of statistical tools is needed to evaluate exactly how many ciphertexts are needed.

   Finally, an important point is that the data complexity of our attacks is very limited since we build the ciphertext sequences in an off-line computation

Table 1. Summary of cryptanalysis results against some SSSC

Type of Attack	Target	Complexity	Data
Distinguisher [4]	KNOT	2^{18}	2^{18}
Key Recovery [12]	KNOT	2^{69}	$2^{36.6}$
Key Recovery [12]	KNOT	2^{62}	$2^{38.6}$
Key Recovery [7]	SSS	10 seconds	10 kBytes
Key Recovery [13]	HBB	2^{12}	2^{12}
Key Recovery (this paper)	MOSQUITO	2^{81}	2^{10}
Key Recovery (this paper)	MOSQUITO	2^{70}	2^{11}
Exhaustive Search	MOSQUITO	2^{96}	96 bits

(described in Section 4.6 and Section 4.7). The data complexity is roughly of $4 \times 96 \times 2 \simeq 2^{10}$ ciphertext bits in the first attack and $8 \times 96 \times 2 \simeq 2^{11}$ for the second attack.

5   Conclusion

All the dedicated Self-Synchronizing Stream Ciphers (SSSC) of the KNOT-MOSQUITO family are subject to differential chosen ciphertext attacks. Our results, combined with previous results on HBB, KNOT and SSS show that it is extremely difficult to design a SSSC resistant against chosen-ciphertext attacks.

Some designers argued [19] that chosen ciphertext attacks should fall outside the security model for SSSC. However, they are taken into account in block cipher-based constructions, and could be more realistic than expected. Even the authors of MOSQUITO [6] stressed out that dedicated SSSC should resist chosen ciphertext attacks.

Since no dedicated SSSC still stands, we believe that block-cipher-based constructions should now be favored if one needs a self-synchronizing algorithm, for practical purpose. An interesting direction would also be to see how much one can "lighten" an existing block cipher (AES for instance), in order to obtain a SSSC faster than the CFB (or OCFB) mode [1,9].

References

1. A. Alkassar, A. Geraldy, B. Pfitzmann, and A.-R. Sadeghi. Optimized Self-Synchronizing Mode of Operation. In M. Matsui, editor, *Fast Software Encryption – 2001*, volume 2355 of *Lectures Notes in Computer Science*, pages 78–91. Springer, 2002.
2. F. Arnault and T. Berger. A new class of stream ciphers combining LFSR and FCSR architectures. In A. Menezes and P. Sarkar, editors, *Progress in Cryptology – INDOCRYPT'02*, volume 2551 of *Lectures Notes in Computer Science*, pages 22–33. Springer, 2002.

3. S. Babbage. Stream Ciphers: What Does the Industry Want ? In *State of the Art of Stream Ciphers* workshop (SASC'04), 2004.
4. J. Daemen. *Cipher and Hash Function Design. Strategies based on Linear and Differential Cryptanalysis.* PhD thesis, Katholieke Universiteit Leuven, march 1995. Chapter 9.
5. J. Daemen, R. Govaerts, and J. Vandewalle. A Practical Approach to the Design of High Speed Self-Synchronizing Stream Ciphers. In *Singapore ICCS/ISITA '92*, pages 279–283. IEEE, 1992.
6. J. Daemen and P. Kitsos. Submission to ECRYPT call for stream ciphers: the self-synchronizing stream cipher Mosquito. eSTREAM, ECRYPT Stream Cipher Project, Report 2005/018, 2005. http://www.ecrypt.eu.org/stream.
7. J. Daemen, J. Lano, and B. Preneel. Chosen Ciphertext Attack on SSS. eSTREAM, ECRYPT Stream Cipher Project, Report 2005/044, 2005. http://www.ecrypt.eu.org/stream.
8. eSTREAM - The ECRYPT Stream Cipher Project http://www.ecrypt.eu.org/stream/.
9. FIPS PUB 81. *DES Modes of Operation*, 1980.
10. P-A. Fouque, G. Martinet, and G. Poupard. Practical Symmetric On-Line Encryption. In T. Johansson, editor, *Fast Software Encryption – 2003*, volume 2887 of *Lectures Notes in Computer Science*, pages 362–375. Springer, 2003.
11. P. Hawkes and G. Rose. Primitive Specification and Supporting Documentation for SOBER-t32. In *First Open NESSIE Workshop*, 2000. Submission to NESSIE.
12. A. Joux and F. Muller. Loosening the KNOT. In T. Johansson, editor, *Fast Software Encryption – 2003*, volume 2887 of *Lectures Notes in Computer Science*, pages 87–99. Springer, 2003.
13. A. Joux and F. Muller. Two Attacks Against the HBB Stream Cipher. In H. Gilbert and H. Handschuh, editors, *Fast Software Encryption – 2005*, volume 3557 of *Lectures Notes in Computer Science*, pages 330–341. Springer, 2005.
14. U. Maurer. New Approaches to the Design of Self-Synchronizing Stream Ciphers. In D.W. Davies, editor, *Advances in Cryptology – Eurocrypt'91*, volume 547 of *Lectures Notes in Computer Science*, pages 458–471. Springer, 1991.
15. J. Mitra. A Near-Practical Attack against B mode of HBB. In *Advances in Cryptology - Asiacrypt'05*, 2005. To appear.
16. F. Muller. Differential Attacks and Stream Ciphers. In *State of the Art in Stream Ciphers*. ECRYPT Network of Excellence in Cryptology, 2004. Workshop Record.
17. National Institute of Standards and Technology (NIST). Advanced Encryption Standard (AES) FIPS Publication 197, November 2001. Available at http://csrc.nist.gov/publications/fips/fips197/fips-197.pdf.
18. B. Preneel, M. Nuttin, R. Rijmen, and J. Buelens. Cryptanalysis of the CFB Mode of the DES with a Reduced Number of Rounds. In D.R. Stinson, editor, *Advances in Cryptology – Crypto'93*, volume 773 of *Lectures Notes in Computer Science*. Springer, 1993.
19. G. Rose, P. Hawkes, G. Paddon, and M. Wiggers de Vries. Primitive Specifications for SSS. eSTREAM, ECRYPT Stream Cipher Project, Report 2005/028, 2005. http://www.ecrypt.eu.org/stream.
20. P. Sarkar. Hiji-Bij-Bij : A New Stream Cipher with a Self-Synchronizing Mode of Operation. In T. Johansson and S. Maitra, editors, *Progress in Cryptology – INDOCRYPT'03*, volume 2904 of *Lectures Notes in Computer Science*, pages 36–51. Springer, 2003.

21. X. Wang, Y. Yin, and H. Yu. Finding Collisions in the Full SHA1. In V. Shoup, editor, *Advances in Cryptology – Crypto'05*, volume 3621 of *Lectures Notes in Computer Science*, pages 17–36. Springer, 2005.

22. X. Wang and H. Yu. How to Break MD5 and Other Hash Functions. In R. Cramer, editor, *Advances in Cryptology – Eurocrypt'05*, volume 3494 of *Lectures Notes in Computer Science*, pages 19–35. Springer, 2005.

23. X. Wang, H. Yu, and Y. Yin. Efficient Collision Search Attacks on SHA0. In V. Shoup, editor, *Advances in Cryptology – Crypto'05*, volume 3621 of *Lectures Notes in Computer Science*, pages 1–16. Springer, 2005.

24. D. Watanabe and S. Furuya. A MAC Forgery Attack on SOBER-128. In B. Roy and W. Meier, editors, *Fast Software Encryption – 2004*, volume 3017 of *Lectures Notes in Computer Science*, pages 472–482. Springer, 2004.

25. B. Zhang, H. Wu, D. Feng, and F. Bao. Chosen Ciphertext Attack on a New Class of Self-Synchronizing Stream Ciphers. In A. Canteaut and K. Viswanathan, editors, *Progress in Cryptology – INDOCRYPT'04*, volume 3348 of *Lectures Notes in Computer Science*, pages 73–83. Springer, 2004.

Distinguishing Attacks
on the Stream Cipher Py*

Souradyuti Paul[1], Bart Preneel[1], and Gautham Sekar[1,2]

[1] Katholieke Universiteit Leuven, Dept. ESAT/COSIC,
Kasteelpark Arenberg 10,
B–3001, Leuven-Heverlee, Belgium
[2] Dept. of Electronics and Instrumentation, Dept. of Physics
Birla Institute of Technology and Science, Pilani, India
{souradyuti.paul, bart.preneel}@esat.kuleuven.be,
gautham.sekar@gmail.com

Abstract. The stream cipher Py designed by Biham and Seberry is a submission to the ECRYPT stream cipher competition. The cipher is based on two large arrays (one is 256 bytes and the other is 1040 bytes) and it is designed for high speed software applications (Py is more than 2.5 times faster than the RC4 on Pentium III). The paper shows a statistical bias in the distribution of its output-words at the 1st and 3rd rounds. Exploiting this weakness, a distinguisher with advantage greater than 50% is constructed that requires $2^{84.7}$ randomly chosen key/IV's and the first 24 output bytes for each key. The running time and the data required by the distinguisher are $t_{ini} \cdot 2^{84.7}$ and $2^{89.2}$ respectively (t_{ini} denotes the running time of the key/IV setup). We further show that the data requirement can be reduced by a factor of about 3 with a distinguisher that considers outputs of later rounds. In such case the running time is reduced to $t_r \cdot 2^{84.7}$ (t_r denotes the time for a single round of Py). The Py specification allows a 256-bit key and a keystream of 2^{64} bytes per key/IV. As an ideally secure stream cipher with the above specifications should be able to resist the attacks described before, our results constitute an academic break of Py. In addition we have identified several biases among pairs of bits; it seems possible to combine all the biases to build more efficient distinguishers.

1 Introduction

The cipher Py, designed by Biham and Seberry [3], was submitted to the ECRYPT project [7] as a candidate for Profile 1 which covers software based stream ciphers suitable for high-speed applications. In the last couple of years a growing interest has been noticed among cryptographers to design fast and

---

* This work was supported in part by the Concerted Research Action (GOA) Mefisto 2000/06 and Ambiorix 2005/11 of the Flemish Government and in part by the European Commission through the IST Programme under Contract IST-2002-507932 ECRYPT.

M.J.B. Robshaw (Ed.): FSE 2006, LNCS 4047, pp. 405–421, 2006.

secure stream ciphers because of weaknesses being found in many *de facto* standards such as RC4 and also due to the failure of the NESSIE project [12] to find a stream cipher that met its very stringent security requirements. The current stream cipher, namely Py, is one of the attempts in this direction.

Py is the most recent addition to the class of stream ciphers whose design principles are motivated by that of RC4 (see [9,10,13,14,15]). Like RC4, Py also uses the technique of random shuffle to update the internal state. In addition, Py uses a new technique of rotating all array elements in every round with a minimal running time. The high performance (it is 2.5 times faster than the RC4 on Pentium III) and its apparent security make this cipher very attractive for selection to the Profile 1 of the ECRYPT project.

This paper identifies several biased pairs of output bits of Py at rounds t and $t + 2$ (where $t > 0$). The weaknesses originate from the non-uniformity of the distributions of carry bits in modular addition used in Py. Using those biases, we have constructed a class of distinguishers. We show that the best of them works successfully with $2^{84.7}$ randomly chosen key/IV's, the first 24 bytes for each key (i.e., a total of $2^{89.2}$ bytes) and running time $t_{ini} \cdot 2^{84.7}$ where t_{ini} is the running time of the key/IV setup of Py. We also show that a simple adjustment to the above distinguisher reduces the number of key/IV's, the data complexity and the running time to $2^{28.7}$, a total of $2^{87.7}$ bytes and $t_r \cdot 2^{84.7}$ respectively, where t_r is the running time of a single round of Py. Note that the allowable key-size and keystream length of Py are 256 bits and 2^{64} bytes respectively. Therefore, these results imply that – even if our attack has a larger total complexity – Py fails to provide the security level expected from an ideal stream cipher with the parameter sizes of Py. Therefore, we believe that our results present a theoretical break of the cipher; see Sect. 9 for an elaborate discussion on this issue. It is important to note that the weaknesses of Py which are described in this paper, still cannot be implemented in practice in view of the its high time complexity. However, the individual distinguishers open the possibility to combine them in order to generate more efficient distinguishers.

2   Description of Py

Py is a synchronous stream cipher which normally uses a 32-byte key (however, the key can be of any size from 1 byte to 256 bytes) and a 16-byte initial value or IV (IV can also be of any size from 1 byte to 64 bytes). The allowable keystream length per key/IV is 2^{64} bytes. Py works in three phases – a key setup algorithm, an IV setup algorithm and a round function which generates two output-words (each output-word is 4 bytes long). The internal state of Py contains two S-boxes Y, P and a variable s. Y contains 260 elements each of which is 32 bits long. The elements of Y are indexed by [-3, -2,..., 256]. P is a permutation of the elements of $\{0, ..., 255\}$. The main feature of the stream cipher Py is that the S-boxes are updated like 'rolling arrays' [3]. The technique of 'rolling arrays' means that, in each round of Py, *(i)* one or two elements of the S-boxes are updated (line 1 and 7 of Algorithm 1) and *(ii)* all the elements are cyclically rotated by one position toward the left (line 2 and 8 of Algorithm 1). In our analysis, we have assumed that, after the key/IV setup, Y, P and the variable s are uniformly distributed and independent. Under this assumption we analyzed the round function of Py (or Pseudorandom Bit Generation Algorithm) which is described in Algorithm 1. See [3] for a detailed description of the key/IV setup algorithms.

The inputs to Algorithm 1 are $Y[-3, ..., 256]$, $P[0, ..., 255]$ and s, which are obtained after the key/IV setup. Lines 1 and 2 describe how P is updated and rotated. In the update stage, the 0th element of P is swapped with another element in P, which is accessed indirectly, using $Y[185]$. The next step involves a cyclic rotation by one position, of the elements in P. This implies that the entry in $P[0]$ becomes the entry in $P[255]$ in the next round and the entry in $P[i]$ becomes the entry in $P[i-1]$ ($\forall i \in \{1, 2, ..., 255\}$). Lines 3 and 4 of Algorithm 1 indicate how s is updated and its elements rotated. Here, the '$ROTL32(s, x)$' function implies a cyclic left rotation of s by x bit-positions. The output-words (each 32-bit) are generated in lines 5 and 6. The last two lines of the algorithm explain the update and rotation of the elements of Y. The rotation of Y is carried in the same manner as the rotation of P.

Algorithm 1. Single Round of Py

Require: $Y[-3, ..., 256]$, $P[0, ..., 255]$, a 32-bit variable s
Ensure: 64-bit random output
    /*Update and rotate P*/
1: swap $(P[0], P[Y[185]\&255])$;
2: rotate (P);
    /* Update s*/
3: $s+ = Y[P[72]] - Y[P[239]]$;
4: $s = ROTL32(s, ((P[116] + 18)\&31))$;
    /* Output 8 bytes (least significant byte first)*/
5: output $((ROTL32(s, 25) \oplus Y[256]) + Y[P[26]])$;
6: output $((\quad\quad s \quad\quad \oplus Y[-1]) + Y[P[208]])$;
    /* Update and rotate Y*/
7: $Y[-3] = (ROTL32(s, 14) \oplus Y[-3]) + Y[P[153]]$;
8: rotate(Y);

2.1 Notation and Convention

As Py uses different types of internal and external states (e.g. integer arrays, 32-bit integer) and they are updated every round, it is important to denote all the states and rounds in a simple but consistent way. In every round of Py, the S-box P and the variable s are updated before the output generation (see Algorithm 1). The other S-box, namely Y, is updated after the output generation.

1. In the beginning of any round i, the components of the internal state are denoted by P_{i-1}, s_{i-1} but Y_i.
2. At the end of any round i, the internal state is updated to P_i, s_i and Y_{i+1}. (If the above two conventions are followed, we have P_i, s_i and Y_i in the

formulas for the generation of the output-words in round i (line 5 and 6 of Algorithm 1)).

3. The nth element of the arrays Y_i and P_i, are denoted by $Y_i[n]$ and $P_i[n]$ respectively. The jth bit of $Y_i[n]$, $P_i[n]$ and s_i are denoted by $Y_i[n]_{(j)}$, $P_i[n]_{(j)}$ and $s_{i(j)}$ respectively (following the convention that the least significant bit is the 0th bit).

4. The output-words generated in line 5 and line 6 of Algorithm 1 are referred to as the '1st output-word' and the '2nd output-word' respectively.

5. $O_{l,m}$ denotes the lth ($l \in \{1, 2\}$) output-word generated in the mth round of Py. $O_{l,m(j)}$ denotes the jth bit of $O_{l,m}$. For example, $O_{1,3(5)}$ denotes the 5th bit of the 1st output-word in round 3.

6. The '+' operator denotes *addition modulo* 2^{32} except when it is used to increment elements of P (particularly in expressions of the form $P_i[n] = P_j[m] + 1$, where '+' denotes *addition over* $\mathbb{Z}$). Similarly, '-' and '$\oplus$' denote *subtraction modulo* 2^{32} and bitwise *exclusive-or*.

7. $P[A]$ denotes the probability of occurrence of the event A.

8. $[a, b]$ denotes the set of all integers between a and b including both.

9. A Pseudorandom Bit Generator will be denoted by PRBG.

2.2   Assumption

We assume that the key setup and the IV setup algorithms of Py are perfect, i.e., after the execution of them, the permutation P, the elements of Y and the s are uniformly distributed and independent. When we are interested in the analysis of the mixing of bits of the internal state by the PRBG, the above assumption is reasonable, particularly when it is difficult to derive any relation between inputs and outputs of the key/IV setup algorithm. Apart from that the assumption is in agreement with a claim made in Sect. 6.4 of [3] that the key/IV setup leaks no statistical information on the internal state.

3   Motivational Observation

Our main observation is that, if certain conditions on the elements of the S-box P are satisfied then the least significant bit (lsb) of the 1st output-word at the 1st round is equal to the lsb of the 2nd output-word at the 3rd round.

Theorem 1. $O_{1,1(0)} = O_{2,3(0)}$ *if the following six conditions on the elements of the S-box P are simultaneously satisfied.*

1. $P_2[116] \equiv -18 (\mathrm{mod}\ 32)$ *(event A)*,
2. $P_3[116] \equiv 7 (\mathrm{mod}\ 32)$ *(event B)*,
3. $P_2[72] = P_3[239] + 1$ *(event C)*,
4. $P_2[239] = P_3[72] + 1$ *(event D)*,
5. $P_1[26] = 1$ *(event E)*,
6. $P_3[208] = 254$ *(event F)*.

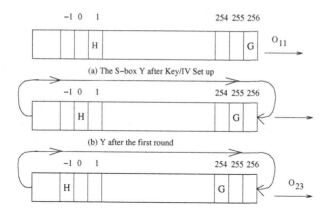

Fig. 1. (a) $P_1[26] = 1$ (condition 5): G and H are used in $O_{1,1}$, (b) Y_2 (i.e., Y after the 1st round), (c) $P_3[208] = 254$ (condition 6): G and H are used in $O_{2,3}$

Proof. The formulas for the $O_{1,1}$, $O_{2,3}$ and s_2 are given below (see Sect. 2):

$$O_{1,1} = (ROTL32(s_1, 25) \oplus Y_1[256]) + Y_1[P_1[26]], \qquad (1)$$

$$O_{2,3} = (s_3 \oplus Y_3[-1]) + Y_3[P_3[208]], \qquad (2)$$

$$s_2 = ROTL32(s_1 + Y_2[P_2[72]] - Y_2[P_2[239]], ((P_2[116] + 18) \bmod 32)). \quad (3)$$

- Condition 1 (i.e., $P_2[116] \equiv -18(\bmod 32)$) reduces (3) to

$$s_2 = s_1 + Y_2[P_2[72]] - Y_2[P_2[239]] \,.$$

- Condition 2 (i.e., $P_3[116] \equiv 7(\bmod 32)$) together with Condition 1 implies

$$s_3 = ROTL32((s_1 + Y_2[P_2[72]] - Y_2[P_2[239]] + Y_3[P_3[72]] - Y_3[P_3[239]]), 25) \,.$$

- Condition 3 and 4 (that is, $P_2[72] = P_3[239] + 1$ and $P_2[239] = P_3[72] + 1$) reduce the previous equation to

$$s_3 = ROTL32(s_1, 25) \,. \qquad (4)$$

From (1), (2), (4) we get:

$$O_{1,1} = (ROTL32(s_1, 25) \oplus Y_1[256]) + Y_1[P_1[26]], \qquad (5)$$

$$O_{2,3} = (ROTL32(s_1, 25) \oplus Y_3[-1]) + Y_3[P_3[208]] \,. \qquad (6)$$

In Fig. 1, conditions 5 and 6 are described. According to the figure,

$$H = Y_1[P_1[26]] = Y_3[-1], \qquad (7)$$

$$G = Y_1[256] = Y_3[P_3[208]] \,. \qquad (8)$$

Applying (7) and (8) in (5) and (6) we get,

$$O_{1,1(0)} \oplus O_{2,3(0)} = Y_1[256]_{(0)} \oplus Y_1[P_1[26]]_{(0)} \oplus Y_3[-1]_{(0)} \oplus Y_3[P_3[208]]_{(0)} = 0 \,.$$

This completes the proof. □

4   Bias in the Distribution of the 1st and the 3rd Outputs

In this section, we shall compute $P[O_{1,1(0)} \oplus O_{2,3(0)} = 0]$ using the results of Sect. 3. We now recall the six events (or conditions) A, B, C, D, E, F as described in Theorem 1. First, we shall compute $P[A \cap B \cap C \cap D \cap E \cap F]$. The elements involved in the calculation of the probability are $P_1[26]$, $P_2[72]$, $P_2[116]$, $P_2[239]$, $P_3[72]$, $P_3[116]$, $P_3[208]$, and $P_3[239]$. Now we observe that Algorithm 1 ensures that all the above elements occupy unique indices in round 1. We calculate the probabilities step by step using Bayes' rule under the assumption described in Sect. 2.2.

1. $P[E] = \frac{1}{256}$,

2. $P[E \cap F] = P[F|E] \cdot P[E] = \frac{1}{255} \cdot \frac{1}{256}$,

3. $P[A \cap E \cap F] = P[A|E \cap F] \cdot P[E \cap F] = \frac{8}{254} \cdot \frac{1}{255} \cdot \frac{1}{256}$,

4. $P[A \cap B \cap E \cap F] = P[B|A \cap E \cap F] \cdot P[A \cap E \cap F] = \frac{8}{253} \cdot \frac{8}{254} \cdot \frac{1}{255} \cdot \frac{1}{256}$,

5. Similarly, $P[A \cap B \cap C \cap E \cap F] = \frac{247}{251 \cdot 252} \cdot \frac{8}{253} \cdot \frac{8}{254} \cdot \frac{1}{255} \cdot \frac{1}{256}$,

6. $P[A \cap B \cap C \cap D \cap E \cap F] \approx \frac{244}{249 \cdot 250} \cdot \frac{247}{251 \cdot 252} \cdot \frac{8}{253} \cdot \frac{8}{254} \cdot \frac{1}{255} \cdot \frac{1}{256} \approx 2^{-41.9}$.

Under the assumption of randomness and uniformity of the distributions of the S-box elements and of s after the key/IV setup, if any of the six events – described in Theorem 1 – does not occur then $P[O_{1,1(0)} \oplus O_{2,3(0)} = 0] = \frac{1}{2}$ (see Appendix A for a justification for that). That is,

$$P[O_{1,1(0)} \oplus O_{2,3(0)} = 0|(A \cap B \cap C \cap D \cap E \cap F)^c] = \frac{1}{2} \,.$$

We denote the event $A \cap B \cap C \cap D \cap E \cap F$ by L and its complement by L^c. Therefore,

$$\begin{aligned} P[O_{1,1(0)} \oplus O_{2,3(0)} = 0] &= P[O_{1,1(0)} \oplus O_{2,3(0)} = 0|L] \cdot P[L] \\ &\quad + P[O_{1,1(0)} \oplus O_{2,3(0)} = 0|L^c] \cdot P[L^c] \\ &= 1 \cdot 2^{-41.91} + \frac{1}{2} \cdot (1 - 2^{-41.91}) \\ &= \frac{1}{2} \cdot (1 + 2^{-41.91}) \,. \end{aligned} \qquad (9)$$

Note that, if Py had been an ideal PRBG then the above probability would have been exactly $\frac{1}{2}$.

5   The Distinguisher

A *distinguisher* is an algorithm which distinguishes a stream of bits from a perfectly random stream of bits, that is, a stream of bits that has been chosen according to the uniform distribution. There are several ways a cryptanalyst may try to distinguish between a string, generated by an insecure pseudorandom bit generator, and one from a perfectly random source. In one case, she selects a *single* key/IV randomly and produce keystream, seeded by the chosen key/IV, long enough to distinguish it from random with high success probability. This attack scenario is rather common and such *distinguisher* is called a *regular distinguisher*. In a different scenario, to build a distinguisher, the adversary may use *many* randomly chosen key/IV's rather than a single key and a few *specified* bytes from each of the keystreams generated by those key/IV's. The *distinguisher*, so constructed, is called a *prefix distinguisher*. A bias present in the output at time t in a single stream may hardly be detected by a *regular distinguisher* but a *prefix distinguisher* can easily discover the anomaly with a few bytes. This fact was nicely demonstrated by Mantin and Shamir [11] to detect a strong bias toward zero in the second output byte of RC4. In addition to that, there exist *hybrid distinguishers* that may fall between the above two extreme cases, that is, the adversary may use *many* key/IV's and for each key/IV she collects *long* keystream. The idea of constructing distinguishers using *many* randomly chosen key/IV's has been a well studied subject. Goldreich has shown that a distribution which is *computationally indistinguishable* from the *uniform distribution* based on a *single sample* is also *computationally indistinguishable* from the *uniform distribution* based on *multiple samples* [8].

The distinguishers that we construct in this section and Sect. 6, using the bias described in Sect. 4, are *prefix distinguishers*. In Sect. 7, we build a *regular distinguisher*; however, the number of outputs needed for this distinguisher exceeds the allowable keystream length per key/IV. In Section 8, we propose a *hybrid distinguisher* mainly to reduce the time cost of our *prefix distinguisher*.

Algorithm 2. *A Distinguisher* separating Py from Random

Require: An n-bit sequence $(z_1, z_2, z_3, \cdots, z_n)$
Ensure: Whether the sequence is random or generated by Py
  1: Compute LLR $= \sum_i \log(\frac{P_0[z_i]}{P_1[z_i]})$;
  2: If LLR ≥ 0 then return 1 (i.e., "The sequence is from Py")
       else 0 (i.e., "The sequence is random");

Algorithm 2. The *prefix distinguisher* that separates Py from random is described in Algorithm 2. The input to the algorithm is a realization of the sequence of binary random variables $(z_1, z_2, z_3, \cdots, z_n)$. The adversary first generates n key/IV pairs $X_1, X_2, X_3, \cdots, X_n$ randomly and then computes $z_i = O_{1,1(0)} \oplus O_{2,3(0)}$ for all X_i, $1 \leq i \leq n$. Using the results obtained by Baignères, Junod and Vaudenay [1], it can be shown that Algorithm 2 is an *optimal distinguisher*. Given a fixed number of samples, an *optimal distinguisher* attains the *maximum advantage*. Note that the random variables z_i's are independent of each other and each of them follows the distribution computed in Sect. 4 (call the distribution D_0). Let the uniform distribution on alphabet $[0, 1]$ be denoted by D_1. In Algorithm 2, $P_0[z_i]$ (shorthand for $P_{D_0}[z_i]$) denotes the probability of occurrence of z_i when chosen according to D_0 (similarly $P_1[z_i]$ and $P_{D_1}[z_i]$).

Let the Algorithm 2, the sequence of variables $(z_1, z_2, z_3, \cdots, z_n)$ and the quantity $\sum_i \log(\frac{P_0[z_i]}{P_1[z_i]})$ be denoted by $\mathcal{F}$, Z and LLR respectively. Now we will compute the *advantage* of $\mathcal{F}$ (the advantage of this distinguisher has been independently calculated by Paul Crowley [5]). The *advantage* of a distinguisher – a measure indicating the efficiency of an algorithm to distinguish a distribution from another – is given by the following formula [1]:

$$\text{Adv}_{\mathcal{F}}^n = \left| P_{D_0^n}[\mathcal{F}(Z) = 1] - P_{D_1^n}[\mathcal{F}(Z) = 1] \right|. \tag{10}$$

Following the results in [1], it can be shown that for large n,

$$P_{D_0^n}[\mathcal{F}(Z) = 1] = P_{D_0}[\text{LLR} \geq 0] \approx \Phi\left(\frac{\sqrt{n}\mu_0}{\sigma_0}\right),$$

$$P_{D_1^n}[\mathcal{F}(Z) = 1] = P_{D_1}[\text{LLR} \geq 0] \approx \Phi\left(\frac{\sqrt{n}\mu_1}{\sigma_1}\right).$$

where Φ is the standard normal distribution function expressed as,

$$\Phi(z) = \frac{1}{\sqrt{2\pi}} \int_{-\infty}^{z} e^{-\frac{1}{2}u^2} \, du.$$

If the two distributions D_0 and D_1 are close (i.e., $\left|P_0[z] - P_1[z]\right| \ll P_1[z]$) then

$$\mu_0 \approx -\mu_1 \approx \frac{1}{2} \sum_{z \in [0,1]} \frac{(P_0[z] - P_1[z])^2}{P_1[z]} \text{ and } \sigma_0^2 \approx \sigma_1^2 \approx \sum_{z \in [0,1]} \frac{(P_0[z] - P_1[z])^2}{P_1[z]}.$$

The above equations suggest that, for a given n, using the known distributions D_0 and D_1, the *advantage* of Algorithm 2 can be computed from (10). Some simple calculations show that, if $P_0[0] - P_1[0] = \frac{1}{M}$, then, to ensure the *advantage* of the distinguisher to be greater than 0.5, the required number of samples is

$$n = 0.4624 \cdot M^2. \tag{11}$$

In the present case $P_0[0] - P_1[0] = \frac{1}{2^{42.9}}$ (see Sect. 4). Therefore, from (11), $n = 0.4624 \cdot (2^{42.9})^2 = 2^{84.7}$ samples (i.e., as many randomly chosen key/IV's) can distinguish Py from random with an advantage that exceeds 0.5. The time cost to build this distinguisher is $t_{ini} \cdot 2^{84.7}$ where t_{ini} is the running time of the key/IV setup of Py. Note that, for each key/IV, we collect the first 24 bytes of the keystream. Therefore, the number of bytes required to establish the distinguisher is $2^{84.7} \cdot 24 = 2^{89.2}$.

6   Biases Among Other Pairs of Bits and Distinguishers

In Sect. 4, we have showed a bias in $(O_{1,1(0)}, O_{2,3(0)})$. In this section, we show that the bias is present in $(O_{1,1(i)}, O_{2,3(i)})$, where $0 \leq i \leq 31$; however, the bias gradually reduces as i increases. From (1) and (2), we get:

$$O_{1,1(i)} = ROTL32(s_1, 25)_{(i)} \oplus Y_1[256]_{(i)} \oplus Y_1[P_1[26]]_{(i)} \oplus c_{1(i)},$$
$$O_{2,3(i)} = s_{3(i)} \oplus Y_3[-1]_{(i)} \oplus Y_3[P_3[208]]_{(i)} \oplus c_{3(i)},$$

where $0 \leq i \leq 31$ and c_1, c_3 are the carry terms in (1) and (2) respectively.

A Special Case. If all the 6 conditions of Theorem 1 are satisfied, $O_{1,1}$ and $O_{2,3}$ can be written in the following form (see Theorem 1):

$$O_{1,1} = (S \oplus G) + H, \tag{12}$$
$$O_{2,3} = (S \oplus H) + G, \tag{13}$$

which implies that

$$O_{1,1(i)} \oplus O_{2,3(i)} = c_{1(i)} \oplus c_{3(i)}, \qquad 0 \leq i \leq 31,$$

where the carries $c_{1(i)}$ and $c_{3(i)}$ can be calculated from the following recursive relations (note that $c_{1(0)} = c_{3(0)} = 0$),

$$c_{1(i)} = c_{1(i-1)}(S_{(i-1)} \oplus G_{(i-1)}) \oplus c_{1(i-1)}H_{(i-1)} \oplus$$
$$H_{(i-1)}(S_{(i-1)} \oplus G_{(i-1)}), \tag{14}$$
$$c_{3(i)} = c_{3(i-1)}(S_{(i-1)} \oplus H_{(i-1)}) \oplus c_{3(i-1)}G_{(i-1)} \oplus$$
$$G_{(i-1)}(S_{(i-1)} \oplus H_{(i-1)}). \tag{15}$$

Computing $P[O_{1,1(i)} \oplus O_{2,3(i)} = 0]$. Note that

$$P[O_{1,1(i)} \oplus O_{2,3(i)} = 0] = P[O_{1,1(i)} \oplus O_{2,3(i)} = 0|L] \cdot P[L]$$
$$+ P[O_{1,1(i)} \oplus O_{2,3(i)} = 0|L^c] \cdot P[L^c]$$
$$= \underbrace{P[c_{1(i)} \oplus c_{3(i)} = 0|L]}_{p_i} \cdot P[L]$$
$$+ \underbrace{P[O_{1,1(i)} \oplus O_{2,3(i)} = 0|L^c]}_{X_i} \cdot P[L^c], \tag{16}$$

where $i \in [0, 31]$ and the event L is $A \cap B \cap C \cap D \cap E \cap F$. Note that four components are involved in (16); they are $P[L]$, $P[L^c]$, p_i and X_i. Next, we show how to determine these four quantities.

1,2. Computing $P[L]$ and $P[L^c]$: the results in Sect. 4 show that $P[L] = 2^{-41.9}$ and $P[L^c] = (1 - 2^{-41.9})$.

3. Computing p_i: now we recursively compute $P[c_{1(i)} \oplus c_{3(i)} = 0|L]$, denoted by p_i in (16) (similarly p_{i-1} should be understood), from the following equation derived directly from (14) and (15).

$$c_{1(i)} \oplus c_{3(i)} = (c_{1(i-1)} \oplus c_{3(i-1)})(S_{(i-1)} \oplus G_{(i-1)} \oplus H_{(i-1)}) \oplus$$
$$S_{(i-1)}(G_{(i-1)} \oplus H_{(i-1)}). \tag{17}$$

Table 1. Truth table for (17). The last column in each row indicates the probability of the occurrence of that row.

$c_{1(i-1)} \oplus c_{3(i-1)}$	$S_{(i-1)}$	$B_{(i-1)}$	$A_{(i-1)}$	$c_{1(i)} \oplus c_{3(i)}$	Probability
0	0	0	0	0	$p_{i-1}/8$
0	0	0	1	0	$p_{i-1}/8$
0	0	1	0	0	$p_{i-1}/8$
0	0	1	1	0	$p_{i-1}/8$
0	1	0	0	0	$p_{i-1}/8$
0	1	0	1	1	
0	1	1	0	1	
0	1	1	1	0	$p_{i-1}/8$
1	0	0	0	0	$(1-p_{i-1})/8$
1	0	0	1	1	
1	0	1	0	1	
1	0	1	1	0	$(1-p_{i-1})/8$
1	1	0	0	1	
1	1	0	1	1	
1	1	1	0	1	
1	1	1	1	1	

Note that the variables G, H, S are uniformly distributed and independent. The truth table for (17) is shown in Table 1. From Table 1, using Bayes' rule, we obtain the following recursion to compute p_i,

$$p_i = \frac{p_{i-1}}{2} + \frac{1}{4}.$$

We already know that $p_0 = 1$ (i.e., $P[O_{1,1(0)} \oplus O_{2,3(0)} = 0|L] = 1$). Therefore, solving the above recurrence relation, finally we get

$$p_i = \frac{1}{2} + \frac{1}{2^{i+1}}, \quad 0 \le i \le 31. \tag{18}$$

4. Computing X_i: according to the results obtained in Appendix A it is reasonable to assume that

$$X_i = \frac{1}{2}, \quad \text{for all } i \in [0, 24] \cup [26, 31].$$

General Expression. Using the above results, recalling (16), we find,

$$P[O_{1,1(i)} \oplus O_{2,3(i)} = 0] = \frac{1}{2}(1 + 2^{-(41.9+i)}), \tag{19}$$

where $i \in [0, 24] \cup [26, 31]$. It is also reasonable to assume (due to the event L' as described in Appendix A) that

$$P[O_{1,1(25)} \oplus O_{2,3(25)} = 0] \ge \frac{1}{2}(1 + 2^{-(41.9+25)})$$

$$\ge \frac{1}{2}(1 + 2^{-66.9}).$$

From (19), one may see that $P[O_{1,1(i)} \oplus O_{2,3(i)} = 0]$ attains the maximum value if $i = 0$. Our distinguisher, described in Sect. 4 and 5, exploits the case if $i = 0$. Equation (19) suggests that many distinguishers can be generated using different $(O_{1,1(i)}, O_{2,3(i)})$'s rather than only $(O_{1,1(0)}, O_{2,3(0)})$, however, the amount of bias decreases as i increases (i.e., we get the most effective distinguisher if $i = 0$). For example, if $i = 1$,

$$P[O_{1,1(1)} \oplus O_{2,3(1)} = 0] = \frac{1}{2}(1 + 2^{-42.91}).$$

For the above case, taking the 1st bits of $O_{1,1}$ and $O_{2,3}$, the number of samples (i.e., the number of key/IV's) required to establish a distinguisher with advantage exceeding 0.5 is $2^{86.7}$ (see (11)). Similarly, if we consider $i = 2$ then the number of required samples is $2^{88.7}$.

7 Generalizing the Bias at Rounds t and $t + 2$: A Distinguisher Using a Single Keystream

Under assumptions similar to those in Sect. 2.2, the results of Sect. 3 and Sect. 4 are valid even if we consider any rounds t and $t + 2$ $(t > 0)$ instead of just rounds 1 and 3. In other words, instead of $(O_{1,1(0)}, O_{2,3(0)})$, one can show that the bias exists even in the distribution of $(O_{1,t(i)}, O_{2,(t+2)(i)})$. Now, we state a theorem which is the generalized version of Theorem 1.

Theorem 2. $O_{1,t(0)} = O_{2,(t+2)(0)}$ *if the following six conditions on the elements of the S-box P are simultaneously satisfied.*

1. $P_{t+1}[116] \equiv -18 (\mathrm{mod}\ 32)$,
2. $P_{t+2}[116] \equiv 7 (\mathrm{mod}\ 32)$,
3. $P_{t+1}[72] = P_{t+2}[239] + 1$,
4. $P_{t+1}[239] = P_{t+2}[72] + 1$,
5. $P_t[26] = 1$,
6. $P_{t+2}[208] = 254$.

Using the above theorem and the techniques used before, it is easy to show that (see (9))

$$P[O_{1,t(0)} \oplus O_{2,(t+2)(0)} = 0] = \frac{1}{2}(1 + 2^{-41.91}).$$

The fact that the above probability is valid, $\forall t > 0$, allows us to generate a *regular distinguisher* with the number of rounds $2^{84.7}$ of a *single keystream* (see Sect. 5 for a definition of a *regular distinguisher*). This means that $2^{84.7} \times 2^3 = 2^{87.7}$ bytes of *a single* stream generated by a randomly chosen key/IV are sufficient to distinguish Py from random with success probability greater than 0.5. The work-load here is also comparable to $2^{87.7}$. However, this attack is rendered ineffective because the amount of required bytes falls outside the allowable keystream length of 2^{64} bytes.

8   A More Efficient Hybrid Distinguisher

The results of Sect. 5 and Sect. 7 lead us in a natural way to build a *hybrid* distinguisher by making a trade-off between the number of key/IV's and output bytes per key/IV. It is apparent from the previous discussion that, to realize our distinguisher, we need $2^{84.7}$ pairs of internal states (recall that the internal state of Py consists of the arrays P, Y and a 32-bit integer s) with each pair being separated by one round. Then, under the assumption that the first state of each pair is randomly generated, those pairs can be used to build a distinguisher. As the allowable number of rounds per key/IV is $2^{64-8} = 2^{56}$, the number of required key/IV's is $2^{84.7-56} = 2^{28.7}$ to construct this *hybrid distinguisher*. The main difference between the *prefix distinguisher* in Sect. 5 and this *hybrid distinguisher* is that the running time to build this *hybrid distinguisher* is much smaller, as it requires the key/IV setup to run only for $2^{28.7}$ times compared to $2^{84.7}$ times for the previous *prefix distinguisher*. Therefore, the time and the data complexity of this distinguisher are $t_r \cdot 2^{84.7}$ and $2^{87.7}$ bytes respectively, where t_r is the running time of a single round of Py. Furthermore, this *hybrid* distinguisher does not breach the cipher specifications.

9   Do Our Distinguishers Break the Cipher Py?

The subject of what constitutes a break of a *practical stream cipher* or a PRBG is a highly contentious issue even if the area is quite well developed in theory. Theoretically, a cryptographically strong pseudorandom bit generator (CSPRBG) is an algorithm $\mathcal{A}$ that, on being given a random seed k as input, generates a sequence of pseudo-random bits $a_1, a_2, a_3, \cdots$. The function $\mathcal{A}$ possesses the following properties (see Blum and Micali [4]):

1. Each bit a_i can be produced in time polynomial in the length of seed k.
2. Given the algorithm $\mathcal{A}$ and the first s output bits generated by an unknown seed k, it is *computationally infeasible* to predict the $s+1$st bit with biased probability. The s is polynomial in the length of the seed k.

We see that, theoretically, a PRBG is studied according to how it behaves when the length of *seed* is increased asymptotically. The major problem in fitting the analyses of practical stream ciphers into the above framework is that, most of the ciphers work with *fixed sized* keys and keystream bits (e.g. Py allows 256-bit key and 2^{64} bytes of keystream per key/IV pair). Such constraints make the asymptotic analyses of practical stream ciphers impossible. For a practical PRBG with a *fixed sized key* (such as Py), given the first s output bits generated by an unknown key/IV, the $s+1$st bit can be predicted with a high probability with running time bounded above by a trivial exhaustive search. As there is no non-trivial upper bound on the running time of a distinguishing attack on a stream cipher (or PRBG) with a fixed sized key, any legal distinguishing attack with running time less than exhaustive search constitutes an academic break of the cipher.[1] Therefore, our attacks from Sect. 5, Sect. 6 and Sect. 8 imply a theoretical break of Py. However, it should be noted that each of the attacks presented in the paper requires a workload larger than 2^{85} and therefore, poses no practical threats to the cipher.

Do Our Distinguishing Attacks on Py Violate the Designers' Claims?
The stream cipher Py is claimed by the designers to have up to 256-bit security (see Appendix A of [3]). In the authors' words, "The security claims are for keys up to 256 bits (32 bytes) and IVs up to 128 bits (16 bytes)". 256-bit is also the category of security level under which Py is included in the ECRYPT project [6]. According to the discussion on the definition of *n-bit security* of a perfectly secure stream cipher, it is clear that this claim is compromised by our attacks.

However, in Sect. 6.1 of [3], the authors claim, "There are no distinguishing attacks that succeed given less than 2^{64} bytes of key stream with a complexity less than of exhaustive search." It is understood from [2], that those 2^{64} bytes, as mentioned in the claim, may be generated by many keys rather than a single key. Under this interpretation, our attacks do not violate this claim, since our best attack requires $2^{87.7}$ bytes of output.

As a result we conclude that two claims, mentioned above, contradict each other with respect to the attacks mentioned in this paper. At this point, we leave it to the reader to decide on the implications of our distinguishers.

10   Future Work

One could try to combine the individual biases of the pairs of bits presented here to develop a more sophisticated distinguisher with fewer output bytes. Paul Crowley has reduced the time and output bytes of our distinguisher to 2^{72} each, by analyzing our observation in Sect. 3 using a Hidden Markov Model [5]. A plausible strategy consists of identifying many more correlations between internal and external states of Py in order to reduce the time and data complexity of the distinguisher.

11   Conclusion and Remarks

The paper presented several weaknesses of the stream cipher Py. We discovered a class of distinguishers for the cipher, the best of which works with $2^{87.7}$ bytes and comparable time. We also showed that the output stream of Py with a recommended keystream length of 2^{64} bytes, contains biases at different points – this fact can be exploited to build more effective distinguishers. These results break the cipher Py academically. However, the data complexity for the best distinguishing attack falls well beyond the time complexity what is feasible today. Therefore, these weaknesses pose no practical threat to the security of the cipher

---

[1] A *legal distinguishing attack* is the one which does not violate the specified parameters of the cipher.

at this moment. However, the shortened version of Py, known as Py6, may contain more serious weaknesses than the ones described here, but the complete description of Py6 has not been provided in [3].

Acknowledgments

We thank Paul Crowley for providing us with useful references. It is a great pleasure to thank Eli Biham for his comments on our work. Thanks are due to Hongjun Wu and Jongsung Kim for numerous valuable discussions. We also acknowledge the constructive comments of the anonymous reviewers of FSE'06.

References

1. T. Baignères, P. Junod and S. Vaudenay, "How Far Can We Go Beyond Linear Cryptanalysis?," *Asiacrypt 2004* (P. Lee, ed.), vol. 3329 of *LNCS*, pp. 432–450, Springer-Verlag, 2004.
2. E. Biham, Personal Communication, Dec. 2005.
3. E. Biham, J. Seberry, "Py (Roo): A Fast and Secure Stream Cipher using Rolling Arrays," eSTREAM, ECRYPT Stream Cipher Project, Report 2005/023, 2005.
4. M. Blum, S. Micali, "How to Generate Cyptographically Strong Sequence of Psudo-random Bits," *Siam Journal of Computing*, vol. 13, No. 4, pp. 850–864, November 1984.
5. P. Crowley, "Improved Cryptanalysis of Py," *Workshop Record of SASC 2006 – Stream Ciphers Revisited*, ECRYPT Network of Excellence in Cryptology, February 2006, Leuven (Belgium), pp. 52–60.
6. Daniel. J. Bernstein, "Comparison of 256-bit stream ciphers at the beginning of 2006," *Workshop Record of SASC 2006 – Stream Ciphers Revisited*, ECRYPT Network of Excellence in Cryptology, pp. 70–83.
7. Ecrypt, `http://www.ecrypt.eu.org`.
8. O. Goldreich, "Lecture Notes on Pseudorandomness–Part-I," Department of Computer Science, Wiezmann Institute of Science, Rehovot, ISRAEL, January 23, 2001.
9. G. Gong, K. C. Gupta, M. Hell, Y. Nawaz, "Towards a General RC4-Like Keystream Generator," *First SKLOIS Conference, CISC 2005* (D. Feng, D. Lin, M. Yung, eds.), vol. 3822 of *LNCS*, pp. 162–174, Springer-Verlag, 2005.
10. Robert J. Jenkins Jr., "ISAAC," *Fast Software Encryption 1996* (D. Gollmann, ed.), vol. 1039 of *LNCS*, pp. 41–49, Springer-Verlag, 1996.
11. I. Mantin, A. Shamir, "A Practical Attack on Broadcast RC4," *Fast Software Encryption 2001* (M. Matsui, ed.), vol. 2355 of *LNCS*, pp. 152–164, Springer-Verlag, 2001.
12. NESSIE: New European Schemes for Signature, Integrity and Encryption, `http://www.cryptonessie.org`.
13. Souradyuti Paul, Bart Preneel, "A New Weakness in the RC4 Keystream Generator and an Approach to Improve the Security of the Cipher," *Fast Software Encryption 2004* (B. Roy, ed.), vol. 3017 of *LNCS*, pp. 245–259, Springer-Verlag, 2004.
14. Hongjun Wu, "A New Stream Cipher HC-256," *Fast Software Encryption 2004* (B. Roy, ed.), vol. 3017 of *LNCS*, pp. 226–244, Springer-Verlag, 2004.
15. Bartosz Zoltak, "VMPC One-Way Function and Stream Cipher," *Fast Software Encryption 2004* (B. Roy, ed.), vol. 3017 of *LNCS*, pp. 210–225, Springer-Verlag, 2004.

A   Uniformity of Bits If L Does Not Occur

We first write the general formula to calculate $Z = O_{1,1} \oplus O_{2,3}$.

$$O_{1,1} = (ROT32(s, 25) \oplus G) + H, \tag{20}$$

$$O_{2,3} = (ROT32(ROT32(s + I - J, r) + K - L, l) \oplus M) + N, \tag{21}$$

where
$s = s_1$, $G = Y_1[256]$, $H = Y_1[P_1[26]]$, $I = Y_2[P_2[72]]$, $J = Y_2[P_2[239]]$, $r = P_2[116] + 18 \bmod 32$, $K = Y_3[P_3[72]]$, $L = Y_3[P_3[239]]$, $l = P_3[116] + 18 \bmod 32$, $M = Y_3[-1]$, $N = Y_3[P_3[208]]$.

Below we isolate 18 cases, divided into 4 groups, where the relation between internal and external states is not trivial. The symbol '/' is used to mean 'or'. Note that the equalities in each group are satisfied if they do not violate the condition of uniqueness of permutation elements of S-box P. The notation $A \Leftrightarrow B$ signifies that the A and B are identical elements in two different rounds of the S-box Y (i.e., $A = B$ but their indices may be changed in different rounds).

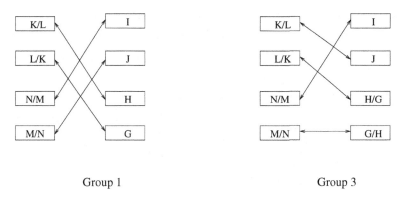

Group 1   Group 3

Fig. 2

1. $I \Leftrightarrow N/M$, $J \Leftrightarrow M/N$, $K \Leftrightarrow G/H$, $L \Leftrightarrow H/G$ (a total of 4 cases). See Fig.2.
2. $I \Leftrightarrow K/L$, $J \Leftrightarrow L/K$, $M \Leftrightarrow H$, $N \Leftrightarrow G$ (a total of 2 cases). See Fig.3.
3. $I \Leftrightarrow N/M$, $J \Leftrightarrow K/L$. The G is identical to one of the remaining two elements (so is the H) (a total of 6 cases). See Fig.2.
4. Similar to the above, $J \Leftrightarrow M/N$, $I \Leftrightarrow L/K$. The G is identical to one of the remaining two elements (so is the H) (a total of 6 cases).

Fact 1. *After the key/IV setup, if the permutation P falls outside all of the 18 cases described above then $O_{1,1}$ and $O_{2,3}$ are independent and uniformly distributed over $[2^{32} - 1, 0]$.*

Now we denote $O_{1,1(0)} \oplus O_{2,3(0)}$ by R_0.

Theorem 3. *After the key/IV setup, let the S-box P be one of the 16 cases described in Groups 1,3 and 4. Then*

$$Prob[R_0 = 0 \,|\, P] = \frac{1}{2}.$$

Proof. Now we prove the theorem by considering Groups 1, 3 and 4 separately.

Group 1. (See Fig. 2). For this group, R_0 can be written in the following form,

$$R_0 = s_{(7)} \oplus s_{(w)} \oplus (G_{(0)} \oplus H_{(0)} \oplus K_{(u)} \oplus L_{(u)})$$
$$\oplus (I_{(w)} \oplus J_{(w)} \oplus M_{(0)} \oplus N_{(0)}) \oplus C.$$

Note that the C is a nonlinear function of several bits of s, M, N, H, G. Now we take three possible subcases.

1. If $u \neq 0$ then R_0 is uniformly distributed since C is independent of $K_{(u)}$ and $L_{(u)}$.
2. If $u = 0$, $w \neq 0$ then R_0 is uniformly distributed since C is independent of $I_{(w)}$ and $J_{(w)}$.
3. If $u = 0$, $w = 0$ then $R_0 = s_{(7)} \oplus s_{(0)}$. Therefore, R_0 is uniformly distributed.

Group 3. (See Fig. 2). R_0 can be written in the following form,

$$R_0 = s_{(7)} \oplus s_{(w)} \oplus (G_{(0)} \oplus N_{(0)}) \oplus (K_{(u)} \oplus J_{(w)})$$
$$\oplus (H_{(0)} \oplus L_{(u)}) \oplus (M_{(0)} \oplus I_{(w)}) \oplus C.$$

Of the 6 cases in Group 3, we are considering *only* the following case where $I \Leftrightarrow M$, $J \Leftrightarrow K$, $G \Leftrightarrow N$ and $H \Leftrightarrow L$. In a similar way as above we divide this case into three subcases.

1. If $u \neq 0$ then C is independent of $L_{(u)}$ and thus R_0 is uniformly distributed.
2. If $u = 0$, $w \neq 0$ then R_0 is uniformly distributed since C is independent of $J_{(w)}$.
3. If $u = 0$, $w = 0$ then $R_0 = s_{(7)} \oplus s_{(0)}$. Therefore, R_0 is uniformly distributed.

All the other 5 cases of this group can be proved in a similar fashion.

Group 4. Proof for this group is similar to that for Group 3.   □

Discussion. From Fact 1 and Theorem 3, it is clear that, if P does not fall within Group 2 then $Prob[R_0 = 0|P] = \frac{1}{2}$. The probability of the occurrence of Group 2 is approximately 2^{-31}. Therefore, for a fraction of $(1 - 2^{-31})$ of all cases, R_0 is uniformly distributed. Sect. 4 shows that, for the event L occurring with probability $2^{-41.9}$, $Prob[R_0 = 0 \,|\, P = L] = 1$.

Therefore, we are able to prove that, for a fraction of $(1 - 2^{-31.001})$ of cases, there exists a bias in R_0 toward zero. It is, however, nontrivial to determine the distribution of R_0 for the remaining fraction of $2^{-31.001}$ of the cases, because of vigorous mixing of bits in a nonlinear way. Our experiments suggest that it is very unlikely that the positive bias generated in a large fraction of $(1 - 2^{-31.01})$

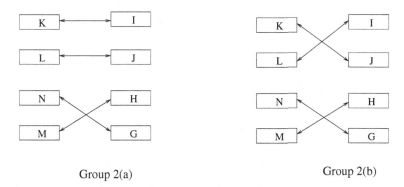

Group 2(a)   Group 2(b)

Fig. 3. Group 2(a): $I \Leftrightarrow K$, $J \Leftrightarrow L$, $M \Leftrightarrow H$, $N \Leftrightarrow G$; Group 2(b): $I \Leftrightarrow L$, $J \Leftrightarrow K$, $M \Leftrightarrow H$, $N \Leftrightarrow G$

of the cases can be compensated by a very minuscule fraction of $2^{-31.001}$ of them. According to a small number of experiments that we carried out, a slight bias toward zero was detected for that remaining fraction of $2^{-31.001}$ also. However, we ignored that bias and assumed R_0 to be uniformly distributed for those cases in building the distinguishers described in the paper.

In addition to the event L, for which R_i is biased toward zero $\forall i \in [1, 31]$ (see Sect. 6), we also identify another event L', for which R_{25} is again biased toward zero (all other R_i's are uniformly distributed individually). The event L' occurs when $P_2[116] \equiv -18 (\mathrm{mod}\, 32)$, $P_3[116] \equiv 7 (\mathrm{mod}\, 32)$, $P_2[72] = P_3[72] + 1$, $P_2[239] = P_3[239] + 1$, $P_1[26] = 1$, $P_3[208] = 254$ (see Group 2(a) of Fig. 3). Using similar arguments as above, it can be shown that R_i is uniformly distributed over $[0, 1]$ for the rest of the cases.

Resynchronization Attacks on WG and LEX[*]

Hongjun Wu and Bart Preneel

Katholieke Universiteit Leuven, ESAT/SCD-COSIC
Kasteelpark Arenberg 10, B-3001 Leuven-Heverlee, Belgium
{wu.hongjun, bart.preneel}@esat.kuleuven.be

Abstract. WG and LEX are two stream ciphers submitted to eStream – the ECRYPT stream cipher project. In this paper, we point out security flaws in the resynchronization of these two ciphers. The resynchronization of WG is vulnerable to a differential attack. For WG with 80-bit key and 80-bit IV, 48 bits of the secret key can be recovered with about $2^{31.3}$ chosen IVs . For each chosen IV, only the first four keystream bits are needed in the attack. The resynchronization of LEX is vulnerable to a slide attack. If a key is used with about $2^{60.8}$ random IVs, and 20,000 keystream bytes are generated from each IV, then the key of the strong version of LEX could be recovered easily with a slide attack. The resynchronization attack on WG and LEX shows that block cipher related attacks are powerful in analyzing non-linear resynchronization mechanisms.

Keywords: cryptanalysis, stream cipher, resynchronization attack, differential attack, slide attack, WG, LEX.

1 Introduction

For the research on stream ciphers, resynchronization atacks have not been studied as thoroughly as the keystream generation algorithm itself. Ten years ago, Daemen, Govaerts and Vandewalle analyzed the weakness of linear resynchronization mechanism with known output Boolean function [5]. Later Golić and Morgari studied linear resynchronization mechanisms with unknown output function [7]. However almost all the stream ciphers proposed recently use non-linear resynchronization mechanisms, so the previous attacks on linear resynchronization mechanisms could no longer be applied. Recently Armknecht, Lano and Preneel applied algebraic attacks and linear cryptanalysis to the resynchronization mechanism and obtained lower bounds for the nonlinearity required from a secure resynchronization mechanism [1]. In this paper, we apply the differential attack and slide attack to stream ciphers with non-linear resynchronization. We show that the cryptanalysis techniques used to attack block ciphers are also useful in the analysis of non-linear resynchronization mechanisms.

---

[*] This work was supported in part by the Concerted Research Action (GOA) Ambiorics 2005/11 of the Flemish Government and in part by the European Commission through the IST Programme under Contract IST-2002-507932 ECRYPT.

M.J.B. Robshaw (Ed.): FSE 2006, LNCS 4047, pp. 422–432, 2006.

WG [11] and LEX [4] are two stream ciphers submitted to eStream, the ECRYPT stream cipher project [6]. The keystream generation algorithms of WG and LEX are quite strong. The keystream generation of WG is based on the WG transformation which has excellent cryptographic properties [8]. The keystream generation of LEX is based on the Advanced Encryption Standard [10]. However, the resynchronization mechanism of WG and LEX are insecure. The resynchronization mechanism of WG is vulnerable to the differential attack [2] and that of LEX is vulnerable to a slide attack [3]. Breaking WG requires $2^{31.3}$ chosen IVs, and breaking the strong version of LEX requires about $2^{60.8}$ random IVs.

This paper is organized as follows. WG and LEX are introduced in Sect. 2. The differential attack on WG is presented in Sect. 3, and the slide attack on LEX is described in Sect. 4. Section 5 concludes this paper.

2 Description of WG and LEX

WG and LEX are described in Sec. 2.1 and 2.2, respectively.

2.1 The Stream Cipher WG

WG is a hardware oriented stream cipher with key length up to 128 bits; it supports IV sizes from 32 bits to 128 bits. The main feature of the WG stream cipher is the use of the WG transformation to generate keystream from an LFSR.

Keystream Generation
The keystream generation diagram of WG is given in Fig. 1. WG has a regularly clocked LFSR which is defined by the feedback polynomial

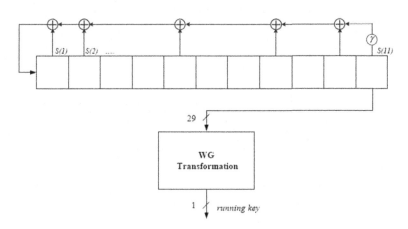

Fig. 1. Keystream generation diagram of WG [11]

$$p(x) = x^{11} + x^{10} + x^9 + x^6 + x^3 + x + \gamma \tag{1}$$

over $GF(2^{29})$, where $\gamma = \beta^{464730077}$ and β is the primitive root of $g(x)$

$$g(x) = x^{29} + x^{28} + x^{24} + x^{21} + x^{20} + x^{19} + x^{18} + x^{17} +$$
$$x^{14} + x^{12} + x^{11} + x^{10} + x^7 + x^6 + x^4 + x + 1. \tag{2}$$

Then the non-linear WG transformation, $GF(2^{29}) \rightarrow GF(2)$, is applied to generate the keystream from the LFSR.

Resynchronization (Key/IV Setup)
The key/IV setup of WG is given in Fig. 2. After the key and IV have been loaded into the LFSR, it is clocked 22 steps. During each of these 22 steps, 29 bits from the middle of the WG transformation are XORed to the feedback of LFSR, as shown in Fig. 2.

One step of the key/IV setup can be expressed as follows:

$$T = S(1) \oplus S(2) \oplus S(5) \oplus S(8) \oplus S(10) \oplus (\gamma \times S(11)) \oplus WG'(S(11)),$$
$$S(i) = S(i-1) \text{ for } i = 11 \cdots 2; \ S(1) = T \ ,$$

where $WG'(S(11))$ denotes the 29 bits extracted from the WG transformation, as shown in Fig. 2.

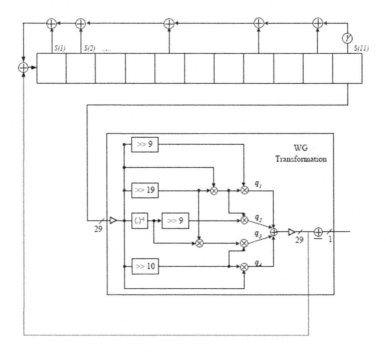

Fig. 2. Key/IV setup of WG [11]

The WG cipher supports several key and IV sizes: the key size can be 80 bits, 96 bits, 112 bits and 128 bits and the IV sizes can be 32 bits, 64 bits, 80 bits, 96 bits, 112 bits, and 128 bits. Slightly different resynchronization mechanisms are used for the different IV sizes. The details are given in Sect. 3.

2.2   The Stream Cipher LEX

LEX is based on the block cipher AES. The keystream bits are generated by extracting 32 bits from each round of AES in the 128-bit Output Feedback (OFB) mode [9]. LEX is about 2.5 times faster than AES. Fig. 3 shows how the AES is initialized and chained. First a standard AES key-schedule for a secret 128-bit key K is performed. Then a given 128-bit IV is encrypted by a single AES invocation: $S = AES_K(IV)$. The S and the subkeys are the output of the initialization process.

S is encrypted by K in the 128-bit OFB mode (for the more secure variant, K is changed every 500 AES encryptions). At each round, 32 bits of the middle value of AES are extracted to form the keystream. The bytes $b_{0,0}$, $b_{0,2}$, $b_{2,0}$, $b_{2,2}$ at every odd round and the bytes $b_{0,1}$, $b_{0,3}$, $b_{2,1}$, $b_{2,3}$ at every even round are selected, as shown in Fig. 4.

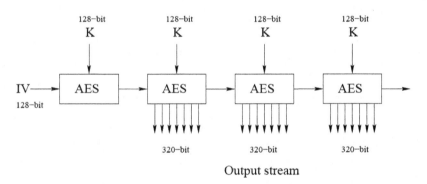

Output stream

Fig. 3. Initialization and stream generation [4]

$b_{0,0}$	$b_{0,1}$	$b_{0,2}$	$b_{0,3}$
$b_{1,0}$	$b_{1,1}$	$b_{0,0}$	$b_{1,3}$
$b_{2,0}$	$b_{2,1}$	$b_{2,2}$	$b_{2,3}$
$b_{3,0}$	$b_{3,1}$	$b_{3,2}$	$b_{3,3}$

Odd rounds   *Even rounds*

Fig. 4. The positions of the output extracted in the even and odd rounds [4]

3   The Differential Attacks on the Resynchronization of WG

The resynchronization of WG can be broken with a chosen IV attack based on differential cryptanalysis. (We remind the readers that the details of the differential attack given in this paper are slightly different from the standard differential attack on a block cipher, such as the generation of the differential pairs and the filtering of the wrong pairs.) WG with a 32-bit IV size is not vulnerable to the attack given in this section (since no special differential can be introduced into this short IV). In Sec. 3.1 the attack is applied to break an WG with an 80-bit key and an 80-bit IV. The attacks on WG with IV sizes larger than 80 bits are given in Sect. 3.2. The attack on a WG with a 64-bit IV size is given in Sec. 3.3.

3.1   An Attack on WG with an 80-Bit Key and an 80-Bit IV

We investigate the security of the key/IV setup of WG with an 80-bit key and an 80-bit IV. For this version of WG, denote the key as $K = k_1, k_2, k_3, \cdots, k_{80}$ and the IV as $IV = IV_1, IV_2, IV_3, \cdots, IV_{80}$. They are loaded into the LFSR as follows:

$$S_{1,\ldots,16}(1) = k_{1,\ldots,16} \qquad S_{17,\ldots,24}(1) = IV_{1,\ldots,8}$$
$$S_{1,\ldots,8}(2) = k_{17,\ldots,24} \qquad S_{9,\ldots,24}(2) = IV_{9,\ldots,24}$$
$$S_{1,\ldots,16}(3) = k_{25,\ldots,40} \qquad S_{17,\ldots,24}(3) = IV_{25,\ldots,32}$$
$$S_{1,\ldots,8}(4) = k_{41,\ldots,48} \qquad S_{9,\ldots,24}(4) = IV_{33,\ldots,48}$$
$$S_{1,\ldots,16}(5) = k_{49,\ldots,64} \qquad S_{17,\ldots,24}(5) = IV_{49,\ldots,56}$$
$$S_{1,\ldots,8}(6) = k_{65,\ldots,72} \qquad S_{9,\ldots,24}(6) = IV_{57,\ldots,72}$$
$$S_{1,\ldots,8}(7) = k_{73,\ldots,80} \qquad S_{17,\ldots,24}(7) = IV_{73,\ldots,80}$$

All the remaining bits of the LFSR are set to zero. Then the LFSR is clocked 22 steps with the middle value from the WG transformation being used in the feedback.

The chosen IV attack on WG goes as follows. For each secret key K, we choose two IVs, IV' and IV'', so that IV' and IV'' are identical in 8 bytes, but differ in two bytes: $IV'_{17,\ldots,24} \neq IV''_{17,\ldots,24}$ and $IV'_{49,\ldots,56} \neq IV''_{49,\ldots,56}$. The differences satisfy $IV'_{17,\ldots,24} \oplus IV''_{17,\ldots,24} = IV'_{49,\ldots,56} \oplus IV''_{49,\ldots,56}$.

Denote the value of $S(i)$ $(1 \leq i \leq 11)$ at the end of the j-th step by $S^j(i)$, and denote loading the key/IV as the 0th step. After loading the key and the chosen IV into LFSR, we know that the differences in $S(2)$ and $S(5)$ are the same, i.e., $S'^0(2) \oplus S''^0(2) = S'^0(5) \oplus S''^0(5)$. We denote this difference as $\triangle_1$, i.e., $\triangle_1 = S'^0(2) \oplus S''^0(2) = S'^0(5) \oplus S''^0(5)$.

We now examine the differential propagation during the 22 steps in the key/IV setup. The complete differential propagation is shown in Table 1, where the differences at the i-th step indicate the differences at the end of the i-th step. The difference $\triangle_2 = (\gamma \times S'^6(11) \oplus WG'(S'^6(11)) \oplus (\gamma \times S''^6(11) \oplus WG'(S''^6(11)) = (\gamma \times S'^0(5) \oplus WG'(S'^0(5)) \oplus (\gamma \times S''^0(5) \oplus WG'(S''^0(5))$. Similarly, we obtain that $\triangle_3 = (\gamma \times S'^0(2) \oplus WG'(S'^0(2)) \oplus (\gamma \times S''^0(2) \oplus WG'(S''^0(2))$.

Table 1. The differential propagation in the key/IV setup of WG

	S(1)	S(2)	S(3)	S(4)	S(5)	S(6)	S(7)	S(8)	S(9)	S(10)	S(11)
step 0	0	$\triangle_1$	0	0	$\triangle_1$	0	0	0	0	0	0
step 1	0	0	$\triangle_1$	0	0	$\triangle_1$	0	0	0	0	0
step 2	0	0	0	$\triangle_1$	0	0	$\triangle_1$	0	0	0	0
step 3	0	0	0	0	$\triangle_1$	0	0	$\triangle_1$	0	0	0
step 4	0	0	0	0	0	$\triangle_1$	0	0	$\triangle_1$	0	0
step 5	0	0	0	0	0	0	$\triangle_1$	0	0	$\triangle_1$	0
step 6	$\triangle_1$	0	0	0	0	0	0	$\triangle_1$	0	0	$\triangle_1$
step 7	$\triangle_2$	$\triangle_1$	0	0	0	0	0	0	$\triangle_1$	0	0
step 8	$\triangle_1\oplus\triangle_2$	$\triangle_2$	$\triangle_1$	0	0	0	0	0	0	$\triangle_1$	0
step 9	0	$\triangle_1\oplus\triangle_2$	$\triangle_2$	$\triangle_1$	0	0	0	0	0	0	$\triangle_1$
step 10	$\triangle_1\oplus\triangle_2\oplus\triangle_3$	0	$\triangle_1\oplus\triangle_2$	$\triangle_2$	$\triangle_1$	0	0	0	0	0	0
step 11	$\triangle_2\oplus\triangle_3$	$\triangle_1\oplus\triangle_2\oplus\triangle_3$	0	$\triangle_1\oplus\triangle_2$	$\triangle_2$	$\triangle_1$	0	0	0	0	0
step 12	$\triangle_1\oplus\triangle_2$	$\triangle_2\oplus\triangle_3$	$\triangle_1\oplus\triangle_2\oplus\triangle_3$	0	$\triangle_1\oplus\triangle_2$	$\triangle_2$	$\triangle_1$	0	0	0	0
step 13	$\triangle_2\oplus\triangle_3$	$\triangle_1\oplus\triangle_2$	$\triangle_2\oplus\triangle_3$	$\triangle_1\oplus\triangle_2\oplus\triangle_3$	0	$\triangle_1\oplus\triangle_2$	$\triangle_2$	$\triangle_1$	0	0	0
step 14	$\triangle_3$	$\triangle_2\oplus\triangle_3$	$\triangle_1\oplus\triangle_2$	$\triangle_2\oplus\triangle_3$	$\triangle_1\oplus\triangle_2\oplus\triangle_3$	0	$\triangle_1\oplus\triangle_2$	$\triangle_2$	$\triangle_1$	0	0
step 15	$\triangle_1\oplus\triangle_2\oplus\triangle_3$	$\triangle_3$	$\triangle_2\oplus\triangle_3$	$\triangle_1\oplus\triangle_2$	$\triangle_2\oplus\triangle_3$	$\triangle_1\oplus\triangle_2\oplus\triangle_3$	0	$\triangle_1\oplus\triangle_2$	$\triangle_2$	$\triangle_1$	0
step 16	$\triangle_1\oplus\triangle_2\oplus\triangle_3$	$\triangle_1\oplus\triangle_2\oplus\triangle_3$	$\triangle_3$	$\triangle_2\oplus\triangle_3$	$\triangle_1\oplus\triangle_2$	$\triangle_2\oplus\triangle_3$	$\triangle_1\oplus\triangle_2\oplus\triangle_3$	0	$\triangle_1\oplus\triangle_2$	$\triangle_2$	$\triangle_1$
step 17	$\triangle_1\oplus\triangle_4$	$\triangle_1\oplus\triangle_2\oplus\triangle_3$	$\triangle_1\oplus\triangle_2\oplus\triangle_3$	$\triangle_3$	$\triangle_2\oplus\triangle_3$	$\triangle_1\oplus\triangle_2$	$\triangle_2\oplus\triangle_3$	$\triangle_1\oplus\triangle_2\oplus\triangle_3$	0	$\triangle_1\oplus\triangle_2$	$\triangle_2$
step 18	$\triangle_3\oplus\triangle_4\oplus\triangle_5$	$\triangle_1\oplus\triangle_4$	$\triangle_1\oplus\triangle_2\oplus\triangle_3$	$\triangle_1\oplus\triangle_2\oplus\triangle_3$	$\triangle_3$	$\triangle_2\oplus\triangle_3$	$\triangle_1\oplus\triangle_2$	$\triangle_2\oplus\triangle_3$	$\triangle_1\oplus\triangle_2\oplus\triangle_3$	0	$\triangle_1\oplus\triangle_2$
step 19	$\triangle_1\oplus\triangle_2\oplus\triangle_3\oplus\triangle_5\oplus\triangle_6$	$\triangle_3\oplus\triangle_4\oplus\triangle_5$	$\triangle_1\oplus\triangle_4$	$\triangle_1\oplus\triangle_2\oplus\triangle_3$	$\triangle_1\oplus\triangle_2\oplus\triangle_3$	$\triangle_3$	$\triangle_2\oplus\triangle_3$	$\triangle_1\oplus\triangle_2$	$\triangle_2\oplus\triangle_3$	$\triangle_1\oplus\triangle_2\oplus\triangle_3$	0
step 20	$\triangle_4\oplus\triangle_6$	$\triangle_1\oplus\triangle_2\oplus\triangle_3\oplus\triangle_5\oplus\triangle_6$	$\triangle_3\oplus\triangle_4\oplus\triangle_5$	$\triangle_1\oplus\triangle_4$	$\triangle_1\oplus\triangle_2\oplus\triangle_3$	$\triangle_1\oplus\triangle_2\oplus\triangle_3$	$\triangle_3$	$\triangle_2\oplus\triangle_3$	$\triangle_1\oplus\triangle_2$	$\triangle_2\oplus\triangle_3$	$\triangle_1\oplus\triangle_2\oplus\triangle_3$
step 21	$\triangle_4\oplus\triangle_5\oplus\triangle_7$	$\triangle_4\oplus\triangle_6$	$\triangle_1\oplus\triangle_2\oplus\triangle_3\oplus\triangle_5\oplus\triangle_6$	$\triangle_3\oplus\triangle_4\oplus\triangle_5$	$\triangle_1\oplus\triangle_4$	$\triangle_1\oplus\triangle_2\oplus\triangle_3$	$\triangle_1\oplus\triangle_2\oplus\triangle_3$	$\triangle_3$	$\triangle_2\oplus\triangle_3$	$\triangle_1\oplus\triangle_2$	$\triangle_2\oplus\triangle_3$
step 22	$\triangle_2\oplus\triangle_3\oplus\triangle_4\oplus\triangle_5\oplus\triangle_6\oplus\triangle_7\oplus\triangle_8$	$\triangle_4\oplus\triangle_5\oplus\triangle_7$	$\triangle_4\oplus\triangle_6$	$\triangle_1\oplus\triangle_2\oplus\triangle_3\oplus\triangle_5\oplus\triangle_6$	$\triangle_3\oplus\triangle_4\oplus\triangle_5$	$\triangle_1\oplus\triangle_4$	$\triangle_1\oplus\triangle_2\oplus\triangle_3$	$\triangle_1\oplus\triangle_2\oplus\triangle_3$	$\triangle_3$	$\triangle_2\oplus\triangle_3$	$\triangle_1\oplus\triangle_2$

From Table 1, we notice that at the end of the 22th step, the difference at $S^{22}(10)$ is $\triangle_2 \oplus \triangle_3$. From the above description of $\triangle_2$ and $\triangle_3$, we know that

$$\triangle_2 \oplus \triangle_3 = ((\gamma \times S'^0(5) \oplus WG'(S'^0(5)) \oplus (\gamma \times S''^0(5) \oplus WG'(S''^0(5))) \oplus$$
$$((\gamma \times S'^0(2) \oplus WG'(S'^0(2)) \oplus (\gamma \times S''^0(2) \oplus WG'(S''^0(2))). \quad (3)$$

This shows that the value of $\triangle_2\oplus\triangle_3$ is determined by $k_{17,\ldots,24}$, $k_{49,\ldots,64}$, $IV'_{9,\ldots,24}$, $IV'_{49,\ldots,56}$, $IV''_{9,\ldots,24}$, $IV''_{49,\ldots,56}$.

From the keystream generation of WG, we know that the first keystream bit is generated from $S^{22}(10)$ (after the key/IV setup, the LFSR is clocked, and

$S^{23}(11)$ is used to generate the first keystream bit). If $\triangle_2 \oplus \triangle_3 = 0$, then the first keystream bits for IV' and IV'' should be the same. This property is applied in the attack to determine whether the value of $\triangle_2 \oplus \triangle_3$ is 0.

Assume that the value of $\triangle_2 \oplus \triangle_3$ is randomly distributed, then $\triangle_2 \oplus \triangle_3 = 0$ with probability 2^{-29}. We thus need to generate about 2^{29} pairs $(\triangle_2, \triangle_3)$ in order to obtain a pair satisfying $\triangle_2 \oplus \triangle_3 = 0$. Note that the key is fixed and that $S'^0(2) \oplus S''^0(2) = S'^0(5) \oplus S''^0(5)$ must be satisfied. Three bytes of IV ($IV'_{9,...,24}$, $IV'_{49,...,56}$) and one-byte difference ($\triangle_1$) can be freely chosen to generate different $(\triangle_2, \triangle_3)$, so there are about $2^{24} \times 255/2 \approx 2^{31}$ available pairs of $(\triangle_2, \triangle_3)$. Hence there is no problem to generate 2^{29} pairs of $(\triangle_2, \triangle_3)$.

Then we proceed to determine which pair $(\triangle_2, \triangle_3)$ satisfies $\triangle_2 \oplus \triangle_3 = 0$. For each pair $(\triangle_2, \triangle_3)$, we modify the values of $IV'_{1,...,8}$ and $IV''_{1,...,8}$, but we ensure that $IV'_{1,...,8} = IV''_{1,...,8}$. This modification does not affect the value of $\triangle_2 \oplus \triangle_3$, but it affects the value of $S^{22}(10)$. We generate keystream and examine the first keystream bits. If the values of the first keystream bits are the same, then the chance that $\triangle_2 \oplus \triangle_3 = 0$ is improved. In that case, we modify $IV'_{1,...,8}$ and $IV''_{1,...,8}$ again and observe the first keystream bits. This process ends when the first keystream bits are not the same or this process is repeated for 40 times. If one $(\triangle_2, \triangle_3)$ passes the test for 40 times, then we know that $\triangle_2 \oplus \triangle_3 = 0$ with probability extremely close to 1. (Each wrong pair could pass this filtering process with probability 2^{-40}. One pair of 2^{29} wrong pairs could pass this process with probability 2^{-11}.) Thus with about $2 \times 2^{29} \times \sum_{i=1}^{40} \frac{i}{2^i} = 2^{31}$ chosen IVs, we can find a pair $(\triangle_2, \triangle_3)$ satisfying $\triangle_2 \oplus \triangle_3 = 0$. Subsequently according to Eqn. (3) and $\triangle_2 \oplus \triangle_3 = 0$, we recover 24 bits of the secret key, $k_{17,...,24}$ and $k_{49,...,64}$.

The above attack can be improved if we consider the differences at $S^{22}(7)$ and $S^{22}(8)$. The differences there are both $\triangle_1 \oplus \triangle_2 \oplus \triangle_3$. If the value of $\triangle_1 \oplus \triangle_2 \oplus \triangle_3$ is 0, then the third and fourth bits of the two keystreams would be the same. If we only observe the third and fourth keystream bits, then $k_{17,...,24}$ and $k_{49,...,64}$ can be recovered with $2 \times 2^{29} \times \sum_{i=1}^{20} (\frac{1}{2^{i-1}} - \frac{1}{2^i}) \times i = 2^{30.4}$ chosen IVs.

In the attack, we observe the first, third and fourth keystream bits, then recovering $k_{17,...,24}$ and $k_{49,...,64}$ requires about $2 \times 2^{28} \times 2^{1.13} = 2^{30.1}$ chosen IVs (the value $2^{1.13}$ is obtained through numerical computation).

By setting the difference at $S^0(3)$ and $S^0(6)$ and observing the second and third bits of the keystream, we can recover another 24 bits of the secret key, $k_{25,...,40}$ and $k_{65,...,72}$. We need $2^{30.4}$ chosen IVs.

So with about $2^{30.1} + 2^{30.4} = 2^{31.3}$ chosen IVs, we can recover 48 bits of the 80-bit secret key. It shows that the key/IV setup of the WG stream cipher is insecure.

3.2 The Attacks on WG with Key and IV Sizes Larger Than 80 Bits

The WG ciphers with the key and IV sizes larger than 80 bits are all vulnerable to the chosen IV attacks. The attacks are very similar to the above attack. We omit the details of the attacks here. The results are given below:

1. For WG with 96-bit key and 96-bit IV, 48 bits of the key can be recovered with complexity about the same as the above attack.
2. For WG with IV sizes larger than 96 bits, 72 bits of the key can be recovered with complexity about 1.5 times that of the above attack.

3.3   The Attacks on WG with 64-Bit IV Size

We use WG with an 80-bit key and a 64-bit IV as an example to illustrate the attack. For WG cipher with an 80-bit key and a 64-bit IV, the key and IV are loaded into the LFSR as follows:

$$S_{1,\ldots,16}(1) = k_{1,\ldots,16} \qquad\qquad S_{1,\ldots,16}(2) = k_{17,\ldots,32}$$
$$S_{1,\ldots,16}(3) = k_{33,\ldots,48} \qquad\qquad S_{1,\ldots,16}(4) = k_{49,\ldots,64}$$
$$S_{1,\ldots,16}(5) = k_{65,\ldots,80} \qquad\qquad S_{1,\ldots,16}(9) = k_{1,\ldots,16}$$
$$S_{1,\ldots,16}(10) = k_{17,\ldots,32} \oplus 1 \qquad S_{1,\ldots,16}(11) = k_{33,\ldots,48}$$

$$S_{17,\ldots,24}(1) = IV_{1,\ldots,8} \qquad\qquad S_{17,\ldots,24}(2) = IV_{9,\ldots,16}$$
$$S_{17,\ldots,24}(3) = IV_{17,\ldots,24} \qquad\qquad S_{17,\ldots,24}(4) = IV_{25,\ldots,32}$$
$$S_{17,\ldots,24}(5) = IV_{33,\ldots,40} \qquad\qquad S_{17,\ldots,24}(6) = IV_{41,\ldots,48}$$
$$S_{17,\ldots,24}(7) = IV_{49,\ldots,56} \qquad\qquad S_{17,\ldots,24}(8) = IV_{57,\ldots,64}$$

In the attack, we introduce differences at $S(2)$ and $S(5)$, but we can only generate about 2^{23} pairs of $(\triangle_2, \triangle_3)$ since we can only modify $IV_{9,\ldots,16}$ and $IV_{33,\ldots,40}$. Thus we can obtain a pair $(\triangle_2, \triangle_3)$ satisfying $\triangle_2 \oplus \triangle_3 = 0$ or $\triangle_1 \oplus \triangle_2 \oplus \triangle_3 = 0$ with probability 2^{-5}. Once we know $\triangle_2 \oplus \triangle_3 = 0$ or $\triangle_1 \oplus \triangle_2 \oplus \triangle_3 = 0$, we can recover 29 bits of information on $k_{17,\ldots,32}$ and $k_{65,\ldots,80}$. It shows that 29 bits of information of the secret key can be recovered with probability 2^{-5}. This attack requires about $2^{25.1}$ chosen IVs.

The attack on WG with 96-bit key and 64-bit IV is similar to the above attack. We introduce differences at $S(2)$ and $S(5)$ or at $S(3)$ and $S(6)$. In the attack 29 bits of information on $k_{17,\ldots,32}$ and $k_{65,\ldots,80}$ can be recovered with probability 2^{-5}, and another 29 bits of information on $k_{33,\ldots,48}$ and $k_{81,\ldots,96}$ can be recovered with probability 2^{-5}.

The attack on WG with 112-bit key and 64-bit IV is also similar. The result is that 29 bits of information on $k_{17,\ldots,32}$ and $k_{65,\ldots,80}$ can be recovered with probability 2^{-5}, 29 bits of information on $k_{33,\ldots,48}$ and $k_{81,\ldots,96}$ can be recovered with probability 2^{-5}, and 29 bits of information on $k_{49,\ldots,64}$ and $k_{97,\ldots,112}$ can be recovered with probability 2^{-5}.

The attack on WG with 128-bit key and 64-bit IV is also similar. The result is that 29 bits of information on $k_{17,\ldots,32}$ and $k_{65,\ldots,80}$ can be recovered with probability 2^{-5}, 29 bits of information on $k_{33,\ldots,48}$ and $k_{81,\ldots,96}$ can be recovered with probability 2^{-5}, 29 bits of information on $k_{49,\ldots,64}$ and $k_{97,\ldots,112}$ can be recovered with probability 2^{-5}, and 29 bits of information on $k_{64,\ldots,80}$ and $k_{113,\ldots,128}$ can be recovered with probability 2^{-5}.

4   A Slide Attack on the Resynchronization of LEX

The security of LEX depends heavily on the fact that only a small amount of information is released for each round (including the input and output) of AES. The slide attack intends to retrieve all the information of one AES round input (or output) in LEX.

Denote $S_i = E_K^i(IV)$, where $E^i(m)$ means that m is encrypted i times, $S_0 = IV$; denote the 320 bits extracted from the i-th encryption with k_i for $i \geq 2$. For two IVs, IV' and IV'', if $k_2' = k_j''$ ($j > 2$), then we know that $S_1' = S_{j-1}''$. Immediately, we know that $S_{j-2}'' = S_0' = IV'$. Note that k_{j-1}'' is extracted from $E_K(S_{j-2}'')$, so k_{j-1}'' is extracted from $E_K(IV')$; this means that we know the input to AES, and we know 32 bits from the output of the first round. In the following, we show that it is easy to recover the secret key from this 32 bits of information of the first round output.

Denote the 16-byte output of the r-th round of AES with $m_{i,j}^r$ ($0 \leq i, j \leq 3$), and denote the 16-byte round key at the end of the r-th round with $w_{i,j}^r$ ($0 \leq i, j \leq 3$). Now if $m_{0,0}^1$, $m_{0,2}^1$, $m_{2,0}^1$, $m_{2,2}^1$ are known, i.e, four bytes of the first round output are known, then we obtain the following four equations:

$$m_{0,0}^1 \oplus w_{0,0}^1 = \text{MixColumn}((m_{0,0}^0 \oplus w_{0,0}^0) \| (m_{1,3}^0 \oplus w_{1,3}^0)$$
$$\| (m_{2,2}^0 \oplus w_{2,2}^0) \| (m_{3,1}^0 \oplus w_{3,1}^0)) \,\&\, 0xFF \tag{4}$$

$$m_{2,0}^1 \oplus w_{2,0}^1 = (\text{MixColumn}((m_{0,0}^0 \oplus w_{0,0}^0) \| (m_{1,3}^0 \oplus w_{1,3}^0)$$
$$\| (m_{2,2}^0 \oplus w_{2,2}^0) \| (m_{3,1}^0 \oplus w_{3,1}^0)) \gg 16) \,\&\, 0xFF \tag{5}$$

$$m_{0,2}^1 \oplus w_{0,2}^1 = \text{MixColumn}((m_{0,2}^0 \oplus w_{0,2}^0) \| (m_{1,1}^0 \oplus w_{1,1}^0)$$
$$\| (m_{2,0}^0 \oplus w_{2,0}^0) \| (m_{3,3}^0 \oplus w_{3,3}^0)) \,\&\, 0xFF \tag{6}$$

$$m_{2,2}^1 \oplus w_{2,2}^1 = (\text{MixColumn}((m_{0,2}^0 \oplus w_{0,2}^0) \| (m_{1,1}^0 \oplus w_{1,1}^0)$$
$$\| (m_{2,0}^0 \oplus w_{2,0}^0) \| (m_{3,3}^0 \oplus w_{3,3}^0)) \gg 16) \,\&\, 0xFF. \tag{7}$$

Each equation leaks one byte of information on the secret key. In the above four equations, 12 bytes of the subkey are involved. To recover all these 12 bytes, we need three inputs to AES and the related 32-bit first round outputs so that we can obtain 12 equations. These 12 equations can be solved with about $\alpha \times 2^{32}$ operations, where α is a small constant. With 96 bits of the key have been recovered, the rest of the 32 bits of AES can be recovered by exhaustive search.

We now compute the number of IVs required to generate three collisions. Suppose that a secret key is used with about $2^{65.3}$ random IVs, and each IV^i is used to generate a 640-bit keystream k_2^i, k_3^i. Since the block size of AES is 128 bits, we know that with high probability there are three collisions $k_2^i = k_3^j$ for different i and j since $\frac{2^{65.3} \times (2^{65.3}-1)}{2} \times 2^{-128} \approx 3$.

The number of IVs could be reduced if more keystream bits are generated from each IV. In [4], it is suggested to change the key every 500 AES encryptions for a strong variant of LEX. Suppose that each IV is applied to generate 500 320-bit outputs, then with $2^{60.8}$ IVs, we could find three collisions $k_2^i = k_x^j$ ($2 < x < 500$)

and recover the key of LEX. For the original version of LEX, the AES key is not changed during the keystream generation. Suppose that each IV is used to generate 2^{50} keystream bytes, then the key could be recovered with about 2^{43} random IVs (here we need to consider that the state update function of LEX is reversible; otherwise, the amount of IV required in the attack could be greatly reduced).

For a secure stream cipher with a 128-bit key and a 128-bit IV, each key would never be recovered faster than exhaustive key search no matter how many IVs are used together with that key. But for LEX each key could be recovered faster than exhaustive search if that key is used together with about 2^{61} random IVs. We thus conclude that LEX is theoretically insecure.

For a stream cipher with 128-bit key and 128-bit IV, if the attacker can choose the IV, then one of 2^{64} keys could be recovered with about 2^{64} pre-computations (based on the birthday paradox). The complexity of such an attack is close to our attack on LEX. However, there are two major differences between these two attacks. One difference is that the attack based on birthday paradox is a chosen IV attack while our attack is a random IV attack. Another difference is that the attack based on birthday paradox results in the recovery of one of n keys, while our attack recovers one particular key. Recovering one of n keys and recovering one particular key are two different types of attacks being used in different scenarios, so it is not meaningful to simply compare their complexities.

5   Conclusion

In this paper, we show that the resynchronization mechanisms of WG and LEX are vulnerable to a differential attack and a slide attack, respectively. It shows that the block cipher cryptanalysis techniques are powerful in analyzing the non-linear resynchronization mechanism of a stream cipher.

The designers of WG recommended to use 44 steps in the initialization to resist a differential attack [12]. It is a small modification to the design to achieve secure key/IV setup. However, it is inefficient. We recommend to change the primitive polynomial tap positions so that the tap distances are coprime, and to generate the first keystream bit from $S(1)$ instead of $S(10)$. Then we expect that WG with 22-step key/IV setup will be able to resist a differential attack.

Acknowledgements

The authors would like to thank the anonymous reviewers of SASC 2006 and FSE 2006 for their helpful comments. Special thanks go to Alex Biryukov for pointing out that the attack on LEX is slide attack and for helpful discussion.

References

1. F. Armknecht, J. Lano, and B. Preneel, "Extending the Resynchronization Attack," *Selected Areas in Cryptography – SAC 2004*, LNCS 3357, H. Handschuh, and A. Hasan (eds.), Springer-Verlag, pp. 19-38, 2004.

2. E. Biham, A. Shamir, "Differential Cryptanalysis of DES-like Cryptosystems," in *Advances in Cryptology – Crypto'90*, LNCS 537, pp. 2-21, Springer-Verlag, 1991.
3. A. Biryukov, D. Wagner, "Slide Attacks," *Fast Software Encryption – FSE'99*, LNCS 1636, pp. 245-259, Springer-Verlag, 1999.
4. A. Biryukov, "A New 128-bit Key Stream Cipher LEX," *ECRYPT Stream Cipher Project Report 2005/013*. Available at http://www.ecrypt.eu.org/stream/
5. J. Daemen, R. Govaerts, J. Vandewalle, "Resynchronization weakness in synchronous stream ciphers," *Advances in Cryptology - EUROCRYPT'93*, Lecture Notes in Computer Science, vol. 765, pp. 159-167, 1994.
6. ECRYPT Stream Cipher Project, at http://www.ecrypt.eu.org/stream/
7. J. D. Golić, G. Morgari, "On the Resynchronization Attack," *Fast Software Encryption – FSE 2003*, LNCS 2887, pp. 100-110, Springer-Verlag, 2003.
8. G. Gong, A. Youssef. "Cryptographic Properties of the Welch-Gong Transformation Sequence Generators," *IEEE Transactions on Information Theory*, vol. 48, No. 11, pp. 2837-2846, Nov. 2002.
9. National Institute of Standards and Technology, "DES Modes of Operation," Federal Information Processing Standards Publication (FIPS) 81. Available at http://csrc.nist.gov/publications/fips/
10. National Institute of Standards and Technology, "Advanced Encryption Standard (AES)," Federal Information Processing Standards Publication (FIPS) 197. Available at http://csrc.nist.gov/publications/fips/
11. Y. Nawaz, G. Gong. "The WG Stream Cipher," *ECRYPT Stream Cipher Project Report 2005/033*. Available at http://www.ecrypt.eu.org/stream/
12. Y. Nawaz, G. Gong. "Preventing Chosen IV Attack on WG Cipher by Increasing the Length of Key/IV Setup," *ECRYPT Stream Cipher Project Report 2005/047*. Available at http://www.ecrypt.eu.org/stream/

Author Index

Lecture Notes in Computer Science

For information about Vols. 1–3984

please contact your bookseller or Springer

Vol. 4033: B. Stiller, P. Reichl, B. Tuffin (Eds.), Performability Has its Price. X, 103 pages. 2006.

Vol. 4032: O. Etzion, T. Kuflik, A. Motro (Eds.), Next Generation Information Technologies and Systems. XIII, 365 pages. 2006.

Vol. 4031: M. Ali, R. Dapoigny (Eds.), Innovations in Applied Artificial Intelligence. XXIII, 1353 pages. 2006. (Sublibrary LNAI).

Vol. 4029: L. Rutkowski, R. Tadeusiewicz, L.A. Zadeh, J. Zurada (Eds.), Artificial Intelligence and Soft Computing – ICAISC 2006. XXI, 1235 pages. 2006. (Sublibrary LNAI).

Vol. 4027: H.L. Larsen, G. Pasi, D. Ortiz-Arroyo, T. Andreasen, H. Christiansen (Eds.), Flexible Query Answering Systems. XVIII, 714 pages. 2006. (Sublibrary LNAI).

Vol. 4026: P.B. Gibbons, T. Abdelzaher, J. Aspnes, R. Rao (Eds.), Distributed Computing in Sensor Systems. XIV, 566 pages. 2006.

Vol. 4025: F. Eliassen, A. Montresor (Eds.), Distributed Applications and Interoperable Systems. XI, 355 pages. 2006.

Vol. 4024: S. Donatelli, P. S. Thiagarajan (Eds.), Petri Nets and Other Models of Concurrency - ICATPN 2006. XI, 441 pages. 2006.

Vol. 4021: E. André, L. Dybkjær, W. Minker, H. Neumann, M. Weber (Eds.), Perception and Interactive Technologies. XI, 217 pages. 2006. (Sublibrary LNAI).

Vol. 4020: A. Bredenfeld, A. Jacoff, I. Noda, Y. Takahashi (Eds.), RoboCup 2005: Robot Soccer World Cup IX. XVII, 727 pages. 2006. (Sublibrary LNAI).

Vol. 4019: M. Johnson, V. Vene (Eds.), Algebraic Methodology and Software Technology. XI, 389 pages. 2006.

Vol. 4018: V. Wade, H. Ashman, B. Smyth (Eds.), Adaptive Hypermedia and Adaptive Web-Based Systems. XVI, 474 pages. 2006.

Vol. 4017: S. Vassiliadis, S. Wong, T.D. Hämäläinen (Eds.), Embedded Computer Systems: Architectures, Modeling, and Simulation. XV, 492 pages. 2006.

Vol. 4016: J.X. Yu, M. Kitsuregawa, H.V. Leong (Eds.), Advances in Web-Age Information Management. XVII, 606 pages. 2006.

Vol. 4014: T. Uustalu (Ed.), Mathematics of Program Construction. X, 455 pages. 2006.

Vol. 4013: L. Lamontagne, M. Marchand (Eds.), Advances in Artificial Intelligence. XIII, 564 pages. 2006. (Sublibrary LNAI).

Vol. 4012: T. Washio, A. Sakurai, K. Nakajima, H. Takeda, S. Tojo, M. Yokoo (Eds.), New Frontiers in Artificial Intelligence. XIII, 484 pages. 2006. (Sublibrary LNAI).

Vol. 4011: Y. Sure, J. Domingue (Eds.), The Semantic Web: Research and Applications. XIX, 726 pages. 2006.

Vol. 4010: S. Dunne, B. Stoddart (Eds.), Unifying Theories of Programming. VIII, 257 pages. 2006.

Vol. 4009: M. Lewenstein, G. Valiente (Eds.), Combinatorial Pattern Matching. XII, 414 pages. 2006.

Vol. 4008: J.C. Augusto, C.D. Nugent (Eds.), Designing Smart Homes. XI, 183 pages. 2006. (Sublibrary LNAI).

Vol. 4007: C. Àlvarez, M. Serna (Eds.), Experimental Algorithms. XI, 329 pages. 2006.

Vol. 4006: L.M. Pinho, M. González Harbour (Eds.), Reliable Software Technologies – Ada-Europe 2006. XII, 241 pages. 2006.

Vol. 4005: G. Lugosi, H.U. Simon (Eds.), Learning Theory. XI, 656 pages. 2006. (Sublibrary LNAI).

Vol. 4004: S. Vaudenay (Ed.), Advances in Cryptology - EUROCRYPT 2006. XIV, 613 pages. 2006.

Vol. 4003: Y. Koucheryavy, J. Harju, V.B. Iversen (Eds.), Next Generation Teletraffic and Wired/Wireless Advanced Networking. XVI, 582 pages. 2006.

Vol. 4001: E. Dubois, K. Pohl (Eds.), Advanced Information Systems Engineering. XVI, 560 pages. 2006.

Vol. 3999: C. Kop, G. Fliedl, H.C. Mayr, E. Métais (Eds.), Natural Language Processing and Information Systems. XIII, 227 pages. 2006.

Vol. 3998: T. Calamoneri, I. Finocchi, G.F. Italiano (Eds.), Algorithms and Complexity. XII, 394 pages. 2006.

Vol. 3997: W. Grieskamp, C. Weise (Eds.), Formal Approaches to Software Testing. XII, 219 pages. 2006.

Vol. 3996: A. Keller, J.-P. Martin-Flatin (Eds.), Self-Managed Networks, Systems, and Services. X, 185 pages. 2006.

Vol. 3995: G. Müller (Ed.), Emerging Trends in Information and Communication Security. XX, 524 pages. 2006.

Vol. 3994: V.N. Alexandrov, G.D. van Albada, P.M.A. Sloot, J. Dongarra, Computational Science – ICCS 2006, Part IV. XXXV, 1096 pages. 2006.

Vol. 3993: V.N. Alexandrov, G.D. van Albada, P.M.A. Sloot, J. Dongarra, Computational Science – ICCS 2006, Part III. XXXVI, 1136 pages. 2006.

Vol. 3992: V.N. Alexandrov, G.D. van Albada, P.M.A. Sloot, J. Dongarra, Computational Science – ICCS 2006, Part II. XXXV, 1122 pages. 2006.

Vol. 3991: V.N. Alexandrov, G.D. van Albada, P.M.A. Sloot, J. Dongarra, Computational Science – ICCS 2006, Part I. LXXXI, 1096 pages. 2006.

Vol. 3990: J. C. Beck, B.M. Smith (Eds.), Integration of AI and OR Techniques in Constraint Programming for Combinatorial Optimization Problems. X, 301 pages. 2006.

Vol. 3989: J. Zhou, M. Yung, F. Bao, Applied Cryptography and Network Security. XIV, 488 pages. 2006.

Vol. 3988: A. Beckmann, U. Berger, B. Löwe, J.V. Tucker (Eds.), Logical Approaches to Computational Barriers. XV, 608 pages. 2006.

Vol. 3987: M. Hazas, J. Krumm, T. Strang (Eds.), Location- and Context-Awareness. X, 289 pages. 2006.

Vol. 3986: K. Stølen, W.H. Winsborough, F. Martinelli, F. Massacci (Eds.), Trust Management. XIV, 474 pages. 2006.